Mammals Biochronology and Paleoecology of the Euro-Mediterranean Quaternary

Mammals Biochronology and Paleoecology of the Euro-Mediterranean Quaternary

Guest Editors

Alessio Iannucci
George E. Konidaris
Dimitris S. Kostopoulos
Joan Madurell-Malapeira
Raffaele Sardella

MDPI

Basel • Beijing • Wuhan • Barcelona • Belgrade • Novi Sad • Cluj • Manchester

Guest Editors

Alessio Iannucci
Department of Geosciences
Eberhard Karls University
of Tübingen
Tübingen
Germany

George E. Konidaris
School of Geology
Aristotle University
of Thessaloniki
Thessaloniki
Greece

Dimitris S. Kostopoulos
School of Geology
Aristotle University
of Thessaloniki
Thessaloniki
Greece

Joan Madurell-Malapeira
Dipartimento di Scienze
della Terra
Università di Firenze
Firenze
Italy

Raffaele Sardella
Dipartimento di Scienze
della Terra
Sapienza Università di Roma
Roma
Italy

Editorial Office

MDPI AG
Grosspeteranlage 5
4052 Basel, Switzerland

This is a reprint of the Special Issue, published open access by the journal *Quaternary* (ISSN 2571-550X), freely accessible at: www.mdpi.com/journal/quaternary/special_issues/2EO2261S80.

For citation purposes, cite each article independently as indicated on the article page online and using the guide below:

Lastname, A.A.; Lastname, B.B. Article Title. *Journal Name* **Year**, *Volume Number*, Page Range.

ISBN 978-3-7258-3752-6 (Hbk)
ISBN 978-3-7258-3751-9 (PDF)
https://doi.org/10.3390/books978-3-7258-3751-9

Contents

quaternary

Editorial

Introduction: Mammals Biochronology and Paleoecology of the Euro-Mediterranean Quaternary

Alessio Iannucci [1,*], George E. Konidaris [2,3], Dimitris S. Kostopoulos [2], Joan Madurell-Malapeira [4,5] and Raffaele Sardella [6]

1 Section of Terrestrial Palaeoclimatology, Department of Geosciences, Eberhard Karls University of Tübingen, Sigwartstr. 10, 72076 Tübingen, Germany
2 School of Geology, Aristotle University of Thessaloniki, 54124 Thessaloniki, Greece; gkonidaris@geo.auth.gr (G.E.K.); dkostop@geo.auth.gr (D.S.K.)
3 Paleoanthropology, Institute for Archaeological Sciences, Department of Geosciences, Eberhard Karls University of Tübingen, Rümelinstr. 23, 72070 Tübingen, Germany
4 Earth Science Department, University of Florence, 50121 Florence, Italy; joan.madurellmalapeira@unifi.it
5 Department of Geology, Faculty of Sciences, Universitat Autònoma de Barcelona, 08193 Cerdanyola del Vallès, Spain
6 Department of Earth Sciences, Sapienza University of Rome, P.le Aldo Moro 5, 00185 Rome, Italy; raffaele.sardella@uniroma1.it
* Correspondence: alessio.iannucci@mnf.uni-tuebingen.de

1. Introduction

The Quaternary is a time of fundamental climatic shifts and environmental changes that highlight the need for a thorough investigation from different perspectives and at multiple scales to disentangle the factors involved in the response of the biota. In turn, recognizing bioevents (e.g., the dispersal or extinction of species) and relating them to geological time is crucial for correlating changes between distant geographic regions. The mammalian fossil record is widely used for biochronological correlations and paleoecological reconstructions of the Euro-Mediterranean region and represents an especially important proxy for inferring the timing, pattern, and paleoenvironmental context of the earliest events of hominin dispersal into Europe. Conducting research on the response of mammals to the paleoenvironmental changes in the Euro-Mediterranean Quaternary and developing a consistent and precise biochronological framework require the thoughtful integration of data and schemes from different geographic areas.

With the aim of fostering progress in this direction, we organized a scientific session on "Mammals Biochronology and Paleoecology of the Euro-Mediterranean Quaternary" during the XXI INQUA Conference, which took place in Rome in July 2023. With attendees from more than ten countries, the session was indeed a stimulating occasion for discussion. This Special Issue gathers selected papers derived from original contributions presented during the session or developed as a result of the contextual scientific exchange. A brief summary of the articles is presented below.

Received: 3 February 2025
Revised: 12 February 2025
Accepted: 15 February 2025
Published: 6 March 2025

Citation: Iannucci, A.; Konidaris, G.E.; Kostopoulos, D.S.; Madurell-Malapeira, J.; Sardella, R. Introduction: Mammals Biochronology and Paleoecology of the Euro-Mediterranean Quaternary. *Quaternary* **2025**, *8*, 12. https://doi.org/10.3390/quat8010012

2. Summary

The articles of this Special Issue cover a large geographical scope, encompassing most of the northern Mediterranean region and adjacent areas, and a wide range of approaches, including biochronological to paleoecological research and more. Small and large mammals are considered, with a preponderance for the latter. Carnivora and hippopotamuses receive particular attention, with dedicated papers and specific digressions in other works. Recurrent themes include the establishment of local biochronological/biostratigraphic schemes or the identification or refinement of the knowledge on faunal dispersal events and their correlation. Paleoecological and paleoenvironmental insights, often with considerations on implications for hominins, are also present in several works. Among the articles, there are some that begin with taxonomic descriptions or specific sites, but even in these cases, the results are contextualized in a broader European dimension.

Spassov [1] presents a critical compilation of large mammalian dispersal events and a correlation of the Villafranchian (ca. 3.6/3.4–1.2 Ma) biochronology between Eastern and Western Europe. Taxa whose dispersal is discussed in detail include *Mammuthus*, *Equus*, *Canis*, *Acinonyx*, *Puma*, *Panthera*, *Pachycrocuta*, *Hippopotamus*, and *Bison*. Significant faunal turnovers occurred: during the final Pliocene (early Villafranchian); at the beginning of the Pleistocene (middle Villafranchian); around 2.2–2.1 Ma, as represented by the fauna of sites such as Slivnitsa, Coste San Giacomo, and Senèze, which the author proposed should be united in a Faunal Unit, marking the beginning of the late Villafranchian; immediately after the latter period, hence within the Olivola Faunal Unit; and at the late Villafranchian–Epivillafranchian transition, which is more elusive to date. In the final part of the article, biochronological correlations based on the reviewed dispersals and faunal turnovers are presented.

Tesakov et al. [2] provide a biochronological scheme based on small mammals, especially arvicolids, for the Plio-Pleistocene of Eastern Turkey and Transcaucasus (Armenia and Azerbaijan). Based on originally collected and reviewed material from the Early Pliocene to the early Middle Pleistocene of sixteen localities, the authors built a stratigraphic chart, enabling correlation with Anatolian and European biozonations. The important events recognized include the following: the Early Pliocene *Promimomys* stage, the late Ruscinian vole radiation, the Villanyian *Mimomys* radiation and the dispersal of *Borsodia*, the *Allophaiomys* datum at the Villanyian–Biharian transition, and the Microtini radiation at beginning of the Biharian.

Konidaris and Kostopoulos [3] establish for the first time a series of Faunal Units for Greece. The updated information on the taxonomy, stratigraphy, and chronology of Late Pliocene to Middle Pleistocene large mammal successions in the country allowed the authors to define eight Faunal Units (Milia, Dafnero, Gerakarou, Tsiotra Vryssi, Krimni, Apollonia, Marathousa, and Apidima) characterized by first and last occurrences of various taxa. Similarities and correlations with respect to Western Europe (Italy, France, and Spain) are then highlighted and discussed, with major large mammalian faunal turnovers recognized at ca. 2.6–2.5 Ma, 2.0 Ma, 1.7–1.6 Ma, and within the Mid-Pleistocene Revolution.

Konidaris et al. [4] describe new material and review previous findings of mammals from the late Early to Middle Pleistocene deposits of the Megalopolis Basin (Greece), which yielded a long and rather continuous faunal sequence. New discoveries include a new late Early Pleistocene site, Choremi 6, and two partial skeletons of *Hippopotamus* and *Palaeoloxodon* from Marathousa 1. Newly discovered small mammals allowed the authors to discuss age constraints for the sites of the Megalopolis Basin, and then their biochronological importance in Greece and Europe. The basin retained freshwater bodies even during glacial stages, as indicated by the occurrence of faunal elements highly dependent on the availability of freshwater (beavers, hippopotamuses, and otters), supporting the view that it acted as a refugium area for several species.

Mecozzi et al. [5] review the Italian fossil record of Middle Pleistocene hippopotamuses, including *Hippopotamus antiquus* and *Hippopotamus amphibius*, also presenting previously unpublished material from several localities. Morphological comparisons supported the view that a replacement between *H. antiquus* and *H. amphibius* occurred at ca. 0.5 Ma (during MIS 13), though it cannot be ruled out that both species might have been present around that time. Biometric analyses confirm the considerations of previous studies, which indicated that *H. antiquus* was generally larger than *H. amphibius*, although the authors observe that remains of *H. antiquus* dated to ca. 0.6 Ma show a reduced size when compared to older fossils of the same taxon.

Bellucci et al. [6] describe the carnivoran record from the Early Pleistocene of Coste San Giacomo (Italy) and discuss biochronological implications. The site, which is dated at ca. 2.2 Ma and represents the reference for the homonymous Faunal Unit, played an important role in our comprehension of the middle to late Villafranchian faunal turnover and in the development of the "Wolf event" concept. In Italy, Coste San Giacomo documents the first clear occurrence of *Canis etruscus* (together with Pantalla), the last occurrence of the hyena *Pliocrocuta perrieri*, and the first of *Homotherium latidens*.

Iannucci [7] shows that suids were one of the rarest group of artiodactyls in the Pleistocene of Europe and describes suid material from Peyrolles (France), dated at ca. 1.47 Ma. The author rejects the hypothesis that suids were absent from Europe between ca. 1.8 Ma and 1.2 Ma. The "suid gap" case is then taken as an example to highlight and discuss the large uncertainty in investigating the late Early Pleistocene, which is a crucial timespan for early *Homo* dispersal.

Madurell-Malapeira et al. [8] review the carnivorans from the late Early Pleistocene of the Vallparadís Section (Spain), encompassing the sites of Cal Guardiola and Vallparadís Estació. Several paleobiological, paleoecological, paleobiogeographical, and biochronological considerations on the identified species are discussed, with special emphasis on their relevance within the context of the Early–Middle Pleistocene Transition. The diverse carnivoran fauna of Vallparadís is then compared to that of other Eurasian and African Pleistocene localities, through cluster and ordination analyses. The results point to an overall separation of African, Asian, and European sites, though the Western Asian localities of Dmanisi and 'Ubeidiya cluster with European localities. The carnivore guild of the Vallparadís Section closely resembles those of other Western European Epivillafranchian localities, especially Untermassfeld and Vallonnet.

Rodríguez-Gómez et al. [9] provide estimates of carrying capacity (considered therein the prey biomass species that can be sustained over time in an ecosystem) and total available biomass based on the herbivore species recognized at the Early Pleistocene sites of Orce, in Spain (Fuente Nueva-3, Barranco León, and Venta Micena). The carrying capacity values are similar for the analyzed faunal assemblages, while Venta Micena appears more productive than the other sites in terms of total available biomass. The estimated values, which represent optimal conditions, are then compared with the relative abundances of species at the sites, allowing the authors to discuss the taphonomical and ecological factors responsible for the observed differences, with emphasis on the implications for hominins.

Fidalgo et al. [10] present a review of Quaternary hippopotamuses from the Iberian Peninsula (Spain and Portugal), covering Early, Middle, and Late Pleistocene localities. More than forty localities with hippopotamuses are considered, denoting a rich regional fossil record, especially for the Early Pleistocene. The authors recognize a delayed appearance of *Hippopotamus* in the fossil record of the Iberian Peninsula (ca. 1.7 Ma) compared to other European Mediterranean regions (2.2–2.0 Ma; France, Italy, and Greece). They hypothesize a brief coexistence between *Hippopatamus antiquus* and *Hippopotamus amphibius* (ca. 0.45 Ma) and suggested that hippopotamus populations went extinct in the Iberian Peninsula between MIS 5 and MIS 3, similar to Italy. Paleoecological considerations, with highlights on pathologies and rare cases of the exploitation of hippopotamuses by hominins, are also discussed.

3. Future Directions

Trans-regional discussions on mammalian biochronology and paleoecology are fundamental to enhance our comprehension of the Quaternary. Several articles of this Special Issue, which was initiated with the aim of promoting such scientific exchange, identified or related their results with/to important moments of faunal turnover that are widely recognized across Europe. However, not all periods and geographic areas are equivalently known, and even for those that we have more data on, there is still much that remains to be understood.

Ultimately, biochronological correlations depend on the reliability of the taxonomic attributions and the dating of the localities from which the faunal elements come from. Therefore, the application of absolute and relative dating methods alternative to and in combination with biochronology—or the cross-correlation of different local/regional biochronological schemes—and the discovery and description of new material—or the reexamination of older/historical collections and refinement of previous taxonomic attributions—will be crucial to improving the correlations at a regional-to-pan-European scale. New systematic fieldwork in well-stratified and dated contexts will ensure reliable paleogeographic, paleoenvironmental, and taphonomic interpretations, while applying modern methods to the revision of older collections (or contextualize the latter in light of recent advancements) will provide further support. Paleoecological information is deeply entwined with biochronology, making integrated studies that address both aspects especially important. Indeed, faunal events that provide data for correlation often document the response of the biota to climatic and environmental changes.

Eventually, a deeper understanding of the recent geological past is critical to gain perspective on ongoing and foreseeable environmental changes, and insights drawn from the mammalian fossil record represent a fundamental source of information.

Funding: Alessio Iannucci is supported by a Humboldt Research Fellowship provided by the Alexander von Humboldt Foundation.

Acknowledgments: We would like to thank the authors who contributed to the session and the Special Issue, and the reviewers who invested their time and effort in evaluating the articles.

Conflicts of Interest: The authors declare no conflicts of interest.

References

1. Spassov, N. The final Pliocene and Early Pleistocene faunal dispersals from East to Europe and correlation of the Villafranchian biochronology between Eastern and Western Europe. *Quaternary* **2024**, *7*, 43. [CrossRef]
2. Tesakov, A.S.; Frolov, P.; Simakova, A.; Yakimova, A.; Titov, V.; Ranjan, P.; Çelik, H.; Trifonov, V. Plio-Pleistocene small mammal-based biochronology of Eastern Anatolia and Transcaucasus. *Quaternary* **2024**, *7*, 42. [CrossRef]
3. Konidaris, G.E.; Kostopoulos, D.S. The Late Pliocene–Middle Pleistocene large mammal faunal units of Greece. *Quaternary* **2024**, *7*, 27. [CrossRef]
4. Konidaris, G.E.; Athanassiou, A.; Tourloukis, V.; Chitoglou, K.; van Kolfschoten, T.; Giusti, D.; Thompson, N.; Tsartsidou, G.; Roditi, E.; Panagopoulou, E.; et al. The late Early–Middle Pleistocene mammal fauna from the Megalopolis Basin (Peloponnese, Greece) and its importance for biostratigraphy and paleoenvironment. *Quaternary* **2024**, *7*, 41. [CrossRef]
5. Mecozzi, B.; Iannucci, A.; Arzarello, M.; Carpentieri, M.; Moncel, M.-H.; Peretto, C.; Sala, B.; Sardella, R. Middle Pleistocene hippopotamuses from the Italian Peninsula: An overview. *Quaternary* **2024**, *7*, 20. [CrossRef]
6. Bellucci, L.; Bona, F.; Conti, J.; Mecozzi, B.; Strani, F.; Sardella, R. The Early Pleistocene carnivoran of Coste San Giacomo (Anagni, central Italy): Biochronological implications. *Quaternary* **2024**, *7*, 57. [CrossRef]
7. Iannucci, A. The occurrence of suids in the post-Olduvai to pre-Jaramillo Pleistocene of Europe and implications for late Villafranchian biochronology and faunal dynamics. *Quaternary* **2024**, *7*, 11. [CrossRef]
8. Madurell-Malapeira, J.; Prat-Vericat, M.; Bartolini-Lucenti, S.; Faggi, A.; Fidalgo, D.; Marciszak, A.; Rook, L. A review on the latest Early Pleistocene carnivoran guild from the Vallparadís Section (NE Iberia). *Quaternary* **2024**, *7*, 40. [CrossRef]

9. Rodríguez-Gómez, G.; Espigares, M.P.; Martínez-Navarro, B.; Ros-Montoya, S.; Guerra-Merchán, A.; Martín-González, J.A.; Campaña, I.; Pérez-Ramos, A.; Granados, A.; García-Aguilar, J.M.; et al. Carrying capacity, available meat and the fossil record of the Orce sites (Baza Basin, Spain). *Quaternary* **2024**, *7*, 37. [CrossRef]
10. Fidalgo, D.; Madurell-Malapeira, J.; Martino, R.; Pandolfi, L.; Rosas, A. An updated review of the Quaternary hippopotamus fossil records from the Iberian Peninsula. *Quaternary* **2024**, *7*, 4. [CrossRef]

quaternary

MDPI

Article

The Final Pliocene and Early Pleistocene Faunal Dispersals from East to Europe and Correlation of the Villafranchian Biochronology between Eastern and Western Europe

Nikolai Spassov

Department of Paleontology and Mineralogy, National Museum of Natural History at the Bulgarian Academy of Sciences, Tsar Osvoboditel Blvd. 1, 1000 Sofia, Bulgaria; nspassov@nmnhs.com

Abstract: The Villafranchian stage in the mammal fauna evolution in Eurasia (ca. 3.6/3.4 Ma—ca. 1.2 Ma) is associated with the beginning of the formation of the modern appearance of the mammal megafauna of today's Palaearctic. The cooling and the aridification starting with the beginning of the Early Pleistocene gradually eliminated the quasi-tropical appearance of the Late Neogene landscapes and fauna of Europe. The time from the Mid-Piacenzian (ca. 3.3–3.0 Ma) to the end of the Early Pleistocene was a time of particularly intense dispersal of species, of faunal exchange between Eurasia and Africa, and of the entry of new mammals into Europe from the East. That is why the correlation of the biochronology of the Villafranchian fauna between Eastern and Western Europe is of particular interest. Accumulated data make possible a more precise correlation of these faunas today. A correlation of selected Eastern European localities with established faunal units and MNQ zones is made in the present work. Usually, the dispersal from Asia or from E. Europe to W. Europe is instantaneous from a geological point of view, but in a number of cases, reaching W. Europe happens later, or some species known to be from Eastern Europe do not reach Western Europe. The main driving forces of the faunal dispersals, which are the key bioevents in the faunal formation, are climate changes, which in turn, affect the environment. We can summarize the following more significant Villafranchian bioevents in Europe: the End Pliocene (Early Villafranchian: MNQ16) turnover related to the first appearance of a number of taxa, for example, felids, canids, proboscideans, and ungulates; the Quaternary beginning turnover. Correlated with this are the beginning of the Middle Villafranchian, which should be placed at about 2.6 Ma; the Coste San Giacomo faunal unit turnover (Senèze and Slivnitsa localities should be included here, and the FU itself, at the very beginning of the late Villafranchian (=MNQ18a)); the *Pachycrocuta* event at the very beginning of the Olivola FU; and the events related to the Late Villafranchian/Epivillafranchian bounfary.

Keywords: Villafranchian events; Plio-Pleistocene turnovers; faunal dispersals; European biochronology; faunal correlations

check for updates

Citation: Spassov, N. The Final Pliocene and Early Pleistocene Faunal Dispersals from East to Europe and Correlation of the Villafranchian Biochronology between Eastern and Western Europe. *Quaternary* **2024**, 7, 43. https://doi.org/10.3390/quat7040043

Academic Editors: Alessio Iannucci, George Konidaris, Dimitris S. Kostopoulos, Joan Madurell Malapeira, Raffaele Sardella and Maria Rita Palombo

Received: 22 May 2024
Revised: 25 September 2024
Accepted: 30 September 2024
Published: 11 October 2024

1. Introduction

The Neogene Arabian desert climax puts an end to the faunal dispersal from Eurasia to Africa at the Pliocene beginning. Then, the warming and the global retreat of deserts during the Mid-Piacenzian (ca. 3.3–3.0 Ma) cause, as it seems, a new, significant bi-directional African–Eurasian dispersals (Böhme et al. [1]). The time from the Mid-Piacenzian (ca. 3.3–3.0 Ma) to the end of the Late Villafranchian (ca. 1.2 Ma) was a time of particularly intense dispersal of species, of faunal exchange between Eurasia and Africa, and of the entry of new mammals into Europe from the East. The cooling and the aridification starting at the Pliocene/Pleistocene boundary and continuing throughout the Early Pleistocene (van Asperen, Kahlke [2]) gradually eliminated the quasi-tropical appearance of the Late Neogene landscapes and fauna of Europe (Spassov [3]). That is why the mammalian fauna of the transition from Pliocene to Pleistocene (usually designated as Villafranchian fauna) is of considerable interest. The Villafranchian Mammal Age is a biochronological unit based on large mammals, and the "Villafranchian" concept is of widespread use within the scientific communities of continental biostratigraphers in Western and Southern Europe (Rook and Martínez-Navarro [4] and references therein), but is also used, and much more widely, for Eurasian faunas (Vangengeim [5]; Sotnikova [6]; Sotnikova, Rook [7]). The Villafranchian stage (Azzaroli [8,9]) in the mammal fauna evolution in Eurasia (covering, according to present-day concepts, the approximate interval between ca. 3.6/3.4 Ma—about 1.2 Ma) is associated with the beginning of the formation of the modern appearance of the mammalian fauna of today's Palaearctic (Spassov [10,11]). Changes in the climate and in the environment during the Plio-Pleistocene and the Early Pleistocene led to faunal changes, showing the evolution of faunas in time, on the basis of which the Villafranchian stage in the evolution of the fauna is subdivided.

Chronostratigraphy and biochronology are of primary importance for relating biologic events to the geologic time scale (Lindsey [12]). Numerous biochronological studies discuss the question of the chronological sub-division of the Villafranchian biochron (Rook and Martinez-Navarro [4] and references therein). Particularly established approaches regarding the biochronological subdivision of the large mammal assemblages are the MN zone system of Mein [13], further developed for the Plio-Pleistocene by Guérin [14,15], and the mammal units system proposed by Azzaroli [9,16] and refined in the 1990s by Torre et al. [17] and Gliozzi et al. [18]. The obtained absolute dates for a number of iconic Villafranchian localities from Southern Europe (Nomade et al. [19]) were of primary importance for refining the chronology of Villafranchian faunas. Among the main events leading to faunal changes in the Villafranchian of Europe are the faunal dispersals, which are the subject of a number of studies, including Lindsay et al. [20]; Azzaroli [16]; Spassov [11,21]; Cregut-Bonnoure [22]; Sardella, Palombo [23]; Palombo et al. [24]; Sotnikova, Rook [7]; Croitor, Brugal [25]; Rook, Martínez -Navarro [4]; O'Regan et al. [26]; Palombo [27,28]; Koufos, Kostopoulos [29]; and Iannucci et al. [30,31]. Changes in the appearance of the European fauna are generally due to the entry of new mammal species from the East. That is why the correlation of the biochronology of the Villafranchian fauna between Eastern and Western Europe is of particular interest. A number of attempts to correlate the megafaunas of Eastern Europe with those of Western Europe have been made to date (e.g., Samson [32]; Vangengeim, Sotnikova [5]; Radulescu, Samson [33]; Spassov [11,21,34]; Palombo et al. [35]; Cregut-Bonnoure [22]; Kostopoulos et al. [36]; Kahlke et al. [37]; Andreescu et al. [38]; and Terhune et al. [39]). The zoogeographic features of the different regions of Europe and even more of the Palaearctic create difficulties in correlating local faunas and show the limitations of known biochronological schemes in this respect. However, the large number of data accumulated recently makes it possible to specify the main faunal events showing the evolution of the Villafranchian fauna of Europe, but also to make more precise correlations (see Palombo [40]), and Konidaris and Kostopoulos [41]) of the main Villafranchian localities from Eastern and Western Europe. These are also the goals of the present study, in connection with which, selected faunal events are discussed below.

2. Results and Discussion

2.1. The Faunal Dispersals as Key Bioevents for the Formation and the Evolution of the Villafranchian Faunas of Europe

The *Mammuthus* event. Lindsey et al. [20] draw attention to data suggesting a similar age of the occurrence of *Equus* and *Mammuthus* in Eurasia. They allow the appearance of both species in Europe, in the interval 3.0–2.6 Ma, and Azzaroli [16] summarizes this dispersal in a common dispersal event, accepted and mentioned repeatedly in the later literature. Recent data on the first appearance of these two genera in Europe show that these two events should most likely be separated because they are not synchronous (Iannucci, Sardella [42]), although some questions in this regard remain (see below). The origin of the genus *Mammuthus* is known to be African. The oldest *Mammuthus* remains in Europe, with a secure dating, are known from Tulucești and Cernătești, Romania (Radulesco, Samson [43]) and from the faunal complex at Ripa Skortselskaya, Moldova (Alekseeva [44]) which is from an equivalent horizon (Skortselskian horizon: Nikiforova et al. [45]), and they represent the first *Mammuthus* species from Eurasia: *M. rumanus* (Markov, Spassov [46];

Lister, van Essen [47]; Markov [48]). The localities have been palaeomagnetically correlated to the mid-Gauss subchron (C2An2n, 3.207–3.116 Ma) (Andreescu et al. [38]). It is also worth mentioning the mandible from Bossilkovtsi, Bulgaria (Markov, Spassov [46]) (probably at least 3 Ma?) and the find from Tsotylio, Greece (ca. 3.2 Ma?) (Kostopoulos and Koulidou [49]) show very primitive morphology and evolutionary stages similar to those of the mentioned finds from Romania, while at the same time, showing the Balkan route of dispersal. Thus, in Eurasia, *Mammuthus* was a newcomer from Africa during the mid-Piacenzian, and its appearance in Europe is confined to 3.2 Ma (Markov [48]; Böhme et al. [1]).

The *Equus* s. lato event. Of the three taxa that have become classic examples of bioevents associated to one degree or another with the beginning of the Pleistocene, *Equus*, *Mammuthus*, and *Canis* (Azzaroli [16]), an example of a very rapid dispersal from the East is *Equus*. *Equus* appeared in N. America at least at ca. 4 Ma (Rook et al. [50]). In China, there are no known finds from earlier than 2.5 million years ago (Sun and Deng [51]); this does not mean that the horse did not arrive there earlier, because in Europe today, there are a number of paleontological sites that indicate an earlier appearance of the monodactyl horse of the continent (the subgeneric taxonomy of *Equus*, especially its early forms, and even the generic taxonomy of the monodactyl Villafranchian horses of Eurasia remains controversial; see Bernor et al. [52] and Eisenmann [53]). At least four European sites with the presence of primitive, stenonoid horses have an age of about 2.6 Ma (Roca-Neyra, El Rincón 1; Huélago, Montopoli); the Asia Minor site of Güliazi is also of this age, and Pardines (France) is a little younger than 2.6 Ma (Iannucci and Sardella [42]). Here we must also add the Bulgarian locality of Varshets, where remains very close to *E. stenonis* from Saint-Vallier were found (Spassov [54]). The age of Varshets, given its fauna, placing it (Spassov [11]) between Saint-Vallier (probably around 2.5 Ma) and Roca-Neyra (about 2.6 Ma), (see: Nomade et al. [19]) should be at least 2.5 Ma. With this, the age of the earliest appearance of monodactyl horses in Europe and Western Asia should seem clear, but there is some, albeit debatable, evidence of an earlier age from Romanian sites. Samson [32] described *Equus (Allohyppus) euxinus* from Maluşteni and redescribed *E. simionecui* from Bereşti. He also mentioned a "*Plesippus (Allohippus) athanasiui*" metapodial from Capeni. This metatarsal III is noted also by Radulescu and Samson [33,55] as *Plesippus athanasiui and Allohippus* cf. *athanasiui*, respectively, and it is discussed (as *Equus* cf. *athanasiui*) by Forstén [56], who presents some basic dimensions of this metapodial. In addition, Radulescu and Samson [55] note the presence of the monodactyl horse also in Tuluceşti. Bereşti is an MN14 locality with an age of more than 4 Ma (Radulescu and Samson [55]; Andreescu et al. [38]), and the presence of a monodactyl horse with cabaloid features (Forstén [57]) there seems illogical.

It is likely that the remains entered the locality deposits accidentally or were collected from the surface/near the locality. According to our observations (2023), the remains of *Equus* from the locality, stored in the collection of the University of Bucharest, with an apparently cabaloid morphology, have a more different fossilization from that of the hipparion remains from the same locality. The Maluşteni locality is almost the same age, or perhaps only slightly younger, and is referred to as MN15a or the MN14/MN15 boundary (Andreescu et al. [38]; Radulescu and Samson [55]; Crespo et al. [58]). At the same time, Crespo et al. [58] noted the presence, albeit single, of *Spermophilus* cf. *nogaici*, which should be an indication of an MNQ16 zone. "*Equus euxinus*" has primitive, stenonoid features. It is not excluded, considering *S.* cf. *nogaici* in the locality, that there is also a later level in Maluşteni, or most likely, separate intrusions of a later (MNQ16) fauna there. The question requires further research. The finds from the Pliocene Capeni locality (with a biochronological age referring to MN15b, according to Radulescu and Samson [55], but possibly younger; see Forstén [56]), and from the final Pliocene Tuluceşti locality (3.2 Ma; see above) were not found by me in the collections of the Institute of Biospeleology and the University of Bucharest during my visit in 2023, and it remains unclear where they are stored. These remains, however, should belong to monodactyl horses. It seems that the question of the first appearance of the monodactyl horses in Europe cannot yet be considered definitively resolved. It is possible that there was an initial dispersal on the continent in the Mid-Piacesian, a little more than three million years ago, and a second mass dispersal as a result of climatic changes that began at the Pliocene–Pleistocene boundary.

The *Canis* event. The term 'wolf event' was proposed by Azzaroli [16] for the faunal turnover, related to the appearance of the *Canis* genus in Europe. This event has been discussed repeatedly in a number of recent studies (see Torre et al. [17]; Turner [59]; Spassov [11]; Sotnikova, Rook [7]; Rook and Martínez-Navarro, [4]; Palombo [27]; Bartolini-Lucenti, Spassov [60]; Iannucci et al. [30]; Iannucci, Sardella [42]; and references therein). The accumulation of data and analysis on this issue shows that there is some diachronism of this phenomenon in Europe, and that it makes sense to distinguish the phenomena of the first appearance of the genus from its mass appearance (see also: Iannucci et al. [31]).

1. The earliest appearance of the genus *Canis* in Europe: Lacombat et al. [61] mention several fragmentary remains of mandibles from Vialette, whose age is estimated (see there) to be 3.14 Ma and refer them to *Canis* sp. (*Canis* has been mentioned in this Late Pliocene locality since Heintz et al. [62]). Following the publication of Lacombat et al. [61], the first occurrence of the genus *Canis* in Europe is generally accepted to be associated with this Late Pliocene locality (Sotnikova, Rook [7]; Rook and Martínez-Navarro, [4]; Palombo [27]). Iannucci et al. [31], however, note the existence of chronological heterogeneity of the fauna preserved in the old collections (Crozatier Museum) labeled as Vialette. At the same time, we [1] (Spassov in Böhme et al.) expressed the opinion that the canid from Vialette most likely represents the genus *Eucyon*, known in Europe from the latest Miocene until the Pleistocene beginning. The most informative is the mandibular fragment (2003-5-401-VIA) with m1-m2. The analysis of a cast, kindly provided to me (2006) by A. Monguillon (Université de Lyon), enabled me to draw some conclusions. The canid is really large, and its dental dimensions are similar to those of a coyote (Lm_1 = 21 mm; p4 = 11.7 × 5 mm; p3 = 10.2 × 4 mm, after cast). This is also reminiscent of the possible *Eucyon*—"Canis" *michauxi* (Martin [63]). The lower carnassial tooth of this enigmatic canid from the Pliocene of Perpignan is not known, but the preserved p3-p4 (incorrectly represented in Martin's [63] figure as p2-p3) are even larger. The talonid of m1 of the specimen 2003-5-401-VIA is heavily eroded and of incomplete relief, but as seen in Figure 1, the hypoconid and the entoconid do not contact at their bases. This morphology should correspond to the state of the talonid in Eucyon; in *Canis* well-developed cristids that contact and fuse at their bases descend from the hypoconid and the entoconid towards each other (Tedford and Qiu [64]).

A potential candidate for the earliest known *Canis* s. str. (for *Canis* (*Xenocyon*) arrival in Europe see below)) from Europe is apparently the left semimandible, designated as *Canis neschersensis* Croizet et Jobert (see Blainville: Ostéographie, Canis, pl. XIII). It is stored in the Laboratory of Paleontology at the National Museum of Natural History, Paris under number MNHN.F.PET2010 from Perrier-Etouaires (Figure 1). This number corresponds to an older number from the laboratory of Comparative Anatomy of the Museum—A.C. 658. The museum catalog noted that the find was from Puy-de-Dôme, Auvergne (Puy-de-Dôme = plateau of Perrier = plateau of Boulade, nota mea: NS), specifically from the volcanic sands of Neschers, and was registered under no. 132 in the catalog of the paleontological collection created by l'Abbé Croizet and brought to the museum in Paris in 1839. A number of paleontological remains found in the volcanic sands of the Perrier Plateau have been well-known since the first half of the 19th century. The attribution of the find to the Etouaires site (Ravin des Etouaires = Perrier-Etouaires) in the latest catalogue of the Paris Museum (see above) should be explained by the grouping made by a number of later authors of some smaller sites in the vicinity to the Ravin des Etouaires and uniting them under the name Les Etouaires, assuming that these remains with similar fossilization are also roughly similar in age (Heintz [65]). Heintz [65] included in this group of localities the locality of Neschers from the volcanic sands of the region (distinguishing it from the Late Pleistocene Neschers fossiliferous formation), at the same time separating these localities from other and younger Perrier plateau sites such as Roca-Neyra, Pardines, and Peyrolles. Nomade et al. [19] provided an age of 2.78 Ma for the pumices of la Côte d'Ardé, which they considered stratigraphically very close to Les Etouaires classical site. At the same time, as mentioned above, the fauna from the old collections marked "Les Etouaires" is to one degree or another composite and includes fossils from other localities, as well. That is why I accept the opinion (Iannucci, Sardella [39]) that the fauna of Les Etouaires, as a whole, is constrained between a plinian fall dated at ca. 3.11 Ma and an overlying debris avalanche, whose pumices (contemporary to the Roca-Neyra fauna) have been dated (Nomade et al. [19]) at ca. 2.60 Ma. As a result, I can agree with the statement that the *C. neschersensis* find should fall, in general, within this time interval (Iannucci et al. [31]). After A. Iannucci (pers. comm.), the mandible of *C. neschersensis* has a different fossilization than some findings of certain Pliocene age (like the suids) from Perrier. It could be coeval with the remains of *C. (Xenocyon) falconeri* from Boulade (also from the Perrier plateau), a locality close to Roca-Neyra and correlated with it in age (2.6 Ma) (Bartolini-Lucenti, Spassov [60]), or it could be younger (?), but it seems to me that it is most likely earlier (see above; Heintz [65]) and could refer to the end of the early Villafranchian (MNQ16).

Following is a short comparative description of MNHN.F.PET2010: measurements: Lp1-m3 (alveolar) = 76.5 mm; m1 = 23.5 × 8.5; mandible height under m1 (labially) = 21.2. The size is slightly inferior to *C. etruscus*—*C. arnensis* and closer to "*C. apolloniensis*." The mandible is distinct from the Villafranchian group of taxa after several morphological characters, namely by the specific premolar features: large premolars, strongly developed on all pm (p2-p4) posterior additional cuspids (strong posterior additional cuspid on p2 including a very strong secondary additional cuspid on p4), and a lack of any gap between them (p4 partially overlaps the base of the anterior paraconid ridge). It is also distinct by some of its plesiomorphic (not wolf-like) characteristics, such as the narrow m1 trigonid with a rather flattened labial surface, the well-developed pre-hypoconid and pre-entoconid on the m1 talonid, and the strongly convex ventral surface of the mandible corpus. The typical *Canis* cristids on the talonid cuspids are well-developed, but the entoconid is narrow and mesiodistally elongated. The m1 talonid is relatively large. The M_2 is relatively short, but broad. Its mesial cuspids show a plesiomorphic condition: They are almost equally large, positioned in a straight line almost perpendicular to the longitudinal axis of the teeth (i.e., the labial one is not mesially placed in relation to the lingual one) (Spassov [66]). The teeth show, on the other hand, some derived characteristics that resemble *C. mosbachensis*, which appeared in Europe towards the end of the early Pleistocene: the paraconid of m1 is oblique and exceeds the height of p4, and p3 is relatively low-set in relation to adjacent premolars. Therefore, a more detailed comparison with this species makes sense: The mandible from Neschers differs from the European representatives of *C. mosbachensis* (as well as from all Villafranchian European *Canis*, as I already noted) by the remarkably short alveolar space, which is why there are no gaps between c1-m2 (due to lack of space, m2 is very high positioned on the surface of the mandible ramus); the labial outline of m1 is flat; the line of the p4 base is horizontal, in an angle with the line of the bases of p2-p3; p2 is with a well-marked posterior additional cuspid, which appears to be absent in the European representatives of the species and very rare in the Asian subspecies; and the mandible corpus is (as noted above) with a strongly convex ventral surface. It is interesting that in some features, such as the oblique m1 paraconid, the height of paraconid of m1 exceeding the height of p4, the relatively convex mandible body, the lack of gaps between teeth, the shape of m2, the position of p4 base in relation of the bases of p2-p3, the strong posterior additional cuspid in p2, the strong secondary additional cuspid in p4, the complex m1 talonid relief, and the mesio-distaly elongated m1 entoconid, the mandible of the *Canis* from Neschers resembles *Canis chihliensis palmidens* (NIH 164, Paris Museum) from Nihewan, China, from the Pleistocene beginning (2.4 Ma). In summary, it can be said that the unclear geological age and the lack of more material, bearing in mind the not small variability in *Canis*, do not give certainty when discussing *C. neschersensis* as the earliest *Canis* in Europe. However, some arguments noted above regarding its probable age (possibly between 2.8 and 2.6/2.4 Ma?), as well as its specific morphology, give reason to also mention, in the discussion of the early *Canis* appearance, this fossil as the first putative representative of the genus *Canis* in Europe.

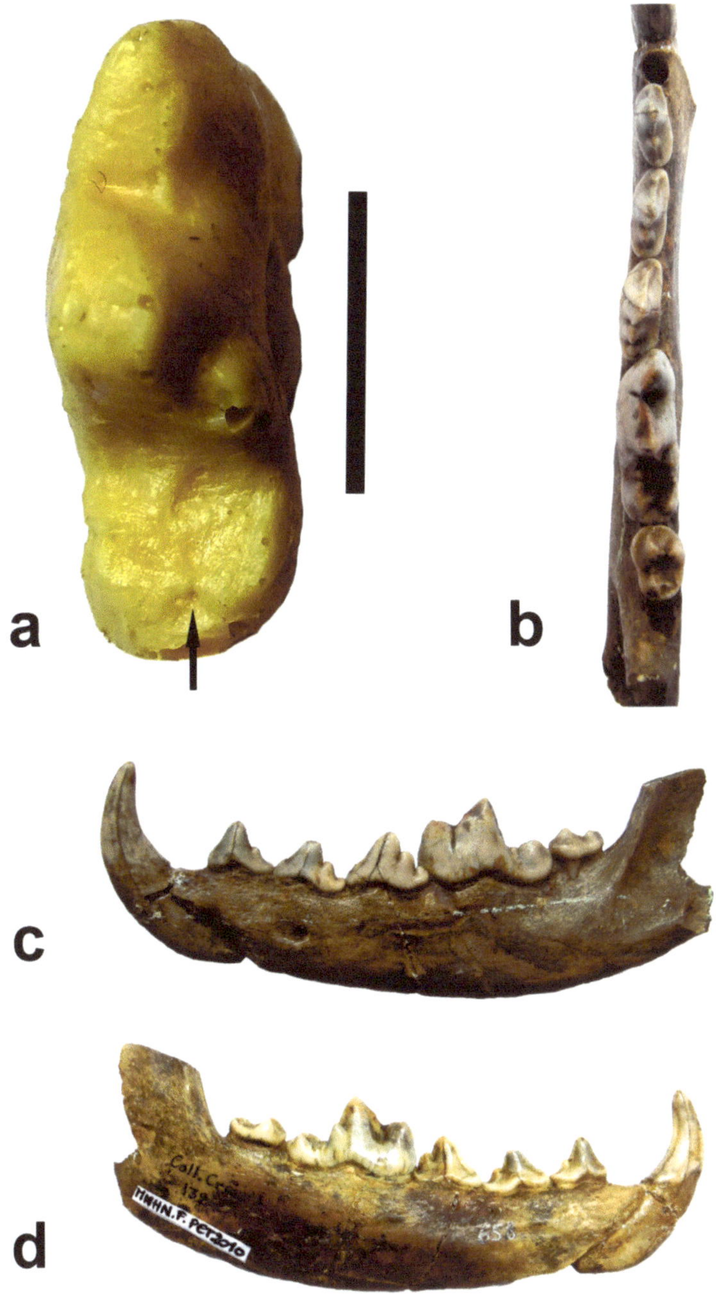

Figure 1. Canids discussed in paragraph 1.1. (**a**): m1 of "*Canis*" from Vialette in occlusal view (cast of mandible 2003-5-401-VIA, Crozatier Museum of Le Puy-en-Velay). The arrow shows the hypoconid and the metaconid (though heavily worn) are not fused at their bases. (**b**–**d**): *Canis neschersensis* (MNHN.F.PET2010) stored in the coll. of the Laboratory. of Paleontology of the National Museum of Natural History, Paris, in occlusal, labial, and lingual views.

Torre [67] and Iannucci et al. [31] noted that the presence of *Canis* (as probable "*Canis* aff. *etruscus*") from "Etouaires" was reported also by Heintz et al. [62] based on fossil material from the Bravard Collection housed in the British Museum and listed by Lydekker with provenance as "Tour de Boulade." Torre [67] (following Lydekker) considers that this fossil must "belong to a large wolf which is very similar to those of the last Glaciation." Here we have to note that from Boulade (near Roca-Neyra), the earliest European remains of *Canis* (*Xenocyon*) are described based on old collections that are now kept at the Claude Bernard University in Lyon (Bartolini-Lucenti, Spassov [60]). The tower "Tour de Boulade" itself is in the same region, above the Parentignat bridge, but at about 4.5 km from the mentioned site of Boulade. It is not clear whether the old designation of the site as "Boulade" has anything to do with "Tour de Boulade" (where Late Pleistocene fauna was indicated; Fosse et al. [68]). At the same time, however, as Sebastien Nomade clarifies (in a letter to S. Bartolini-Lucenti, 2021), the site La Boulade, where the remains of the mentioned *C.* (*Xenocyon*) *falconeri* should come from, is about 2.5 km away from Roca-Neyra, and stratigraphically and geochronologically, La Boulade and Roca-Neyra are both dated between 2.59 and 2.60 Ma. A comparison of a photograph from the cast of the mandible, stored in the collections of the British Museum (owned by the University of Florence and kindly provided to me by Saverio Bartolini-Lucenti) shows the following: The mandible is larger overall than a recent European *C. lupus* and has more elongated m2, but at the same time, shows considerable similarities with the wolf. From C. (Xenocyon) *falconeri* (IGF 865, coll. of the University of Florence), it is distinguished by the more robust m1 (especially the paraconid part); the shorter premolar row and smaller p1; and the larger i3. The articular process is wolf-like in shape, not as in *C.* (*Xenocyon*), and is rather more robust. Evolved wolves reach (albeit much later in time) the apomorphic features of the carnassial tooth of *C.* (*X.*) *falconeri*. However, the m1 hypoconid of the mentioned mandible is not as centrally positioned on the talonid and is not as significantly larger than the entoconid, as it is in a *C.* (*X.*) *falconeri* sample from Boulade (Roca-Neyra).

In relation to the time of the first appearance of *Canis* in Europe, some other unclear cases deserve to be noted. Among them are a mandible from Csarnota final Pliocene (Hungary), and a mandible fragment with p3 and p4 is mentioned as "*Canis* sp." in Capeni (Radulescu, Samson [55]; see also above for the remains of *Equus* also noted in this locality). At the same time, Marciszak et al. [69] mentioned, but did not describe the presence of *Canis* in the Late Pliocene Weze 2 locality (ca. 2.8–2.6 Ma). Argant [70] reported from Saint Vallier (with a probable age of almost 2.5 Ma; Nomade et al. [19]) the presence of scarce long bone fragments and an occipital skull fragment that resembled *Canis*. At the same time, the presence of "*Canis*" (?) is also established in Chilhac (Monguillon-Douillet A., [71]), whose age is determined at ca. 2.36 Ma (Nomade et al. [19]). In relation to all this, we must not forget the presence of *Canis* at Neschers. But it seems more likely to me that the remains (or most of them) refer to *Eucyon*, which must have been displaced by *Canis* in its later mass appearance in Coste San Giaccomo, Slivnitsa, and Senèze, than to belong to *Canis* s. str. Three separate teeth from the close-in-age locality of Varshets, Bulgaria (MNQ17, ca. 2.5 Ma), have been assigned to this genus (Spassov [11]). These cases cannot be resolved without a special, additional comparison.

2. The mass appearance of the genus: In Western Europe (France, Italy, Spain), the *Canis* mass appearance occurred in the period of about 2.2–1.98 Ma (Palombo [27]), but this can be said for all of Europe in general. We can call this dispersal, using Azzaroli's concept [16], *Canis* event s. str. or mass *Canis* appearance in Europe. The earliest *Canis* mass arrival on the continent must be related to the Coste San Giacomo unit. We can claim that the following localities (starting from east to west) record practically the same wave of dispersal of the "wolf-like" *Canis* to Europe: Slivnitsa, Bulgaria (see below: Spassov [11,72]), Coste San Giacomo (2.2 Ma: Florindo et al. [73]) and Quercia (2.2–2.1 Ma: Iannucci [74], Italy), and Senèze, France (its base age is of ca. 2.2 and the youngest fossils ranged in age between 2.10 and 2.08 Ma after a new investigation in. prep.: Eric Delson, pers. comm.). It seems that these localities, which record

the first secure mass invasion of the genus (Spassov [11,72,75]; Rook, Mart1F31nez-Navarro [4]; Iannucci [74]) have a very similar/close age and should be placed in the C. San Giacomo Unit. As it seems, two species entered practically simultaneously from the East to Europe (*C. etruscus* and *C. arnensis* are apparently present simultaneously in the locality of Slivnitsa; Spassov [72]), and two species of *Canis* are detected in Senèze after a new investigation (chapter in press on Senèze carnivores: A. Argant, pers. comm.). Slivnitsa and Senèze have very similar fauna and should be of the same/very close age (Spassov [75]). They show other mass dispersals, as well (see below). The Slivnitsa faunal event must, therefore, correlate (Spassov [34]) with the climatochronologic zone SCT10 of Zubakov and Borzenkova [76]. This zone, documented in Georgia and the Azov region, has an estimated age predating the Olduvai warming (between the Reunion subchron [2.193 Ma] and the beginning of the Olduvai event [ca. 1.95 Ma]).

Canis (assigned to *C.* cf. *senesensis*) is also mentioned for the Khapry faunal complex, southern Russia (especially for the Liventsovka and Khapry s. str. localities; see Titov [77]). After the statement of several authors (e.g., Sotnikova et al. [78]; Sotnikova, Rook [7]) these *Canis* finds must be correlated biostratigraphically to MNQ17, and along the Northern Black Sea coast, the genus should have already appeared in the Middle Villafranchian (MNQ17). Spassov [11] cast doubt on the dating of these finds, suggesting that they could originate from the upper and younger faunal levels of the multilayer Liventsovka section. The occurrence of three and possibly six equid *Equus* species also suggests a heterogeneity in age of the Khaprovskiy (Khapry) complex (Eisenmann [53]). Titov [77] estimated that the Khapry Sands, which extend for more than 120 km and up to 2 km width, could have accumulated for as long as 400,000 years and have an age from the early MNQ17 (excluding the earliest Middle Villafranchian) to the pre-Olduvai part of the Matuyama chron, between 2.6/2.5 and 2.2 Ma. According to Iannucci and Sardella [42], the occurrence of *Pachycrocuta brevirostris* there makes even wider the time limits during which the sands of the Khapry complex (having negative magnetic polarity and, thus, a pre-Olduvai age) were accumulated (between 2.6 and not less than 2.0 Ma). Considering the possible upper limit of the age of the Khapry complex (2.2–2.1/2.0 Ma), it is very likely that the appearance of *Canis* there also coincides with the C. San Giacomo Faunal Unit and represents a manifestation of the same *Canis* event traced west of the Black Sea in that time (Spassov [11]; see also below). Ultimately, even if the mass *Canis* arrival took place in the Azov region in the late NMQ17 (?), to the west of the Azov Sea, this penetration was recorded at the geologic time of Slivnitsa and Senèze (the Coste San Giacomo Unit).

Large felid events (*Acinonyx* and *Puma* dispersal in the latest Pliocene [MNQ16] and the *Panthera* dispersal in the Early Pleistocene [beginning of the Late Villafranchian—MNQ18a]).*Acinonyx* is thought to have originated in Africa, where the oldest remains (3.85–3.60 Ma) are likely from Laetoli (Werdelin, Dehghani [79]). The oldest remains of Eurasian *Acinonyx* are known from the MN16 of Perrier (Puy de Dôme), where the genus and probably the species A. *pardinensis* was found in two different, but very close sites: La Côte d'Ardé (with *A. p. pardinensis*) and Les Étouaires (with *A. p. arvernensis*), the former being the type locality according to Hemmer et al. [80]. Interestingly, the new dating of La Côte d'Ardé, by which Nomade et al. [19] dated, in fact, the very nearby site of Les Étouaires s. str., is 2.78 Ma. This means that the two disputed taxa are practically of the same age (above, we drew attention to the assemblage of fauna from different localities, often attributed to the Les Étouaires locality). This casts doubt on the subspecific status of these taxa. It seems that the fossil Eurasian cheetah is known from a large number of localities from the Early Villafranchian (only in Europe) until the Middle Pleistocene and from S.-W. Europe to China (Cherin et al. [81]).

The Les Étouaires locality (s. str.?) also seems to be associated with the earliest European record of the Eurasian *Puma pardoides* and probably also with the earliest saber-toothed cat *Megantereon* (Hugueney et al. [82]; Hemmer et al. [83]; Hemmer, Kahlke [84]; Cherin et al. [85]). The Eurasian fossil *Puma* is found with certainty in more than 10

Villafranchian localities known from S.-W. Europe till Mongolia (Cherin et al. [85]). Hemmer et al. [83] suggested, based on some fossil material, that the genus originated in Africa (Laetoli), but this suggestion was not supported by the analyses of Werdelin, Dehghani [79]. The origin of the genus is more likely to be related to Central Asia, and its earliest known occurrence (3.1 Ma) appears to be in Shamar, Northern Mongolia (Sotnikova [86]; Hemmer [80]). According to the available data, both *Puma* and *Acinonyx* appeared in Europe at the very end of the Pliocene (MNQ16b) in the Perrier localities ("Les Etouaires").

Much later, the genus *Panthera* appeared in Eastern Europe. The oldest finds of the genus are probably those from the Laetoli upper unit (3.85–3.60 Ma) (*Panthera* sp.; see Werdelin, Dehghani [79]). After Hemmer and Kahlke [84], the earliest Eurasian "jaguars" (sensu Hemmer [87]) (i.e., the early representatives of *Panthera* in Eurasia) obviously descended from an African *Panthera* population. The same author believes that the Eurasian fossil representatives of the genus do not differ significantly from the modern jaguar and should be included in the same species [84,88] (Hemmer [88]; Hemmer, Kahlke [84]; and references therein). This opinion is not accepted by all authors (see Jiangzuo [89]), and until recently, the question about the taxonomic appartenance of the Villafranchian Eurasian *Panthera* representatives remained controversial. For a clearer characterization of phylogenetic lines, we can enrich the taxon *Panthera* with one more clade, the subgenus *Onca*, in which jaguar-like cats can be divided into *P.* (*Onca*) *toscana*—the earliest (Late Villafranchian) Eurasian fossil jaguars; *P.* (*Onca*) *gombaszogensis*—the later representatives of Eurasian jaguars; and *P.* (*Onca*) *onca*—the American jaguars (for the correct transcription of the name as *P. gombaszogensis*, not *P. gombaszoegensis*, according to ICZN rules, see Wagner [90]).

The smallest and most ancient form of the European jaguar *P.* (*Onca*) *gombaszogensis toscana* is recorded, apparently, for the first time in Slivnitsa, Bulgaria (Spassov [11,72]). *P. gombaszogensis* has also been recorded in the Villafranchian fauna of Tegelen, but the Tegelen large mammal fauna probably does not represent a single time frame, and a large part of the fauna originates, it seems, from sediments deposited a little later, during the Olduvai Subchrone (Hoek Ostende [91]). The presence of the species (see Iannucci et al. [31]) at Puebla de Valverde (2.13–1.98 Ma) is uncertain. A little later than the appearance in Slivnitsa is the finding of *P. gombaszogensis* in localities such as Gerakarou 1, Greece (Koufos [92]), and in Olivola, Italy (Torre et al. [17]; Gliozzi et al. [18]; Iannucci et al. [31]). Gerakarou 1, where *Pachycrocuta brevirostris* is also present, should have an age later than Slivnitsa and related to the boundary between the C. San. Giacomo Unit (MNQ18a) and the Olivola Unit (MNQ18 b) (Spassov [11]; see Section 2.2. and Section 2.3).

***Pachycrocuta brevirostris* appearance.** The first appearance of this giant hyena in Europe is considered by a number of authors as a faunal event because of the high impact of this giant supercarrion eater hyaenid in the Early Pleistocene faunal assemblages, but often also because of the "simultaneous" appearance with another hipercarnivore, *Panthera gombaszogensis*, in Europe (Martínez-Navarro [93]; Rook, L., Martínez-Navarro [4]; Croitor et al. [94]; Iannucci et al. [31]). The dispersal of this felid, however, is earlier, as noted above. At the same time, other species noted their mass distribution in Eastern or Western Europe at that time (see Section 2.2). The so-called *Pachycrocuta brevirostris* event is frequently considered to mark the beginning of the Late Villafranchian [4,93,95] (Martínez-Navarro [93]; Rook and Martínez-Navarro [4]; Sianis et al. [95]), but according to the understanding expressed in this study, this beginning is marked by the C. San Giacomo-Slivnitsa–Senèze turnover (see Section 2.2). One of the earliest occurrences of *Pachycrocuta* in Europe is likely from the Iberian Peninsula: Based on some *P. brevirostris* remains from the localities P-1 and SCC-1 (~2.12–1.92 Ma) in the Fonelas area, as well as from the locality of Almenara-Casablanca 1 (pre-Olduvai age), Madurell-Malapeira et al. [96] have placed the *P. brevirostris* event before the Olivola Faunal Unit. These rare finds, however, apart from probably being close in age to the beginning of the Olivola FU, are poorly informative and can generally be considered rather tentatively (Iannucci et al. [30]). The giant hyena has been recorded in Gerakarou-1, Greece (Koufos [92,97]), and its presence in this locality must be among the first appearances of this carnivore on the continent (Spassov [11]; Iannucci et al. [30]). The age of Gerakarou should be approximately at the boundary between the C. San Giacomo and Olivola faunal units, slightly later than that of Slivnitsa (which we include in Coste San Giacomo FU) and similar to the very beginning of Olivola FU (Spassov [11]). In Olivola, Italy, *P. brevirostris* is well-known (Gliozzi et al. [18]; Iannucci et al. [30]) and pointed out as one of the main faunal elements initiating the Late Villafranchian faunal dispersal events (Azzaroli [9]). The Olivola locality has not been directly dated. The lowest possible placement of Olivola is around 2.1 Ma (Napoleone et al. [98]), and it is usually placed between 2.1–1.9 Ma (~2 Ma: Iannucci et al. [31]). Ultimately, the beginning of the *Pachycrocuta brevirostris* event in Europe should be dated, according to existing data (at least west of the Azov Sea), to about 2 or a slightly more than 2 Ma (it is not clear whether the species appeared in Khapry, in the Azov region, shortly before).

***Hippopotamus* event.** The dispersal of hippos into Europe is considered as an event of biostratigraphic importance and related to climatic changes (Rook, L., Martínez-Navarro [4]). The first appearanceof the *Hippopotamus* in Europe is now attested at 2.2 Ma (Coste San Giaccomo, Italy; see Bellucci et al. [99]; Fidalgo et al. [100]; Iannucci et al. [31]) and is possibly also from Elis (Greece), with unclear age. The age of Elis is mentioned as "Middle Villafranchian" (?) (Reimann, Strauch [101]) and "Lowest Pleistocene," at ~2.1 Ma (Athanassiou [102]; Athanassios Athanassiou, pers. comm.). The *Hippopotamus* incisor from Coste San Giacomo was collected during field collection, ex situ, causing some authors to doubt the age of the finding (Marra et al. [103]; Mecozzi [104]). However, field activities and excavations at Coste San Giacomo have pointed out that the vertebrate assemblage comes from a single fossiliferous level (Iannucci et al. [31]). After Mazza and Rustioni [105], a first phalange from Senèze (FSL 211082), previously ascribed to *Equus* sp., must be attributed to *Hippopotamus*, a claim based on which Iannucci et al. [31] accept the presence of the hippopotamus (also mentioned in the older literature) also at Senèze. However, two posterior first phalanges cataloged as FSL 211,082 are discussed as *Allohippus senezensis* by Eisenmann and Delson in a publication currently in preparation (E. Delson, pers. comm.) In sum, it seems that there is no evidence of the *Hippopotamus* in the Senèze assemblage.

The question of how, from where, and under what climatic conditions the *Hippopotamus* colonized Europe is of interest. Despite their aquatic lifestyle, hippopotamuses cannot swim (Mazza [106]). There are no reliable data for a land bridge via the Gibraltar Straits or via Sicilia during the Villafranchian (Spassov [34] and references therein), which means that the hippo most likely could not have passed from North Africa to Europe via these hypothetical dispersal routes (especially Gibraltar) discussed by a number of authors (see Fidalgo et al. [100] and references therein). The logical route of dispersal, also supported by the earliest finds in the area, is that via the Levantine corridor (see Made et al. [107]; Fidalgo et al. [100]). The route from the east to the continent is also the route of almost all migrations of large mammals to Europe (Spassov [11,34]). This entry route for the hippopotamus is supported by the registered Meria cooling documented, as already mentioned, in the Black Sea region in the time between the Reunion subchron and the beginning of the Olduvai event showing signs of a temporary closing of the Bosphorus (Zubakov and Borzenkova [76]).

Iannucci et al. [31] note that the appearance of the hippopotamus in Europe documented an African dispersal of a species linked to humid conditions in a context that is generally deemed to denote the spread of open-adapted faunal elements of mainly Asian affinities. But did the appearance of the hippopotamus really occur in humid conditions at that time on the continent, and does this appearance really contradict the spread of open-adapted faunal elements from the East to Europe? In fact, the hippo cannot stand the strong sun and spends most of the day in water to stay cool and hydrated. At the same time, the presence of water bodies does not mean wet conditions. The tropical savannah conditions in which *H. amphibius* lives today are characterized as usually being very dry. This animal does not need humid conditions, but rivers and patches of water scattered throughout in open landscapes (Parker [108]), conditions that do not contradict a relatively cool environment in southern Europe ~ 2.2 Ma ago.

The *Bison* (*Eobison*)—*B.* (*Bison*) replacement and the Lower Epivillafranchian boundary. The end of the Villafranchian shows, according to a number of researchers, a transitional fauna between the Villafranchian one and that which is characteristic of the Middle Pleistocene. This episode in the history of the megafauna is referred to in the literature in different ways: Latest Villafranchian sensu Koufos, [109], Final Villafranchian sensu Spassov [11], and the term Epivillafranchian, proposed at the beginning of the 1960s, which has acquired the widest distribution and gradually become established as the name of this biochron (Kahlke [110]; Kahlke et al. [37] and references therein). Its time span is between the Late Villafranchian s. str. and the Galerian biochrons (1.2–0.9 Ma). In general, it is defined as the time between the *Praemegaceros verticornis*—*Bison menneri* first occurrence and the *Crocuta crocuta* first occurrence] (Kahlke [110]; Bellucci et al. [111] and references therein). The disappearance of the primitive bison *Bison* (*Eobison*) and its replacement by the evolved bison of the subgenus *B.* (*Bison*) seems to be among the important faunal events reflecting climatic and landscape changes (van Asperen, Kahlke [2]) and, thus, indicating the boundary between the Late Villafranchian and the Epivillafranchian.

The putative ancestor of the bison is *Leptobos* (Cherin et al. [112] and references therein), or it is its possible sister species, bearing in mind the new data (Akbar Khan et al. [113]) about the rather early appearance of the primitive bison. While *Leptobos* has been present in Europe since at least 3.3 Ma and throughout the Villafranchian mammal age (Masini et al. [114]; Mead et al. [115]), *Bison* probably appeared in Asia and entered Europe much later. Its first occurrence is registered in the Upper Siwaliks of Pakistan with *Bison* (*Eobison*) cf. *sivalensis* in the Early Villafranchian, between 3.3–2.6 Ma (Akbar Khan et al. [113]) and reaching the boundaries of Europe (Dmanisi) ~1.76–1.8 Ma ago with *B.* (*E.*) *georgicus* (Burchak-Abramovich and Vekua, [116]; Sorbelli et al. [117]). The last primitive bison that should be included in the subgenus *B.* (*Eobison*) are known from Pirro, Italy (*B. degiulii*), and Apollonia, Greece, (*B.* cf. *degiulii*) (with probable ages of about 1.3 and about 1.2 Ma). They are distinguished by an increasing size compared to earlier representatives of the subgenus and by more massive metapodials (Sorbelli et al. [117]). The same authors note that the progressive climatic deterioration and the spread of open environments led to an increase in metapodial stoutness and in the body size of these forms. Increasing size is a natural process in the evolution of many taxa, but it is difficult to agree that the massiveness of the metapodials of these bisons is due to inhabitance in more open and arid landscapes. It has long been known, as a result of a number of studies, that active running in steppe landscapes and on terrains harder than forest landscapes lead to the exact opposite result—to a reduction in the massiveness and elongation of the metapodials, both in equids and in artiodactyls (Gromova [118]; Eisenmann [119]; Scott [120]). The massiveness of the metapodials of the last representatives of *B.* (*Eobison*) should be due to the increase in size and, hence, the weight of these forms, which at the intensification of the cursoriality in open, plain landscapes, leads for biomechanical reasons to the shortening of metapodials (Gambaryan [121]), and as a consequence of this, to a change in their proportions and an increase in their massiveness.

2.2. Major Villafranchian Events. A Summary

The data on the first appearance of different taxa in the Villafranchian of Europe are constantly replenished, and today, it is clear that the processes are more complex and often diachronic (Iannucci et al. [31]). There is hardly any doubt that the main driving forces of mass dispersals are more significant climatic and, hence, environmental changes. Some of the more significant dispersals are briefly summarized below.

End Pliocene (Early Villafranchian: MNQ16) events. As noted above, the time from the Mid-Piacenzian (~3.3–3.0 Ma) marks the beginning of a particularly intense dispersal of species, of a new faunal exchange between Eurasia and Africa, as well as of the entry of new mammals into Europe from the East in the Early Villafranchian (for the first appearances of individual faunal elements, see above). At that time, *Leptobos* appeared in Southern Europe, and this seems to be the time of origin of the genus. The appearance of *Mammuthus* in Europe is associated with this time. Its earliest finds are usually associated with the species *M. rumanus*, and its first secure occurrence (Romania and Moldova) is about 3.2 Ma ago. The first appearance of the monodactyl horses in Europe is a question we cannot yet accept as definitively resolved. It is possible that there was a first (unsuccessful?) dispersal on the continent in the Mid-Piacenzian, a little more than three million years ago, and a subsequent mass dispersal during climatic changes that began at the Pliocene–Pleistocene boundary.

The site of Les Étouaires marks the first occurrence of several taxa of large carnivores. The "Les Étouaires" faunal assemblage is, as it seems, complex and made of various outcrops and levels. They do, however, appear to be relatively close in age within the MNQ16 zone (see above). The earliest remains of *Acinonyx pardinensis* are from Les Étouaires and from the neighboring La Côte d'Ardé site, accurately dated after the deposits from the base of the Les Étouaires fluvial sequence at Côte d'Ardé (2.78 Ma after Nomade et al. [18]). The Les Étouaires locality itself is also correlated with the age of these deposits (see the corresponding passage above). From these levels also appear to be the remains of the earliest *Puma* in Europe. The first occurrence of *Canis* s. str. (possibly with *C. neschersensis*) is not entirely clear. The age of the type locality of "Neschers" is not clear, but we have some reason, as we noted earlier, to assume an age close to Étouaires. This locality also appears to be associated with the first (certain) occurrence of *Pliotragus ardeus* (Cregut-Bonnoure [22]; Croitor et al. [94]). Its occurrence in localities of earlier age, for example, in the Late Miocene of Stratzing, Austria, can most likely be explained by the presence of higher stratigraphical levels there (Gentry [122]).

The Quaternary beginning and the Middle Villafranchian faunal events. The first certain appearance of the monodactyl horses (*Equus* or *Allohypus*, according to different taxonomic interpretations) is related to the Pliocene/Pleistocene boundary. A number of localities mentioned above, and with an age very close to the beginning of the Quaternary (close to or synchronous with the Gauss–Matuyama transition (i.e., 2.588 Ma), show the first certain mass appearance of the horse. As such, these mass dispersal event localities, such as Roca-Neyra, for example (2.60 Ma) should mark the beginning of the Middle Villafranchian and the beginning of MNQ17, not the end of MNQ16 and the Early Villafranchian. A little later, around the beginning of the Quaternary, the suid *Sus strozzi* appeared in Europe, commonly found there in the first part of the Early Pleistocene of Europe (Cherin et al. [123]). Of all the earliest finds, the most securely dated seems to be that of Saint-Vallier (Iannucci et al. [31]), a type locality of the MNQ17 zone, with an age older than 2.4 Ma, perhaps close to 2.5 Ma (Nomade et al. [19]). It seems that the Pliocene/Pleistocene boundary is also associated with the first occurrence of *C.* (*Xenocyon*) *falconeri* in Europe, according to remains from France with a probable age of 2.6 Ma (Bartolini-Lucenti, Spassov [60]), and from Poland at 2.4–2.2 Ma (Marciszak [69]). The replacement of the Archaic *Mammuthus rumanus* with *M. meridionalis* probably also occurred soon after the beginning of the Quaternary and in the Middle Villafranchian: The earliest *M. meridionalis* apparently appear in Khapry, S. Russia (Titov [77]).

Slivnitsa–Coste San Giacomo–Senèze turnover. Although the appearance of *Canis* in Europe takes place probably, as we noted already, at the end of the Pliocene (or around the Pliocene-Pleistocene boundary), the mass appearance of the "wolf-like" *Canis* s. str., which we can associate with the "wolf event" of Azzaroli [16], becomes later. This wave of dispersal from the East is associated with the time between 2.2–2.1 Ma. Coste San Giacomo (CSG), Italy (Napoleone et al. [98]), is associated with an age of 2.2 Ma. The base age of Senèze (France) was also found to be ca. 2.20 Ma after a new investigation (in prep.: E. Delson, pers. comm.) (2.21 to 2.09 after Nomade et al. [19]). In CSG, the gomphothere *Anancus* (relic from there?) is still found, but in both localities, as well as in the apparently similar in age Slivnitsa (Bulgaria), "wolf-like *Canis*" appear, such as *C. arnensis*, which is similar in size to the coyote (Brugal, Boudadi-Maligne [124]) or slightly larger. Two species of Canis are present in Senèze, according to the findings from a new investigation (chapter in prep. on Senèze carnivores: A. Argant, pers. comm.), and apparently also in

Slivnitsa (Spassov [11,72]). We correlate the Slivnitsa faunal event (Spassov [34]) with the climatochronologic zone SCT10 of Zubakov and Borzenkova [76]. This zone, documented in Georgia and the Azov region, has an estimated age between the Reunion subchron and the beginning of the Olduvai event. During this period, which corresponds to the Meria cooling documented in the Black Sea region, there are indications that some Aegean islands close to the Anatolian coast were connected with Asia Minor (Dermitzakis [125]) and that the Black Sea was a freshwater sea; this suggests a temporary closing of the Bosphorus (Zubakov and Borzenkova [76]). This explains the new wave of faunal dispersal from the East during that time. With Slivnitsa, a Villafranchian Balkan faunal unit can be designated, but it would be most logical that Senèze and Slivnitsa, together with Coste San Giacomo, should be united in a common faunal unit (CSG Unit) and separated in the MNQ18a zone (sensu Spassov [11,34,126]), thus marking the beginning of the Late Villafranchian. The MNQ18a zone and CSG unit in the range indicated above are characterized in part by the mass appearance of *Canis* and by the first appearance of *Panthera* on the continent (Slivnitsa), by the first appearance of *Ovis* and *Hemitragus* (Slivnitsa and Senèze) (Spassov [11,34,72,75]), as well as by the first appearance of the hippo in Coste San Giacomo (Bellucci et al. [99]; Fidalgo et al. [100]). The dominance of bovid species over cervids (Slivnitsa) at that time is also an indication of the cooling and opening of the landscape.

In the context of the above, some data and ideas about the earliest possible entry of the human genus into Europe are of interest. In recent years, an increasing number of discoveries have supported the idea that the earliest human occupation of Europe was via the Bosphorus/Peri-Pontic pathway only (Strait et al. [127]) and took place during the Villafranchian, significantly predating 1 Ma. The time of the earliest human appearance in Europe could be related to conditions of increasing aridification and to a domination of open/mosaic landscapes.

The time of Slivnitsa (Meria cooling), which should be the time interval between the Reunion subchron and the beginning of the Olduvaian event, has been proposed for the time of the earliest possible appearance of the genus *Homo* on the continent (Spassov [34] and references therein). The discovery of lithic artefacts at the site of Kermek (Azov Sea region, S. Russia) with an age of ca. 2 Ma (Shchelinsky et al. [128]; M. Gurova, Inst. of Archaeology, Sofia, pers. comm.) seems to confirm the above hypothesis about the time of the first (probably unsuccessful) appearance of *Homo* in Europe.

The Olivola FU beginning. The age of the Italian locality of Olivola, which provided the name of the Olivola Faunal Unit (Gliozzi et al. [18]), has not been directly dated and is usually placed at ca. 2 Ma; its lowest possible placement is around 2.1 Ma (Napoleone et al. [98]). The very beginning of the Olivola Faunal Unit (the Coste San Giacomo/Olivola FU boundary) is related to the appearance of *Pachycrocuta* in Europe and the wide spread of *Canis*. The first occurrence of *Pachycrocuta* seems to be immediately after that of the mass appearance of *Canis* in the Coste San Giacomo FU. Apparently, the appearance and spread of these carnivores is related to the dispersal of a number of ungulates, as a result of the change (the opening) of the landscape. After some investigations (Croitor et al. [94]), the dominant ruminant species in Eastern Europe (after the example of the Dacian Basin) before the *Pachycrocuta* event (*Pliotragus ardeus, Gazellospira torticornis, Rucervus radulescui, Metacervocerus rhenanus,* and *Mitilanotherium inexspectatum*) became extinct during the Pachycrocuta faunal turnover and were replaced by a more cold-adapted assemblage of ruminants (*Megalovis latifrons, Eucladoceros* sp., *Dama* sp., and *Praemegaceros obscurus*). The mentioned turnover is also manifested in Western Europe, where the time span of the Olivola FU shows the peak of bovid diversity and the disappearance of *Gazella borbonica* (Masini et al. [114]).

The Late Villafranchian/Epivillafranchian boundary. If we assume that the Epivillafranchian represents a separate episode of the development of the European megafauna, then the end of the Late Villafranchian in a narrow sense should end at the boundary with the Epivillafranchian. The boundary between these two episodes of faunal evolution is marked by new climatic changes leading to a new faunal turnover. Among the many faunal changes (some of which are controversial for taxonomic reasons; Bellucci et al. [111]), we can mention again the disappearance of *Bison* (*Eobison*), whose last occurrence is at Pirro and probably also at Apollonia (Sorbelli et al. [117]), as well as the first occurrence of *Praemegaceros verticornis* and *Bison menneri*, and perhaps also of *Megaloceros savini* (Kahlke [110]; Bellucci et al. [111] and references therein). Also worth mentioning here is the appearance of the evolved form of *Sus strozzi* (Iannucci [74]), which is possibly a separate subspecies (considered by some to be an ancient form of the later *Sus scrofa*).

2.3. Correlation of the Villafranchian Biochronology between Eastern and Western Europe (Correlation of Selected Eastern European Localities)

The study of the Villafranchian faunistic complexes from Eastern Europe and Central Asia is of great significance for the elucidation of the origin and evolution of this fauna in Europe as a whole. There is a logical trend to correlate stratigraphically the biocomplexes in Eastern and Southeastern Europe and even in Central Asia with those earlier defined in W. Europe by using the same biostratigraphic criteria. The dispersal of Villafranchian faunal elements through Europe is usually an instantaneous event from a geological point of view, but in some cases, W. Europe is reached later, and some species known in Eastern Europe never even reach Western European territories (Vangengeim [129]; Spassov [11]; Cregut-Bonnoure [22]). The environmental differences and the zoogeographic features of the fauna spread from the Urals to the most western and southwestern territories of the continent create a number of difficulties in the biochronological correlation. I have specifically focused on this issue in previous studies, in which I attempted such a correlation (Spassov [11,75]). Many new data (the discovery of new paleontological localities, new mass fossil material, and the absolute dating of a number of key deposits) make it possible to refine this correlation (Figure 2).

Early Villafranchian. The Balkan localities of mammalian megafauna from the early Villafranchian are few. The Romanian localities with *Mammuthus rumanus*, Cernăteşti and Tulucesti, as well as the Moldovian locality of Ripa Skortselskaya, must be placed by geochronology at 3.2–3.1 Ma (Radulescu et al. [55]; Nikiforova et al. [45]; Andreescu et al. [38]). The *M. rumanus* mandible from Bossilkovtsi (Bulgaria), originating from Pliocene sands (M. Böhme, University of Tubingen: pers. comm.) in its morphological features, is similar to the stage of this species from the indicated Romanian localities, and the Bossilkovtsi locality should be close in age. The archaic *Mammuthus* maxilla from Tsotylio, Greece, is also probably of a similar age (Kostopoulos, Koulidou [49]).

Middle Villafranchian. The Khapry faunal complex (Southern Russia), in which fauna is revealed in at least seven palaeontological localities, has a wide possible temporal span of ~400,000 years and occupies a time interval from the beginning of the Middle Villafranchian (not including its beginning) to ~2.2 Ma (Titov [77]). Given the presence of *P. brevirostris*, we can place this faunal assemblage at approximately between 2.5 and 2.1 Ma (see the discussion in the chapter The *Canis* Event). The earliest Middle Villafranchian locality on the Balkans appears to be Varshets, Bulgaria. According to its fauna, the presence of *Nyctereutes tingi* (Spassov [11; Tamvakis et al. [130]), the presence of the primitive *Gazellospira* sp. (known in Roca-Neyra) (Spassov [54]), and the evolutionary stage of *Martes* (Marciszak et al. [131]) shows a transition from the fauna of Roca-Neyra to that of Saint-Vallier (the benchmark locality of the MNQ 17 zone), with which there is a great similarity (Spassov [11,34]). Given the age of Roca-Neyra (2.6 Ma) and the assumed age of Saint-Vallier (over 2.4 and under/close to 2.5 Ma) (Nomade et al. [19]), we can assume that the age of Varshets is around 2.5 Ma. The Dafnero site, with several localities, is a typical MNQ17 site, which is somewhat later than Varshets after its fauna (Koufos, Kostopoulos [29,132]; Spassov [11]), which is confirmed by the results obtained for its absolute age (2.4–2.3 Ma, probably closer to 2.3) (Benammi et al. [133]). The Greek localities Sesklo and Volakas should also be of similar age (Benammi et al. [133]; Koufos, Kostopoulos [29]; Spassov [11]).

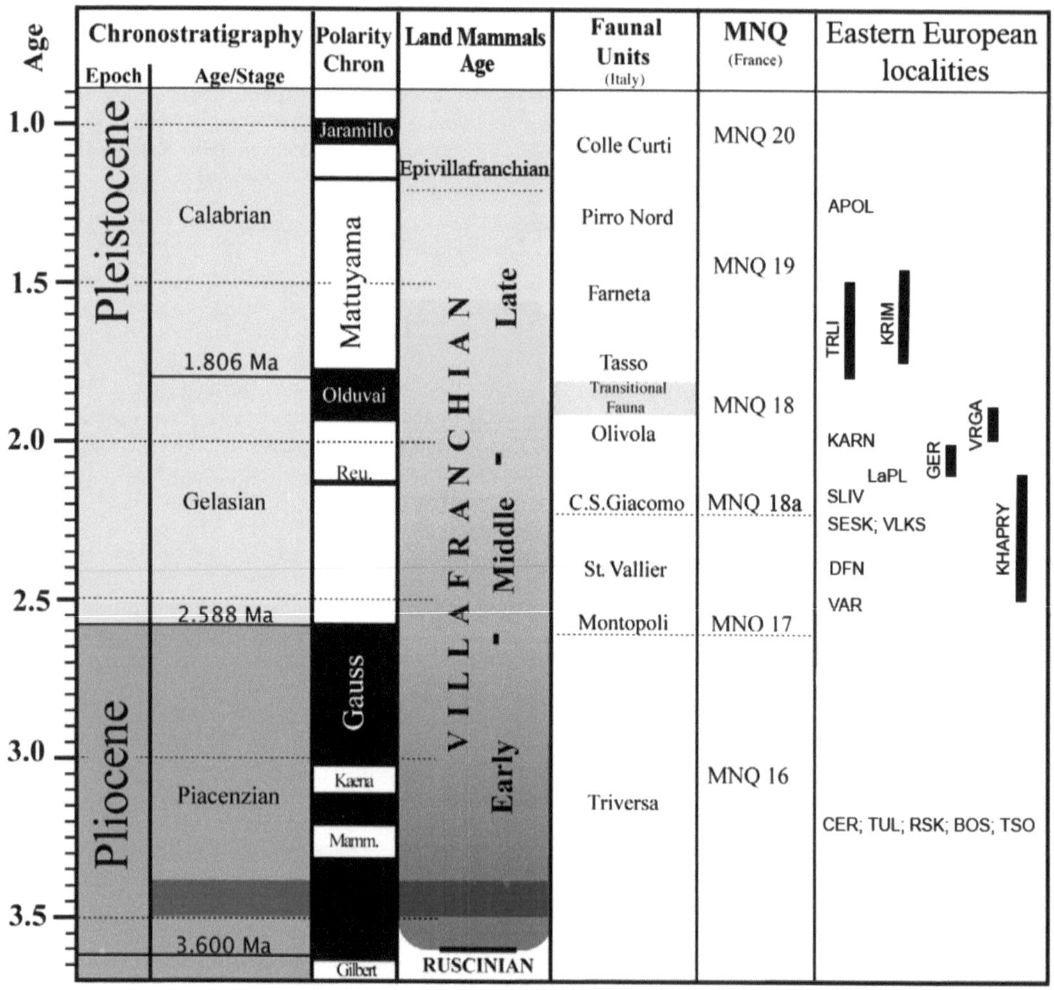

Figure 2. Correlation of the Villafranchian biochronology between Eastern and Western Europe (geological age and biochronological position of selected Eastern European Villafranchian localities). The biochronology table and the correlation between the faunal units and the MNQ zones is based on Nomade et al. [18] with some original modifications. The additional column on the far right presents the position of the Eastern European localities, discussed in Section 2.3. Abbreviations of polarity subchrons: Reu.—Réunion; Mamm.—Mammoth. Localities abbreviations: CER—Cernăteşti; TUL—Tulucesti; RSK—Ripa Skortselskaya; BOS—Bossilkovtsi; TSO—Tsotylio; DFN—Dafnero; SESK—Sesklo; VAR – Varshets; VLKS—Volakas; SLIV—Slivnitsa; LaPI—La Pietris; GER—Gerakarou; KARN—Karnezeika; VGRA—Vale Graunceanului; KRIM—Krimni; TRLI—Trlica; APOL—Apollonia.

Late Villafranchian. The biochronological position of Slivnitsa has been discussed in detail above. Slivnitsa should be the earliest Balkan locality, marking the beginning of the Late Villafranchian (MNQ18a), and deserves to represent a Balkan analogue of Coste San Giacomo Unit, in which unit (CSG), we have reason to place this locality. La Pietris, Romania, does not have a very clear position. The site still has *Nyctereutes megamastoides*, which is absent in Olivola (but this may also be due to geographical differences). Characteristic artiodactyls are *Pliotragus ardeus* (Radulescu et al. [55]) and the cervid Rucervus (*Arvernoceros*) *radulescui*

(Terhune et al. [39]), which appear to be generally typical of "pre-Pachycrocuta event" faunas (Terhune et al. [39]; Croitor et al. [94]). The site is perhaps only slightly later than Slivnitsa and similar in age to La Puebla (Spassov [11]), placed today (Nomade et al. [19]) at ~2.1–2.0 Ma. Of fairly close age is probably the Romanian site Valea Graunceanului (Radulescu et al. [55]), which has a largely similar fauna, but also *Smutsia* (Pholidota) and possibly also *Pachycrocuta* (?) (Terhune at al. [39]). This site, which presents apparently wetter and forested conditions, should be later than Slivnitsa, but probably also a little later than La Pietris (Spassov [11]), with an age perhaps corresponding to the beginning of the Olduvai subchron (?), which, however, somewhat contradicts the presence of some "pre-Pachycrocuta event" ungulates. The Geek locality of Gerakarou-1 was especially discussed in the analysis of the *Pachycrocuta* event as a locality that possibly marked the first appearance of this hyena on the continent (see above). It marks the beginning of the Late Villafranchian in Greece (Koufos [97]; Konidaris, Kostopoulos [41,134]) and shows many similarities with Slivnitsa. At the same time, *Pachycrocuta* already occurs in Gerakarou-1, and it should be placed a little later than this Bulgarian locality (Spassov [11]), most likely at the very beginning of the Olivola FU (MNQ18b, according to my understanding), which is why we place it here at ca. 2.1–2 Ma. Almost of the same age (~2 Ma) should be the pre-Olduvai locality of Karnezeika in Southern Greece (Sianis et al. [95]). The fauna from the Late Villafranchian locality of Krimni-3 (Northern Greece) is placed within rather wide limits (1.8–1.5 Ma) and apparently shows the last occurrence of *Palaeotragus* and *Sus strozzi* s. str. in the southern Balkans, as well as the presence of the giant ostrich *Pachystruthio dmanisensis* (Konidaris et al. [135]). The presence in this fauna of *Stephanorhinus* cf. *hundsheimensis* contradicts this dating because this rhino is apparently known in Europe and in Anatolia from about the beginning of the Epivillafranchian (Pandolfi, Erten [136]), and I place this locality within the indicated limits somewhat tentatively. The fauna from the Trlica cave (Montenegro), TRL11-10 level, has also been determined in similar, broad limits—1.8–1.5 Ma (Vislobokova et al. [137]). The Greek locality of Apollonia has an important biochronological significance. It is positioned by Koufos and Kostopoulos [29] in the Epivillafranchian. At the same time, the presence of the primitive bison *B.* (*Eobison*) (Sorbelli et al. [117]) seems to confirm to one degree or another my statement (Spassov [11]) that the site has significant similarities with Pirro (Pirro FU), which has an age of ca. 14–1.2 Ma (Napoleone et al. [98]). However, Apollonia shows a transitional fauna to the Epivillafranchian one. In this sense, placing the locality within ~ 1.3–1.1 Ma (Konidaris, Kostopoulos [41,134]) or at the boundary of the Late Villafranchian s. str./Epivillafranchian, at the time of the beginning of a new strong aridification (see Spassov [34]), i.e., at 1.2 Ma or slightly before (not later than this date) seems logical.

Funding: This study research received support from the Bulgarian National Foundation (Contract No. KP-06-H44/6 from 14 July 2021).

Data Availability Statement: The considered material is curated and available for study in the public institutions mentioned in the text, and the data analyzed in this work are available either directly within this article or from the cited references.

Acknowledgments: I am very grateful to Angélique Monguillon (Lyon) for the casts of "*Canis*" remains from Vialette sent years ago, as well as to Marius Robu and Alex Petculescu ("Emil Racovita" Institute of Speleology, Bucharest) and to Stefan Vasile (University of Bucharest) for kindly providing access to the collections of these institutions in Bucharest. I would like to sincerely thank for the helpful discussions and for sending information and photographs (some of them unpublished) Claire Sagne (Laboratory de Paléontologie—MNHN, Paris [CR2P]), Eric Delson (American Museum of Natural History, New York), Saverio Bartolini-Lucenti (University of Florence, Firenze), Alessio Iannucci (Eberhard Karls University Tübingen), Alain Argant (Aix Marseille Université, Aix-en-Provence), Athanassios Athanassiou (Hellenic Ministry of Culture, Athens), and Georgi N. Markov (National Museum of Natural History-BAS, Sofia). I would like to thank the guest editor of the special issue, Alessio Iannucci, as well as the two anonymous reviewers for their comments and suggestions, which greatly contributed to the improvement of the manuscript.

Conflicts of Interest: The author declare no conflicts of interest.

References

1. Böhme, M.; Spassov, N.; Majidifard, M.; Gärtner, A.; Kirscher, U.; Marks, M.; Dietzel, C.; Uhlig, G.; El Atfy, K.; Begun, D.B.; et al. Neogene hyperaridity in Arabia drove the directions of mammalian dispersal between Africa and Eurasia. *Commun. Earth Environ.* **2021**, *2*, 85. [CrossRef]
2. Van Asperen, E.; Kahlke, R.-D. Dietary traits of the late Early Pleistocene Bison menneri (Bovidae, Mammalia) from its type site Untermassfeld (Central Germany) and the problem of Pleistocene 'wood bison'. *Quat. Sci. Rev.* **2017**, *177*, 299–313. [CrossRef]
3. Spassov, N. The Mammalian Megafauna from the Late Villafranchian Localities Varshets and Sivnitsa (Bulgaria) and the Biochronology of the Villafranchian in S.-E. Europe. Ph.D. Thesis, National Museum Natural History, Bulgarian Academy of Sciences, Sofia, Bulgaria, 1999; p. 284. (In Bulgarian).
4. Rook, L.; Martínez-Navarro, B. Villafranchian: The long story of a Plio-Pleistocene European large mammal biochronologic unit. *Quat. Int.* **2010**, *219*, 134–144. [CrossRef]
5. Vangengeim, E.; Sotnikova, M. Paleozoogeographic characteristics of the fauna. In *Biostratigraphy of the Late Pliocene–the Early Pleistocene of Tajikistan*; Nikiforova, K., Vangengeim, E., Eds.; Nauka: Moscow, Russia, 1988; pp. 103–114.
6. Sotnikova, M. *Carnivore Mammals of the Pliocene and the Early Pleistocene*; Nauka: Moskow, Russia, 1989; p. 122.
7. Sotnikova, M.; Rook, L. Dispersal of the Canini (Mammalia, Canidae: Caninae) across Eurasia during the Late Miocene to Early Pleistocene. *Quat. Int.* **2010**, *212*, 86–97. [CrossRef]
8. Azzaroli, A. Villafranchian correlations based on large mammals. *G. Geol.* **1970**, *35*, 111–131.
9. Azzaroli, A. The villafranchian stage in Italy and the Plio–Pleistocene boundary. *G. Geol.* **1977**, *41*, 61–79.
10. Spassov, N. The Villafranchian mammalian fauna and its investigation in Europe, on the Balkans and in Bulgaria. *Geol. Balc.* **1997**, *27*, 83–94.
11. Spassov, N. The Plio-Pleistocene vertebrate fauna in South-Eastern Europe and the megafaunal migratory waves from the east to Europe. *Rev. Paleobiol.* **2003**, *22*, 197–229.
12. Lindsay, E. Chronostratigraphy, Biochronology, Datum Events, Land Mammal Ages, Stage of Evolution, and Appearance Event Ordination. *Bull. Am. Mus. Nat. Hist.* **2003**, 212–230. [CrossRef]
13. Mein, P. Résultats du groupe de travail des Vertébrés. In *Report on Activity on the RCMNS Working Group (1971–1975); VIth Congress of the Regional Committee on Mediterranean Neogene Stratigraphy*; Bratislava, Czechoslovakia, 1975; pp. 78–81.
14. Guérin, C. Premiere Biozonation du Pleistocene Europeen, principal resultat biostratigraphique de l'etude des Rhinocerotidae du miocene terminal au pleistocene superieur d'Europe Occidentale. *Geobios* **1982**, *15*, 593–598. [CrossRef]
15. Guérin, C. Biozones or Mammal Units? Methods and limits in Biochronology. In *European Neogene Mammal Chronologt*; Lindsay, E.H., Fahlbusch, V., Mein, P., Eds.; Plenum Press: New York, NY, USA, 1990; pp. 119–130.
16. Azzaroli, A. Quaternary mammals and the end-villafranchian dispersal Event—A turning point in the history of Eurasia. *Palaeogeogr. Palaeoclimatol. Palaeoecol.* **1983**, *44*, 117–139. [CrossRef]
17. Torre, D.; Ficcarelli, G.; Masini, F.; Rook, L.; Sala, B. Mammal dispersal events in the early Pleistocene of Western Europe. *Cour. Forschungsinstitut Senckenberg* **1992**, *153*, 51–58.
18. Gliozzi, E.; Abbazzi, L.; Argenti, P.; Azzaroli, A.; Caloi, L.; Capasso Barbato, L.; Di Stefano, G.; Esu, D.; Ficcarelli, G.; Girotti, O.; et al. Biochronology of selected mammals, molluscs and ostracods from the Middle Pliocene to the Late Pleistocene in Italy. The state of the art. *Riv. Ital. Paleontol. Stratigr.* **1997**, *103*, 369–388. [CrossRef]
19. Nomade, S.; Pastre, J.; Guillou, H.; Faure, M.; Guérin, C.; Delson, E.; Debard, E.; Voinchet, P.; Messager, E. 40Ar/39Ar constraints on some French Landmark Late Pliocene to Early Pleistocene large mammalian paleofaunas, paleoenvironmental and paleoecological implications. *Quat. Geochronol.* **2014**, *421*, 2–15. [CrossRef]
20. Lindsay, E.; Opdyke, N.; Johnson, N. Pliocene dispersal of the horse Equus and late Cenozoic mammalian dispersal events. *Nature* **1980**, *287*, 135–138. [CrossRef]
21. Spassov, N. The Late Villafranchian and biochronology of South-East Europe: Faunal migrations, environmental changes and possible first appearance of the genus *Homo* on the continent. In *On Late Plio/Pleistocene Extinctions and Evolution in the Palearctic: The Vatera Site, Proceedings of the Ist International Workshop, Vatera, Lesvos, Greece, 29 September–2 October 1999*; 2002; Annales Geologiques des Pays Hellenique, 1 Série, A; Volume 39, pp. 221–243.
22. Cregut-Bonnoure, É. Apport des Caprinae et Antilopinae (Mammalia, Bovidae) à la biostratigraphie du Pliocène terminal et du Pléistocène d'Europe. *Quaternaire* **2007**, *18*, 73–97. [CrossRef]
23. Sardella, R.; Palombo, M. The Pliocene-Pleistocene boundary: Which significance for the so called "Wolf event"? evidences from Western Europe. *Quaternaire* **2007**, *18*, 65–71. [CrossRef]
24. Palombo, M.; Sardella, R.; Novelli, M. Carnivora dispersal in Western Mediterranean during the last 2.6Ma. *Quat. Int.* **2008**, *179*, 176–189. [CrossRef]
25. Croitor, R.; Brugal, J.-P. Ecological and evolutionary dynamics of the carnivore community in Europe during the last 3 million years. *Quat. Int.* **2010**, *212*, 98–108. [CrossRef]
26. O'Regan, H.; Turner, A.; Bishop, L.; Elton, S.; Lamb, A. Hominins without fellow travellers? First appearances and inferred dispersals of Afro-Eurasian large-mammals in the Plio-Pleistocene. *Quat. Sci. Rev.* **2011**, *30*, 1343–1352. [CrossRef]
27. Palombo, M. Deconstructing mammal dispersals and faunal dynamics in SW Europe during the Quaternary. *Quat. Sci. Rev.* **2014**, *96*, 50–71. [CrossRef]

28. Palombo, M. Discrete dispersal bioevents of large mammals in Southern Europe in the post-Olduvai Early Pleistocene: A critical overview. *Quat. Int.* **2017**, *431*, 3–19. [CrossRef]
29. Koufos, G.; Kostopoulos, D. The Plio-Pleistocene Large Mammal Record of Greece: Implications for Early Human Dispersals into Europe. In *Paleoanthropology of the Balkans and Anatolia*; Harvati, K., Roksandic, M., Eds.; Vertebrate Paleobiology and Paleoanthropology Series; Springer Science+Business Media: Dordrecht, The Netherlands, 2016; pp. 269–281.
30. Iannucci, A.; Mecozzi, B.; Sardella, R.; Iurino, D. The extinction of the giant hyena *Pachycrocuta brevirostris* and a reappraisal of the Epivillafranchian and Galerian Hyaenidae in Europe: Faunal turnover during the Early–Middle Pleistocene Transition. *Quat. Sci. Rev.* **2021**, *272*, 107240. [CrossRef]
31. Iannucci, A.; Mecozzi, B.; Sardella, R. Beware of the "Wolf event"—Remarks on large mammal dispersals in Europe and the late Villafranchian faunal turnover. *Alp. Mediterr. Quat.* **2023**, *36*, 75–90.
32. Samson, P. Les Equidés fossils de Roumanie (Pliocène moyen—Pléistocène supérieur). *Geol. Romana* **1975**, *14*, 165–352.
33. Radulescu, C.; Samson, P. Biochronology and evolution of the Early Pliocene to the Early Pleistocene mammalian faunas of Romania. *Boll. Della Soc. Paleontol. Ital.* **2001**, *40*, 285–291.
34. Spassov, N. Southeastern Europe as a Route for the Earliest Dispersal of Homo Toward Europe: Ecological Conditions and the Timing of the First Human Occupation of Europe. In *Paleoanthropology of the Balkans and Anatolia*; Harvati, K., Roksandic, M., Eds.; Vertebrate Paleobiology and Paleoanthropology Series; Springer Science+Business Media: Dordrecht, The Netherlands, 2016; pp. 281–290. [CrossRef]
35. Palombo, M.; Valli, A.; Kostopoulos, D.; Alberdi, M.; Spassov, N.; Vislobokova, I. Similarity Relationships between the Pliocene to Middle Pleistocene Large Mammal Faunas of Southern Europe from Spain to the Balkans and the North Pontic Region. *Cour. Forsch. Inst. Senckenberg* **2006**, *256*, 329–347.
36. Kostopoulos, D.; Palombo, M.; Alberdi, M.; Valli, A. Pliocene to Pleistocene large mammal diversity and turnover in North Mediterranean region: The Greek Peninsula with respect to the Iberian and Italian ones. *Geodiversitas* **2007**, *29*, 401–419.
37. Kahlke, R.-D.; García, N.; Kostopoulos, D.; Lacombat, F.; Lister, A.M.; Mazza, P.; Spassov, N.; Titov, V. Western Palaearctic palaeoenvironmental conditions during the Early and early Middle Pleistocene inferred from large mammal communities, and implications for hominin dispersal in Europe. *Quat. Sci. Rev.* **2011**, *30*, 1368–1395. [CrossRef]
38. Andreescu, I.; Codrea, V.; Lubenescu, V.; Munteanu, T.; Petculescu, A.; Stiucă, E.; Terzea, E. New developments in the upper pliocene–pleistocene stratigraphic units of the dacian basin (eastern paratethys), Romania. *Quat. Int.* **2013**, *284*, 15–29. [CrossRef]
39. Terhune, C.; Curran, S.; Croitor, R.; Dragușin, V.; Gaudin, T.; Petculescu, A.; Robinson, C.; Robu, M.; Werdelin, L. Early Pleistocene fauna of the Oltet¸ River Valley of Romania: Biochronological and biogeographic implications. *Quat. Int.* **2020**, *533*, 14–33. [CrossRef]
40. Palombo, M.R. Large mammal faunal dynamics in southwestern Europe during the late Early Pleistocene: Implications for the biochronological assessment and correlation of mammalian faunas. *Alp. Mediterr. Quat.* **2016**, *29*, 143–168.
41. Konidaris, G.E.; Kostopoulos, D.S. The Late Pliocene–Middle Pleistocene Large Mammal Faunal Units of Greece. *Quaternary* **2024**, *7*, 27. [CrossRef]
42. Iannucci, A.; Sardella, R. What Does the "Elephant-Equus" Event Mean Today? Reflections on Mammal Dispersal Events around the Pliocene-Pleistocene Boundary and the Flexible Ambiguity of Biochronology. *Quaternary* **2023**, *6*, 16. [CrossRef]
43. Radulesco, C.; Samson, P. Review of the Villafranchian s.s. faunas of Romania. *Alp. Mediterr. Quat.* **1995**, *8*, 377–382.
44. Alexeeva, L. On the new theriocomplex in Northern Black Sea zone. *Bull. Aacademy Sci. Georgian SSR* **1977**, *86*, 485–488. (In Russian)
45. Nikiforova, K.; Alexandrova, L.; Troubikhin, V.; Tchepalyga, A. Correlation of Pliocene and Eopleistocene deposits of the European part of the and Romania. The continental Upper Pliocene of the Black-Caspian region. *Izv. Akad. Nauk. USSR Ser. Geol.* **1986**, 5–17. (In Russian)
46. Markov, G.; Spassov, N. Primitive mammoths from Northeast Bulgaria in the context of the earliest mammoth migrations in Europe. In *Advances in Vertebrate Paleontology "Hent to Pantha"*; Petculescu, A., Stiuca, E., Eds.; Institute of Speleology of the Romanian Academy: Bucharest, Romania, 2003; pp. 53–58.
47. Lister, A.M.; van Essen, H.E. Mammuthus rumanus the earliest mammoth in Europe. In *Advances in Vertebrate Paleontology "Hent to Pantha"*; Petculescu, A., Stiuca, E., Eds.; Institute of Speleology of the Romanian Academy: Bucharest, Romania, 2003; pp. 47–52.
48. Markov, G. *Mammuthus rumanus*, early mammoths, and migration out of Africa: Some interrelated problems. *Quat. Int.* **2012**, *276*, 23–26. [CrossRef]
49. Kostopoulos, D.; Koulidou, I. An early mammoth maxilla from north-western Greece. *Quat. Int.* **2015**, *379*, 155–163. [CrossRef]
50. Rook, L.; Bernor, R.; Avilla, L.S.; Cirilli, O.; Flynn, L.; Jukar, A.; Sanders, W.; Scott, E.; Wang, X. Mammal Biochronology (Land Mammal Ages) Around the World from Late Miocene to Middle Pleistocene and Major Events in Horse Evolutionary History. *Front. Ecol. Evol.* **2019**, *7*, 278. [CrossRef]
51. Sun, B.; Deng, T. The Equus Datum and the Early Radiation of Equus in China. *Front. Ecol. Evol.* **2019**, *7*, 429. [CrossRef]
52. Bernor, R.; Cirilli, O.; Jukar, A.M.; Potts, R.; Buskianidze, M.; Rook, L. Evolution of Early Equus in Italy, Georgia, the Indian Subcontinent, East Africa, and the Origins of African Zebras. *Front. Ecol. Evol.* **2019**, *7*, 166. [CrossRef]
53. Eisenmann, V. The equids from Liventsovka and other localities of the Khaprovskii Faunal Complex, Russia: A revision. *Geobios* **2022**, *70*, 17–33. [CrossRef]
54. Spassov, N. Brief Review of the Pliocene Ungulate Fauna of Bulgaria. *Quat. Hors-Sér.* **2005**, *2*, 201–212.

55. Radulescu, C.; Samson, P.; Petculescu, A.; Stiucă, E. Pliocene Large Mammals of Romania. *Coloq. Paleontol.* **2003**, *1*, 549–558.
56. Forstén, A. A review of *Equus stenonis* Cocchi (Perissodactyla, Equidae) and related forms. *Quat. Sci. Rev.* **1999**, *18*, 1373–1408. [CrossRef]
57. Forstén, A. Mitochondrial-DNA time-table and the evolution of Equus: Comparison of molecular and paleontological evidence. *Ann. Zool. Fenn.* **1991**, *28*, 301–309.
58. Crespo, V.; Vasile, S.; Petculescu, A.; Rătoi, B.; Haiduc, B. The Early Pliocene small mammals (Eulipotyphla, Rodentia, Lagomorpha) from Berești and Mălușteni (eastern Romania): A fresh look at old collections. *Earth Environ. Sci. Trans. R. Soc. Edinb.* **2023**, *114*, 49–63. [CrossRef]
59. Turner, A. Villafranchian-Galerian larger carnivores of Europe dispersions and extinctions. *Cour. Forschungsinst. Senckenberg* **1992**, *153*, 153–160.
60. Bartolini-Lucenti, S.; Spassov, N. Cave canem! The earliest *Canis* (*Xenocyon*) (Canidae, Mammalia) of Europe: Taxonomic affinities and paleoecology of the fossil wild dogs. *Quat. Sci. Rev.* **2022**, *276*, 107315. [CrossRef]
61. Lacombat, F.; Abbazzi, L.; Ferretti, M.; Martínez-Navarro, B.; Moullé, P.-E.; Palombo, M.; Rook, L.; Turner, A.; Valli, A.-M. New data on the Early Villafranchian fauna from Vialette (Haute-Loire, France) based on the collection of the Crozatier Museum (Le Puy-en-Velay, Haute-Loire, France). *Quat. Int.* **2008**, *179*, 64–71. [CrossRef]
62. Heintz, E.; Guérin, C.; Martin, R.; Prat, A. Principaux gisements villafranchiens de France: Listes fauniques et biostratigraphie. *Mémoires Bur. Rech. Géol. Min.* **1974**, *78*, 169–182.
63. Martin, R. Trois nouvelles espèces de Caninae (Canidae, Carnivora) des gisements plio-villafranchiens d'Europe. *Doc. Lab. Géol. La Fac. Sci. Lyon. Notes Mém.* **1973**, *57*, 87–96.
64. Tedford, R.; Qiu, Z. A new canid genus from the Pliocene of Yushe, Shanxi Province. *Vertebr. PalAsiatica* **1996**, *34*, 27–40.
65. Heintz, E. *Les Cervides Villafranchiens de France et d'Espagne*; Voll. I; Nouv. Ser., Ser. C; Muséum National d'Histoire Naturelle: Paris, France, 1970; Volume 34, p. 303.
66. Spassov, N. A New Late Villafranchian Locality of vertebrate Fauna—Slivnitsa (Bulgaria) and the Carnivore Dispersal Events in Europe on the Pliocene/Pleistocene Boundary. *Hist. Nat. Bulg.* **1998**, *9*, 101–113.
67. Torre, D. The Ruscinian and the Villafranchian dogs of Europe. *Boll. Della Soc. Paleontol. Ital.* **1979**, *18*, 162–165.
68. Fosse, P.; Madelaine, S.; Oberlin, C.; Cretin, C.; Bonnet-Jacquement, P. Un lion des cavernes (*Panthera* (*Leo*) *spelaea*) exploité au Dryas récent: Les données du gisement du Peyrat (Saint-Rabier, Dordogne, France). *Paleo* **2017**, *28*, 91–115. [CrossRef]
69. Marciszak, A.; Kropczyk, A.; Gornig, W.; Kot, M.; Nadachowski, A.; Lipecki, G. History of Polish Canidae (Carnivora, Mammalia) and Their Biochronological Implications on the Eurasian Background. *Genes* **2023**, *14*, 539. [CrossRef]
70. Argant, A. Les carnivores du gisement Pliocène final de Saint-Vallier (Drôme, France). *Geobios* **2004**, *37*, 133–182. [CrossRef]
71. Monguillon-Douillet, A. Les carnivores du site plio-pleistocène de Chilhac (Haute-Loire, France). In *Etude Paléontologique, Biochronologique, Paléoécologique et Taphonomique*; Unpublished Mémoire de D.E.A. Quaternaire et Géologie; Muséum National d'Histoire Naturelle: Paris, France, 2000.
72. Spassov, N. The position of jackals in the *Canis* Genus and life-history of the golden jackal (*Canis aureus* L.) in Bulgaria and the Balkans. *Hist. Nat. Bulg.* **1989**, *1*, 44–56.
73. Florindo, F.; Marra1, F.; Angelucci, D.; Biddittu, I.; Bruni, L.; Florindo, F.; Gaeta, M.; Guillou, H.; Jicha, L.; Macrì, P.; et al. Environmental evolution, faunal and human occupation since 2 Ma in the Anagni basin, central Italy. *Sci. Rep.* **2021**, *11*, 7056. [CrossRef] [PubMed]
74. Iannucci, A. Sus strozzii (Suidae, Mammalia) from the historical locality of Quercia (Early Pleistocene, Italy). *Geobios* **2023**, *77*, 27–36. [CrossRef]
75. Spassov, N. Biochronology and zoogeographic affinities of the Villafranchian faunas of South Europe. *Historia naturalis bulgarica.* **2000**, *12*, 89–128.
76. Zubakov, V.; Borzenkova, I. Global Palaeoclimate of the Late Cenozoic. Developments in palaeontology and stratigraphy. *Amst. Elsevier* **1990**, *12*, 456.
77. Titov, V. *Late Pliocene Large Mammals from Northeastern Sea of Azov Region*; Matishov, G.G., Ed.; SSC RAS Publishing: Rostov-on-Don, Russia, 2008; p. 264. (In Russian)
78. Sotnikova, M.; Bajgusheva, V.; Titov, V. Carnivores of the Khapry faunal assemblage and their stratigraphic implications. *Stratigr. Geol. Correl.* **2002**, *10*, 375–390.
79. Werdelin, L.; Dehghani, R. Carnivora. In *Paleontology and Geology of Laetoli: Human Evolution in Context*; Harrison, T., Ed.; Vertebrate Paleobiology and Paleoanthropology Series; Springer: Dordrecht, The Netherlands, 2011; pp. 189–232. [CrossRef]
80. Hemmer, H.; Kahlke, R.-D.; Vekua, A.K. The cheetah *Acinonyx pardinensis* (Croizet et Jobert, 1828) s.l. at the hominin site of Dmanisi (Georgia)—A potential prime meat supplier in Early Pleistocene ecosystems. *Quat. Sci. Rev.* **2011**, *30*, 2703–2714. [CrossRef]
81. Cherin, M.; Iurino, D.; Sardella, R.; Rook, L. *Acinonyx pardinensis* (Carnivora, Felidae) from the Early Pleistocene of Pantalla (Italy): Predatory behavior and ecological role of the giant PlioePleistocene cheetah. *Quat. Sci. Rev.* **2014**, *87*, 82–97. [CrossRef]
82. Hugueney, M.; Guérin, C.; Poidevin, J.L. Découverte de *Trogontherium minus* Newton 1890 (Rodentia: Castoridae) dans le Villafranchien inférieur de Perrier- Étouaires (Puy-de-Dôme, France): Implications phylogénétiques. *C. R. Acad. Sci. Ser. II* **1989**, *309*, 763–768.

83. Hemmer, H.; Kahlke, R.-D.; Vekua, A. The Old World puma—*Puma pardoides* (Owen, 1846) (Carnivora: Felidae)—In the Lower Villafranchian (Upper Pliocene) of Kvabebi (East Georgia, Transcaucasia) and its evolutionary and biogeographical significance. *Neues Jahrb. Geol. Paläontol. Abh.* **2004**, *233*, 197–231. [CrossRef]
84. Hemmer, H.; Kahlke, R.-D. New results on felids from the Early Pleistocene site of Untermassfeld. In *The Pleistocene of Untermassfeld near Meiningen*; Kahlke, R.-D., Ed.; Part 5; Verlag des Römisch-Germanischen Zentralmuseums: Weimar, Germany, 2022; pp. 1465–1566.
85. Cherin, M.; Iurino, D.; Sardella, R.; Rook, L. Earliest occurrence of *Puma pardoides* (Owen, 1846) (Carnivora, Felidae) at the Plio/Pleistocene transition in western Europe: New evidence from the Middle Villafranchian assemblage of Montopoli, Italy. *C. R. Palevol* **2013**, *12*, 165–171. [CrossRef]
86. Sotnikova, M. Verhnepliotsenovye Carnivora Tsentralnoy Azii (The Late Pliocene carnivores of Central Asia). *Izv. Akad. Nauk. SSSR. Seriya Geol.* **1976**, *11*, 133–137. (In Russian)
87. Hemmer, H. Zur Charakterisierung und stratigraphischen Bedeutung von *Panthera gombaszoegensis* (Kretzoi, 1938). *Neues Jahrb. Geol. Paläontol. Monatshefte* **1971**, *12*, 701–711.
88. Hemmer, H. Die Feliden aus dem Epivillafranchium von Untermaßfeld. In *Das Pleistozän von Untermaßfeld bei Meiningen (Thüringen). Teil 3*; Kahlke, R.-D., Ed.; Des Römisch-Germanischen Zentralmuseums: Mainz, Germany, 2001; Volume 40, pp. 699–782.
89. Jiangzuo, Q.; Liu, J. First record of the Eurasian jaguar in southern Asia and a review of dental differences between pantherine cats. *J. Quat. Sci.* **2020**, *35*, 817–830. [CrossRef]
90. Wagner, J. A short nomenclatural note to correct spelling of the species-group name "*gombaszögensis*". *Quat. Hors-Ser.* **2011**, *4*, 245.
91. van den Hoek Ostende, L.; de Vos, J. A century of research on the classical locality of Tegelen (province of Limburg, The Netherlands). *Cour. Forsch.-Inst. Senckenberg* **2006**, *256*, 291–304.
92. Koufos, G. The Pleistocene carnivores of the Mygdonia basin (Macedonia, Greece). *Ann. Paléontol.* **1992**, *78*, 205–257.
93. Martínez-Navarro, B. Early Pleistocene faunas of Eurasia and hominid dispersals. In *Out of Africa I: The First Hominin Colonization of Eurasia*; Fleagle, J., Shea, J., Grine, F., Baden, A., Leakey, R., Eds.; Vertebrate Paleobiology and Paleoanthropology Series; Springer: Dordrecht, The Netherlands, 2010; pp. 207–224.
94. Croitor, R.; Robinson, C.; Curran, S.; Terhune, C.; Drăgușin, V.; Pobiner, B.; Popescu, D.; Petculescu, A. Early pleistocene ruminants (Artiodactyla, Mammalia) from the Dacian Basin (South Romania) before and after the *Pachycrocuta* event: Implications for hominin dispersals in Western Eurasia. *Hist. Biol.* **2023**, *36*, 485–533. [CrossRef]
95. Sianis, P.; Athanassiou, A.; Roussiakis, S.; Iliopoulos, G. Carnivora from the Early Pleistocene locality of Karnezeika (Southern Greece). *Geobios* **2023**, *79*, 43–59. [CrossRef]
96. Madurell-Malapeira, J.; Ros-Montoya, S.; Espigares, M.; Alba, D.; Aurell-Garrido, J. Villafranchian large mammals from the Iberian Peninsula: Paleobiogeography, paleoecology and dispersal events. *J. Iber. Geol.* **2014**, *40*, 167–178. [CrossRef]
97. Koufos, G. The Villafranchian carnivoran guild of Greece: Implications for the fauna, biochronology and paleoecology. *Integr. Zool.* **2014**, *9*, 444–460. [CrossRef]
98. Napoleone, G.; Albianelli, A.; Azzaroli, A.; Bertini, A.; Magi, M.; Mazzini, M. Calibration of the Upper Valdarno basin to the Plio-Pleistocene for correlating the Apennine continental Sequences. *Alp. Mediterr. Quat.* **2003**, *16*, 131–166.
99. Bellucci, L.; Mazzini, I.; Scardia, G.; Bruni, L.; Parenti, F.; Segre, A.G.; Segre Naldini, E.; Sardella, R. The site of Coste San Giacomo (Early Pleistocene, Central Italy): Palaeoenvironmental analysis and biochronological overview. *Quat. Int.* **2012**, *267*, 30–39. [CrossRef]
100. Fidalgo, D.; Galli, E.; Madurell-Malapeira, J.; Rosas, A. Earliest Pleistocene European hippos: A review. *Comun. Geol.* **2021**, *108*, 65–69.
101. Reimann, C.; Strauch, F. Ein *Hippopotamus*-Schädel aus dem Pliozän von Elis (Peloponnes, Griechenland). *Neues Jahrb. Geol. Paläontol. Abh.* **2008**, *249*, 203–222. [CrossRef]
102. Athanassiou, A. The fossil record of continental hippopotamids (Mammalia: Artiodactyla: Hippopotamidae) in Greece. In *Fossil Vertebrates of Greece, 2*; Vlachos, E., Ed.; Springer Nature: Cham, Switzerland, 2022; pp. 281–289.
103. Marra, F.; Pandolfi, L.; Petronio, C.; Di Stefano, G.; Gaeta, M.; Salari, L. Reassessing the sedimentary deposits and vertebrate assemblages from Ponte Galeria area (Rome, central Italy): An archive for the Middle Pleistocene faunas of Europe. *Earth-Sci. Rev.* **2018**, *139*, 104–122. [CrossRef]
104. Mecozzi, B.; Iannucci, A.; Mancini, M.; Tentori, D.; Cavasinni, C.; Conti, J.; Messina, M.Y.; Sarra, A.; Sardella, R. Reinforcing the idea of an early dispersal of *Hippopotamus amphibius* in Europe: Restoration and multidisciplinary study of the skull from the Middle Pleistocene of Cava Montanari (Rome, central Italy). *PLoS ONE* **2023**, *18*, e0293405. [CrossRef]
105. Mazza, P.; Rustioni, M. The fossil bear from Senèze (Southern France). *Rend. Lincei* **1994**, *5*, 17–26. [CrossRef]
106. Mazza, P. If Hippopotamuses cannot swim, how did they colonize islands? *Lethaia* **2014**, *47*, 494–499. [CrossRef]
107. Made, J.; Sahnouni, M.; Boulaghraief, K. *Hippopotamus* gorgops from El Kherba (Algeria) and the context of its biogeography. In Proceedings of the II Meeting of African Prehistory, Burgos, Spain, 15–16 April 2015; Sahnouni, M., Semaw, S., Rios Garaizar, J., Eds.; Consorcio CENIEH: Burgos, Spain, 2017; pp. 135–169.
108. Parker, J. What Biome Do Hippos Live In? *Build a Stash Newsletter*. 16 March 2023. Available online: https://www.buildastash.com/post/what-biome-do-hippos-live-in (accessed on 10 February 2024).
109. Koufos, G. The Villafranchian mammalian faunas and biochronology of Greece. *Boll. Della Soc. Paleontol. Ital.* **2001**, *40*, 217–223.

110. Kahlke, R.-D. Late Early Pleistocene European large mammals and the concept of an Epivillafranchian biochron. In *Late Neogene and Quaternary Biodiversity and Evolution: Regional Developments and Interregional Correlations, Proceedings of the 18th International Senckenberg Conference (VI International Palaeontological Colloquium in Weimar, 2004)*; Volume II; Kahlke, R.-D., Maul, L.C., Mazza, P., Eds.; 2007; Volume 259, pp. 265–278, Courier Forschungsinstitut Senckenberg.
111. Bellucci, L.; Sardella, R.; Rook, L. Large mammal biochronology framework in Europe at Jaramillo: The Epivillafranchian as a formal biochron. *Quat. Int.* **2014**, *389*, 84–89. [CrossRef]
112. Cherin, M.; D'Allestro, V.; Masini, F. New Bovid Remains from the Early Pleistocene of Umbria (Italy) and a Reappraisal of *Leptobos merlai*. *J. Mamm. Evol.* **2019**, *26*, 201–224. [CrossRef]
113. Akbar Khan, M.; Kostopoulos, D.; Akhtar, M.; Nazir, M. Bison remains from the Upper Siwaliks of Pakistan. *Neues Jahrb. Geol. Paläontologie Abh.* **2010**, *258*, 121–128. [CrossRef]
114. Masini, F.; Palombo, M.R.; Rozzi, R. A reappraisal of the Early to Middle Pleistocene Italian Bovidae. *Quat. Int.* **2013**, *288*, 45–62. [CrossRef]
115. Mead, J.; Jin, C.; Wei, G.; Sun, C.; Wang, Y.; Swift, S.; Zheng, L. New data on Leptobos crassus (Artiodactyla, Bovidae) from Renzidong Cave, Early Pleistocene (Nihewanian) of Anhui, China, and an overview of the genus. *Quat. Int.* **2014**, *354*, 139–146. [CrossRef]
116. Burchak-Abramovich, N.I.; Vekua, A.K. On a new pleistocene bovine from Eastern Georgia. In *Paleotheriology*; Tatarinov, L., Ed.; Nauka: Moscow, Russia, 1994; pp. 253–261. (In Russian)
117. Sorbelli, L.; Cherin, M.; Kostopoulos, D.; Sardella, R.; Mecozzi, B.; Plotnikov, V.; Prat-Vericat, M.; Azzara, B.; Bartolini-Lucenti, S.; Madurell-Malapeira, J. Earliest bison dispersal in Western Palearctic: Insights from the Eobison record from Pietrafitta (Early Pleistocene, central Italy). *Quat. Sci. Rev.* **2023**, *301*, 107923. [CrossRef]
118. Gromova, V. *Hippariony (Rod Hipparion) po Materialam Taraklii, Pavlodara Idrugim*; Trudy Paleontologicheskogo Instituta. Publ. House of the Academy of Sciences: Moscow, Russia, 1952; Volume 36, 480p. (In Russian)
119. Eisenmann, V. What metapodial morphometry has to say about some Miocene hipparions. In *Paleoclimate and Evolution, with Emphasis on Human Origins*; Vrba, E.S., Denton, G.H., Partridge, T.C., Burckle, L.H., Eds.; Yale University Press: New Haven, CT, USA, 1995; pp. 148–163.
120. Scott, R. The Comparative Paleoecology of Late Miocene Eurasian Hominoids. Ph.D. Thesis, The University of Texas at Austin, Austin, TX, USA, 2004; p. 458.
121. Gambaryan, P. *The Running of Mammals*; Publishing House Nauka: Leningrad, Russia, 1972; p. 334.
122. Gentry, A. An Ovibovine (Mammalia, Bovidae) from the Neogene of Stratzing, Austria. *Ann. Naturhistorische MuseumWien* **2001**, *102*, 189–199.
123. Cherin, M.; Sorbelli, L.; Crotti, M.; Iurino, D.; Sardella, R.; Souron, A. New material of *Sus strozzii* (Suidae, Mammalia) from the Early Pleistocene of Italy and a phylogenetic analysis of suines. *Quat. Sci. Rev.* **2018**, *194*, 94–115. [CrossRef]
124. Brugal, J.-P.; Boudadi-Maligne, M. Quaternary small to large canids in Europe: Taxonomic status and biochronological contribution. *Quat. Int.* **2011**, *243*, 171–182. [CrossRef]
125. Dermitzakis, M. The evolution of the Aegeis during the Late Cenozoic. *Geol. Balc.* **1990**, *20*, 3–16. [CrossRef]
126. Spassov, N. Villafranchian Succession of Mammalian Megafaunas from Bulgaria and the Biozonation of South-East Europe. In *Actes du Congres Biochrom '97*; Aguilar, J.-P., Legendre, S., Michaux, J., Eds.; Mémoires et travaux de l'Institut de Montpellier de l'École Pratique des Hautes Études; Institut de Montpellier: Montpellier, France, 1997; Volume 21, pp. 669–676.
127. Strait, D.; Orr, C.; Hodgkins, J.; Spassov, N.; Gurova, M.; Miller, C.; Tzankov, T. The Human Fossil Record of Bulgaria and the Formulation of Biogeographic Hypotheses. In *Paleoanthropology of the Balkans and Anatolia*; Harvati, K., Roksandic, M., Eds.; Vertebrate Paleobiology and Paleoanthropology Series; Springer Science+Business Media: Dordrecht, The Netherlands, 2016; pp. 187–211.
128. Shchelinsky, V.; Gurova, M.; Tesakov, A.; Titov, V.; Frolov, P.; Simakova, A. The early Pleistocene site of Kermek in western Ciscaucasia (Southern Russia): Stratigraphy, biotic record and lithic industry (preliminary results). *Quat. Int.* **2016**, *393*, 51–69. [CrossRef]
129. Vangengeim, E. *Paleontological Justification of the Quaternaqr Stratigraphy of North Asia*; Nauka: Moscow, Russia, 1977; p. 171. (In Russian)
130. Tamvakis, A.; Savvidou, A.; Spassov, N.; Youlatos, D.; Merceron, G.; Kostopoulos, D. New insights on Early Pleistocene *Nyctereutes* from the Balkans based on material from Dafnero-3 (Greece) and Varshets (Bulgaria). *Palaeoworld* **2023**, *32*, 555–572. [CrossRef]
131. Marciszak, A.; Lipecki, G.; Spassov, N. *Martes wenzensis* Stach, 1959 within the early history of the genus *Martes* Pinel, 1792. *Acta Geol. Pol.* **2024**, *74*, e4. [CrossRef]
132. Koufos, G.; Kostopoulos, D. New Carnivore Material from the Plio-Pleistocene of Macedonia (Greece) with the description of a new Canid. *Münch. Geowiss. Abh.* **1997**, *34*, 33–63.
133. Benammi, M.; Aidona, E.; Merceron, G.; Koufos, G.; Kostopoulos, D. Magnetostratigraphy and Chronology of the Lower Pleistocene Primate Bearing Dafnero Fossil Site, N. Greece. *Quaternary* **2020**, *3*, 22. [CrossRef]
134. Konidaris, G.; Kostopoulos, D. The Early Pleistocene large mammal succession of Greece: Implications for pan-European biogeographic correlations. In Proceedings of the XXI INQUA Congress, Rome, Italy, 13–20 July 2023; Session 135. Mammals, Biochronology and Palaeoecology of the Euro-Mediterrane an Quaternary; Abstracts of Papers. Sapienza University of Rome: Rome, Italy, 2023. Available online: https://inquaroma2023.exordo.com/programme/presentations (accessed on 2 January 2024).

135. Konidaris, G.; Amanatidou, M.; Chitoglou, K.; Fragkioudakis, E.; Gerakakis, N.; Giannakou, V.; Gkeme, A.; Kalaitzi, C.; Ladas, A.; Tsakalidis, C.; et al. Krimni-3: A New Lower Pleistocene Vertebrate Site in Mygdonia Basin. In Proceedings of the 16th International Congress of the Geological Society of Greece, Patras, Greece, 17–19 October 2022. Bulletin of the Geological Society of Greece, Sp. Publ. 10, Ext. Abs. GSG 2022, 138.
136. Pandolfi, L.; Erten, H. *Stephanorhinus hundsheimensis* (Mammalia, Rhinocerotidae) from the late early Pleistocene deposits of the Denizli Basin (Anatolia, Turkey). *Geobios* **2017**, *50*, 65–73. [CrossRef]
137. Vislobokova, I.; Agadzhanyan, A.; Lopatin, A. The case of Trlica TRL11–10 (Montenegro): Implications for possible early hominin dispersals into the Balkans in the middle of the Early Pleistocene. *Quat. Int.* **2020**, *554*, 15–35. [CrossRef]

quaternary

MDPI

Article

Plio-Pleistocene Small Mammal-Based Biochronology of Eastern Anatolia and Transcaucasus

Alexey S. Tesakov [1,*], Pavel Frolov [1], Alexandra Simakova [1], Albina Yakimova [1], Vadim Titov [2], Pranav Ranjan [1], Hasan Çelik [3] and Vladimir Trifonov [1]

[1] Geological Institute, Russian Academy of Sciences, Pyzhevsky 7, 119017 Moscow, Russia; pavlentiy987@mail.ru (P.F.); simak2001@mail.ru (A.S.); albina.yakimova@ginras.ru (A.Y.); pranavranjan42@gmail.com (P.R.); trifonov@ginras.ru (V.T.)

[2] Southern Scientific Centre, Russian Academy of Sciences, Chekhova 41, 344006 Rostov-on-Don, Russia; vvtitov@yandex.ru

[3] Geological Engineering Department, Engineering Faculty, Fırat University, Elâzığ 23119, Türkiye; hasancelik@firat.edu.tr

* Correspondence: tesak@ginras.ru

Abstract: The known Plio-Pleistocene mammalian record, mainly represented by small mammals, and its biotic and geological context in the vast region of Eastern Turkey and Transcaucasus provides a sound base for regional biochronology. Recently obtained faunal associations and the main evolutionary lineages found in the region support direct correlations to the European (ELMA/MN/MQ) and the Eastern European (faunal complexes/MQR-MNR) biochronological systems. Important data on palynology, aquatic and terrestrial mollusks, and magnetostratigraphy integrate the reviewed material into a robust local biochronology. The range of standard biochrons of Early Pliocene through late Early Pleistocene and the regional Anatolian zones M-P are reliably detected. The Early Pleistocene time range (zone P) is refined based on rhizodont lagurines *Borsodia* and Euro-Asian larger voles *Mimomys* ex gr. *pliocaenicus*. The successive zone R for Early Pleistocene faunas with early rootless Microtini is proposed.

Keywords: mammals; biochronology; eastern Turkey; transcaucasus; pliocene; early pleistocene; middle pleistocene

Citation: Tesakov, A.S.; Frolov, P.; Simakova, A.; Yakimova, A.; Titov, V.; Ranjan, P.; Çelik, H.; Trifonov, V. Plio-Pleistocene Small Mammal-Based Biochronology of Eastern Anatolia and Transcaucasus. *Quaternary* **2024**, *7*, 42. https://doi.org/10.3390/quat7040042

Academic Editors: Raffaele Sardella and Joan Madurell Malapeira

Received: 1 June 2024
Revised: 2 September 2024
Accepted: 20 September 2024
Published: 29 September 2024

1. Introduction

Fossil mammals and particularly smaller forms, such as rodents, are well known for their importance in biochronology because of their rapid evolution, wide distribution, and abundant fossil record, e.g., [1,2]. Following the pioneering works on Neogene and early Quaternary mammals by Fikret Ozansoy [3] and others, Otto Sickenberg and his colleagues continued studies on fossil mammalian faunas of Turkey [4–7]. These studies paved the way for later studies of primarily Neogene mammals and mammalian biochronology. Large mammals of the Pliocene and Pleistocene of Anatolia, particularly proboscideans, carnivorans, and ungulates, are constantly in the focus of recent studies [8–10].

After several decades of research, the Plio-Pleistocene small mammals of Asiatic Turkey are relatively well studied [11–15] and many others. Important new data on this and adjacent regions accumulated recently [16–22].

The last decade brought informative new fossil material on small mammals, particularly from the Plio-Pleistocene of east Anatolia and adjacent areas of Transcaucasus [23–29]. Though not exceptionally rich and continuous, this record presents a contribution to the knowledge of the small mammal history and enables a needed modern review of regional mammalian biostratigraphy and its integration with the European (MN/Q/ELMA) [1], Eastern European (faunal complexes/MQR-MNR) [30], and Anatolian [31] biochronological systems. The goal of this contribution is a review of new data relevant to the update and revision of the regional mammalian Plio-Pleistocene biochronology of Eastern Turkey and Transcaucasus.

2. Material and Methods

This study is based on fossil material from East Anatolia and adjacent regions of Transcaucasus in Armenia and Azerbaijan both originally collected in the field, studied in scientific collections, or reviewed from the literature. The bulk of data from Eastern Anatolia were collected by the joint Russian–Turkish cooperation project of the Geological Institute RAS in Moscow and the Firat University of Elâzığ. Data on the localities Tekman, Kümbetli, Pekecik B and C, Agri-East, and Duzdag are published for the first time. The geographic positions of the studied sites are shown in Figure 1. All key fossil collections are stored in the institution's acronym as GIN, Geological Institute of the Russian Academy of Sciences in Moscow; EU, Firat University of Elâzığ; MTA, General Directorate of Mineral Research and Exploration, Ankara.

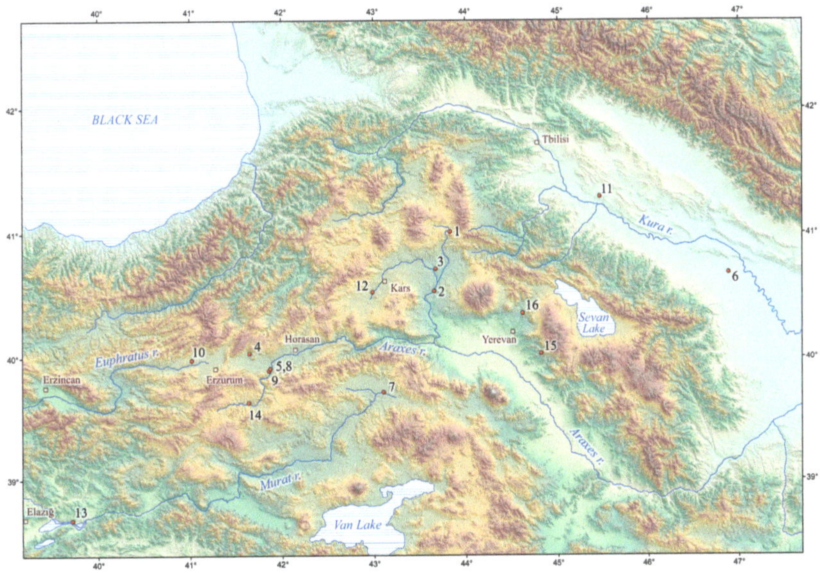

Figure 1. Schematic map showing the geographic position of the reviewed Pli-Pleistocene localities. 1. Krasar, 2. Haykadzor, 3. Demirkent, 4. Pasinler-A, 5. Pekecik C, 6. Duzdag, 7. Agri-East, 8. Pekecik B, 9. Pekecik A, 10. Paşayurdu, 11. Kushkuna, 12. Kümbetli, 13. Karangibaşi, 14. Tekman, 15. Jradzor, 16. Nurnus.

Mostly arvicolids (Arvicolinae, Cricetidae, Rodentia) are included in the review because this is the dominant and most important for biostratigraphical studies of small mammal group for the Plio-Pleistocene of the region. The utilized terminology of the European mammal biochronology follows Fejfar et al. [1] for mammal ages and MQ zones, and the regional zonation for the Pliocene follows Ünay et al. [31] and de Bruijn et al. [32].

In the text, most generalized Early Pliocene voles are referred to as *Promimomys*, diagnosed by low hypsodonty, the undifferentiated thickness of the enamel wall, lack of the external cement, predominantly two enamel islets in M3, compact anteroconid of m1 with integral T4–T5 and with a deep enamel islet. Generic differentiation of Early Pliocene voles based on merely the shape of anteroconid, with its triangular shape indicating *Promimomys* and trifoliate shape (with developed incoming angles BRA3 and LRA4) signaling *Mimomys*, is discouraged. Many of thus understood primitive voles of the Ruscinian are, in fact, unrelated to true Villanyian *Mimomys* and may represent stem groups of multiple genera of Plio-Pleistocene arvicolines (see the discussion in [19] (p. 314)).

3. Results

Plio-Pleistocene biotic record.

In this section we review the most important mammalian associations of the studied region. Their correlation with standard and regional stratigraphic charts is shown in Figure 2. Necessary short no tes on the most important correlative faunas of the neighbouring regions are given.

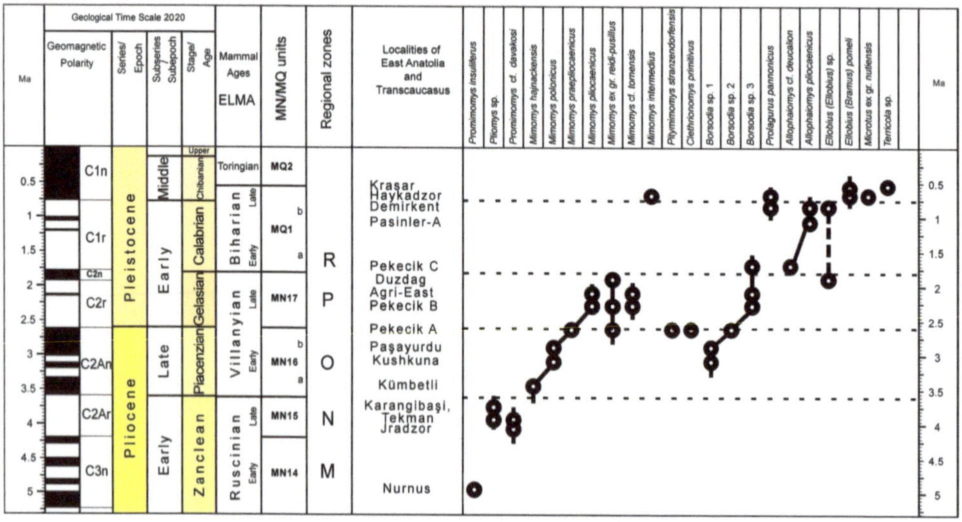

Figure 2. Stratigraphic chart showing the sequence of mammalian localities, occurrences of forms of Arvicolinae and their phyletic connections. Regional zones are after [31].

3.1. Early Pliocene (Early Ruscinian)

Nurnus

Located in central Armenia in an abandoned diatomite pit 6 km north of Yerevan (40°21′27.9″ N 44°37′08.1″ E), this site provided the only fauna of the earliest Pliocene age in the region. The scanty material, including bones of hipparionine horse, rhino, and fish imprints, has long been attributed to the Pliocene. It was the excavation and research by Melik-Adamyan [24] that ultimately refined the age of the fauna as Early Pliocene based on small mammal remains. Still, insufficiently published fauna includes a pika, close to *Ochotona mediterranensis* Suata-Alpaslan, 2015 from İğdeli, a rabbit *Trischizolagus gambariani* Melik-Adamyan, 1986, and a primitive arvicoline *Promimomys* cf. *insuliferus* shortly described as *Polonomys* sp. in [23] The evolutionary level of this vole places this fauna in the unit MN14.

Correlative faunas. In contrast to the detailed sequential Ruscinian record of the Ptolemais Basin in Greece [33], mammalian faunas of the Early Pliocene age are relatively rare and patchy in Anatolia. The oldest faunas of this age document the primitive vole *Promimomys* cf. *insuliferus* in İğdeli [34] and Nasrettinhoca 2 [22]. Faunas from Hacisam and Aşaği Page in central and eastern Anatolia appear to be close in age, based on the morphology of voles with a poorly differentiated triangular anteroconid [15]. The fauna of Çeştepe in Kazan Basin near Ankara is presumably close in age [20]. The oldest known early Ruscinian faunal association of Dinar-Akçaköy, according to the recent revision [35], includes the primitive vole *Promimomys enginae* Suata-Alpaslan, 2015. This form shows some characters, such as squared anteroconid, known in later European voles of the genus

Germanomys. The presence of these voles enables the attribution of these faunas to the unit MN14 and zone M of the Anatolian regional biozonation.

3.2. *Early Pliocene (Late Ruscinian)*

3.2.1. Jradzor

The site is located in central Armenia (40°1′42.53″ N 44°48′53.92″ E). Smaller vertebrate fauna originated from an extensive stratigraphic sequence of lake sediments strongly influenced by pyroclastic material. Paleomagnetic and radiometric analyses provide a rare case of accurate placement of fossiliferous beds to upper Lower Pliocene, in the time range between 3.9 and 4.1 Ma [36]. Illustrated voles *Promimomys* sp. and *Mimomys* cf. *davakosi* (level JZ-3, ca. 4.1–4.0 Ma) belong to a medium-size generalized form *Promimomys* ex gr. *moldavicus-davakosi*, more advanced in hypsodonty than in lower Pliocene levels of the Ptolemais sequence [33]. Of the preliminary illustrated coeval forms from Jradzor [36], especially important are the leporid *Trischizolagus* cf. *dumitrescuae*, the glirid *Myomimus* cf. *maritsensis*, the murid *Occitanomys* sp., the gerbil *Pseudomeriones* cf. *tchaltensis*, and the cricetid *Neocricetodon* sp. The fauna is correlated with early Late Ruscinian, MN15.

3.2.2. Karangibaşi

The site belongs to the Mio-Pliocene Çaybağı Formation and is located 16 km west of Palu town on the right bank of the Murat River (38°39′51.80″ N 39°43′07.12″ E). Small mammal fauna includes *Amblycoptus* sp., Desmaninae gen., *Ochotona* sp., Leporidae gen., cf. *Propliomys*. Scanty arvicoline remains to represent a brachiodont vole resembling late Early Pliocene *Propliomys* [37].

3.2.3. Tekman

The fossil locality is situated on the right bank of the Araxes River near the bridge on the road from Haciömer to Tekman. The site was reported by Sickenberg et al. [5] with the original faunal list including *Promimomys* sp. and *Castillomys* sp. Subsequently, a more detailed revised species list, strangely with reference to the same source, was published by Suata-Alpaslan [38]. This presumably revised list includes the vole *Mimomys* sp. and murids *Occitanomys* sp. and *Orientalomys*. The same site was reported under the name Cęvırme, referred to the Pliocene Işıklar Formation in a publication devoted to fossil *Capoeta* fish from the original collections in Tübingen University [39]. As a result of our studies, a small outcrop of lacustrine deposits at the bank of the Araxes River (39°37′48.97″ N 41°38′43.37″ E) yielded a very small collection of micromammals, including Ochotonidae gen., *Promimomys* cf. *moldavicus-davakosi* and *Propliomys* sp. The low hypsodonty suggests an Early Pliocene age; the presence of pliomyoid vole points to the late Ruscinian, MN15 biochronological unit.

Correlative faunas. The most studied of coeval small mammalian faunas in north-central Anatolia is the reference fauna of Çalta [11,12]. The single arvicolid form in this locality was originally referred to *Mimomys gracilis* Kretzoi, 1959 [11], and later to *Mimomys davakosi* van de Weerd, 1979 [12]. The morphology of this vole with the deep Mimomys reentrant (BRA3), which insulates late in the wear process [11], indicates an early stage of the pliomyine evolutionary trend. Early morphological stages of late Early Pliocene *Propliomys* voles of the peri-Black Sea region, for example *Propliomys destinatus* (Tesakov, 2005) [40] from Odessa Catacombs or *Mimomys moldavicus*/*Propliomys* from Dranic-2 [41], show high similarity to the vole from Çalta.

Recently a small late Early Pliocene (MN15) cricetine assemblage was described from Afşar 1 in western Anatolia [42]. Arvicolines were referred to as *Mimomys* cf. *gracilis* and *Pliomys* sp. This is well supported by the morphology of M3. The late Early Pliocene *Propliomys* was also reported from central Anatolia in Ortalica [15]. The locality of Ericek in western Anatolia [19] documents a large vole, "*Mimomys*" *occitanus* Thaler, 1955. It can represent either a late Ruscinian early member of a large *Mimomys* group related to *M. pliocaenicus* of Plio-Pleistocene or an endemic short-lived lineage of the *Promimomys* evolutionary grade. The biochron of Late Ruscinian is also characterized in Anatolia by faunas of Tozaklar and Zirnak ("*Mimomys*" cf. *occitanus*), Sürsürü (medium-size *Promimomys* with pliomyine variability), Taşova (primitive *Promimomys*) [15].

3.3. *Late Pliocene (Early Villanyan)*

3.3.1. Kümbetli

A roadside 5 m thick section of sandy-silty fluvial deposits of the Pliocene infill of the Kars sedimentary basin (40°31′51.47″ N 42°59′22.43″ E) yielded microvertebrate remains, including fragmentary shells of turtles, bones of anurans, and few molars of a larger vole *Mimomys hajnackesis* (=*hassiacus*). This is the first record of early Late Pliocene arvicolines in eastern Turkey. It may represent the MN16a unit in the region.

3.3.2. Kushkuna

This site is located in western Azerbaijan, at the left bank of the Kura River valley (41°17′36.71″ N 45°28′21.90″ E). The site exposes a thick sequence of the marine deposits of the Caspian Sea Akchagylian transgression of the Late Pliocene age [43]. Small mammal assemblage includes a large form *Mimomys polonicus* and a scanty smaller vole *Borsodia* sp. 1 (reported as *B.* ex gr. *steklovi-novoasovica* [44]. The evolutionary level of the larger vole *Mimomys* and the geological constraints of the section indicate the correlation of this micromammal fauna to the Late Pliocene, the biostratigraphic unit MN16b, and the regional Anatolian zone O.

3.3.3. Paşayurdu

Recently studied section of Late Pliocene paleo-delta fluvial deposits at the margin of the Erzurum intermontane Basin (39°58′44.82″ N 41°01′19.72″ E) between the villages Paşayurdu and Çigdemli. This site yielded a small mammal association with *Mimomys* cf. *polonicus* and *Borsodia* sp. [28]. The hypsodonty level of the larger *Mimomys* matches that of the type *M. polonicus* Kowalski, 1961 Rębielice Królewski in Poland. *Borsodia* from Paşayurdu is characterized by uniformly thick enamel, low dentine tracts (HH-index of m1, ca. 1.6), and a very short posterior lobe of M3. All these characters can be regarded as primitive. This record of *Borsodia* appears to be the oldest known in Anatolia. The fauna of small mammals from Pashayurdu-Çigdemli dates between 3.0 and 2.6 Ma. The lower age limit is controlled by the lower boundary of the MN16b biochronological zone, which is estimated to be close to 3.0 Ma [1].

Correlative faunas. The earlier part of the Early Villanyian (MN16a) in central Anatolia is represented by the fauna from Kadiözü, with large brachyodont *Mimomys* cf. *hajnackensis* (reported as *Mimomys* sp.) [15], and by the faunas from Hoyhoytepe and Mercan 1, also with *M.* cf. *hajnackensis* [22].

The properly described fauna of Afşar 2 in western Anatolia [42] includes larger *Mimomys* cf. *hajnackensis* (reported as *M. hassiacus*), medium-size *Mimomys* cf. *stehlini* (as *M. gracilis*), and *Propliomys graecus* (as *Pliomys*). Lower hypsodonty levels of the voles indicate a correlation with the MN16a biochronological unit.

The later part of the Late Pliocene (MN16b) in the published faunas is difficult to recognize because of the lack of conspicuous records of large *Mimomys* of the *M. polonicus* grade. The late Early Villanyian age may be presumed for Yenice-1 with rather advanced M3 of *Propliomys graecus* [15].

3.4. *Early Pleistocene (Late Villanyian)*

Pekecik A. The small mammal fauna of the lignite-rich lacustrine deposits near the village of Pekecik (39°53′35.79″ N 41°51′41.09″ E) at the right bank of the Araxes River in the Horasan Basin (Yolüstü or Horasan Formation) was first briefly reported by Ünay and de Bruijn [15] with *Mimomys pliocaenicus*, *Borsodia* sp., and *Clethrionomys* sp. The fauna was dated to the late Villanyan. The revision of the original collection (MTA) and additional material collected from this level led to a revised list including *Allocricetus* sp., *Mimomys praepliocaenicus* Rabeder, 1981 (HH-index ca. 3), *Mimomys reidi* Hinton, 1910, *Borsodia* sp. 2 (HH-index ca. 2.5), *Pitymimomys stranzendorfensis* Rabeder, 1981, and *Clethrionomys primitivus* Popov, 2000 [32] (Simakova et al., 2021). The paleomagnetic constraints of normally magnetized host deposits indicate the chronological position of this fauna in the uppermost part of the Gauss Chron and an age slightly older than 2.6 Ma. The taxonomic content of the fauna matches the MN17a unit. This part of the section contains direct palynological evidence (brackish-water dinocysts) of the deep inland penetration of the Akchagylian marine transgression into the Horasan Basin. The maximal stage of the transgression is recorded close to the Pliocene–Pleistocene boundary. It gives an additional stratigraphic constraint to the Pekecik sequence [32,35].

Pekecik B. Higher in the section of Horasan Formation in Pekecik (39°54′18.79″ N 41°52′21.48″ E), several successive levels document *Mimomys pliocaenicus*, *Mimomys* ex gr. *reidi-pusillus*, *Mimomys* cf. *tornensis*, and *Borsodia* sp. 3.

Agri-East. The volcano–sedimentary infill of the Agri intermontane basin in the sections at the eastern margins of the town of Agri (39°42′53.33″ N 43°06′21.48″ E) yielded a micromammalian fauna similar to Pekecik B with *Mimomys pliocaenicus*, *Mimomys* cf. *tornensis*, and hypsodont *Borsodia* sp. 3.

The coeval faunas of Pekecik B and Agri-East correspond to the early-middle Gelasian and the middle part of late Villanyian, MN17b.

Duzdag. This section is situated in the central part of the Kura River valley in Azerbaijan (40°40′56.66″ N 46°53′40.09″ E). This several hundred-meters-thick section is one of the most studied sequences documenting the Plio-Pleistocene of Akchagylian and Apsheronian marine transgressions in the Caspian region [45]. The early Apsheronian stratigraphic member was recently regionally constrained to the late Gelasian time slice between the Reunion and Olduvai subchrons [36]. Fluvial freshwater deposits in this part of the section yielded micromammal remains, including *Ellobius (Ellobius)* sp. similar to *E. kujalnikensis* Topachevsky, 1965, and *Mimomys* ex gr. *reidi-pusillus*.

Correlative faunas. The older part of Villanyian is documented by the mammalian fauna of Sarikol Tepe in central Anatolia [8]. This fauna was the first to show the presence of rhizodont lagurines *Borsodia* in Anatolia. *Borsodia* sp. from this site is more advanced in hypsodonty and shows deeper incurving reentrants than *Borsodia* sp. 1 from late early Villanyian Paşayurdu fauna [28]. The review in [15] described several faunas referred to the late Villanyan. The older faunas (Ziyaret, Sivricek, Karasapaca, Çatakli) show the presence of hypsodont *Mimomys pliocaenicus*. The assemblage of Ziyaret shows a probable taphonomic admixture of more primitive *M. hajnackensis*. *Mimomys* sp. 2 from Ziyaret and *Borsodia* sp. from Çatakli may represent *Clethrionomys* sp.

The later faunas of Havutçulu, Şevketin Daği, and Kartaltepe, along with large hypsodont *Mimomys*, document the first appearance of endemic Aegean–Anatolian rootless lagurine voles *Kalymnomys* [15,46]. *Kalymnomys* datum preceding the migrational appearance of rootless Microtini of the *Allophaiomys* grade and other Lagurini (*Lagurodon arankae* Kretzoi, 1954, *Prolagurus*) is recorded in the diverse fauna of Biçakçi in western Anatolia. This fauna includes hypsodont *Mimomys pliocenicus* (HH-index 5–6), *Mimomys tornensis* Janossy et van der Meulen, 1975, *Pitymimomys pitymyoides* (Rabeder, 1981), *Borsodia* gr. *newtoni-arankoides*, *Kalymnomys* sp., and *Clethrionomys kretzoii* (Kowalski, 1958) [22].

3.5. Early Pleistocene (Early Biharian)

Pekecik C. The high level of the Pekecik sequence some 300 m above the Pekecik A level (39°54′18.08″ N 41°52′28.83″ E) yielded an assemblage of *Beremendia fissidens*, *Allophaiomys* cf. *deucalion*, and *Borsodia* sp. 3 resembling *B.* ex gr. *newtoni-arankoides*.

Pasinler-A. Early Pleistocene lacustrine deposits north of Pasinler produce a small material of *Allophaiomys* cf. *pliocaenicus* [25]. This vole, showing an undifferentiated enamel band and early stage of anteroconid complication in m1, fits the evolutionary stage of Microtini dated to the late Early Pleistocene between 1 and 0.8 Ma. Additional material (2017) of the third upper molar of this vole, with shallow postero-lingual reentrant

LRA4, supports the conclusions. The specific fauna of freshwater mollusks of the Pasinler Formation [25] with sculptured pyrgulids is very close in composition with the assemblage of the late Early Pleistocene Ani Formation of the Shirak Basin at the Turkish–Armenian boundary [28]. A similar molluscan assemblage was recovered from the uppermost levels of the Pekecik sequence above the datum of *Allophaiomys* cf. *deucalion*.

Demirkent. This reference section (40°42′53.83″ N 43°40′22.01″ E) documents the penetration of the Akchagylian transgression of the Caspian Sea into the Shirak Basin as evidenced by brackish-water dinocyst assemblage of the Akchagylian type [27]. The locustrine deposits of the lower part of the section are correlated to the Late Pliocene (Piacenzian) based on palynology and normal magnetization of the deposit. Higher in the section, fluvio-lacustrine deposits of the Ani Formation yielded a very small material of micromammals, including *Prolagurus pannonicus* and Microtini cf. *Allophaiomys* sp., and cf. *Ellobius* sp. This late early Biharian association indicates the late Early Pleistocene age of this part of the section.

Correlative faunas. Important early Biharian faunas in central and western Anatolia include Hamamayaği, Değirmendere, Kürttepe, and Kemalpaşa Mahallesi-I [15,47]. The post-Villanyian age is indicated by the presence of primitive rootless Microtini, *Allophaiomys*, or larger endemic *Tibericola* voles. The latter may represent basal stages of the snow voles *Chionomys*. Most of these faunas also represent rootless endemic lagurines *Kalymnomys*. One of the youngest Early Pleistocene small mammal faunas of central Anatolia comes from the Early Palaeolithic site Dursunlu [48]. The vole fauna includes advanced *Allophaiomys nutiensis* Chaline, 1972, lagurine *Lagurodon arankae*, water vole *Mimomys intermedius* (Newton, 1881), and a mole vole *Ellobius* (*Bramus*) sp. This fauna correlates to micromammals of the Ani and Pasinler Formations, the late early Biharian, late Calabrian, and late Early Pleistocene age.

3.6. Early Middle Pleistocene (Late Biharian)

Haykadzor. Fluviatile deposits of the Arapi formation in the Shirak Basin in western Armenia (40°32′16.04″ N 43°39′21.05″ E) yielded arvicoline fauna with *Ellobius* (*Bramus*) *pomeli* Tesakov, 1916, *Prolagurus pannonicus transylvanicus* Terzea, 1989, *Mimomys intermedius*, and *Microtus* ex gr. *nutiensis* [19]. This locality is bracketed between the Matuyama-Brunhes paleomagnetic reversal below (0.78 Ma) and the overlying ashes of the Aragats volcano radiometrically dated to 0.6–0.7 Ma. This position gives the time range of 0.78–0.6 Ma [19]. The biochronological position of this fauna matches the Cromerian of NW Europe, Tiraspolian mammal unit of Eastern Europe, late Biharian, early Cibanian, and early Middle Pleistocene age.

Krasar. Fluviatile deposits of the Upper Akhuryan Basin in western Armenia (41°00′45.22″ N 43°49′52.28″ E). Arvicoline assemblage from Krasar includes *Ellobius pomeli* and *Terricola* sp. [19]. The fauna of this locality is close in age or slightly younger than Haykadzor.

Correlative faunas. Emirkaya 2 is a presumably early Middle Pleistocene fauna from a fissure filling in Central Anatolia [49] as indicated by the co-occurrence of *Mimomys intermedius* (reported as *M. savini*), and *Microtus* and *Terricola*. The presence of *Arvicola* may indicate a mixing of early and late Middle Pleistocene material.

4. Discussion

4.1. Regional Anatolian Biozonation

The regional biochronological zonation of Neogene of Anatolia was proposed by Ünay, de Bruijn, Sarac, and Hordijk [15,31,32]. The lowering of the Pliocene–Pleistocene boundary to 2.6 Ma by the International Commission on Stratigraphy (2009) transferred the uppermost zone P of this scheme, originally correlated to MN17, into the Early Pleistocene. This Early Pleistocene zone P is characterized by *Mimomys pliocaenicus*/*ostramosensis* and *Tibericola sakaryaensis* from the reference localities Hamamayaği and Yağmurlu [31]. The latter fauna likely belongs to an older Pliocene time [50]. In fact, in its original content zone P corresponds to two different zonal units. The lower subunit is characterized by the faunas with dominant Late Villanyan hypsodont rhizodont arvicolines of the *Mimomys* group, such as the *Mimomys pliocaencus* group, co-occurring with advanced *Borsodia*, and first primitive *Kalymnomys* in western Anatolia (e.g., Bıçakçı, Kartaltepe 1, Havutçulu). In eastern Anatolia, the proper match is the new fauna of Agri-East without *Kalymnomys*. The upper subunit includes later faunas with dominant first primitive (*Allophaiomys* grade) arvicolines with rootless molars (e.g., Hamamayaği, Değirmendere, Pekecik C). Rhizodont voles are nearly missing. This later unit corresponds to early Biharian time. We propose to restrict the usage of zone P to the older late Villanyian faunas with the type fauna of Bıçakçı [18]. The immediately chronologically subsequent zone is expedient to denote as zone R with the type fauna of Hamamayaği. More data are needed to extend this zonation into the early Middle Pleistocene.

4.2. East European Biochronology

The regional biochronological system of MNR/MQR units of Eastern Europe [30,40] was established based on the principle of concurrent range zones of several self-checking lineages of arvicoline rodents. In the time periods of a stable, well-differentiated, and homogeneous fauna of small mammals, such as in the Late Pliocene and early Early Pleistocene, this zonation is easily applicable to the fossil record of all Europe and Western Asia, giving a reliable increase in accuracy and more details in the biostratigraphic subdivision of continental deposits [18,51,52]. The eastern Anatolian record shows the stable presence of the most important small mammal groups and the identical sequence of evolutionary events as compared to the East European record. More data are needed to resolve endemic or universal lineages of rhizodont lagurines, *Borsodia*. Some morphological characters, such as more robust molars and more uniformly thick enamel, may indicate an independent development of a *Borsodia* lineage in the eastern Anatolian region in the Late Pliocene and earliest Pleistocene. No reliable evidence on the origin of Aegean–Anatolian endemic *Kalymnomys* lagurines of the Early Pleistocene is yet available.

4.3. European Biochronology

The overviewed sequence of small mammal faunas in the Anatolia and western Transcaucasus shows a very important and growingly reliable information basis for regional stratigraphy and elaboration of the bio-climatic history of the region in Pliocene and Early Pleistocene. The unique geological structure of the region features numerous intermontane basins experiencing long-term subsiding and accumulating long sedimentary sequences of continental deposits. The paleontological record of these basins, most importantly, the record of land mammals, thus can be studied in its continuous geological content. Ruscinian arvicoline faunas of Asia Minor and Transcaucasus, still insufficiently known, present the common evolutionary trends known elsewhere in the Holarctic Realm: brachiodont *Promimomys* grade voles gradually increase their hypsodonty, and by the second half of Early Pliocene show evidence of phyletic radiation and genus-level differentiation. Villanyian arvicoline faunas show stages of radiation of mimomyian voles and some endemic Anatolian lineages (*Borsodia*). Presumably aquatic, large voles of the *Mimomys hajnackensis-pliocaenicus* group are very widespread in the region. This group most efficiently provides interregional and intracontinental biostratigraphic correlations in Piacenzian and Gelasian. Biharian arvicoline faunas of the region document the global *Allophaiomys* dispersal and parallel endemic evolution of other lineages of rootless voles. The present-day state of the art in the understanding of the history of micromammalian faunas in the westernmost part of Asia enables bright perspectives of a further increase in our knowledge.

5. Conclusions

The fossil record of Eastern Anatolia and Transcaucasus enables a recognition of a dozen consecutive biochronological levels from the Early Pliocene to the early Middle Pleistocene. In contrast to the older Neogene fossil record [32], the Plio-Pleistocene biotic record of Anatolia shows a low level of endemicity, which makes it important and applicable for the extensive regions of Northern Eurasia. The region seems to be easily penetrated by continental-wide migration waves of Arvicolinae. The local development of vole faunas shows the main evolutionary events of the group: the Early Pliocene *Promimomys* stage, late Ruscinian vole radiation, Villanyian *Mimomys* radiation, and dispersal of *Borsodia*, the *Allophaiomys* datum at the Villanyian-Biharian transition, and the beginning of the Biharian Microtini radiation. The availability of radiometric age control of some faunas provides precious calibration points of regional and continental biostratigraphic data. The reviewed record is thus easily correlated both to the standard European biochronology (ELMA, MN/MQ units) and regional sequences of Western [2,51,53] and Eastern Europe [52] and Western and central Anatolia [31]. The position of the region at the crossroads between Eastern Europe and the Caucasus and South and Western Europe makes it crucial for studies in the historical biogeography, faunistics, and elaboration and calibration of the Eurasian continental biochronology.

Author Contributions: Conceptualization, A.S.T. and V.T. (Vadim Titov); field studies, A.S.T., P.F., A.S., H.Ç., V.T. (Vladimir Trifonov); biotic data, P.F., A.S., A.Y. and P.R.; geological context, A.S.T., H.Ç. and V.T. (Vladimir Trifonov); data curation, A.S.T.; funding administration, A.S.T., H.Ç. and V.T. (Vladimir Trifonov). All authors have read and agreed to the published version of the manuscript.

Funding: This study was carried out with the financial support of the Russian Science Foundation grant no. 22-17-00249, "Pliocene—Early Pleistocene restructuring of the structural pattern of the Arabian–Caucasian region and its influence on paleogeography, dynamics of biota, and dispersal routes of ancient humans".

Data Availability Statement: All additional information relevant to the presented new material on the geological record of the studied region is available from the authors upon request.

Acknowledgments: We thank Engin Ünay (Sivas) for providing access to the fossil collections in her care and for discussions of the fossil record of micromammals. We thank Mehmet Cihat Alçiçek and Hülya Alçiçek (Pamukkale), Serdar Mayda (Izmir), and Yusuf Carginoglu (Elâzığ) for cooperation and friendly assistance in the laboratory and the field. We thank members of the administrative, field, and laboratory staff at the Geological Institute in Moscow for their great support and cooperation.

Conflicts of Interest: The authors declare no conflicts of interest. The funders had no role in the design of the study, in the collection, analyses, or interpretation of data, in the writing of the manuscript, or in the decision to publish the results.

References

1. Fejfar, O.; Heinrich, W.-D.; Lindsay, E.H. Updating the Neogene rodent biochronology in Europe. *Meded. Ned. Inst. Voor Toegepaste Geowetenschappen TNO* **1998**, *60*, 533–554.
2. Agustí, J.; Lozano-Fernández, I.; Oms, O.; Piñero, P.; Furió, M.; Blain, H.A.; Lopez-García, J.M.; Martínez-Navarro, B. Early to Middle Pleistocene rodent biostratigraphy of the Guadix-Baza basin (SE Spain). *Quat. Int.* **2015**, *389*, 139–147. [CrossRef]
3. Ozansoy, F. Sur les gisements continentaux et les mammifères du Neogene et du Villafranchien d'Ankara (Turquie). *C. R. Acad. Sci. Paris* **1955**, *240*, 992–994.
4. Sickenberg, O.; Tobien, H. New Neogene and Lower Quaternary vertebrate faunas in Turkey. *Newsl. Stratigr.* **1971**, *1*, 51–61. [CrossRef]
5. Sickenberg, O.; Becker-Platen, J.D.; Benda, L.; Berg, D.; Engesser, B.; Gazlry, W.; Heissig, K.; Hünermann, K.A.; Sondaar, P.Y.; Schmidt-Kittler, N.; et al. Die Gliederung des höheren Jungtertiärs und Altquartärs in der Türkei nach Vertebraten und ihre Bedeutung für die internationale Neogen-Stratigraphie. *Geol. Jahrb. B* **1975**, *15*, 1–167.
6. Dayan, E. Über pleistozäne Elefanten-Funde im Umland von Erzurum in Ostanatolien, Ein Beitrag zur Fundgeschichte des Elephas trogontherii im Becken von Pasinler. *Stuttg. Beiträge Naturkunde. Geol. Paläontologie* **1989**, *147*, 1–21.
7. Sen, S.; Bouvrain, G.; Geraads, D. Pliocene vertebrate locality of Çalta, Ankara, Turkey. 12. Paleoecology, biogeography and biochronology. *Geodiversitas* **1998**, *20*, 497–510.
8. Kostopulos, D.; Sen, S. Late Pliocene (Villafranchian) mammals from Sarikol Tepe, Ankara, Turkey. *Mitt. Bayer. Staatsslg. Paläont. Hist. Geol.* **1999**, *39*, 165–202.
9. Albayrak, E.; Lister, A.M. Dental remains of fossil elephants from Turkey. *Quat. Int.* **2012**, *276–277*, 198–211. [CrossRef]
10. Albayrak, E. A new primitive elephantid from Turkey. *Quat. Int.* **2017**, *445*, 43–49. [CrossRef]
11. Sen, S. La fauna de Rongeurs pliocenes de Çalta (Ankara, Turquie). *Bull. Mus. Nat. Hist. Natur.* **1977**, *61*, 89–171.

12. Sen, S. Pliocene vertebrate locality of Çalta, Ankara, Turkey. 4. Rodentia and Lagomorpha. *Geodiversitas* **1998**, *20*, 359–378.
13. Sen, S.; de Bonis, L.; Dalfes, N.; Geraads, D.; Jaeger, J.-J.; Mazin, J.-M. Première decouvérte d'un site à mammifères pléistocénes dans une fissure karstique en Anatolie centrale. *C.R. Acad. Sci. Paris Sér. II* **1991**, *313*, 127–132.
14. Storch, G. Eine jungpleistozäne/altholozäne Nager Abfolge von Antalya, SW-Anatolien (Mammalia, Rodentia). *Z. Säugetierkd* **1988**, *53*, 76–82.
15. Ünay, E.; de Bruijn, H. Plio-Pleistocene rodents ans lagomorphs from Anatolia. *Meded. Ned. Inst. Voor Toegepaste Geowetenschappen TNO* **1998**, *60*, 431–466.
16. Erdal, O.; Erturaç, M.K.; Dalfes, H.N.; Sen, S. Rodents from the Middle Pleistocene of Niksar Basin (Tokat-Turkey): Implications on palaeoenvironment and the North Anatolian Fault. *Neues Jahrbuch Geol. Paläontol. Abh.* **2018**, *289*, 77–111. [CrossRef]
17. Erten, H.; Sen, S.; Sagular, E.k. Miocene and early Pleistocene rodents (Mammalia) from Denizli Basin (southwestern Turkey) and a new species of fossil Mus. *J. Vert. Paleont.* **2015**, *36*, e1054036. [CrossRef]
18. van den Hoek Ostende, L.W.; Diepeveen, F.; Tesakov, A.; Saraç, G.; Mayhew, D.; Alçiçek, M.C. On the brink: Micromammals from the latest Villanyian from Biçakçi (Anatolia). *Geol. J.* **2015**, *50*, 230–245. [CrossRef]
19. van den Hoek Ostende, L.W.; Gardner, J.D.; van Bennekom, L.; Alçiçek, M.C.; Murray, A.; Wesselingh, F.P.; Alçiçek, H.; Tesakov, A. Ericek, a new Pliocene vertebrate locality in the Çameli Basin (southwestern Anatolia, Turkey). *Palaeobiodivers. Palaeoenviron. Quat. Int.* **2015**, *95*, 305–320. [CrossRef]
20. Sen, S.; Delfino, M.; Kazanci, N. Çeştepe, a new early Pliocene vertebrate locality in Central Anatolia and its stratigraphic context. *Annal. Paléontol.* **2017**, *103*, 149–163. [CrossRef]
21. Erturaç, M.K.; Erdal, O.; Sunal, G.; Tüysüz, O.; Sen, S. Quaternary evolution of the Suluova Basin: Implications on tectonics and palaeonvironments of the Central North Anatolian Shear Zone. *Canad. J. Earth Sci.* **2019**, *56*, 1239–1261. [CrossRef]
22. Syromyatnikova, E.V.; Tesakov, A.S.; Mayda, S.; Kaya, T.; Saraç, G. Plio-pleistocene amphibians and reptiles from Central Turkey: New faunas and faunal records with comments on their biochronological position based on small mammals. *Foss. Impr.* **2019**, *75*, 343–358. [CrossRef]
23. Melik-Adamyan, G. New data on Pliocene land vertebrates of Armenian SSR. *Proc. Acad. Sci. Armen. SSR* **1986**, *86*, 135–139.
24. Melik-Adamyan, G. *Stratigraphy and Paleogeography of Pliocene and Lower Neopleistocene of Central and Northwestern Armenia Based on Fauna of Land Vertebrates*; Abridged Thesis. Cand. Sci.; Institute of Geological Sciences: Yerevan, Armenia, 2003; 29p.
25. Vasilyan, D.; Schneider, S.; Bayraktutan, M.S.; Sen, S. Early Pleistocene freshwater communities and rodents from the Pasinler Basin (Erzurum Province, north-eastern Turkey). *Turk. J. Earth Sci.* **2014**, *23*, 1–15. [CrossRef]
26. Tesakov, A.S.; Simakova, A.N.; Frolov, P.D.; Sytchevskaya, E.K.; Syromyatnikova, E.V.; Foronova, I.V.; Shalaeva, E.A.; Trifonov, V.G. Early-Middle Pleistocene environmental and biotic transition in north-western Armenia, southern Caucasus. *Palaeontol. Electron.* **2019**, *22.2.25A*, 1–39. [CrossRef]
27. Simakova, A.N.; Tesakov, A.S.; Çelik, H.; Frolov, P.D.; Shalaeva, E.A.; Sokolov, S.A.; Trikhunkov, Y.I.; Trifonov, V.G.; Bachmanov, D.M.; Latyshev, A.V.; et al. Caspian-type dinocysts in NE Turkey mark deep inland invasion of the Akchagylian brackish-water basin during the terminal Late Pliocene. *Quat. Int.* **2021**, *605–606*, 329–348. [CrossRef]
28. Çelik, H.; Trifonov, V.G.; Tesakov, A.S.; Sokolov, S.A.; Frolov, P.D.; Simakova, A.N.; Shalaeva, E.A.; Belyaeva, E.V.; Yakimova, A.A.; Zelenin, E.A.; et al. Late Pliocene Gilbert type delta and Early Pleistocene drainage system changes in the Erzurum Basin, NE Turkiye. *Stratigr. Geol. Correl.* **2023**, *31*, 700–720. [CrossRef]
29. Trifonov, V.G.; Tesakov, A.S.; Simakova, A.N.; Gaydalenok, O.V.; Frolov, P.D.; Bylinskaya, M.E.; Trikhunkov, Y.I.; Bachmanov, D.M.; Çelik, H.; Hessami, K. Geological and biotic context of the Plio-Pleistocene evolution of the Caucasus-Caspian Region (Akchagylian transgression). *Quat. Int.* **2024**, *686–687*, 120–141. [CrossRef]
30. Tesakov, A.S.; Frolov, P.D.; Titov, V.V.; Dickinson, M.; Meijer, T.; Parfitt, S.A.; Preece, R.C.; Penkman, K.E.H. Aminostratigraphical test of the East European Mammal Zonation for the late Neogene and Quaternary. *Quat. Sci. Rev.* **2020**, *243*, 106434. [CrossRef]
31. Ünay, E.; de Bruijn, H.; Saraç, G. A preliminary zonation of the continental Neogene of Anatolia based on rodents. *Deinsea* **2003**, *10*, 539–548.
32. de Bruijn, H.; Ünay, E.; Hordijk, K. A review of the Neogene succession of the Muridae and Dipodidae from Anatolia, with special reference to taxa known from Asia and/or Europe. In *Fossil Mammals of Asia: Neogene Biostratigraphy and Chronology*; Wang, X., Flynn, L., Fortelius, M., Eds.; Columbia University Press: New York, NY, USA, 2013; pp. 566–582.
33. Hordijk, K.; de Bruijn, H. The succession of rodent faunas from the Mio/Pliocene lacustrine deposits of the Florina-Ptolemais-Servia Basin (Greece). *Hell. J. Geosci.* **2009**, *44*, 21–103.
34. Suata-Alpaslan, F.; Ünay, E.; Ay, F. İğdeli (Gemerek, Sivas) lokalitesi erken Pliyosen fauna istifinin Rodentia ve Lagomorpha (Mammalia) fosilleri: Biyokronolojik ve paleobiyocografik anlamlari. *Cumhur. Üniversitesi Fen-Edeb. Fakültesi Fen Bilim. Derg.* **2010**, *31*, 1–29.
35. Suata-Alpaslan, F. Promimomys enginae nov. sp. (Arvicolidae, Mammalia) from the Early Pliocene locality Dinar-Akcakoy (Afyon, Turkey). In Proceedings of the 15th International Multidisciplinary Scientific Geoconference SGEM2015. Science and Technologies in Geology, Exploration and Mining, Albena, Bulgaria, 18–24 June 2015; Čepek, A., Barta, G., Medak, D., Reinartz, P., Eds.; SGEM: Sofia, Bulgaria, 2015; Volume 1, pp. 385–394.
36. Lazarev, S.; Maul, L.C.; Kuiper, K.; Becker, D.; Bukhsianidze, M.; Hovakimyan, H.; Sahakyan, L.; Vasilyan, D. Pliocene volcanic terrain of the South Caucasus as a missing link in Eurasian palaeobiogeography: Dating and palaeoenvironments of vertebrate assemblages of the Jradzor locality, Armenia. *Palaeogeogr. Palaeoclimatol. Palaeoecol.* **2023**, *625*, 1–27. [CrossRef]

37. Trifonov, V.G.; Çelik, H.; Simakova, A.N.; Bachmanov, D.M.; Frolov, P.D.; Trikhunkov, Y.I.; Tesakov, A.S.; Titov, V.V.; Lebedev, V.A.; Ozherelyev, D.V.; et al. Pliocene–Early Pleistocene history of the Euphrates valley applied to Late Cenozoic environment of the northern Arabian Plate and its surrounding, eastern Turkey. *Quat. Int.* **2018**, *493*, 137–165. [CrossRef]
38. Suata-Alpaslan, F. Anadolu Erken Pliyosen küçük memeli topluluğu: Paleontolojik ve paleoekolojik anlamları. *Kültür Varlıkları Müzeler Genel Müdürlüğü* **2012**, *154*, 1–13.
39. Ayvazyan, A.; Vasilyan, D.; Böhme, M. Possible species-flock scenario for the evolution of the cyprinid genus Capoeta (Cypriniformes: Cyprinidae) within late Neogene lake systems of the Armenian Highland. *PLoS ONE* **2019**, *14*, e0215543. [CrossRef] [PubMed]
40. Tesakov, A.S. Pliocene voles (Pliomys, Arvicolinae, Rodentia) from Odessa Catacombs. *Russ. J. Theriol.* **2005**, *4*, 123–135. [CrossRef]
41. Radulescu, C.; Samson, P. Pliocene and Early Pleistocene arvicolids (Rodentia, Mammalia) of the Dacic Basin, Romania. *Acta Zool. Cracov.* **1996**, *39*, 401–406.
42. Skandalos, P.; Lansing, K.; Demirel, F.A.; Alçiçek, M.C.; Mayda, S.; Dieleman, F.; van den Ostende, L.W.H. Early Pliocene Arvicolinae and Cricetinae from the locality of Afsąr, western Turkey. *Turk. J. Earth Sci.* **2023**, *32*, 27–62. [CrossRef]
43. Lazarev, S.; Kuiper, K.F.; Oms, O.; Bukhsianidze, M.; Vasilyan, D.; Jorissen, E.L.; Bouwmeester, M.J.; Aghayeva, V.; van Amerongen, A.J.; Agustí, J.; et al. Five-fold expansion of the Caspian Sea in the Late Pliocene: New and revised magnetostratigraphic and 40Ar/39Ar age constraints on the Akchagylian stage. *Glob. Planet. Change* **2021**, *206*, 103624. [CrossRef]
44. Tesakov, A.S. *Biostratigraphy of the Middle Pliocene–Eopleistocene of Eastern Europe*; Nauka: Moscow, Russia, 2004; Transactions of the Geological Institute, 554; pp. 1–247. (In Russian)
45. Lebedeva, N.A. *Correlation between Anthropogenic Strata of the Ponto-Caspian*; Nauka: Moscow, Russia, 1978; 136p.
46. Sarica, N. The Plio-Pleistocene age of Büyük Menderes and Gediz grabens and their tectonic signifcance on N-S extensional tectonics in West Anatolia: Mammalian evidence from the continental deposits. *Geol. J.* **2000**, *35*, 1–24. [CrossRef]
47. Ünay, E.; Emre, Ö.; Erkal, T.; Keçer, M. The rodent fauna from the Adapazarý pull-apart basin (NW Anatolia): Its bearings on the age of the North Anatolian fault. *Geodin. Acta* **2001**, *14*, 169–175. [CrossRef]
48. Ünay, E. Allophaiomys from Turkey. *Paludicola* **1998**, *2*, 106–109.
49. Montuire, S.; Sen, S.; Michaux, J. The Middle Pleistocene mammalian fauna from Emirkaya-2, Central Anatolia (Turkey): Systematics and paleoenvironment. *N. Jahrb. Geol. Paläontol. Abh.* **1994**, *193*, 107–144. [CrossRef]
50. Sen, S.; Karadenizli, L.; Antoin, P.; Saraç, G. Late Miocene—early Pliocene rodents and lagomorphs (Mammalia) from the southern part of Çankırı Basin, Turkey. *J. Paleontol.* **2019**, *93*, 173–195. [CrossRef]
51. Mayhew, D.F. Revised biostratigraphic scheme for the Early Pleistocene of the UK based on arvicolids (Mammalia, Rodentia). *Geol. J.* **2015**, *50*, 246–256. [CrossRef]
52. Borodin, A.V.; Strukova, T.V.; Markova, E.A. Calabrian (Eopleistocene) micromammal assemblages from the lacustrine and fluvial deposits of the Southern Trans-Urals and chronological position of some regional stratigraphic units. *Quat. Int.* **2019**, *534*, 89–102. [CrossRef]
53. Piñero, P.; Agustí, J.; Oms, O. The late Neogene rodent succession of the Guadix–Baza Basin (south-eastern Spain) and its magnetostratigraphic correlation. *Palaeontology* **2018**, *61*, 253–272. [CrossRef]

quaternary

MDPI

Article

The Late Pliocene–Middle Pleistocene Large Mammal Faunal Units of Greece

George E. Konidaris [1,2,*] and Dimitris S. Kostopoulos [3]

1 Paleoanthropology, Institute for Archaeological Sciences, Department of Geosciences, Eberhard Karls University of Tübingen, Rümelinstr. 23, 72070 Tübingen, Germany
2 Senckenberg Centre for Human Evolution and Paleoenvironment, Eberhard Karls University of Tübingen, Rümelinstr. 23, 72070 Tübingen, Germany
3 School of Geology, Aristotle University of Thessaloniki, 54124 Thessaloniki, Greece; dkostop@geo.auth.gr
* Correspondence: georgios.konidaris@uni-tuebingen.de

Abstract: Located at the eastern corner of Mediterranean Europe, Greece occupies a critical position for mammal dispersals to/from Europe, Asia, and Africa and constitutes a potential passageway towards Western Europe. During recent decades, numerous fieldwork campaigns in several Pliocene–Pleistocene sites have greatly enriched the fossil record and provided valuable taxonomic and biostratigraphic data. However, a fully developed reference biochronological unit scheme for the Greek record that could contribute to correlations at a continental scale is still pending. In this article, we provide the updated Late Pliocene to Middle Pleistocene large mammal succession, and we introduce the Faunal Units (FUs) of Greece. We define eight FUs, the Milia, Dafnero, Gerakarou, Tsiotra Vryssi, Krimni, Apollonia, Marathousa, and Apidima FUs (from the oldest to the youngest), which are determined by a set of first and last local occurrences. The results form the basis for discussion of already set turnovers, dispersals, and extinction/immigration events and showcase the importance of the local record for the investigation of the European terrestrial ecosystems. By developing the first detailed biochronological scheme for the Pliocene–Pleistocene of Southeastern Europe, this study comprises the basis for an expanded Balkan faunal unit scale and a reference framework for future investigations.

Keywords: biochronology; biostratigraphy; Villafranchian; Galerian; faunal units; Pliocene; Pleistocene; Greece

Citation: Konidaris, G.E.; Kostopoulos, D.S. The Late Pliocene–Middle Pleistocene Large Mammal Faunal Units of Greece. *Quaternary* **2024**, 7, 27. https://doi.org/10.3390/quat7020027

Academic Editor: Maria Rita Palombo

Received: 3 April 2024
Revised: 31 May 2024
Accepted: 7 June 2024
Published: 12 June 2024

1. Introduction

As one of the most likely passageways from Africa/Levant and as the main gate from Asia to Europe, the Balkans have long been considered crucial in understanding and developing the Neogene/Quaternary European continental biostratigraphy and biochronology (e.g., [1–7]). Greece occupies a critical position at the southernmost tip of the Balkans and at the eastern corner of the Mediterranean Europe, and the region's fossil record is of fundamental importance and interest in this discussion.

Research of fossil vertebrate in Greece has a long tradition for almost two centuries, with numerous sites and finds (e.g., [8,9]) that systematically add to the knowledge of the evolution of the Quaternary continental and insular faunas of Europe. During the last two decades, numerous fieldwork campaigns at multiple sites (e.g., Dafnero, Sesklo, Vatera, Volakas, Libakos, Mygdonia Basin, and Megalopolis Basin) by various paleontological teams have greatly enriched the Late Pliocene to Middle Pleistocene fossil record in Greece and provided valuable taxonomic, biostratigraphic, paleoenvironmental, and taphonomic data. Although basic, local biochronological–biostratigraphic schemes have already been proposed and revised on many occasions (e.g., [6,10–13]), they remain atypical, and no systematic effort has been made so far to manage the available data towards the creation of a formal and broader local/regional biochronological unit scheme that could contribute to correlations at a continental scale.

Based on previous studies and new data, we provide the updated Late Pliocene to Middle Pleistocene mammal succession in Greece by means of faunal units and discuss key taxa in both the regional and pan-European frame. Using available local chronological markers and biochronological evidence, we time-calibrate the local faunal scale and correlate it with well-dated faunas of Western Europe. The results form the basis for a discussion of already set mammal turnovers, dispersals, and extinction/immigration events and showcase the importance of the local record for the investigation of the European terrestrial ecosystems. We expect this effort to provide a basis for an expanded Balkan Pliocene–Pleistocene faunal unit scale and a reference framework for future investigations.

2. Methods

Our approach focuses on the Villafranchian, Galerian, and Aurelian European Land Mammal Ages (ELMAs) as redefined by Rook and Martínez-Navarro, Bellucci et al., and Palombo [14–16]. The Villafranchian is subdivided into early (ca. 3.5–2.6 Ma), middle (ca. 2.6–2.1 Ma), and late (ca. 2.1–1.2 Ma) periods, while we also recognize the Epivillafranchian (ca. 1.2–0.8 Ma) as a distinct transitional ELMA between those of the Villafranchian and Galerian [17,18]. Our analysis is based on the biochron concept of faunal units (FUs) as introduced by Azzaroli [19], reviewed by Azzaroli et al. [20] and Gliozzi et al. [21], and discussed by many later authors (e.g., [22–24]). A FU corresponds to an assemblage interval biochron sensu Walsh [25]; it represents a local/regional biochron of the lowest rank, including a typical local/regional taxa association or well recognized evolutionary stage(s) of particular phyletic lineage(s) and is defined by clear bioevents (paleobiological events sensu Walsh [25]) such as the first or last local historical appearances of one or more taxa (e.g., [21,22,25]).

Our dataset includes Greek mammal assemblages spanning from ~3.5 to ~0.12 Ma (Late Pliocene to Middle/Late Pleistocene) that include more than three large mammal taxa identified at least at the genus level. Faunal lists of local faunal assemblages (LFAs) derive from a wide range of literature references (e.g., [6,8,9,13,26,27] and references cited). Published and unpublished data recovered from recently described sites and faunas (such as GAS, SES-L, DFN3, KZ, AK, TSR, PLN, KMN, KYP-3, 4, and MAR-1; Table 1), updated lists of already known LFAs (such as DFN, APL, APD, and LIB), and new chronological calibrations (such as for DFN/DFN3, TSR, and MAR-1) are all involved in the analysis in order to achieve the maximum possible information and time control. In some cases, published taxonomic identifications have been updated and/or reviewed according to the current knowledge to reach a basic taxonomic consensus.

A hierarchical cluster analysis based on presence/absence data at the genus level was applied to a subset of the 23 most complete Greek LFAs (number of recorded genera $\geq$5) and 43 genera of large mammals (Supplementary Material) in order to test their best possible grouping(s) by means of faunal synthesis consistency and correlate clusters with particular time slices. The analysis was performed using PAST [28]. The Q-mode dendrogram [following UPGMA: unweighted pair group method with arithmetic mean algorithm] was obtained by applying several similarity indices, which provided consistent final groupings; the results of the Raup–Crick similarity index, which is used to statistically test presence/absence data through a randomization method [29], were selected for presentation. We excluded, however, a posteriori Aghia Kyriaki (AK; N genera = 6) as it stands as an outlier in most of the runs.

Distinct Greek faunal units (GFUs) have been defined within each cluster based on a species-level analysis on the whole set of data. Local first and last historical appearances (hereinafter referred to as first and last local occurrences, FLOs and LLOs, respectively) of large mammal taxa have been used for the definition of each GFU. Fossiliferous sites included in this study along with their location and selected references are provided in Table 1.

Table 1. Fossiliferous sites included in the study including their abbreviations, locations, and selected references.

Abbreviation	Fossiliferous Site	Location	Selected References
ALK-VOL	Halykes-Volos	Thessaly	[26,30]
AK	Aghia Kyriaki	Aetoloakarnania	[31]
APD-A, B	Apidima Caves A, B	Peloponnese	[5,32]
APL	Apollonia-1	Mygdonia Basin, Macedonia	[33–36]
APO	Apolakkia	Rhodes Island	[26,37,38]
DAM	Damatria	Rhodes Island	[11,26]
DFN, DFN3	Dafnero 1, 3	Haliakmon valley	[39–41]
GAS	Gephyra	Axios valley	[42]
GER	Gerakarou-1	Mygdonia Basin	[6,43]
KAF	Kaiafas	Peloponnese	[11]
KAL, KLT	Kalamoto 1, 2	Mygdonia Basin	[44]
KPT	Kapetanios	Haliakmon valley	[45]
KRI, KRM KMN	Krimni 1, 2, 3	Mygdonia Basin	[46]
KSR	Kastritsi	Thessaly	[47]
KYP-3, 4	Kyparissia 3, 4	Megalopolis Basin, Peloponnese	[48–50]
KZ	Karnezeika	Peloponnese	[51,52]
LIB	Libakos	Haliakmon valley	[45]
MAR	Marathoussa	Mygdonia Basin	[53]
MAR-1, 2	Marathousa 1, 2	Megalopolis Basin	[50,54,55]
MIL	Milia	Haliakmon valley	[56,57]
MKN	Makinia	Aetoloakarnania	[47]
SES-L	Sesklo (Lower Level)	Thessaly	[58,59]
SES-U	Sesklo (Upper Lever)	Thessaly	[58,59]
PEC	Petralona Cave	Chalkidiki peninsula	[60–62]
PLN	Platanochori-1	Mygdonia Basin	[63]
POL	Polylakkos	Haliakmon valley	[45]
PYR	Pyrgos	Peloponnese	[26]
RVL	Ravin of Voulgarakis	Mygdonia Basin	[43]
RIZ	Riza	Mygdonia Basin	[43]
TB-2–5	Tourkovounia 2–5	Attica	[64,65]
TSR:	Tsiotra Vryssi	Mygdonia Basin	[63,66]
VAT	Vatera	Lesvos Island	[67,68]
VOL	Volakas	Drama Basin	[11,69]
VSL	Vassiloudi	Mygdonia Basin	[43]

3. Results

The cluster analysis provided three main groups of LFAs (similarity index ≥ 0.5) that roughly correspond to recognized ELMAs. Galerian and Aurelian LFAs (KYP-3, 4, MAR-1, APD-A, B; cluster "A" in Figure 1) are clearly separated from those of the Villafranchian–Epivillafranchian (cluster "B" in Figure 1) based on the *Homo* + *Palaeoloxodon* + *Cervus* s. str. + *Dama* + *Hippopotamus* large mammal association. Subcluster "a1" can be dated between 780 and ca. 450 ka as it includes the sites KYP-4 (lower part of the Middle Pleistocene <780 ka [70,71], MAR-1 (ca. 450 ka, correlated to Marine Isotope Stage 12 [72]), and KYP-3 (intermediate stratigraphic position between KYP-4 and MAR-1 [70]). Subcluster "a2" is represented in the analysis only by APD-A and B LFAs and corresponds broadly to the late Middle Pleistocene; the crania of *Homo* from Apidima Cave A provided ages of ca. 210 and 170 ka [5], and a comparable or slightly younger age can be assumed for Apidima Cave B.

Cluster "B" incorporates Villafranchian and Epivillafranchian LFAs (Figure 1). Within subcluster "b1", the LFA grouping largely follows the informal scheme of the early ("b1.a"), middle ("b1.b"), and early late ("b1.c") Villafranchian subdivision, whereas late late Villafranchian ("b2.a") and latest Villafranchian–Epivillafranchian ("b2.b") assemblages are incorporated in subcluster "b2". The "b1.a" subcluster includes the LFAs of MIL, SES-L, and APO, characterized by the co-occurrence of *Anancus* + *Hipparion* in contrast to any other group of LFAs included in the analysis. Vlachos et al. [57] correlated the best-documented Milia (MIL) faunal assemblage with MN16a, suggesting an approximate age of 3.0–2.8 Ma [73]. We find the link of this cluster with "b1.c" artificial and therefore of very low credibility; it is mainly based on the common presence of *Sus* and *Gazella*, though species-level taxonomy is sharply different in these two sets of faunas.

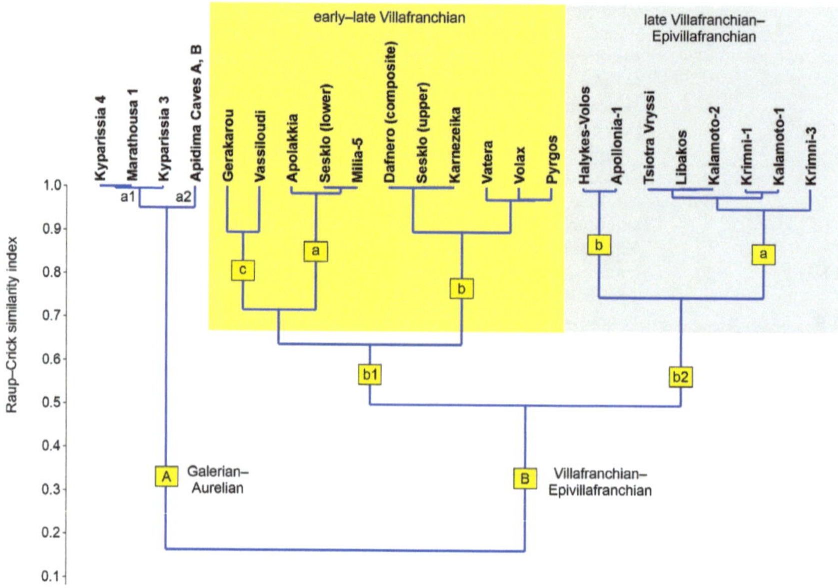

Figure 1. Hierarchical cluster analysis at the genus level showing the faunal similarities among selected Villafranchian to Aurelian large mammal assemblages of Greece.

Subcluster "b1.b" includes the LFAs of DFN/DFN3, SES-U, VAT, VOL, KZ, and PYR (Figure 1). With the exception of Pyrgos (PYR), this faunal assemblage is characterized by the association of *Nyctereutes* + *Equus* + *Palaeotragus* + *Gazella* + *Gazellospira* + *Gallogoral*, while *Paradolichopithecus* is also present in three out of the remaining five LFAs. Based on magnetochronology of the DFN/DFN3 sites [39], this group of LFAs is dated around 2.4–2.3 Ma. The PYR LFA, rather, has to be excluded from this subcluster and is likely associated with "b1.c", as suggested by the lack of basic taxa of this association (e.g., *Gallogoral*, *Nyctereutes*) and evidenced by the species-level taxonomy of its equid content [74].

The subcluster "b1.c" includes the LFAs of GER and VSL (Figure 1) and is characterized by the association of *Equus* + *Sus* + *Gazella* + *Leptobos*. Although *Canis* is recorded only in GER LFA, we suggest it is part of this association, and its absence from VSL is due to the limited material from this LFA. Both GER and VSL belong to the same stratigraphic levels of Mygdonia Basin, below the TSR LFA which provided a maximum age of 1.78 Ma [66]. Hence, we estimate an age of about 2.0 Ma for this faunal group, confirmed by the common presence of *Pliocrocuta* and *Pachycrocuta* in GER.

Within "b2", the subcluster "b2.a" includes the LFAs of TSR, LIB, KLT, KRI, KAL, and KMN (Figure 1). The faunal assemblage is characterized by the co-occurrence of *Mammuthus* + *Equus* + *Stephanorhinus* + *Hippopotamus* + *Dama*-like deer and, although not present in every single LFA, by *Canis* + *Pachycrocuta* and the coexistence of *Leptobos* and *Eobison*. Based on TSR [66], this cluster of LFAs is dated between 1.78 and ca. 1.5 Ma.

The subcluster b2.b includes the LFA of APL and most likely ALK-VOL (Figure 1). At the genus level, it is defined by the association of *Equus* + *Canis* + *Hemitragus* and likely *Pachycrocuta*, *Mammuthus*, and *Pontoceros*. The large mammal association at the genus level appears quite similar to that of the previous assemblage, but key taxa such as *Palaeotragus* and *Leptobos* are missing. Although no geochronological evidence is available for this assemblage, it is safely placed between 1.5 and 0.8 Ma and likely around 1.2 Ma [6].

4. Discussion

4.1. Establishment of the Greek Faunal Units

Based on the cluster analysis results and the local stratigraphic, biostratigraphic, biochronologic, and geochronologic evidence, the entire set of Greek Late Pliocene to Middle Pleistocene LFAs were reclassified chronologically based on the degree of faunal consistency (Figure 2, fourth column). A species-level analysis within and between successive LFA clusters allow eight GFUs to be recognized, which are (from the base to the top): the Milia FU, Dafnero FU, Gerakarou FU, Tsiotra Vryssi FU, Krimni FU, Apollonia FU, Marathousa FU, and Apidima FU (Figure 2, fifth column). Each GFU is defined by a set of FLOs and LLOs (Figure 3).

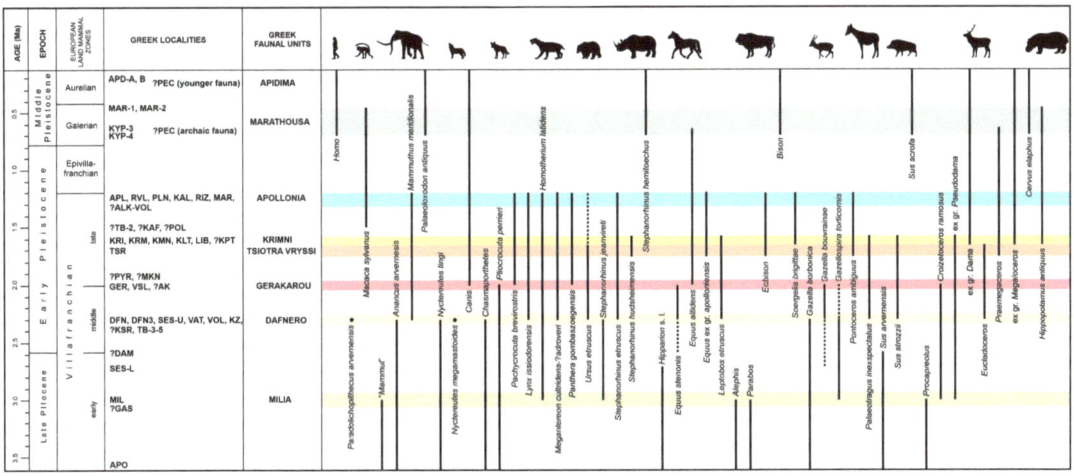

Figure 2. The Late Pliocene–Middle Pleistocene (Villafranchian to Aurelian) large mammal faunal units of Greece, the examined fossiliferous localities, and biochronological range of selected large mammals (silhouette images from PhyloPic, phylopic.org).

The lack of solid, local evidence concerning early to early/middle Villafranchian LFAs from Greece and the rather sparse record allow only the Milia FU to be recognized within this interval at about 3.0 Ma, including MIL and possibly GAS LFAs (e.g., [42,56,75]). The Milia FU is marked by the LLOs of "*Mammut*", *Alephis*, *Parabos*, and *Procapreolus* and the FLOs of *Ursus*, *Homotherium latidens*, *Stephanorhinus jeanvireti*, *Gazellospira torticornis*, *Croizetoceros ramosus*, and cervids of the "*Pseudodama*" lineage (Figure 2). *Mammuthus* might also firstly occur in this FU [76], though data are rather sporadic and not well time-controlled. The older Apolakkia (APO) LFA from Rhodes Island bears very few large mammals and mostly micromammals, the latter indicating a latest Ruscinian (MN15) age [26,37,38]. It is therefore quite possible that at least one more FU can be established below the Milia FU, but data are currently insufficient. Due to the absence of *Equus*, the Sesklo lower level (SES-L) LFA is certainly dated before the Pliocene/Pleistocene boundary [58]. It shares with Milia several common taxa at the genus level (Figure 2), but the large and advanced *Plesiohipparion* and the possible presence of *Gazella bouvrainae* [58] indicate a younger (MN16b) age, and the assemblage might represent a distinct FU between those of Milia and Dafnero, occupying the uppermost part of the early Villafranchian.

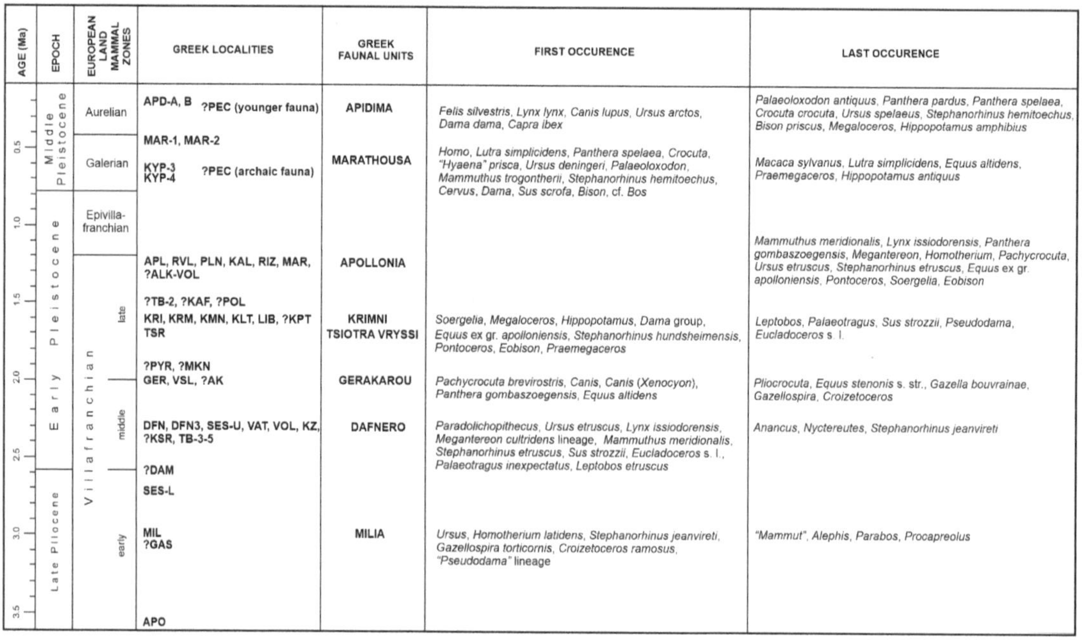

Figure 3. Biostratigraphic table of the large mammal localities and faunal units of Greece with the first and last local occurrences of the various taxa.

Apart from the well-known LFAs of DFN/DFN3, SES-U, VAT, VOL, and KZ (e.g., [6,40,51,58,67,69]), the Dafnero FU may also include TB 3-5 and possibly KSR, both known for limited large mammal remains [11,47,65]. The Dafnero FU is characterized by the presence of *Paradolichopithecus arvernensis* and *Nyctereutes megamastoides*; the LLOs of *Anancus, Nycteureutes tingi, Chasmaporthetes, Stephanorhinus jeanvireti,* and *Gazella borbonica;* and the FLOs of *Lynx issiodorensis, Megantereon cultridens* lineage, *Ursus etruscus, Mammuthus meridionalis, Stephanorhinus etruscus, Sus strozzii, Palaeotragus inexpectatus, Eucladoceros* s. l., and *Leptobos etrusus.* The Damatria (DAM) LFA from the upper part of the homonymous Damatria Formation in Rhodes Island is more difficult to assess within this scheme, due to the puzzling evidence arising from the combination of small and large mammals, which are broadly assigned to MN16 [11,26,77]. Local biostratigraphic evidence places the Plio-Pleistocene boundary in the lower part of the overlying Kritika Formation ([78] and references therein). Hence, the presence of *Equus* and *Sus strozzii* (if this latter is confirmed) suggest that the DAM LFA slightly predates or roughly corresponds to the early/middle Villafranchian transition and could potentially be included in the Dafnero FU, which is suggested to last from ca. 2.3 to ca. 2.5 Ma.

The succeeding Gerakarou FU at around 2.0 Ma includes GER, VSL, AK, and possibly PYR and MKN LFAs (e.g., [6,26,31,47] and references cited). The FLOs of *Canis* s. str. along with *Panthera gombaszoegensis, Pachycrocuta brevirostris,* and *Equus altidens* are found here, whereas *Pliocrocuta perrieri, Equus stenonis* s. str., *Croizetoceros ramosus, Gazella bouvrainae,* and most likely *Gazellospira* appear for the last time in the Greek record (Figure 2).

The Tsiotra Vryssi FU follows at about 1.8–1.6 Ma, represented only by the homonymous LFA from Mygdonia Basin (e.g., [63,66]). The faunal assemblage incorporates most taxa that first occurred in the previous GFU but also includes the FLOs of *Stephanorhinus hundsheimensis, Equus* ex. gr. *apolloniensis, Praemegaceros, Eobison,* and *Pontoceros* (and perhaps a smaller-sized *Megantereon*), indicating the beginning of an important renewal of the mammal community. At the same time, canids related to the *Canis mosbachensis* lineage, alongside *Canis (Xenocyon)* [79], also appear in this GFU.

The following Krimni FU, centered around 1.6–1.5 Ma, includes the LFAs of KRI, KRM, KMN, LIB, and KLT and possibly KPT, POL, TB-2, and KAF, the latter two attributed mainly based on their micromammal content (e.g., [11,44–46,64,65]). This GFU is characterized by the LLOs of *Sus strozzii*, *Palaeotragus*, *Leptobos*, and *Eucladoceros* s. l. and by the FLOs of *Soergelia*, the *Megaloceros* lineages, *Hippopotamus*, and perhaps *Macaca sylvanus*, while cervids of the ex gr. "*Pseudodama*" are replaced by those of true *Dama*.

The Apollonia FU includes APL, RVL, PNT, KAL, MAR, and RIZ LFAs from Mygdonia Basin (e.g., [6,33,43,44,63]) and most likely ALK-VOL from Thessaly [26,30,59]. No new FLOs occur in this assemblage, yet the absence of middle ("archaic") Villafranchian relics (i.e., *Palaeotragus*, *Leptobos*) and the presence of mostly late Villafranchian newcomers indicate the more "modern" characteristics of the fauna. On the other hand, this GFU documents the LLOs of several large mammal taxa, such as *Mammuthus meridionalis*, *Pachycrocuta*, *Lynx issiodorensis*, *Homotherium*, *Megantereon*, *Panthera gombaszoegensis*, *Ursus etruscus*, *Stephanorhinus etruscus*, *Equus* ex gr. *apolloniensis*, *Eobison*, *Soergelia*, and *Pontoceros*. The age of the Apollonia FU is currently estimated at about 1.2 Ma, but it is also possible that the herein included LFAs may cover a wider time span from ca. 1.5 up to 0.9 Ma; alternatively, the scarcity/absence of Greek LFAs from this time interval may obscure the potential presence of an additional "ghost" GFU.

The overlain Marathousa FU, dated between <0.80 and ca. 0.45 Ma, includes LFAs from the Megalopolis Basin in Peloponnese (e.g., KYP-3, 4, MAR-1, 2; [48–50,54,55,80]) and most probably the older (attributed to the early Middle Pleistocene) faunal assemblage from Petralona cave in Chalkidiki Peninsula (e.g., [60,61]). This GFU encompasses the Galerian faunas and is marked by an important reorganization of the fauna including the FLOs of *Homo*, *Palaeoloxodon antiquus*, *Lutra simplicidens*, *Stephanorhinus hemitoechus*, *Bison*, *Sus scrofa*, the *Cervus elaphus* lineage, and possibly *Bos*, while, considering the Middle Pleistocene fauna of Petralona this FU may also include the FLOs of *Ursus deningeri*, *Panthera spelaea*, *Crocuta crocuta*, and "*Hyena*" *prisca* (the latter as *Pliocrocuta perrieri* in [62]). In addition, the presence of a *Panthera* sp., possibly *Panthera pardus*, at KYP-3 [48] could also potentially represent the FLO of this species, as could also be the case for *Mammuthus trogontherii* from Loussika and other localities (e.g., [81]). On the other hand, the Marathousa FU documents the LLOs of *Macaca sylvanus*, *Equus altidens*, *Hippopotamus antiquus*, and *Praemegaceros* as well as of *Lutra simplicidens*.

The uppermost, Apidima FU, represents the establishment of modern faunal elements at the end of the Middle Pleistocene but can practically be extended to the Late Pleistocene as the large mammal fauna does not evidence any renewal during the latter epoch, besides several disappearances towards its end. In addition to the LFA of the Apidima cave complex (APD-A, B) several other LFAs can be included in this FU, for instance, the caves of Petralona (younger fauna), Vraona, Diros, Agios Georgios, Almopia, Kalamakia, Lakonis, Klissoura, Franchthi, Mavri Spilia, and Melitzia as well as Penios valley (e.g., [82–93]). The possibly oldest LFAs of this FU, APD-A and B [5], mark the beginning of the faunal modernization documenting *Capra ibex*, *Dama dama*, *Lynx lynx*, *Vulpes vulpes*, and *Felis sylvestris* ([32]; the red fox and the wildcat could have already been present in the Marathousa FU) while other modern species such as *Ursus arctos* and *Canis lupus* are present in the other LFAs. The Apidima FU includes the LLOs of *Palaeoloxodon antiquus*, *Panthera pardus*, *Panthera spelaea*, *Crocuta crocuta*, *Ursus spelaeus*, *Stephanorhinus hemitoechus*, *Bison priscus*, *Megaloceros*, and *Hippopotamus amphibius*.

4.2. The Greek Faunal Units in the European context

Based on available data from Italy, France, and Spain [14,21,73,94,95], the Milia GFU roughly corresponds to the Triversa FU of Italy and Viallete LFA level in France (Figure 4). The Dafnero GFU is correlated to the St. Vallier FU of Italy and France and the Huelago/Fuenta Nueva-1 LFA level in Spain (Figure 4). The Gerakarou GFU is placed between the Costa St. Giacomo and Olivola FU of Italy, Senèze LFA level in France, and La Puebla de Valverde LFA level in Spain (Figure 4). The Tsiotra Vryssi GFU corresponds to the Tasso FU or to the late Tasso/early Farneta FUs of Italy and the Villanueva di Pitamo LFA level in Spain (Figure 4). The Krimni GFU corresponds to the Farneta FU in Italy and the Venta Micena/Fuente Nueva-2 LFA level in Spain (Figure 4). The Apollonia GFU is placed between the Pirro Nord and Colle Curti FUs of Italy and correlates with the Fuente Nueva-3/Barranco León 5/Sima del Elefante LFA level in Spain (Figure 4). The Marathousa GFU represents a unit within the Isernia and Fontana Ranuccio FUs in Italy and also corresponds to the Gran Dolina TD8, Sima de los Huesos, L'Escale and Arago CM LFAs in Spain and France, respectively (Figure 4). Finally, the Apidima FU can be correlated to several European localities collectively dated to the end of the Middle–Late Pleistocene.

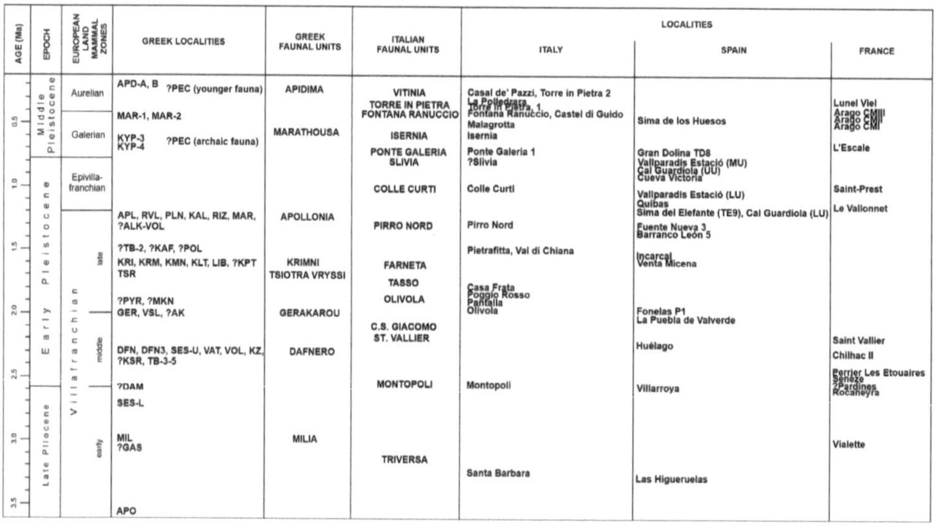

Figure 4. Correlation of the large mammal localities and faunal units of Greece with those of Italy, Spain, and France (data from [14,16,95,96]).

Several authors discuss the problem of diachrony/asynchrony in transcontinental correlations based on mammals and the establishment of mammal dispersal bioevents (e.g., [23,97,98]). The progressive and often longitudinally or latitudinally directed (N↔S, E↔W) environmental changes, the geographical proximity/distance to neighboring regions that are potential reservoirs of new taxa (such as Africa and Asia for Europe), and the changing geographical barriers/routes to and from target regions (such as the three peninsulas of Southern Europe) accentuate local diversification and create basic conditions for diachronicity. On the other hand, discontinuities and gaps in the local faunal record (sampling effect), spatiotemporal imbalances in the distribution of LFAs at local, regional and interregional scales, asymmetries in the density of available geochronologic age estimates, and long-lasting taxonomic inconsistencies (e.g., discussion on species-level systematics of European *Canis, Eucladoceros, Megantereon, "Pseudodama"*, and *Equus* species-level systematics) inhibit or even prohibit the recognition of mammal dispersal events and the ascertainment/control of their possible diachronicity. At present it seems rather unreasonable to try to interpret the S. European Pliocene to Pleistocene data at a finer scale than the temporal resolution corresponding to a full FU (100–200 kyr).

Even under these unfavorable conditions, the Greek record provides supporting evidence for already well-known and long-established mammal dispersal events such as those of *Equus* at 2.6–2.5 Ma (as redefined by [98]) and *Pachycrocuta* at ~2.0 Ma [99]. The former is recorded between SES-L and DAM LFAs (or between Milia and Dafnero GFUs). From the Balkan to the Iberian Peninsulas at least, the *Equus* event coincides with the decline of hipparionine horses [58,100], the replacement of *Sus arvernensis* by *Sus strozii* (e.g., [101]), and the appearance of *Mammuthus meridionalis* (e.g., [102]).

The arrival and rapid expansion of *Pachycrocuta* in Europe is associated with the last occurrences of *Pliocrocuta*, both events recorded in the Gerakarou GFU, which also documents the occurrence of *Panthera gombaszoegensis* [103] that entered Europe roughly at the same time as the giant hyena (e.g., [104]). Although the simultaneous and widely discussed *Canis* event ("wolf event" of [105]) is largely abolished due to the diachronous early evidence of wolf-like taxa across Eurasia (e.g., [14,106–109]), the south European record rather agrees with the firm establishment of cursorial pack-hunting canids (*Canis* ex gr. *etruscus* and *Canis arnensis*) between 2.1 and 2.0 Ma (e.g., [110,111]). It is immediately followed (2.0–1.8 Ma) by the last European occurrences of *Gazella* and *Gazellospira* among herbivores. At the same time *Equus altidens* makes its first appearance in Eastern Europe/the Balkans (e.g., [112,113]).

The next complex bioevent of similar duration starts at 1.7–1.6 Ma. All across Southern Europe, *Leptobos* was replaced by *Eobison*, cervids of the *Praemegaceros*-lineage expanded, *Stephanorhinus hundsheimensis* first occurred in the East [14,46,114], and *Pontoceros* expanded from the north peri-Pontic to the south Balkan territories. At the same time or slightly after (1.6–1.5 Ma), a smaller *Megantereon* replaced the previous larger, one and *Hippopotamus* was firmly established in the northern Mediterranean area (although the genus had sporadically occurred earlier; see [115] and references therein).

A major large mammal turnover is evident in the beginning of the Middle Pleistocene as part of the faunal reorganization close the Early/Middle Pleistocene transition that is associated with the period generally called the "Mid-Pleistocene Revolution" (e.g., [96] and references therein). This event encompasses at the (South) European scale the disappearance of several Villafranchian–Epivillafranchian taxa (e.g., *Mammuthus meridionalis*, *Pachycrocuta*, *Ursus etruscus*, *Megantereon*, *Panthera gombaszoegensis*, and *Soergelia*) and the arrival of several Galerian newcomers (e.g., *Palaeoloxodon*, *Mammuthus trogontherii*, *Crocuta*, *Panthera spelaea*, *Panthera pardus*, *Sus scrofa*, and *Stephanorhinus hemitoechus*). At the Greek scale, this turnover is noticeable in the transition from the Apollonia FU to the Marathousa FU.

5. Conclusions

Low-rank biochronological scales such as FUs have a restricted significance, representing a summary/synthesis of local to regional (bio-)chronological ordering and evolution of the mammalian network under the pressure of climatic/environmental oscillations. In light of the common and multifactorial diachrony/asynchrony phenomena across wide physical and ecological boundaries, the role of FUs appears, however, to be crucial for understanding the dynamics of mammalian dispersal in larger geographical scales and for monitoring bioevents of broader significance (e.g., [14,16,98,116,117]). As local biochronological resolution is based on continuous or adjacent and therefore relatively well-correlated sedimentary sequences, FU scales may provide complementary data to interregional/continental scale correlations, especially in the discontinuous and highly fragmentary terrestrial environments. As such, they facilitate the detection, definition, and control of important bioevents, which in turn lead to the establishment of higher-rank biochronological units/scales.

Previous and ongoing investigations in the Greek Pliocene and Pleistocene allow us to generate a first attempt for a local faunal unit scale (GFU) representing the evolution of large mammal communities from Mid-Pliocene to the Late Pleistocene. Each one of the eight recognized Greek faunal units is defined by a set of first and last local occurrence data, though the latest Pliocene and latest Early Pleistocene are still poorly resolved. Additionally, both the local record as well as those from the south of Western Europe (Italy, France, Spain) suggest that the temporal resolution cannot be improved by current knowledge below a critical interval of 100–200 kyrs, which equals the duration of a full FU. Comparison of the GFU scheme with Western European ones indicates a good correspondence in several dispersal events of particular large mammal taxa. Nevertheless, both the early to middle Villafranchian transition at ~2.6 Ma and the Epivillafranchian faunal reorganization at ca.

1 Ma, thoroughly discussed in the literature (e.g., [14,95,118]), are not expressed as clearly in the large mammal record of Greece as in other Southern European countries due to inadequate data.

Supplementary Materials: The following supporting information can be downloaded at: https://www.mdpi.com/article/10.3390/quat7020027/s1, Table S1: Presence/absence data (genus/locality) used for the hierarchical cluster analysis shown in Figure 1.

Author Contributions: Conceptualization, methodology, investigation, resources, D.S.K. and G.E.K.; validation, formal analysis, data curation, visualization, G.E.K.; writing—original draft preparation, D.S.K.; writing—review and editing, G.E.K. All authors have read and agreed to the published version of the manuscript.

Funding: G.K. is supported by the Deutsche Forschungsgemeinschaft (DFG Project no. 463225251, "MEGALOPOLIS").

Data Availability Statement: All data used in the present analysis are from published resources or provided as Supplementary Material.

Acknowledgments: We thank all colleagues that throughout the years enriched and studied the large mammal fossil record of Greece and for the various discussions that improved several perspectives of the article. We also thank the four reviewers for their comments and suggestions that improved the manuscript.

Conflicts of Interest: The authors declare no conflicts of interest.

References

1. Spassov, N. Biochronology and zoogeographic affinities of the Villafranchian faunas of Bulgaria and South Europe. *Hist. Nat. Bulg.* **2000**, *12*, 89–128.
2. Spassov, N. Southeastern Europe as a route for the earliest dispersal of *Homo* toward Europe: Ecological conditions and the timing of the first human occupation of Europe. In *Paleoanthropology of the Balkans and Anatolia*; Harvati, K., Roksandic, M., Eds.; Springer: Dordrecht, The Netherlands, 2016; pp. 281–290.
3. Kostopoulos, D.S.; Vassiliadou, K.; Koufos, G.D. The beginning of Pleistocene in the Balkan area according to the mammal record; palaeozoogeographical approach. *Ann. Géologiques Des Pays Helléniques* **2002**, *39*, 253–278.
4. Harvati, K.; Panagopoulou, E.; Runnels, C. The paleoanthropology of Greece. *Evol. Anthropol.* **2009**, *18*, 131–143. [CrossRef]
5. Harvati, K.; Röding, C.; Bosman, A.M.; Karakostis, F.A.; Grün, R.; Stringer, C.; Karkanas, P.; Thompson, N.C.; Koutoulidis, V.; Moulopoulos, L.A.; et al. Apidima Cave fossils provide earliest evidence of *Homo sapiens* in Eurasia. *Nature* **2019**, *571*, 500–504. [CrossRef] [PubMed]
6. Koufos, G.D.; Kostopoulos, D.S. The Plio-Pleistocene large mammal record of Greece: Implications for early human dispersals into Europe. In *Paleoanthropology of the Balkans and Anatolia*; Harvati, K., Roksandic, M., Eds.; Springer: Dordrecht, The Netherlands, 2016; pp. 269–280.
7. Tourloukis, V.; Harvati, K. The Palaeolithic record of Greece: A synthesis of the evidence and a research agenda for the future. *Quat. Int.* **2018**, *66*, 48–65. [CrossRef]
8. Vlachos, E. (Ed.) *Fossil Vertebrates of Greece Vol. 1: Basal Vertebrates, Amphibians, Reptiles, Afrotherians, Glires, and Primates*; Springer: Cham, Switzerland, 2022.
9. Vlachos, E. (Ed.) *Fossil Vertebrates of Greece Vol. 2: Laurasiatherians, Artiodactyles, Perissodactyles, Carnivorans, and Island Endemics*; Springer: Cham, Switzerland, 2022.
10. Koufos, G.D.; Kostopoulos, D.S. Biochronology and succession of the Plio-Pleistocene macromammalian localities of Greece. In *Actes Congres BIOCHROM '97*; Aguilar, J.P., Legendre, S., Michaux, E., Eds.; Mémoires et Travaux de l'Ecole Pratique des Hautes Etudes; Institut de Montpellier: Montpellier, France, 1997; Volume 21, pp. 619–634.
11. Koufos, G.D. The Villafranchian mammalian faunas and biochronology of Greece. *Boll. Della Soc. Paleontol. Ital.* **2001**, *40*, 217–223.
12. Koufos, G.D. The Neogene mammal localities of Greece: Faunas, chronology and biostratigraphy. *Hell. J. Geosci.* **2006**, *41*, 183–214.
13. Koufos, G.D. Neogene and Quaternary continental biostratigraphy of Greece based on mammals. In Proceedings of the Proceedings of the 14th International Congress, Thessaloniki, Greece, 25–27 May 2016.
14. Rook, L.; Martínez-Navarro, B. Villafranchian: The long story of a Plio-Pleistocene European large mammal biochronologic unit. *Quat. Int.* **2010**, *219*, 134–144. [CrossRef]
15. Bellucci, L.; Sardella, R.; Rook, L. Large mammal biochronology framework in Europe at Jaramillo: The Epivillafranchian as a formal biochron. *Quat. Int.* **2015**, *389*, 84–89. [CrossRef]
16. Palombo, M.R. Twenty years after: Reflections on the Aurelian European Land Mammal Age. *Alp. Mediterr. Quat.* **2018**, *31*, 177–180.

17. Kahlke, R.D. The Early Pleistocene (Epivillafranchian) faunal site of Untermassfeld (Thuringia, Central Germany). Synthesis of new results. *Eraul* **2000**, *92*, 123–138.
18. Kahlke, R.D. Late Early Pleistocene European large mammals and the concept of an Epivillafranchian biochron. In *Late Neogene and Quaternary Biodiversity and Evolution: Regional Developments and Interregional Correlations. Vol. II, Proceedings of the 18th International Senckenberg Conference (VI International Palaeontological Colloquium in Weimar)*; Kahlke, R.D., Maul, L.C., Mazza, P., Eds.; Courier Forschungsinstitut Senckenberg: Frankfurt, Germany, 2007; Volume 259, pp. 265–278.
19. Azzaroli, A. The Villafranchian stage in Italy and the Plio–Pleistocene boundary. *G. di Geol.* **1977**, *41*, 61–79.
20. Azzaroli, A.; De Giuli, C.; Ficcarelli, G.; Torre, D. Table of the stratigraphic distribution of terrestrial mammalian faunas in Italy from the Pliocene to the early Middle Pleistocene. *Geogr. Fis. Din. Quat.* **1982**, *5*, 55–58.
21. Gliozzi, E.; Abbazzi, L.; Argenti, P.; Azzaroli, A.; Caloi, L.; Capasso Barbato, L.; Di Stefano, G.; Esu, D.; Ficcarelli, G.; Girotti, O.; et al. Biochronology of selected mammals, molluscs and ostracods from the middle Pliocene to the late Pleistocene in Italy. The state of the art. *Riv. Ital. di Paleontol. Stratigr.* **1997**, *103*, 369–388.
22. Palombo, M.R.; Sardella, R. Biochronology and biochron boundaries: A real dilemma or a false problem? An example based on the Pleistocene large mammalian faunas from Italy. *Quat. Int.* **2007**, *160*, 30–42. [CrossRef]
23. Palombo, M.R. Discrete dispersal bioevents of large mammals in Southern Europe in the post-Olduvai Early Pleistocene: A critical overview. *Quat. Int.* **2017**, *431*, 3–19. [CrossRef]
24. Lindsay, E. Chronostratigraphy, biochronology, datum events, land mammal ages, stage of evolution, and appearance event ordination. *Bull. Am. Mus. Nat. Hist.* **2003**, *279*, 212–230. [CrossRef]
25. Walsh, S.L. Fossil datum and paleobiological event terms, paleontostratigraphy, chronostratigraphy, and the definition of Land Mammal "Age" boundaries. *J. Vertebr. Paleontol.* **1998**, *18*, 150–179. [CrossRef]
26. van der Meulen, A.J.; van Kolfschoten, T. Review of the late Turolian to early Biharian mammal faunas from Greece and Turkey. *Mem. Della Soc. Geol. Ital.* **1986**, *31*, 201–211.
27. Koufos, G. Updating the fauna and age of the Neogene–Quaternary Large Mammal Sites of Greece. *SSRN* **2023**. [CrossRef]
28. Hammer, Ø.; Harper, D.A.T.; Ryan, P.D. PAST: Paleontological statistics software package for education and data analysis. *Palaeontol. Electron.* **2001**, *4*, 1–9.
29. Hammer, Ø.; Harper, D.A.T. *Paleontological Data Analysis*; Blackwell Publishing: Oxford, UK, 2006.
30. Athanassiou, A. First results of the palaeontological study of the fissure fillings of the area of Halykes, Magnesia. *Bull. de la Soc. Speleol. de Grece* **1994**, *21*, 318–329.
31. Parparousi, E.M.; Kostopoulos, D.S.; Roussiakis, S.; Iliopoulos, G. The Villafranchian mammal assemblage of Aghia Kyriaki, Aetoloakarnania, Greece: A preliminary palaeontological study. In Proceedings of the Conference "Paleontology and Stratigraphy in Greece in the 21st Century" of the Hellenic Committee for Paleontology and Stratigraphy of the Geological Society of Greece, Athens, Greece, 3 June 2022; pp. 68–69.
32. Tsoukala, E. Quaternary large mammals from the Apidima Caves (Lakonia, S. Peloponnese, Greece). *Beiträge Zur Paläontologie* **1999**, *24*, 207–229.
33. Kostopoulos, D.S. The Plio-Pleistocene artiodactyls (Vertebrata, Mammalia) of Macedonia 1. The fossiliferous site "Apollonia-1", Mygdonia basin of Greece. *Geodiversitas* **1997**, *19*, 845–875.
34. Koufos, G.D. New material and revision of the Carnivora, Mammalia from the Lower Pleistocene locality Apollonia 1, Greece. *Quaternary* **2018**, *1*, 6. [CrossRef]
35. Konidaris, G.E.; Kostopoulos, D.S.; Koufos, G.D. *Mammuthus meridionalis* (Nesti, 1825) from Apollonia-1 (Mygdonia Basin, Northern Greece) and its importance within the Early Pleistocene mammoth evolution in Europe. *Geodiversitas* **2020**, *42*, 69–91. [CrossRef]
36. Gkeme, A.G.; Koufos, G.D.; Kostopoulos, D.S. Reconsidering the equids from the Early Pleistocene fauna of Apollonia 1 (Mygdonia Basin, Greece). *Quaternary* **2021**, *4*, 12. [CrossRef]
37. van de Weerd, A.; Reumer, J.W.F.; de Vos, J. Pliocene mammals from the Apolakkia Formation (Rhodes, Greece). *Proc. K. Ned. Akad. Van Wet. Ser. B* **1982**, *85*, 89–112.
38. Theodorou, G.; Spjeldnaes, N.; Hanken, N.M.; Lauritzen, S.E.; Velitzelos, E.; Athanassiou, A.; Roussiakis, S. Description and taphonomic investigations of Neogene Proboscidea from Rhodos, Greece. *Ann. Géologiques Des Pays Helléniques* **2000**, *38*, 133–156.
39. Benammi, M.; Aidona, E.; Merceron, G.; Koufos, G.D.; Kostopoulos, D.S. Magnetostratigraphy and chronology of the Lower Pleistocene primate bearing Dafnero fossil site, N. Greece. *Quaternary* **2020**, *3*, 22. [CrossRef]
40. Kostopoulos, D.S.; Aidona, E.; Benammi, M.; Gkeme, A.; Grasset, L.; Guy, F.; Koufos, G.D.; Kynigopoulou, Z.; Le Maitre, A.; Novello, A.; et al. The Lower Pleistocene primate-bearing fossil site of Dafnero (W. Macedonia, Greece): New data from classic and innovative approaches. In Proceedings of the 15th international Congress of the Geological Society of Greece, Athens, Greece, 22–24 May 2019.
41. Kostopoulos, D.S.; Guy, F.; Kynigopoulou, Z.; Koufos, G.D.; Valentin, X.; Merceron, G. A 2Ma old baboon-like monkey from Northern Greece and new evidence to support the *Paradolichopithecus–Procynocephalus* synonymy (Primates: Cercopithecidae). *J. Hum. Evol.* **2018**, *121*, 178–192. [CrossRef]
42. Crégut-Bonnoure, E.; Tsoukala, E. The Pliocene Artiodactyla and Proboscidea (Mammalia) from Gephyra (lower Axios valley, Macedonia, Greece). Discovery of a new boselaphine. *Quat. Int.* **2017**, *445*, 200–214. [CrossRef]

43. Koufos, G.D.; Syrides, G.E.; Kostopoulos, D.S.; Koliadimou, K.K. Preliminary results about the stratigraphy and the palaeoenvironment of Mygdonia Basin, Macedonia, Greece. *Geobios* **1995**, *18*, 243–249. [CrossRef]
44. Tsoukala, E.; Chatzopoulou, K. A new Early Pleistocene (latest Villafranchian) site with mammals in Kalamotó (Mygdonia Basin, Macedonia, Greece)—Preliminary report. *Mitteilungen Der Komm. Für Quartärforschung Der Osterr. Akad. Der Wiss.* **2005**, *14*, 213–233.
45. Steensma, K.J. Plio-/Pleistozäne Großsäugetiere (Mammalia) aus dem Beckem von Kastoria/Grevena, Südlich von Neapolis-NW-Griechenland. Ph.D. Thesis, Technische Universität Clausthal, Clausthal, Germany, 1988.
46. Kostopoulos, D.S.; Konidaris, G.E.; Amanatidou, M.; Chitoglou, K.; Fragkioudakis, E.; Gerakakis, N.; Giannakou, V.; Gkeme, A.; Kalaitzi, C.; Tsakalidis, C.; et al. The new fossil site Krimni-3 in Mygdonia Basin and the first evidence of a giant ostrich in the Early Pleistocene of Greece. *Paläontologische Z.* **2023**, *97*, 147–161. [CrossRef]
47. Symeonidis, N.; Theodorou, G.; Schütt, H.; Velitzelos, E. Palaeontological and stratigraphical observations in the area of Achaia and Etoloacarnania (W. Greece). *Ann. Géologiques Des Pays Helléniques* **1985/1986**, *33*, 329–365.
48. Athanassiou, A. Pleistocene vertebrates from the Kyparíssia lignite mine, Megalopolis Basin, S. Greece: Rodentia, Carnivora, Proboscidea, Perissodactyla, Ruminantia. *Quat. Int.* **2018**, *497*, 198–221. [CrossRef]
49. Athanassiou, A.; Michailidis, D.; Vlachos, E.; Tourloukis, V.; Thompson, N.; Harvati, K. Pleistocene vertebrates from the Kyparíssia lignite mine, Megalopolis Basin, S. Greece: Testudines, Aves, Suiformes. *Quat. Int.* **2018**, *497*, 178–197. [CrossRef]
50. Konidaris, G.E.; Athanassiou, A.; Tourloukis, V.; Chitoglou, K.; van Kolfschoten, T.; Giusti, D.; Thompson, N.; Tsartsidou, G.; Roditi, E.; Panagopoulou, E.; et al. The late Early–Middle Pleistocene mammal fauna from the Megalopolis Basin (Peloponnese, Greece) and its importance for biostratigraphy and palaeoenvironment. *Quaternary* **2024**. *under review*.
51. Sianis, P.D. Study of the Villafranchian Large Mammal Fauna from Karnezeika (Argolis, Peloponnese, Greece): Systematics, Taphonomy, Palaeoecology. Ph.D. Thesis, University of Patras, Patras, Greece, 2023.
52. Sianis, P.D.; Athanassiou, A.; Roussiakis, S.; Iliopoulos, G. Carnivora from the Early Pleistocene locality of Karnezeika (Southern Greece). *Geobios* **2023**, *79*, 43–59. [CrossRef]
53. Koufos, G.D.; Vassiliadou, K.V.; Koliadimou, K.K.; Syrides, G.E. Early Pleistocene small mammals from Marathoussa, a new locality in the Mygdonia basin, Macedonia, Greece. *Deinsea* **2001**, *8*, 49–102.
54. Konidaris, G.E.; Athanassiou, A.; Tourloukis, V.; Thompson, N.; Giusti, D.; Panagopoulou, E.; Harvati, K. The skeleton of a straight-tusked elephant (*Palaeoloxodon antiquus*) and other large mammals from the Middle Pleistocene butchering locality Marathousa 1 (Megalopolis Basin, Greece): Preliminary results. *Quat. Int.* **2018**, *497*, 65–84. [CrossRef]
55. Konidaris, G.E.; Tourloukis, V.; Boni, G.; Athanassiou, A.; Giusti, D.; Thompson, N.; Syrides, G.; Panagopoulou, E.; Karkanas, T.; Harvati, K. Marathousa 2: A new Middle Pleistocene locality in the Megalopolis Basin (Greece) with evidence of hominin exploitation of megafauna (*Hippopotamus*). *PaleoAnthropology* **2023**, *2023*, 34–55.
56. Tsoukala, E.; Mol, D. The Proboscidea of the Early Villafranchian site of Milia (Grevena, Macedonia, Greece). *Quat. Int.* **2016**, *406*, 4–24. [CrossRef]
57. Vlachos, E.; Tsoukala, E.; Crégut-Bonnoure, E.; Guérin, C.; Mol, D. The Paradise Lost of Milia (Grevena, Greece; Late Pliocene, early Villafranchian, MN15/MN16a): Faunal composition and diversity. *Quaternary* **2018**, *1*, 13. [CrossRef]
58. Athanassiou, A. A Villafranchian *Hipparion*-bearing mammal fauna from Sésklo (E. Thessaly, Greece): Implications for the question of *Hipparion–Equus* sympatry in Europe. *Quaternary* **2018**, *1*, 12. [CrossRef]
59. Athanassiou, A. Contribution to the Study of the Fossil Mammals of Thessaly. Ph.D. Thesis, National and Kapodistrian University of Athens, Athens, Greece, 1996.
60. Tsoukala, E. Contribution to the Study of the Pleistocene Fauna of Large Mammals (Carnivora, Perissodactyla, Artiodactyla) from Petralona Cave, Chalkidiki (N. Greece). Ph.D. Thesis, Aristotle University of Thessaloniki, Thessaloniki, Greece, 1989.
61. Tsoukala, E.; Guérin, C. The Rhinocerotidae and Suidae of the Middle Pleistocene from Petralona Cave (Macedonia, Greece). *Acta Zool. Bulg.* **2016**, *68*, 243–264.
62. Baryshnikov, G.F.; Tsoukala, E. New analysis of the Pleistocene carnivores from Petralona cave (Macedonia, Greece) based on the collection of the Thessaloniki Aristotle University. *Geobios* **2010**, *43*, 389–402. [CrossRef]
63. Konidaris, G.E.; Tourloukis, V.; Kostopoulos, D.S.; Thompson, N.; Giusti, D.; Michailidis, D.; Koufos, G.D.; Harvati, K. Two new vertebrate localities from the Early Pleistocene of Mygdonia Basin (Macedonia, Greece): Preliminary results. *Comptes Rendus Palevol* **2015**, *14*, 353–362. [CrossRef]
64. van der Meulen, A.J.; Doukas, C.S. The Early Biharian rodent fauna from Tourkobounia 2 (Athens, Greece). *Lynx New Ser.* **2001**, *32*, 255–277.
65. Symeonides, N.; de Vos, J. Großsäuger-Funde aus den pleistozänen Spaltenfüllung von Tourkovounia in Athen. *Ann. Géologiques des Pays Helléniques* **1976**, *28*, 135–143.
66. Konidaris, G.E.; Kostopoulos, D.S.; Maron, M.; Schaller, M.; Ehlers, T.A.; Aidona, E.; Marini, M.; Tourloukis, V.; Muttoni, G.; Koufos, G.D.; et al. Dating of the Lower Pleistocene vertebrate site of Tsiotra Vryssi (Mygdonia Basin, Greece): Biochronology, magnetostratigraphy, and cosmogenic radionuclides. *Quaternary* **2021**, *4*, 1, Correct in *Quaternary* **2022**, *5*, 21. [CrossRef]
67. de Vos, J.; van der Made, J.; Athanassiou, A.; Lyras, G.; Sondaar, P.Y.; Dermitzakis, M.D. Preliminary note on the Late Pliocene fauna from Vatera (Lesvos, Greece). *Ann. Géologiques Des Pays Helléniques* **2002**, *39*, 37–70.
68. Lyras, G.A.; van der Geer, A.E. The Late Pliocene vertebrate fauna of Vatera (Lesvos Island, Greece). *Cranium* **2007**, *24*, 11–24.

69. Kostopoulos, D.S. The Plio-Pleistocene Artiodactyls of Macedonia, Greece: 2. The fossiliferous locality of Volakas, VOL (Volakas Basin, Drama, NE Greece). *Paleontol. Evol.* **1998**, *30–31*, 83–92.
70. Karkanas, P.; Tourloukis, V.; Thompson, N.; Giusti, D.; Tsartsidou, G.; Athanassiou, A.; Konidaris, G.; Roditi, E.; Panagopoulou, E.; Harvati, K. The Megalopolis Paleoenvironmental Project (MegaPal). In *Human Evolution at the CROSSROADS: Research in Greece and Beyond*; Harvati, K., Ioannidou, M., Eds.; Tübingen University Press: Tübingen, Germany, in press.
71. van Kolfschoten, T.; Konidaris, G.E.; Doukas, C.; Athanassiou, A.; Tourloukis, V.; Panagopoulou, E.; Karkanas, P.; Harvati, K. Voles (Rodentia, Mammalia) as a proxy to date the site Kyparissia 4 (Megalopolis Basin, Greece). In *Human Evolution at the CROSSROADS: Research in Greece and Beyond*; Harvati, K., Ioannidou, M., Eds.; Tübingen University Press: Tübingen, Germany, in press.
72. Tourloukis, V.; Muttoni, G.; Karkanas, P.; Monesi, E.; Scardia, G.; Panagopoulou, E.; Harvati, K. Magnetostratigraphic and chronostratigraphic constraints on the Marathousa 1 Lower Palaeolithic site and the Middle Pleistocene deposits of the Megalopolis basin, Greece. *Quat. Int.* **2018**, *497*, 154–169. [CrossRef]
73. Nomade, S.; Pastre, J.F.; Guillou, H.; Faure, M.; Guérin, C.; Delson, E.; Debard, E.; Voinchet, P.; Messager, E. $^{40}Ar/^{39}Ar$ constraints on some French landmark Late Pliocene to Early Pleistocene large mammalian paleofaunas: Paleoenvironmental and paleoecological implications. *Quat. Geochronol.* **2014**, *21*, 2–15. [CrossRef]
74. Gkeme, A.G. Study of the Quaternary Equids of Greece: Systematics, Biostratigraphy, Phylogeny, Palaeoecology. Ph.D. Thesis, Aristotle University of Thessaloniki, Thessaloniki, Greece, 2023.
75. Crégut-Bonnoure, E.; Tsoukala, E. The Late Pliocene Bovidae and Cervidae (Mammalia) of Milia (Grevena, Macedonia, Greece). *Quat. Int.* **2017**, *445*, 215–249. [CrossRef]
76. Kostopoulos, D.S.; Koulidou, I. An early mammoth maxilla from north-western Greece. *Quat. Int.* **2015**, *379*, 155–163. [CrossRef]
77. Benda, L.; Meulenkamp, J.E.; van de Weerd, A. Biostratigraphic correlations in the Eastern Mediterranean Neogene. 3. Correlation between mammal, sporomorph and marine microfossil assemblages from the Upper Cenozoic of Rhodos, Greece. *Newsl. Stratigr.* **1977**, *6*, 117–130. [CrossRef]
78. Schneider, S.; Linse, U. Pliocene–Pleistocene Unionida from Rhodes (Dodecanese, Greece): Insights into the evolution of Eastern Mediterranean freshwater mussels. *Arch. Für Molluskenkd.* **2023**, *152*, 43–69. [CrossRef]
79. Karakosta, A.; Konidaris, G.E.; Kostopoulos, D.S.; Koufos, G.D.; Harvati, K. Preliminary study of the canids from the Lower Pleistocene site of Tsiotra Vryssi (Mygdonia Basin, Greece). In *Human Evolution at the CROSSROADS: Research in Greece and Beyond*; Harvati, K., Ioannidou, M., Eds.; Tübingen University Press: Tübingen, Germany, in press.
80. Konidaris, G.E.; Athanassiou, A.; Panagopoulou, E.; Harvati, K. First record of *Macaca* (Cercopithecidae, Primates) in the Middle Pleistocene of Greece. *J. Hum. Evol.* **2022**, *162*, 103104. [CrossRef]
81. Athanassiou, A. A skeleton of *Mammuthus trogontherii* (Proboscidea, Elephantidae) from NW Peloponnese, Greece. *Quat. Int.* **2012**, *255*, 9–28. [CrossRef]
82. Symeonides, N.; Bachmayer, F.; Zapfe, H. Ergebnisse weiterer Grabungen in der Höhle von Vraona (Attika, Griechenland). *Ann. Géologiques Des Pays Helléniques* **1980**, *30*, 291–299.
83. Tsoukala, E.S. The Pleistocene large mammals from the Agios Georgios cave, Kilkis (Macedonia, N. Greece). *Geobios* **1992**, *25*, 415–433. [CrossRef]
84. Rabeder, G. Jungpleistozäne und Frühholozäne Säugetier-Reste aus der Höhle von Vraona auf Attika, Griechenland. *Ann. Géologiques Des Pays Helléniques* **1995**, *36*, 47–59.
85. Giannopoulos, V.I. Contribution to the Study of Recent and Old Environments of the Most Important Greek Caves. Ph.D. Thesis, National and Kapodistrian University of Athens, Athens, Greece, 2000.
86. Panagopoulou, E.; Karkanas, P.; Tsartsidou, G.; Kotjabopoulou, E.; Harvati, K.; Ntinou, M. Late Pleistocene archaeological and fossil human evidence from Lakonis Cave, Southern Greece. *J. Field Archaeol.* **2004**, *29*, 323–349. [CrossRef]
87. Athanassiou, A. The Late Pleistocene fauna of Peneiós Valley (Lárissa, Thessaly, Greece): New collected material. In Proceedings of the 9th Annual Meeting of the European Association of Vertebrate Palaeontologists (Heraklion, Crete), Program and Abstracts, Heraklion, Crete, 14–19 June 2011; Volume 14.
88. Stiner, M.C.; Munro, N.D. On the evolution of diet and landscape during the Upper Paleolithic through Mesolithic at Franchthi Cave (Peloponnese, Greece). *J. Hum. Evol.* **2011**, *60*, 618–636. [CrossRef] [PubMed]
89. Darlas, A.; Psathi, E. The Middle and Upper Paleolithic on the western coast of the Mani Peninsula (Southern Greece). In *Paleoanthropology of the Balkans and Anatolia: Human Evolution and Its Context*; Harvati, K., Roksandic, M., Eds.; Springer: Dordrecht, The Netherlands, 2016; pp. 95–117.
90. Starkovich, B.M. Paleolithic subsistence strategies and changes in site use at Klissoura Cave 1 (Peloponnese, Greece). *J. Hum. Evol.* **2017**, *111*, 63–84. [CrossRef]
91. Tzortzi, M.; Sianis, P.D.; Iliopoulos, G.; Darlas, A. Carnivores from the late Pleistocene locality Melitzia Cave (Mani peninsula, Greece). *Hist. Biol.* **2024**, *36*, 266–284. [CrossRef]
92. Tourloukis, V.; Thompson, N.; Garefalakis, C.; Karkanas, P.; Konidaris, G.E.; Panagopoulou, E.; Harvati, K. New Middle Palaeolithic sites from the Mani Peninsula, Southern Greece. *J. Field Archaeol.* **2016**, *41*, 68–83. [CrossRef]

93. Tsoukala, E.; Chatzopoulou, K.; Rabeder, G.; Pappa, S.; Nagel, D.; Withalm, G. Paleontological and stratigraphical research in Loutra Arideas Bear Cave (Almopia Speleopark, Pella, Macedonia, Greece). In Proceedings of the 2th International Cave Bear Symposium, Thessaloniki and Aridea (Pella, Macedonia), Greece, 2–5 November 2006; Tsoukala, E., Rabeder, G., Eds.; Scientific Annals. School of Geology Aristotle University of Thessaloniki (AUTH): Thessaloniki, Greece, 2006; Volume 98, pp. 41–67.
94. Palombo, M.R.; Valli, A.M.F. Remarks on the biochronology of mammalian faunal complexes from the Pliocene to the Middle Pleistocene in France. *Geol. Romana* **2003**, *37*, 145–163.
95. Madurell-Malapeira, J.; Ros-Montoya, S.; Espigares, M.P.; Alba, D.M.; Aurell-Garrido, J. Villafranchian large mammals from the Iberian Peninsula: Paleobiogeography, paleoecology and dispersal events. *J. Iber. Geol.* **2014**, *40*, 167–178. [CrossRef]
96. Palombo, M.R. Deconstructing mammal dispersals and faunal dynamics in SW Europe during the Quaternary. *Quat. Sci. Rev.* **2014**, *96*, 50–71. [CrossRef]
97. Alroy, J.; Bernor, R.L.; Fortelius, M.; Werdelin, L. The MN System: Regional or continental? *Mitteilungen Der Bayer. Staatssamml. Für Paläontologie Und Hist. Geol.* **1998**, *38*, 243–258.
98. Iannucci, A.; Sardella, R. What does the "Elephant-*Equus*" event mean today? Reflections on mammal dispersal events around the Pliocene-Pleistocene boundary and the flexible ambiguity of biochronology. *Quaternary* **2023**, *6*, 16. [CrossRef]
99. Martínez-Navarro, B. Early Pleistocene faunas of Eurasia and hominin dispersals. In *Out of Africa I. The First Hominin Colonization of Eurasia*; Fleagle, J.G., Shea, J.J., Grine, F.E., Baden, A.L., Leakey, R.E., Eds.; Springer: Dordrecht, The Netherlands, 2010; pp. 207–224.
100. Cirilli, O.; Pandolfi, L.; Alba, D.M.; Madurell-Malapeira, J.; Bukhsianidze, M.; Kordos, L.; Lordkipanidze, D.; Rook, L.; Bernor, R.L. The last Plio-Pleistocene hipparions of Western Eurasia. A review with remarks on their taxonomy, paleobiogeography and evolution. *Quat. Sci. Rev.* **2023**, *306*, 107976. [CrossRef]
101. Cherin, M.; Sorbelli, L.; Crotti, M.; Iurino, D.A.; Sardella, R.; Souron, A. New material of *Sus strozzii* (Suidae, Mammalia) from the Early Pleistocene of Italy and a phylogenetic analysis of suines. *Quat. Sci. Rev.* **2018**, *194*, 94–115. [CrossRef]
102. Lister, A.M.; Sher, A.V.; van Essen, H.; Wei, G. The pattern and process of mammoth evolution in Eurasia. *Quat. Int.* **2005**, *126–128*, 49–64. [CrossRef]
103. Koufos, G.D. The Pleistocene carnivores of the Mygdonia basin (Macedonia, Greece). *Ann. De Paléontologie* **1992**, *78*, 205–257.
104. Konidaris, G.E. Guilds of large carnivorans during the Pleistocene of Europe: A community structure analysis based on foraging strategies. *Lethaia* **2022**, *55*, 1–18. [CrossRef]
105. Azzaroli, A. Quaternary mammals and the "end-Villafranchian" dispersal event—A turning point in the history of Eurasia. *Palaeogeogr. Palaeoclimatol. Palaeoecol.* **1983**, *44*, 117–139. [CrossRef]
106. Sardella, R.; Palombo, M.R. The Pliocene-Pleistocene boundary: Which significance for the so called "Wolf event"? Evidences from western Europe. *Quaternaire* **2007**, *18*, 65–71. [CrossRef]
107. Sotnikova, M.; Rook, L. Dispersal of the Canini (Mammalia, Canidae: Caninae) across Eurasia during the Late Miocene to Early Pleistocene. *Quat. Int.* **2010**, *212*, 86–97. [CrossRef]
108. Iannucci, A.; Mecozzi, B.; Sardella, R. Beware of the "Wolf event"—Remarks on large mammal dispersals in Europe and the late Villafranchian faunal turnover. *Alp. Mediterr. Quat.* **2023**, *36*, 75–90.
109. Tong, H.; Chen, X.; Zhang, B.; Rothschild, B.; White, S.; Balisi, M.; Wang, X. Hypercarnivorous teeth and healed injuries to *Canis chihliensis* from Early Pleistocene Nihewan beds, China, support social hunting for ancestral wolves. *PeerJ* **2020**, *8*, e9858. [CrossRef] [PubMed]
110. Sardella, R.; Bellucci, L.; Bona, F.; Cherin, M.; Iurino, D.A.; Rook, L. Before and after the earliest *Homo* dispersal in Europe: Evidence from the early Pleistocene sites of the Italian Peninsula. In *European Early Pleistocene Biogeography and Ecology Based on the Mammal Record: Case Studies and Preliminary Syntheses*; Kostopoulos, D.S., Konidaris, G., Tesakov, A., van den Hoek Ostende, L.W., Rook, L., Eds.; Comptes Rendus Palevol: Amsterdam, The Netherlands, 2018; Volume 17, pp. 287–295.
111. Bartolini Lucenti, S.; Rook, L. A review on the Late Villafranchian medium-sized canid *Canis arnensis* based on the evidence from Poggio Rosso (Tuscany, Italy). *Quat. Sci. Rev.* **2016**, *151*, 58–71. [CrossRef]
112. Gkeme, A.; Koufos, G.; Kostopoulos, D. The Early Pleistocene stenonoid horse from Libakos (Western Macedonia, Greece): Biochronological and palaeoecological implications and dispesal events. In Proceedings of the 15th Congress of the Regional Committee on Mediterranean Neogene Stratigraphy (RCMNS), Athens, Greece, 3–6 September 2017; p. 67.
113. Koufos, G.D.; Vlachou, T.D.; Gkeme, A.G. The fossil record of equids (Mammalia: Perissodactyla: Equidae) in Greece. In *Fossil Vertebrates of Greece Vol. 2: Laurasiatherians, Artiodactyles, Perissodactyles, Carnivorans, and Island Endemics*; Vlachos, E., Ed.; Springer: Cham, Switzerland, 2022; pp. 351–401.
114. Kostopoulos, D.S.; Maniakas, I.; Tsoukala, E. Early bison remains from Mygdonia Basin (Northern Greece). *Geodiversitas* **2018**, *40*, 283–319. [CrossRef]
115. Fidalgo, D.; Madurell-Malapeira, J.; Martino, R.; Pandolfi, L.; Rosas, A. An updated review of the Quaternary hippopotamus fossil records from the Iberian Peninsula. *Quaternary* **2024**, *7*, 4. [CrossRef]
116. Bell, C.J.; Lundelius, E.L.J.; Barnosky, A.D.; Graham, R.W.; Lindsay, E.H.; Ruez, D.R.J.; Semken, H.A.J.; Webb, S.D.; Zakrzewski, R.J. The Blancan, Irvingtonian, and Rancholabrean Mammal Ages. In *Late Cretaceous and Cenozoic Mammals of North America*; Woodburne, M.O., Ed.; Columbia University Press: New York, NY, USA, 2004; pp. 232–314.

117. Palombo, M.R. Biochronology of terrestrial mammals and Quaternary subdivisions: A case study of large mammals from the Italian peninsula. *Il Quat.* **2009**, *22*, 291–306.
118. Magri, D.; Palombo, M.R. Early to Middle Pleistocene dynamics of plant and mammal communities in South West Europe. *Quat. Int.* **2013**, *288*, 63–72. [CrossRef]

quaternary

MDPI

Article

The Late Early–Middle Pleistocene Mammal Fauna from the Megalopolis Basin (Peloponnese, Greece) and Its Importance for Biostratigraphy and Paleoenvironment

George E. Konidaris [1,*], Athanassios Athanassiou [2], Vangelis Tourloukis [1,3], Krystalia Chitoglou [1], Thijs van Kolfschoten [4,5], Domenico Giusti [1], Nicholas Thompson [1], Georgia Tsartsidou [2], Effrosyni Roditi [1], Eleni Panagopoulou [2], Panagiotis Karkanas [6] and Katerina Harvati [1]

1 Paleoanthropology, Institute for Archaeological Sciences and Senckenberg Centre for Human Evolution and Palaeoenvironment, Eberhard Karls University of Tübingen, 72070 Tübingen, Germany; vangelis.tourloukis@ifu.uni-tuebingen.de (V.T.); krystalia.chitoglou@uni-tuebingen.de (K.C.); domenico.giusti@uni-tuebingen.de (D.G.); nicholas-charles.thompson@uni-tuebingen.de (N.T.); effrosyni.roditi@student.uni-tuebingen.de (E.R.); katerina.harvati@ifu.uni-tuebingen.de (K.H.)

2 Ephorate of Palaeoanthropology–Speleology, Hellenic Ministry of Culture, 11636 Athens, Greece; aathanas@geol.uoa.gr (A.A.); gtsartsidou@ymail.com (G.T.); elenipanagopoulou@yahoo.com (E.P.)

3 Department of History and Archaeology, School of Philosophy, University of Ioannina, 45110 Ioannina, Greece

4 Faculty of Archaeology, Leiden University, 2300 RA Leiden, The Netherlands; t.van.kolfschoten@arch.leidenuniv.nl

5 Institute of Cultural Heritage, Shandong University, Qingdao 250100, China

6 M.H. Wiener Laboratory for Archaeological Science, American School of Classical Studies at Athens, 10676 Athens, Greece; tkarkanas@ascsa.edu.gr

* Correspondence: georgios.konidaris@uni-tuebingen.de

check for updates

Citation: Konidaris, G.E.; Athanassiou, A.; Tourloukis, V.; Chitoglou, K.; van Kolfschoten, T.; Giusti, D.; Thompson, N.; Tsartsidou, G.; Roditi, E.; Panagopoulou, E.; et al. The Late Early–Middle Pleistocene Mammal Fauna from the Megalopolis Basin (Peloponnese, Greece) and Its Importance for Biostratigraphy and Paleoenvironment. *Quaternary* **2024**, 7, 41. https://doi.org/10.3390/quat7040041

Academic Editors: Yolanda Fernández Jalvo and Maria Rita Palombo

Received: 30 May 2024
Revised: 6 August 2024
Accepted: 19 September 2024
Published: 24 September 2024

Abstract: Recent investigations in the upper Lower–Middle Pleistocene deposits of the Megalopolis Basin (Greece) led to the discovery of several sites/findspots with abundant faunal material. Here, we provide an updated overview including new results on the micro- and macro-mammal fauna. Important new discoveries comprise partial hippopotamus skeletons from Marathousa 1 and the new Lower Pleistocene site Choremi 6, as well as a second partial elephant skeleton from Marathousa 1, including a complete tusk and the rarely found stylohyoideum. Based on the first results from the newly collected micromammals, we discuss age constraints of the sites, and we provide biostratigraphic/biochronologic remarks on key mammal taxa for the Middle Pleistocene of Greece and southeastern Europe. The presence of mammals highly dependent on freshwater for their survival, together with temperate-adapted ones in several stratigraphic layers of the basin, including those correlated with glacial stages, when conditions were colder and/or drier, indicate the capacity of the basin to retain perennial freshwater bodies under milder climatic conditions, even during the harsher glacial periods of the European Middle Pleistocene, and further support its refugial status. Yet, the smaller dimensions of the Megalopolis hippopotamuses may represent a response to the changing environmental conditions of the epoch, not optimal for hippopotamuses. Overall, the Megalopolis Basin comprises a unique fossil record for southeastern Europe and provides valuable insights into the Middle Pleistocene terrestrial ecosystems of Europe, and hominin adaptations in particular.

Keywords: Marathousa; Choremi; Kyparissia; *Palaeoloxodon*; *Hippopotamus*; *Mimomys*; *Arvicola*; Pleistocene; Galerian; Greece

1. Introduction

The Megalopolis Basin (Arcadia, Peloponnesus, Greece; Figure 1a) is a tectonic half-graben that was filled from the Neogene onwards by continental deposits of fluviolacustrine origin. During the late Early and the Middle Pleistocene (ca. 900 to 150 ka), the basin hosted a large, shallow lake, which resulted in a stratigraphic sequence composed mainly of lacustrine sediments with lignite seams (Figure 1c) [1–3]. Since 1969, the mining of the lignite resulted in the exposure of long and high fossiliferous sections that offer the unique opportunity to study the stratigraphy and the paleoenvironment of the Pleistocene paleolake and its environs.

The existence of fossil remains in the Megalopolis Basin, especially of "giant" bones that we now know belong to elephants and hippopotamuses, is known since antiquity, when these were attributed to mythical beings (see, e.g., [4,5]). The first systematic excavations by T. Skouphos (University of Athens) in 1902 unearthed an important number of fossils and highlighted the wealth of the basin in fossil vertebrates and its potential for further paleontological research [6,7]. A part of this first collected material was studied in detail in a series of articles by I. Melentis during the 1960s [4,8–15]. Yet, as is common for historical collections, those specimens lack precise stratigraphic information, and the assemblage does not represent a single, stratified accumulation. Since then, several field surveys and excavations of Pleistocene fossils took place in the basin sporadically throughout the years, and numerous vertebrates' remains have been collected and studied (e.g., [2,16–20]). Among other vertebrate remains collected was a surface-collected hominin molar ([17], see the Appendix by Marinos), which was recently attributed to the Neanderthal lineage [21].

Recent investigations in the basin were conducted by a joint team of the Ephorate of Paleoanthropology–Speleology, the University of Tübingen, and the American School of Classical Studies at Athens, in three phases: the field survey of 2012–2013, the systematic excavation of the site Marathousa 1 in 2013–2019, and the field survey of 2018–2022 (for further details, see, e.g., [22–25]). Aiming primarily to locate sites with paleoanthropological/archaeological interest from securely stratified contexts and to perform detailed paleoenvironmental analyses with chronological control, these investigations targeted the Choremi Formation (Fm) and, in particular, its lower part, the Marathousa Member (Mb). The Marathousa Mb, dated to late Early Pleistocene–Middle Pleistocene, is formed by cyclic sedimentation of lignite and clastic (detrital intervals) sediments, mostly of lacustrine origin, with the lignites thought to correspond to interglacial (warm) stages and the detrital intervals to glacial (cold) stages (Figure 1c) [2,3,26,27]. Three main lignite seams are identified (lignite seams I, II, and III, abbreviated as LI, LII, and LIII, respectively) and subdivided into individual seams within them (LIa–c, LIIa–b, and LIIIa–c), of which LIb encompasses the Matuyama/Brunhes geomagnetic polarity reversal, at 0.78 Ma, and is correlated with the warm Marine Isotope Stage (MIS) 19 (Figure 1c) [3,28].

The aims of the present study are the following: (1) to present an overview of the micro- and macro-mammalian fauna, including updates and new results; (2) to provide age constraints based on the newly collected micromammal fauna; (3) to provide biostratigraphic/biochronological remarks on the first and last occurrences of Middle Pleistocene mammals in Greece, as well as in the wider region of southeastern Europe; (4) to discuss the potential link between the harsher environmental conditions during the glacial stages of the Middle Pleistocene and the smaller size of the Megalopolis hippopotamuses; (5) to highlight the importance of the Megalopolis Basin as a glacial refugial area for several mammals, including elephants, hippopotamuses, and macaques.

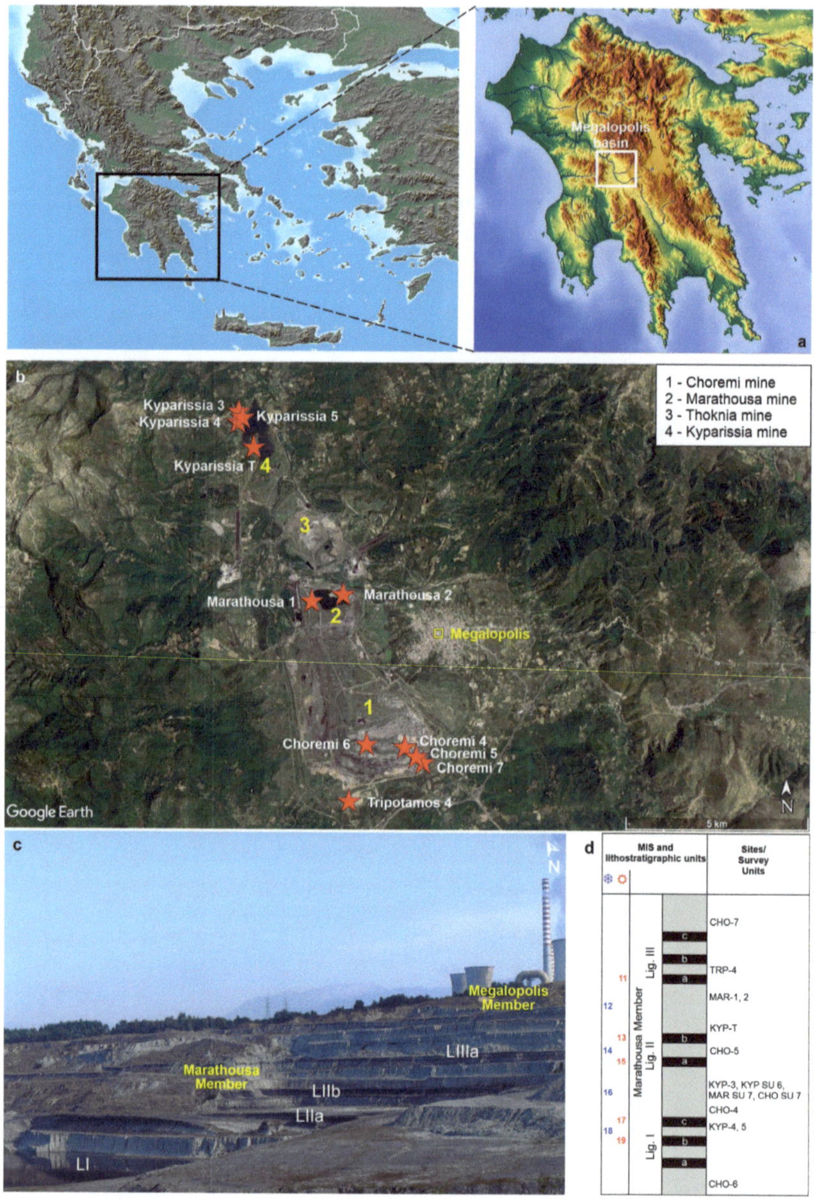

Figure 1. (**a**) Geographic position of the Megalopolis Basin (maps taken from Copernicus Land Monitoring Service: https://land.copernicus.eu/ and https://maps-for-free.com/ (accessed on 30 April 2024)). (**b**) Geographic position of the investigated sites within the lignite mines (numbered 1–4) of the Megalopolis Basin (satellite image from Google Earth). (**c**) Panoramic view of the eastern quarry of the Marathousa mine, indicating the Marathousa and Megalopolis Members of the Choremi Formation, and the lignite seams. (**d**) Simplified stratigraphic column of the Marathousa Member showing the lignite seams and the intercalating detrital intervals, their correlation to the Marine Isotope Stages (MIS) following the age model proposed by Tourloukis et al. [3], and the stratigraphic position of the sites and survey units following Karkanas et al. [25,28].

2. Materials and Methods

The material was recovered during the field investigations (field surveys and excavation) that took place in the Megalopolis Basin during 2012–2022. It is stored at the Ephorate of Paleoanthropology–Speleology in Athens. Systematic excavation was conducted at the site of Marathousa 1 from 2014 to 2019. The rest of the finds (stratified or surface finds, as noted accordingly in the article) were collected from sites (with a sufficient concentration of fossils; Figure 1b) or from findspots during the targeted survey units that systematically prospected the natural or artificial profiles of the mines. A survey unit (SU; CHO, KYP, and MAR in the Choremi, Kyparissia, and Marathousa mines, respectively; Figure 1b) is an arbitrary unit of observation delimited by natural, stratigraphic, or artificial boundaries, e.g., the start/end of a tier inside the mine [22,25]. All this material was collected from exposed artificial profiles, without the means of systematic excavation. The geodetic coordinates of survey units, collected or not collected specimens, and clear stratigraphic contacts were recorded by means of a Differential GPS in RTK mode, using the Hellenic Geodetic Reference System 1987 (HGRS87) and the local datum provided by the Megalopolis DEH.

In the following section, the fossiliferous sites and survey units are presented in chronological order (besides those that correspond roughly to the same stratigraphic position). Their stratigraphic position within the basin and their correlation to MIS are shown in Figure 1d. For the stratigraphic placement of the new find-bearing horizons, the first stratigraphic data from Karkanas et al. [25] were used, while for their age correlation, we employed the age model proposed by Tourloukis et al. [3]. This age model matches the second-option model proposed by Okuda et al. [27], and it was essentially (but preliminary) confirmed by both individual dating assays from the newly identified sites throughout the basin sequence (e.g., [25,29]), as well as by the correlation of the sequence with the Oxygen Isotope Curve (ongoing study).

Sediment samples (ca. 485 kg, dried weight) for small vertebrates were collected from the sites Kyparissia 4 (209 kg, including the test samples studied by Kolfschoten et al. [30]), Kyparissia 3 (ca. 91 kg), and Choremi 7 (ca. 162 kg), as well as from KYP SU 6 (23 kg) during the fieldwork seasons of 2019–2022. Samples were air-dried, soaked in a weak water solution of H_2O_2 (about 1% H_2O_2), wet sieved using a 0.5 mm mesh, and then left to air-dry. Sorting for vertebrate remains was conducted under a stereoscope. Here, we investigated only Rodentia, with a special focus on arvicolines. The study of the entire small mammal fauna, including Eulipotyphla, is pending. We focused on the m1, which is the most diagnostic and informative dental element in arvicolines. Measurements and terminology of arvicoline molars follow van der Meulen [31]. The La/Li index was calculated based on Luzi and Lopez Garcia [32]. Photographs and measurements (in mm, 0.01 accuracy) were taken with a Leica MZ 16A.

Upper premolars/molars are indicated with the upper-case letters P and M, respectively, and the corresponding lower ones with p and m. Dental and postcranial measurements of the large mammal specimens, as well as the updated faunal lists for the sites and SUs, are provided in the Supplementary Materials (Tables S1 and S2).

3. Fossiliferous Sites—Selected Survey Units and Findspots

3.1. *Choremi 6 (CHO-6)*

CHO-6 (Figures 1b and 2) was discovered during the field survey of 2021, when several *Hippopotamus* bones were found stratified in grey clay, some of them partially exposed on the dried surface due the low water level of that year (Figure 2). CHO-6 is located just below LIa (Figure 1d); therefore, it predates the Matuyama/Bruhnes boundary and has an age older than MIS 19 [3,25]. As such, it is dated toward the end of the Lower Pleistocene and represents the oldest known fossiliferous site of the Choremi Fm.

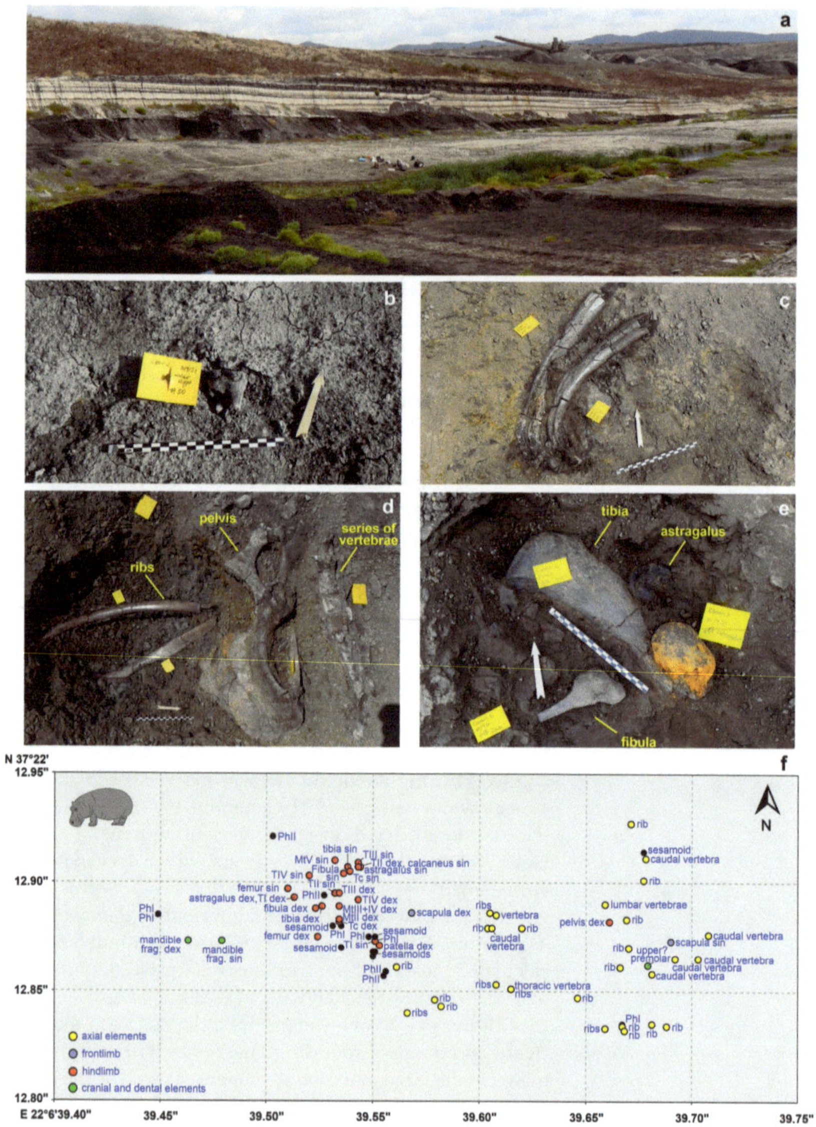

Figure 2. (**a**) Panoramic view of the site Choremi 6. (**b–e**) Fossils belonging to the *Hippopotamus antiquus* skeleton in situ at Choremi 6. (**f**) Distribution map showing the position of the hippopotamus remains (recorded with the use of a Differential GPS—WGS84 datum).

The proximate spatial distribution of the hippopotamus skeletal elements, the lack of duplication, and the consistency in size and ontogenetic age indicate the presence of a single, adult individual (all epiphyses of the long limb bones, the heads of the ribs, and the plates of the vertebrae are fused). Despite the absence of anatomical connection between the bones, the skeleton is not dissociated, and several originally articulated bones were found in very close spatial association (Figure 2). Moreover, most of the available bones belong to the posterior part of the skeleton (femur, patella, tibia, tarsals, metatarsals, pelvis, and lumbar vertebrae), and only few (scapula and mandible) are from the anterior part, most elements of which had already been removed before the discovery of the site (Figures 2 and 3).

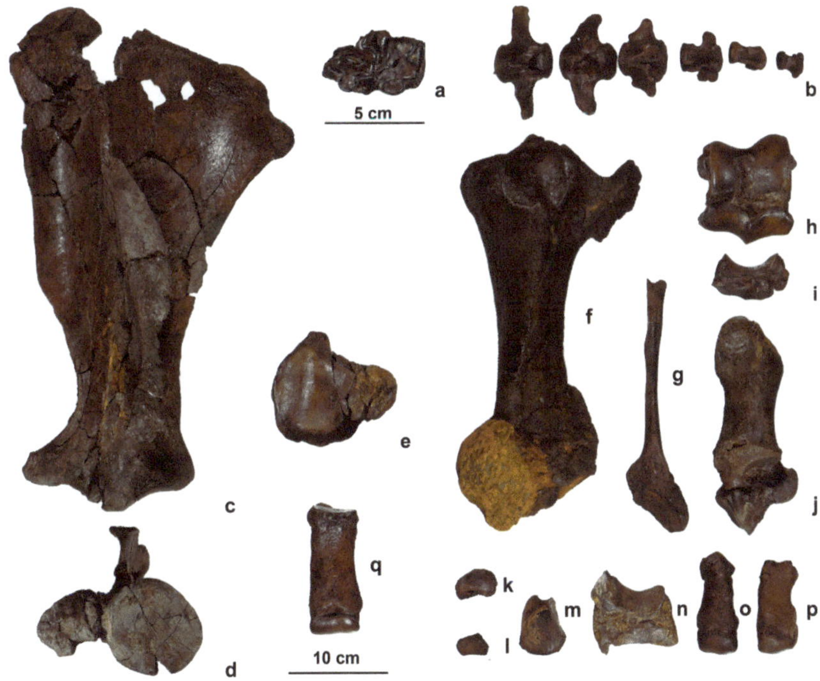

Figure 3. *Hippopotamus antiquus* remains from Choremi 6. (**a**) Lower molar (m1?) in occlusal view; (**b**) series of caudal vertebrae (CHO-6-36, 37, 38, 39, 40, and 41); (**c**,**d**) left scapula (CHO-6-49) in lateral (**c**) and distal (**d**) view; (**e**) right patella (CHO-6-10) in caudal view; (**f**) left tibia (CHO-6-42) in dorsal view (note the presence of a limonite concretion on the distal end); (**g**) left fibula (CHO-6-43) in medial view; (**h**) left astragalus (CHO-6-7) in dorsal view; (**i**) left navicular (central tarsal, CHO-6-15) in lateral view; (**j**) left calcaneus (CHO-6-8) in medial view; (**k**) left ectocuneiform (tarsal I, CHO-6-17) in proximal view; (**l**) left mesocuneiform (tarsal II, CHO-6-25) in proximal view; (**m**) left entocuneiform (tarsal III, CHO-6-11) in distal view; (**n**) left cuboid (tarsal IV, CHO-6-13) in medial view; (**o**) left Mt II (CHO-6-20) in dorsal view; (**p**) left MT V (CHO-6-24) in dorsal view; (**q**) right MT III (CHO-6-21) in dorsal view.

In the biplots comparing variables of postcranial elements, the CHO-6 specimens are distinct from *Hippopotamus amphibius*, and due to their large size, they are plotted well within the sample of *Hippopotamus antiquus*, in most cases toward its lower range (Figure 4; Table S1). Compared to the other known hippopotamuses from the Megalopolis Basin (Kyparissia 1, Kyparissia 4, Kyparissia T, and Marathousa 1), the CHO-6 hippopotamus is the largest one, which is also reflected in the estimated body size of ca. 2400 kg (Table 1).

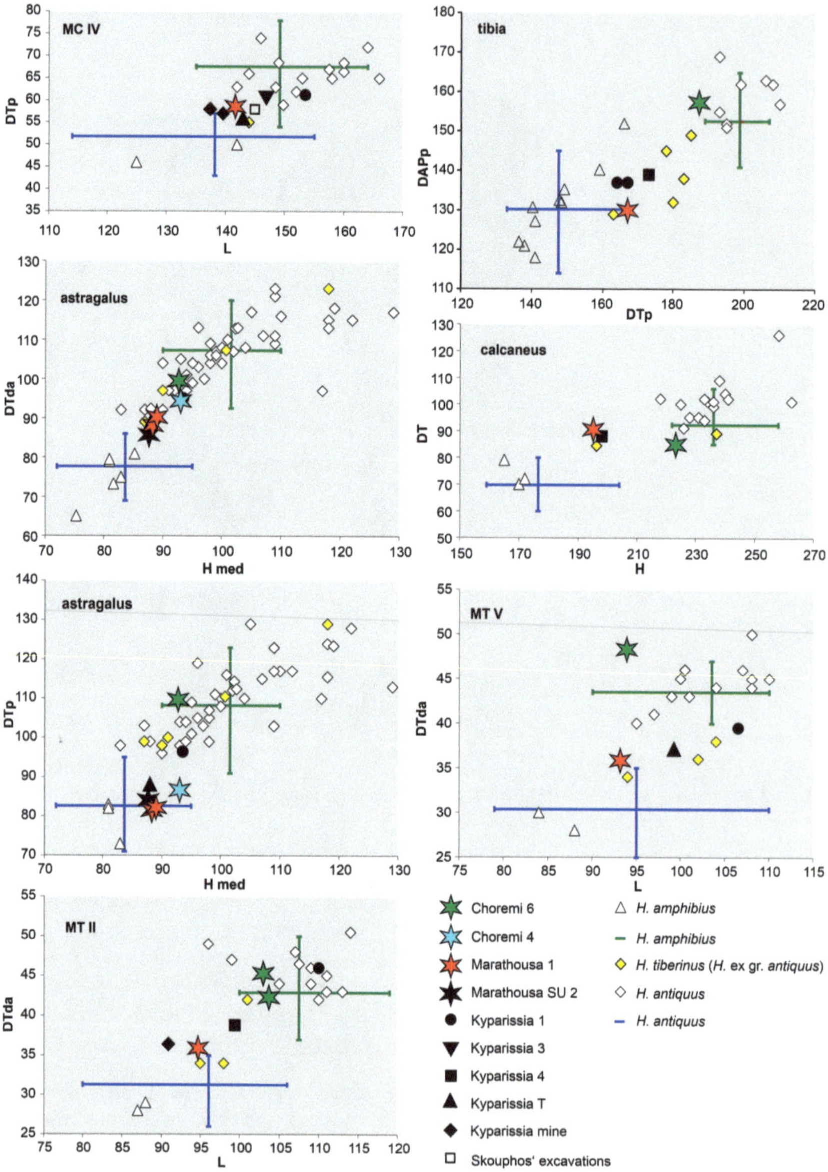

Figure 4. Metrical comparison of *Hippopotamus* postcranial specimens from Megalopolis Basin with *Hippopotamus antiquus*, *Hippopotamus tiberinus* (*H.* ex gr. *antiquus*), and *Hippopotamus amphibius*. Data from [12,20,33–37]. The blue and green lines correspond to the ranges of *H. amphibius* and *H. antiquus* given by [38]. Abbreviations: a, articular facet; d, distal; DAP, anteroposterior diameter; DT, transverse diameter; H, height; L, length; med, medial; p, proximal.

Besides this individual, a large-sized cervid is represented by a right and left hemimandible and a maxilla fragment. All bear the deciduous dentition and the erupting first molar and belong to the same individual. The dimensions of the m1 are similar to those of the giant deer *Praemegaceros verticornis* from Kyparissia 4, yet in the absence of comparative material and of more diagnostic specimens, the CHO-6 specimen is attributed to Cervidae indet. (large-sized). Additionally, an avian large-sized distal phalanx and a turtle shell fragment were found.

Table 1. Estimated body masses of *Hippopotamus* from several sites of the Megalopolis Basin (based on equations for humerus (H), femur (F), and tibia (T) in [39]) compared to *Hippopotamus amphibius* and *Hippopotamus antiquus* (data from [40]). Measurements for Kyparissia 1 and Kyparissia 4 were taken from [20]. Number in parentheses denotes the number of specimens measured.

Site/Species	Measurement	Body Mass (kg)
Choremi 6	T2 (1)	2366
Kyparissia 4 *	F5, T2, T4 (2)	2111
Kyparissia 1 **	H5, T2, T4 (5)	1893
Marathousa 1 ***	H5, T2 (2)	1738
H. amphibius (range/mean)		1048–2090/1476
H. antiquus (range/mean)		1602–4694/3197

* Unknown whether femur and tibia belong to the same individual. ** Minimum of two individuals. *** Skeletally immature individual.

3.2. *Kyparissia 4 (KYP-4)*

KYP-4 (Figure 1b) was discovered in 2007 and since then has provided a rich and diverse assemblage of large mammals and other vertebrates [19,20]. The site lies above the limestone basement and is placed stratigraphically within clastic sediments close to the top of LI (Figure 1d) [25]. The site was revisited during the field survey of 2019 and subsequently in 2020–2022, further enriching the faunal assemblage in both the number of specimens and species. Of importance are the abundance of cranial and postcranial remains of the giant deer *Praemegaceros verticornis* (including three partial skeletons), a partial skeleton of *Hippopotamus antiquus*, and the first evidence for the presence of macaques at the site. Furthermore, the discovery of lithic artefacts provides evidence for the presence of humans.

The updated large mammal fauna includes the following taxa ([19,20,41,42] and ongoing study): *Castor fiber* (beaver), *Macaca sylvanus* (macaque), *Vulpes* sp. (fox), *Felis* sp. (wild cat), *Palaeoloxodon antiquus* (elephant), *Stephanorhinus* cf. *hemitoechus* (rhinoceros), *Equus* spp. (horse), *Hippopotamus antiquus* (hippopotamus), *Sus scrofa* (wild boar), *Bison* sp. (bison), *Dama* sp. (fallow deer), *Cervus elaphus* (red deer), and *Praemegaceros verticornis* (giant deer). A detailed study of the newly collected materials is in progress.

In addition to the large mammals and aiming to provide a secure biochronological framework of the site, sampling for micromammals was conducted in two phases. The first phase (test sampling) took place in 2019–2020 and showed the potential of the site for the preservation and diversity of micromammals [30], while during the second phase in 2021–2022, the systematic sampling significantly increased the available material. The KYP-4 rodent assemblage is dominated by the genus *Mimomys*. There is a clear size difference among the *Mimomys* molars, allowing the distinction of two taxa, the larger *Mimomys*, attributed to *Mimomys* cf. *savini*, and the smaller one attributed to *Mimomys* sp. (Figure 5a–d). Some rooted molars without crown cementum but with "*Mimomys*-enamel differentiation" are ascribed to the genus *Pliomys* (Figure 5e,f), yet their specific attribution is pending. Moreover, some unrooted molars show morphological features (e.g., in the m1, five more or less closed triangles and an *arvalis*-type anterior cap) that indicate the presence of the subgenus *Microtus* (*Microtus*). Additionally, two murid taxa are recognized, one with smaller dimensions and morphology that are compatible with *Mus spretus*, preliminary assigned here to *Mus* cf. *spretus*, and the other one with larger dimensions, attributed to *Apodemus* sp., which could possibly belong to *Apodemus sylvaticus*. Finally, the presence of the shrew *Sorex* sp. is also recognized.

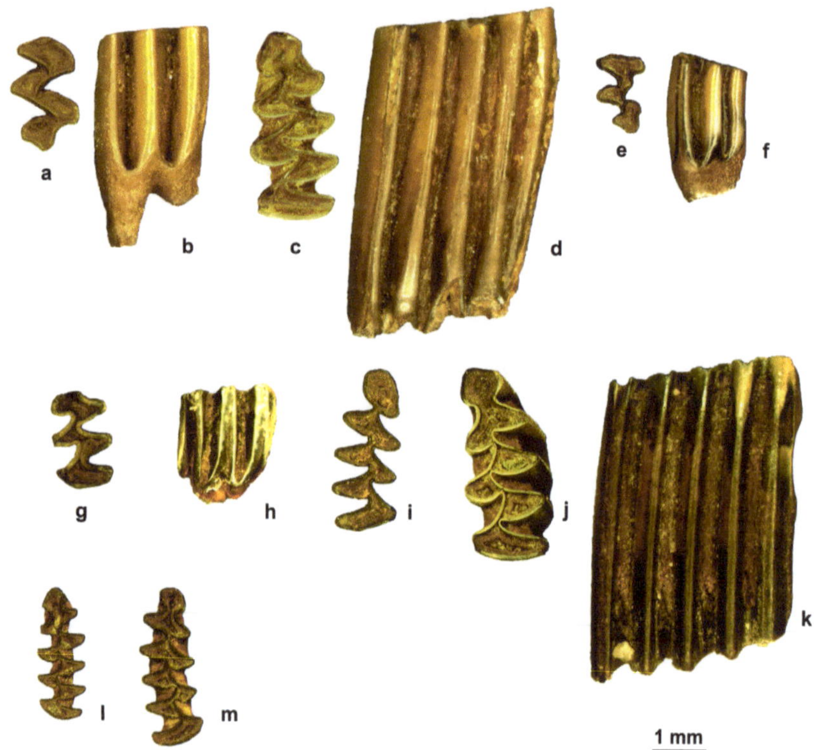

Figure 5. Small mammals from several sites and findspots of the Megalopolis Basin. (**a**,**b**) Right M2 (KYP-4-MM008) of *Mimomys* sp. (smaller size) from Kyparissia 4 in occlusal (**a**) and lingual (**b**) view; (**c**,**d**) left m1 (KYP-4-MM017) of *Mimomys* cf. *savini* from Kyparissia 4 in occlusal (**c**) and lingual (**d**) view; (**e**,**f**) left M3 (KYP-4-MM015) of *Pliomys* sp. from Kyparissia 4 in occlusal (**e**) and lingual (**f**) view; (**g**,**h**) left m2 (KYP-3-MM001) of *Mimomys* sp. (smaller size) from Kyparissia 3 in occlusal (**g**) and lingual (**h**) view; (**i**) right m1 (KYP SU 6-MM002) of *Pliomys* cf. *episcopalis* from Kyparrisia SU 6 in occlusal view; (**j**,**k**) left m1 (CHO-7-MM135) of *Arvicola mosbachensis* from Choremi 7 in occlusal (**j**) and lingual (**k**) view; (**l**) left m1 (CHO-7-MM66) of *Microtus* sp. (cf. *Microtus arvalis*) from Choremi 7 in occlusal view; (**m**) right m1 (CHO-7-MM214) of *Microtus* (*Terricola*) *subterraneus* from Choremi 7 in occlusal view.

3.3. Choremi 4 (CHO-4)

CHO-4 (Figures 1b and 6) was discovered during the field survey of 2018 in Choremi Survey Unit 6 (CHO SU 6). Mammal fossils at CHO-4 were found within a mollusk-rich layer stratigraphically positioned between LIc and LIIa (Figures 1d and 6a–c); therefore, a correlation to MIS 16 is possible (the detrital interval 3, in [3]). As such, CHO-4 stands stratigraphically higher and is younger than CHO-6 and KYP-4. The identified fossils include remains of turtle, beaver (cheek teeth, tibia fragment, and astragalus), deer, and hippopotamus. The latter is represented by an atlas, an astragalus, an ectocuneiform, and a proximal phalanx, all of which were found in very close spatial proximity (Figure 6c–h). The astragalus shows different proportions than the CHO-6 one, especially in terms of the DT proximal, and is similar to the astragali from KYP-T and MAR-1 (Figure 4). Even though most of the variables fall in the lower range of *H. antiquus*, it is closer to the mean values of *H. tiberinus* (*H.* ex gr. *H. antiquus*, according to [43]), except for the DT proximal, which is lower than in the latter species (Figure 4). In addition, a large bovid astragalus

(Figure 6i–l) found on the surface close to the CHO-4 section is morphologically compatible with an attribution to *Bison*, following the criteria in [44].

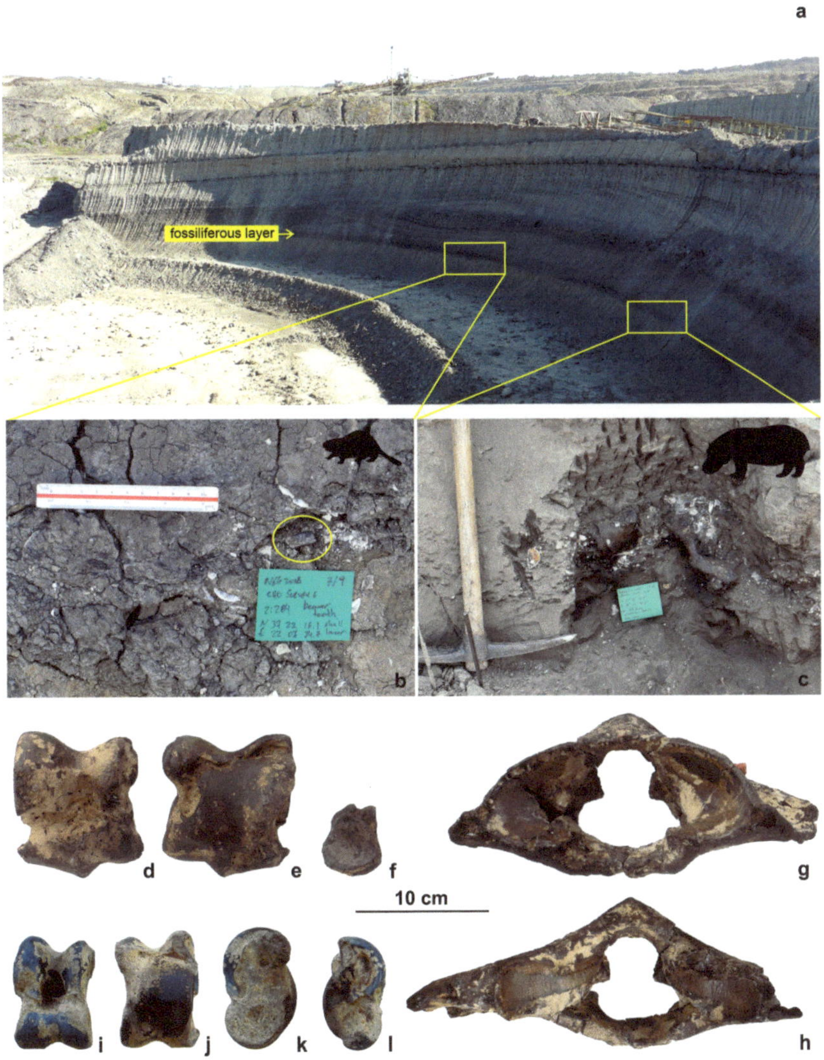

Figure 6. (**a**) Panoramic view of the Choremi 4 section showing the fossiliferous layer. (**b**) A tooth of *Castor fiber* in situ within a shell-rich layer. (**c**) In situ concentration of hippopotamus bones. (**d**,**e**) Right astragalus of *Hippopotamus antiquus* in dorsal (**d**) and plantar (**e**) view. (**f**) Right entocuneiform (tarsal III) in proximal view. (**g**,**h**) Atlas of *H. antiquus* in cranial (**g**) and caudal (**h**) view. (**i**–**l**) Right astragalus of *Bison* sp. found at the surface near the Choremi 4 section in dorsal (**i**), plantar (**j**), medial (**k**), and lateral (**l**) view.

3.4. Kyparissia 3 (KYP-3) and Kyparissia Survey Unit 6 (KYP SU 6)

KYP-3 (Figure 1b) was discovered in 2006 and studied in [19,20]. KYP-3 is located stratigraphically above KYP-4 and is placed in the upper part of the clastic unit underlying LII (Figure 1d) [19,25]. Similar to CHO-4, a correlation to MIS 16 is possible [3]. The site was revisited during the field survey of 2022 and resulted in the discovery of a partial skeleton of the elephant *Palaeoloxodon antiquus* (a third upper molar and several bones, mainly ribs), as well as artiodactyls (Hippopotamidae, Cervidae, and Suidae), turtles, and birds, in direct spatial and stratigraphic association with lithic artefacts and, thus, human presence [25]. The large mammal fauna includes the following taxa ([19,20]; ongoing study): Hyaenidae indet., *Felis* sp., *Panthera* sp. (*Panthera ?pardus*), *Palaeoloxodon antiquus, Sus scrofa, Hippopotamus antiquus, Dama* sp., *Cervus elaphus*, and *Praemegaceros verticornis*. A detailed study of the new materials is in progress.

The samples for micromammals collected from Kyparissia 3 were very poor and yielded only one arvicoline molar, a rooted m2 (Figure 5g,h), which, although not diagnostic, its dimensions denote the presence of a *Mimomys* species, smaller than *M. savini*. On the other hand, test sampling from an adjacent section to KYP-3 (named as KYP SU 6), of comparable stratigraphic position, proved to be much richer and promising for future investigations. From KYP SU 6, rooted arvicoline molars preserving crown cementum belong to *Mimomys*. The size difference among the molars indicates the presence of the larger *M.* cf. *savini* and of a smaller sized one, attributed to *Mimomys* sp. A third arvicoline is identified by a m1 (Figure 5i), which preserves no cement in the synclines, thick enamel with positive differentiation, a small T6, very shallow BRA4, and a rounded AC (anteroconid). These traits are similar to those of *Pliomys episcopalis*, yet due to the single status of the specimen and its larger size, we prefer to ascribe it to *Pliomys* cf. *episcopalis*. Larger-sized *Pliomys* cf. *episcopalis* were also found in Choremiou Section 3, a feature that was interpreted as an endemic feature [2].

3.5. *Marathousa Survey Unit 7 (MAR SU 7)*

MAR SU 7 was investigated in the field seasons of 2018 and 2019 and comprises a long section along the terrace at the northern part of the Marathousa mine, just below LIIa (Figures 1d and 7a,b), thus possibly corresponding to MIS 16 [3]. Several fossils were discovered within two stratigraphically close clastic layers rich in mollusks and organic material. Of interest are fossils of a small-sized cervid and of a mustelid. To the former belong a radius and a metatarsal (Figure 7i–l; Table S1), whose dimensions are smaller than corresponding specimens from Kyparissia 4, Marathousa 1, and Marathousa 2, referred to *Dama* sp. [24,42,45], as well as from "*Cervus*" *peloponnesiacus* [17], and indicate a *Capreolus*-sized cervid (e.g., [46]), which is a rare faunal element in the Megalopolis Basin [19]. The mustelid is represented by an almost complete humerus characterized by strong dorsoventral curvature of the shaft, a laterally expanding supracondylar crest, large deltoid tuberosity, strong lateral curvature of the deltoid ridge, and broad distal epiphysis (Figure 7e–h). These traits, combined with its dimensions (Table S1), are similar to the humerus of the otter *Lutra simplicidens* from Voigtstedt (Germany [47]), and it is, therefore, attributed to this species. This is the second finding of *L. simplicidens* in the Megalopolis Basin, previously known only from MAR-1 [45]. From these mollusk-rich layers, fossils of *Castor fiber* (calcaneus, distal radius; Figure 7c,d), *Cervus elaphus* (metacarpal, radius), turtle shells, and bird bones were also found.

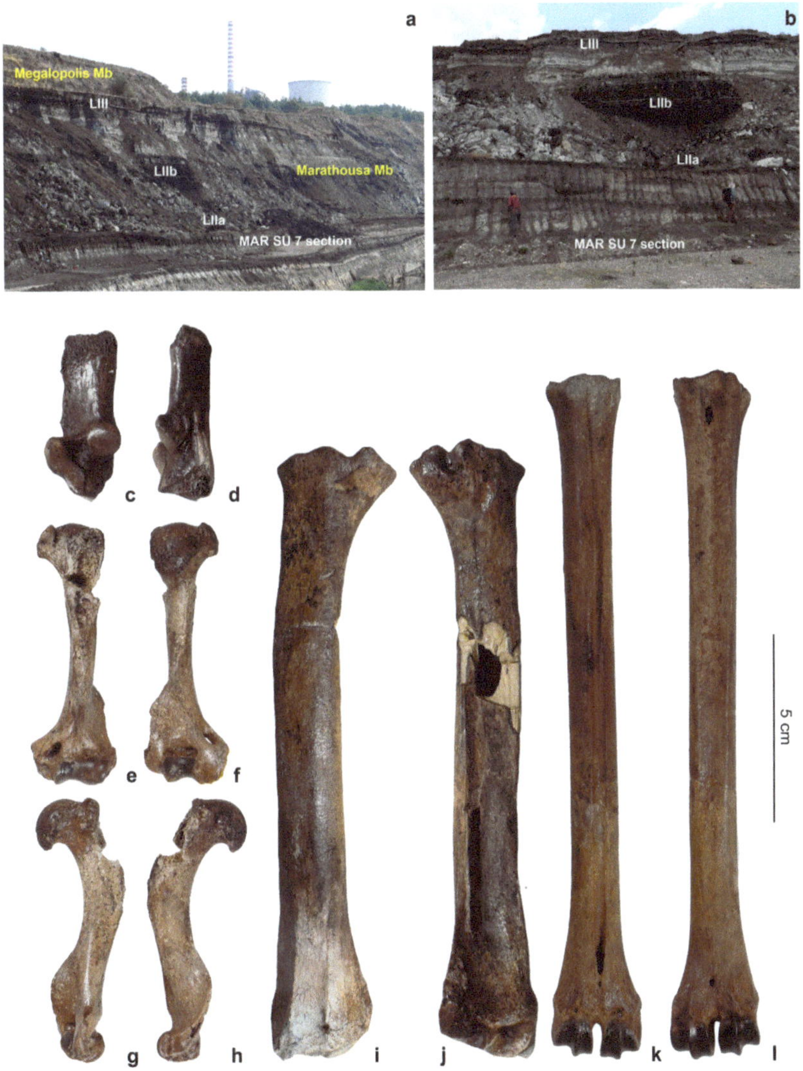

Figure 7. (**a**) Panoramic view of the northern profile in the Marathousa mine, indicating the Marathousa and Megalopolis Members, the lignite seams, and the location of the MAR SU 7. (**b**) Details of the profile showing the MAR SU 7 section. (**c**,**d**) Left calcaneus of *Castor fiber* in medial (**c**) and plantar (**d**) view. (**e–h**) Left humerus of *Lutra simplicidens* in cranial (**e**), caudal (**f**), medial (**g**), and lateral (**h**) view. (**i**,**j**) Left radius of a *Capreolus*-sized cervid in dorsal (**i**) and volar (**j**) view. (**k**,**l**) Metacarpal of a *Capreolus*-sized cervid in dorsal (**k**) and volar (**l**) view.

3.6. *Kyparissia T (KYP-T)*

KYP-T (Figure 1b) was discovered during the field survey of 2012 close to the western margin of the Megalopolis Basin and was systematically excavated in the same year [20]. Stratigraphically, KYP-T was preliminarily placed on the top of LII (Figure 1d) [20], based on the presence of a thin calcareous layer that was identified as the guide horizon for LII, according to [48]. Recent stratigraphic studies at Kyparissia position KYP-3 and KYP-4 lower in the stratigraphic sequence of the Megalopolis Basin, namely, in the upper part of the clastic unit underlying LII, and close to the top of LI, respectively [25]. Thus, KYP-T may actually be older, as it lies at the same altitude as KYP-4. However, the current conditions at KYP-T (collapsed section and growth of dense vegetation) do not allow for new, detailed stratigraphic observations.

The faunal content of KYP-T consists mainly of several *Hippopotamus antiquus* specimens and scarce cervid, avian, and chelonian finds [20]. Those of *Hippopotamus* are rather small sized and may belong to a single individual.

3.7. Marathousa 1 (MAR-1)

MAR-1 (Figure 1b) was discovered during the field survey of 2013 and was systematically excavated until 2019 [22,23,49,50]. The locality is situated between LIIb and LIIIa (Figure 1d), is radiometrically, magnetostratigraphically, and biochronologically dated to ca. 450 ka, and is correlated to the glacial Marine Isotope Stage (MIS) 12 [3,28,45,51–53]. MAR-1 has yielded a stratified and exceptionally well-preserved archaeological and paleontological assemblage, including lithic and bone artefacts, micro- and macro-fauna (insects, ostracods, mollusks, fishes, amphibians, reptiles, birds, and mammals), and micro- and macro-flora ([45,52,54–60] and ongoing studies). Additionally, it preserves evidence of hominin butchering of elephant and other large mammal carcasses, in accordance with the traits of the associated lithic artefacts and the presence of use-wear traces involving butchering activities [45,59,61]. As such, it provides important insights into hominin subsistence strategies and, in particular, megafauna exploitation.

The micro- and macro-mammal fauna of MAR-1 was studied in [52] and [45], respectively. The latter study included the mammal remains from the 2013–2016 collection. The subsequent excavation seasons of 2017–2019 significantly enriched the collection from both areas. In Area A, several more bones of the *Palaeoloxodon* skeleton were unearthed, and these will be treated in detail separately. In Area B, the mammal assemblage increased significantly in the number of specimens, including a complete mandible of the macaque *Macaca sylvanus* cf. *pliocena*, which represents a new entry in the faunal list [57], a partial skeleton of *Hippopotamus*, several elephant remains belonging to a partial skeleton, as well as several remains of beavers, carnivorans, deer, and bovids.

***Hippopotamus* skeleton from Area B.** The skeleton of *Hippopotamus* was found during the mechanical expansion in the southern part of Area B in 2019, within the sedimentary unit UB2b (dark grey, laminated, organic-rich silty sand, and sand rich in shell fragments, interpreted to represent relatively high-energy fluviatile flows entering the margins of the lake within the zone of lake-surface fluctuation [28]) and was subsequently systematically excavated (Figure 8). Several hippopotamus teeth and postcranial remains collected in this southern part of the trench during the initial opening of the site in 2013 belong most possibly to this skeleton, as they fully match in the status of bone preservation, dimensions, and ontogenetic age, while there is no duplication of skeletal elements (Figure 8). The spatial distribution of the recorded skeletal remains shows that the skeleton, although dissociated, retained in proximity anatomically connected bones, indicating minimal disturbance (Figure 8a). Moreover, most of the recorded bones belong to the posterior part of the skeleton (sacrum, lumbar vertebra, femur, tibia, and tarsals). Considering all available specimens, the skeleton is represented by dental and all types of axial and appendicular (both front- and hind-limb) elements, indicating that there was no selective bone survival.

The hippopotamus remains belonged to a skeletally immature individual with unfused heads of ribs and plates of vertebrae, unfused distal epiphysis of radius, and partially fused ilium, head, and distal epiphysis of femur, and proximal epiphysis of tibia (Figure 8). This stage of epiphyseal fusion is compatible with the slightly worn m2 and the minimally worn M3 (Figure 9a,b,d,e), which is equivalent to a mean age of 17 years (with a lifespan mean 43) of *H. amphibius* (group X [62]), indicating an early prime individual.

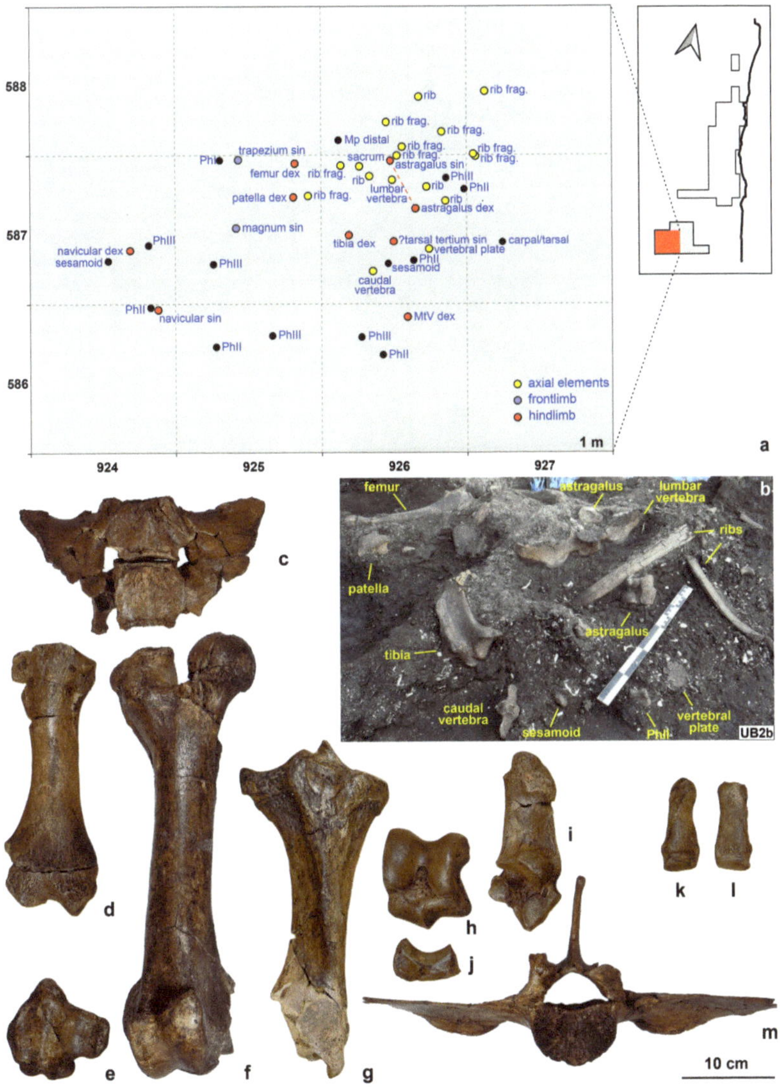

Figure 8. *Hippopotamus antiquus* skeleton from Marathousa 1 (Area B; UB2b). (**a**) Distribution map showing the position (recorded with the use of a total station) of the hippopotamus remains; (**b**) hippopotamus fossils in situ; (**c**) sacrum (MAR-1B-926/587-17) in ventral view; (**d**) right radius (MAR-1B-1) in dorsal view; (**e**) right patella (MAR-1B-925/587-2) in caudal view; (**f**) right femur (MAR-1B-925/587-1) in cranial view; (**g**) right tibia (MAR-1B-926/587-10) in dorsal view; (**h**) right astragalus (MAR-1B-926/587-4) in dorsal view; (**i**) right calcaneus (MAR-1B-2) in medial view; (**j**) right navicular (central tarsal, MAR-1B-924/587-3) in lateral view; (**k**) right MT II (MAR-1B-4) in dorsal view; (**l**) right MT V (MAR-1B-926/586-1) in dorsal view; (**m**) lumbar vertebra (MAR-1B-926/587-9) in cranial view.

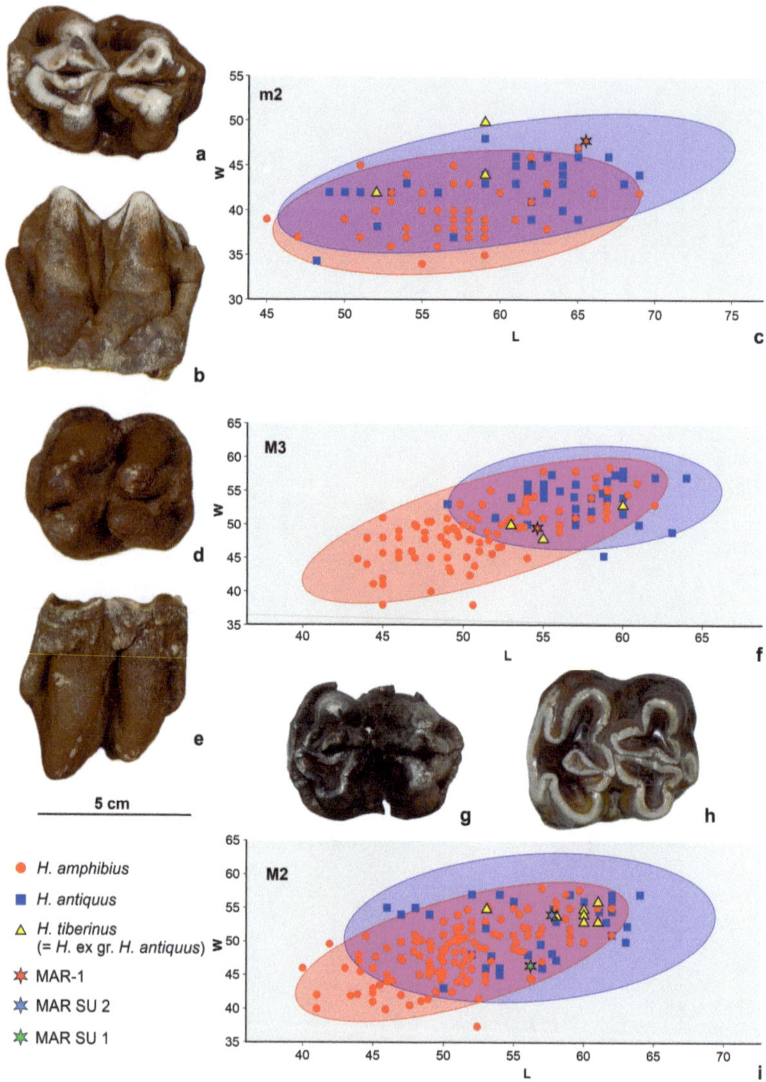

Figure 9. (**a**,**b**) Left m2 (MAR-1B-8) of *Hippopotamus antiquus* from Marathousa 1 (Area B), most possibly belonging to the hippopotamus skeleton, in occlusal (**a**) and buccal (**b**) view. (**c**) Biplot comparing *Hippopotamus* m2 from various localities. (**d**,**e**) Left M3 (MAR-1B-7) of *H. antiquus* from Marathousa 1 (Area B), most possibly belonging to the hippopotamus skeleton, in occlusal (**d**) and buccal (**e**) view. (**f**) Biplot comparing *Hippopotamus* M3 from various localities. (**g**) Left M2 of *H. antiquus* from MAR SU 1 in occlusal view. (**h**) Left M2 of *H. antiquus* from MAR SU 2 in occlusal view. (i) Biplot comparing *Hippopotamus* M2 from various localities. Biplots with 95% confidence ellipses for *Hippopotamus amphibius* and *H. antiquus*, performed with the software package PAST v. 4.16 [63]. Data are from [33,64–68].

For the taxonomic attribution, in the absence of a skull, we focus our comparisons on the dimensions and proportions of the molars and of the more informative postcranial bones (Table S1). In the biplots for the m2 and M3 (Figure 9c,f), *H. amphibius* and *H. antiquus* are largely overlapping, and only in the case of M3 *H. antiquus* presents generally larger dimensions. The MAR-1 molars plot toward the spectrum of the larger dimensions. The m2 stands at the upper ranges of both species but within the 95% confidence ellipses of *H. antiquus*, whereas the M3 at the lower values of *H. antiquus*, and close to *H. tiberinus*. The postcranial bones are smaller than those of CHO-6 and *H. antiquus* in general, and for most skeletal elements, they are close to specimens from Kyparissia (KYP-1, KYP-4, and KYP-T; Figure 4). Despite being largely within the variation of *H. amphibius*, their best fit is *H. tiberinus*; yet, it should be noted again that the MAR-1 bones belonged to a skeletally immature individual and, therefore, in the case of long limb bones where the epiphyses were not completely fused (radius, femur, and tibia), the measurements are a slight underestimate of the expected ones if the epiphyses were fully fused.

Palaeoloxodon **skeleton from Area B.** Besides the elephant skeleton of Area A, a second partial skeleton was unearthed in the southern part of Area B in 2019, within the sedimentary unit UB4c (dark grey, massive organic, and intra-clast-rich silty sand, interpreted to represent subaerial-originated dilute mudflows and hyper-concentrated flows plunged into the lake margin [28]). The discovered skeletal elements include a tusk, a stylohyoid bone, three cervical vertebrae (including the axis), two ribs, and a distal fragment of the humerus (Figure 10). All these remains were found in close spatial proximity (Figure 10a,b) and possibly belonged to the same individual. The left tusk (MAR-1B-928/587-48) is completely preserved, as the pulp cavity is open and filled with sediment. It is long (length = 3.4 m), robust toward its proximal end (circumference = 560 mm, maximum diameters 176 × 171 mm; 56 cm from the base), and shows the typical slight curvature and weak torsion of *Palaeoloxodon antiquus* (Figure 10a; Table S1). The large dimensions of the tusk indicate most possibly a male individual, which agrees with the large size of the axis (height = 320 mm; comparable to the height of 333 mm of the male individual from Area A [45]).

The reversed-Y-shaped left stylohyoid (MAR-1B- 928/587-32) is almost complete (Figure 10f–h; Table S1), missing only the ventral part of the inferior ramus. The distal ends of the superior and posterior rami are completely preserved, showing the typically rugged edges [69]. The superior ramus is elongate and slender, the posterior ramus is rather short and moderately flattened at its proximal part, and the inferior ramus is slender and shows a weak deflection ($z = 8°$). The angle between the inferior and posterior rami is narrow ($\alpha = 49°$ and $\beta = 39°$; measurements according to [70]). In these traits, the MAR-1 stylohyoid is different from those of the mammoth *Mammuthus trogontherii* (including that from Loussika [71]) and matches better with those of *P. antiquus* [70]. Within *P. antiquus*, the MAR-1 specimen fits best with the proportions shown in the stylohyoid from La Polledrara (MIS 9; Italy), rather than those of Neumark Nord 1 (MIS 5e; Germany), which generally have rather short and stout superior ramus, relatively longer posterior ramus, and a differently angled inferior ramus in regard to posterior ones [70]. To investigate whether these traits have evolutionary significance (and in turn, biochronological value) or represent intraspecific variability requires the discovery of additional specimens. However, the presence of this rare and fragile bone at MAR-1 not only highlights the exceptional preservation of the faunal assemblage at this site, but also enriches the available sample of this important bone for phylogenetic and taxonomic studies on proboscideans [72].

Whether the elephant remains that were found in previous years (proximal tibia, partial ribs and vertebrae, and several other bone fragments) belong to this individual is possible but not certain, especially because a second individual is present in Area B (and thus a third one for MAR-1, collectively), identified by a fragment of an axis (MAR-1B-923/586-1).

Other new faunal remains. Two hemimandibles, five diaphyseal/distal fragments of tibiae (Figure 11a–c; Table S1), and several other postcranial elements show morphology (e.g., lower incisors with smooth enamel on the outer side and triangular cross-section, and high-crowned cheek teeth with deep striids) and dimensions similar to the beaver *Castor fiber* [73,74], a species already reported from the locality [45]. A carnivoran fourth metatarsal from Area A (Figure 11s; Table S1) resembles in morphology and dimensions, those of a wolf-sized canid and is attributed to *Canis* sp., a taxon previously known only from Area B [45].

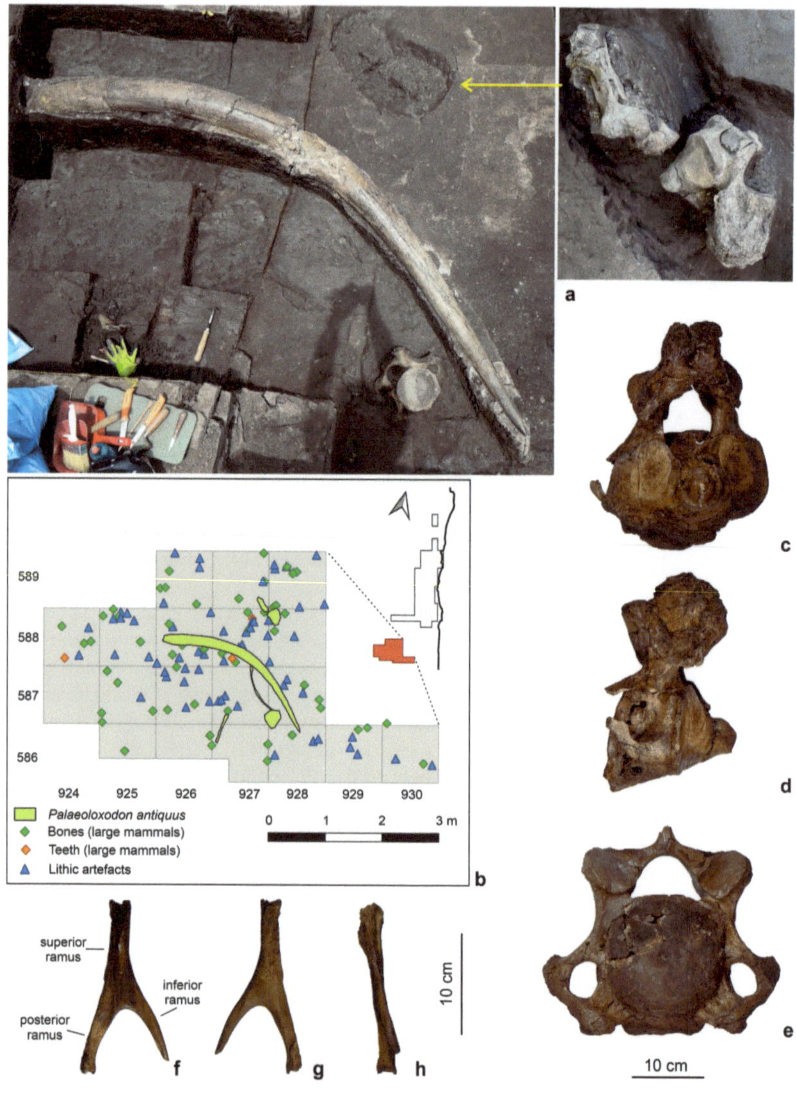

Figure 10. *Palaeoloxodon antiquus* skeleton from Marathousa 1 (Area B; UB4c). (**a**) Tusk and cervical vertebrae (including the axis) in situ. (**b**) Distribution map of the southwestern part of the trench at Area B showing the position (recorded with the use of a total station) of the elephant remains, stratigraphically and spatially associated with lithic artefacts and other faunal remains. (**c**,**d**) Axis (MAR-1B-928/588-39) in cranial (**c**) and right lateral (**d**) view; (**e**) cervical vertebra (MAR-1B-928/586-16) in cranial view; (**f**–**h**) left stylohyoid (MAR-1B-928/587-32) in medial (**f**), lateral (**g**), and anterior (**h**) view.

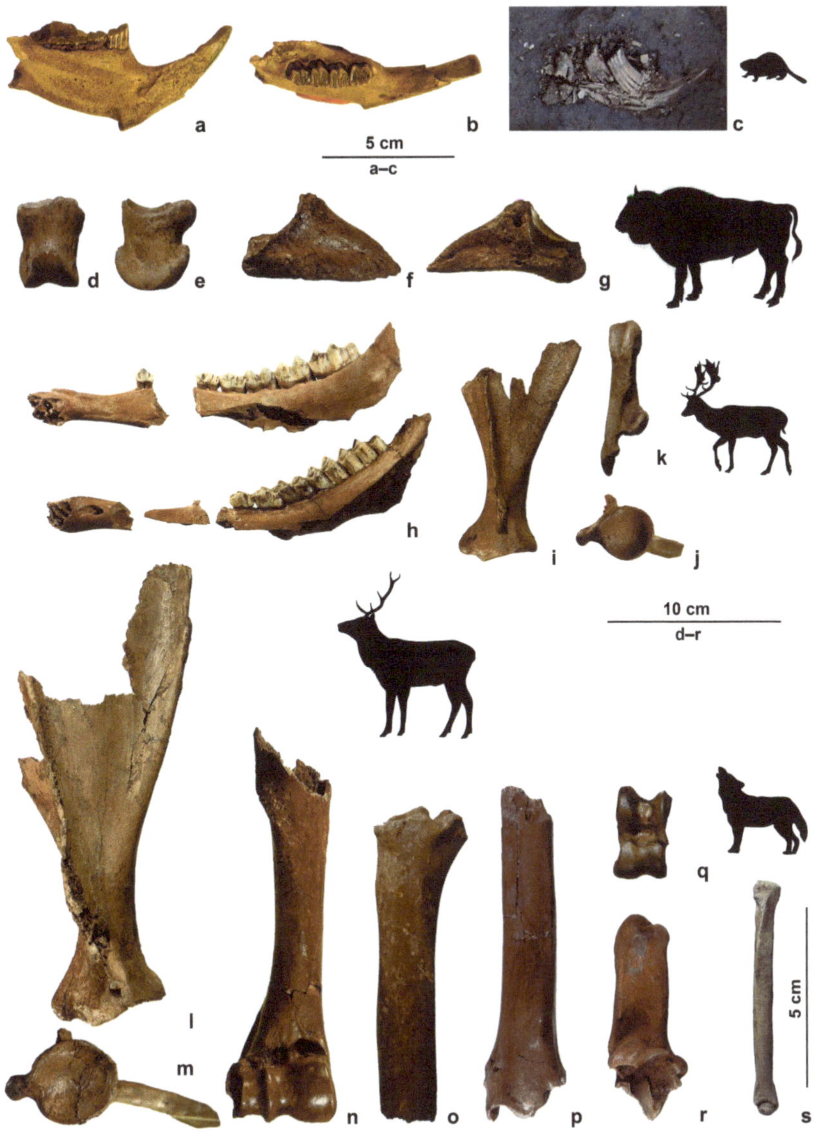

Figure 11. Mammal remains from Marathousa 1. (**a**,**b**) Left hemimandible with the incisor and p4–m3 of *Castor fiber* (surface find at Area B) in medial (**a**) and dorsal (**b**) view. (**c**) Left hemimandible (MAR-1A-941/677-41) of *Castor fiber* in situ at Area A. (**d**,**e**) Second phalanx of the front limb (MAR-1B-930/593-46) of *Bison* sp. in dorsal (**d**) and lateral (**e**) view. (**f**,**g**) Third phalanx of the front limb (MAR-1B-931/594-42) of *Bison* sp. in lateral (**f**) and medial (**g**) view. (**h**) Refitting mandibular fragments (MAR-1B-934/594-71, 932/598-51, 933/595-60, 931/593-38, and 931/596-39) of the same individual of *Dama* sp., above the right hemimandible in medial view, and below the left hemimandible in lateral view. (**i**,**j**) Left scapula fragment (MAR-1B-934/597-35) of *Dama* sp. in lateral (**i**) and distal (**j**) view. (**k**) Left calcaneus (MAR-1B-928/588-21) of *Dama* sp. in plantar view. (**l**,**m**) Left scapula fragment (MAR-1A-935/672-20) of *Cervus elaphus* in lateral (**l**) and distal (**m**) view. (**n**) Distal fragment of left humerus (MAR-1B-931/564-58) of *C. elaphus* in cranial view. (**o**) Proximal fragment of left radius (MAR-1B-926-587-22) in dorsal view. (**p**) Distal fragment of left tibia (MAR-1B-925/586-7) in dorsal

view. (**q**) Right astragalus (MAR-1B-926/586-4) of *C. elaphus* in dorsal view. (**r**) Left calcaneus (MAR-1-939/635-1) of *C. elaphus* in medial view. (**s**) Left MT IV (MAR-1A-941/672-44) of *Canis* sp. in dorsal view. Silhouette images from PhyloPic, phylopic.org.

In the cervid assemblage, two size groups are identified, the medium-sized one belonging to a fallow deer (*Dama* sp.) and the larger one to the red deer (*Cervus elaphus*). Additional material further enriched the specimens of both taxa. In Konidaris et al.'s study [45], a right hemimandible fragment of *Dama* sp. was described, which was found in two refitting pieces (MAR-1B-934/594-71 and 932/598-51). During the subsequent excavations, the left hemimandible bearing the p3–m3 of the same individual was also found (the wear of the teeth and the preservation are identical) in three pieces that again refit (MAR-1B-933/595-60, 931/593-38, and 931/596-39; Figure 11h). Moreover, to the previous published fallow deer material, we add here further dental specimens (P2, m2, and m3), a posterior part of a cranium preserving the occipital condyles, a sacrum, a distal fragment of a scapula, two calcanei, and a proximal phalanx (Figure 11i–k; Table S1). Additional cervid material that shows morphological and metrical traits compatible with an attribution to *Cervus elaphus* [46,75,76] includes isolated cheek teeth (dp3, p3, p4, and m1), a scapula, a distal humerus, a proximal radius, a distal tibia, two astragali, a calcaneus, and a second phalanx (Figure 11l–r; Table S1).

One second and one third phalanx (both of the front limb; Figure 11d–g; Table S1) belong to a bovine [77]. Their morphology and proportions allow for an attribution to *Bison* [44]. A specific attribution is not possible based on the available material, yet similar to the already known second phalanx of the hindlimb [45], an attribution to *Bison schoetensacki* is possible. Finally, a fragment of a fairly worn cheek tooth shows a bunodont pattern with plicated enamel, consistent with an attribution to *Sus*, a genus reported for the first time from the locality.

3.8. *Marathousa 2 (MAR-2)*

MAR-2 was identified during the targeted field survey of 2018, ca. 1.5 km east of MAR-1 and at a similar stratigraphic position (between LII and LIII; Figure 1d); therefore, a comparable age and a correlation to a glacial stage, ca. 450 ka (MIS 12), was suggested [24]. Fossils were collected from exposed sections in two defined areas, Area A and Area B, both yielding mainly hippopotamus bones. In particular, at Area A, the spatial association of the bones (vertebrae and ribs) and the consistency in size and ontogenetic age indicate the presence of a single, skeletally immature, hippopotamus individual (perhaps of close ontogenetic age with the MAR-1 hippopotamus that also has unfused vertebral plates). Additionally, a partial hemimandible of a fallow deer and a femur diaphysis of a beaver were also found. From Area B, isolated hippopotamus ribs and teeth were recovered, of which a large-sized canine permits the attribution to *H. antiquus*. The taphonomic study revealed the presence of cut marks in the spine of a thoracic vertebra of the hippopotamus from Area A, indicating butchering activities, further supported by the discovery of a spatially and stratigraphically associated lithic artefact. This represents one of the very few examples of hippopotami carcass processing during the Early–Middle Pleistocene of western Eurasia [24].

3.9. *Tripotamos 4 (TRP-4)*

TRP-4 (Figure 1b) was discovered during the field survey of 2020 in the southwestern margin of the mine, when lithic and faunal remains were found within a light-colored silty-sand layer positioned possibly above LIIIa, thus overlying MAR-1 (Figure 1d) [25]. The study of the large mammals is pending. The material consisted mostly of bone fragments and isolated teeth. Artiodactyls are represented by *Dama* sp. (hemimandible and other dental and postcranial remains), *Cervus elaphus* (a proximal part of an antler preserving two basal tines), and a bovine (several isolated, hypsodont, molars), while a canine fragment of small dimensions belongs to *Hippopotamus* (Figure 12a–g). The presence of an elephantid is confirmed by several lamellar fragments (Figure 12h).

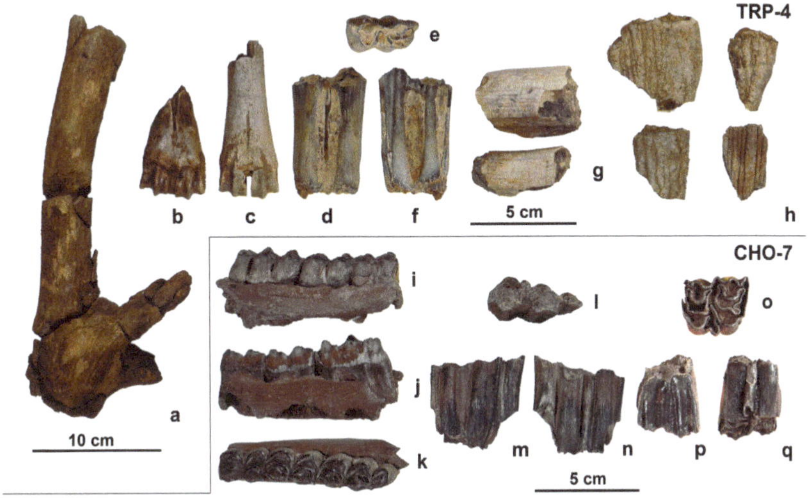

Figure 12. Mammal remains from Tripotamos 4 (TRP-4) and Choremi 7 (CHO-7). (**a**) Antler fragment (TRP-4-F2) of *Cervus elaphus* in lateral view; (**b**) distal metacarpal (TRP-4-F57) of *Dama* sp. in dorsal view; (**c**) distal metacarpal (TRP-4-F47) of *Dama* sp. in dorsal view; (**d–f**) lower molar (TRP-4-F2, surface) of Bovini in lingual (**d**), occlusal (**e**), and buccal (**f**) view; (**g**) fragment of upper canine (TRP-4-F70) of *Hippopotamus* sp.; (**h**) lamellar fragment (TRP-4-F36, F44, F58, and F73) of Elephantidae indet.; (**i–k**) right hemimandible fragment with m1–m3 (CHO-7-F125) of *Cervus elaphus* in lateral (**i**), lingual (**j**), and occlusal (**k**) view; (**l–n**) right m3 (CHO-7-F147) of *Bos* sp. in occlusal (**l**), lingual (**m**), and buccal (**n**); (**o–q**) upper molar (CHO-7-F68) of *Bos* sp. in occlusal (**o**), buccal (**p**), and lingual (**q**) view.

3.10. Choremi 7 (CHO-7)

CHO-7 was discovered during the field survey of 2021 in the southeastern margin of the mine (Figure 1b) and yielded both faunal and cultural material. CHO-7 stands at the highest stratigraphic position of all investigated sites, located above the entire lignite sequence of the Marathousa Mb (above the LIIIc; Figure 1d), just below the overlying Megalopolis Mb; therefore, it comprises the youngest of the investigated sites [25].

The systematic sampling for micromammals from the different layers of the site provided abundant specimens, and the first results of their study are presented herein. The assemblage is dominated by the genus *Microtus*. The occlusal surface of the m1 is characterized by five closed triangles and an *arvalis*-type mesial cap (Figure 5l). The morphology is different than *Microtus agrestis* and indicates the presence of *Microtus arvalis*. In the M2 of the former species, there is an additional closed loop at the distal end of the tooth [78], and this loop is not observed in any of the studied specimens. Additionally, based on the La/Li index, i.e., T4 width/T5 width [32], the two species (*M. arvalis*/*M. agrestis*) can be distinguished, except for an overlap in the larger values of *M. agrestis* and the lower ones of *M. arvalis*. In the CHO-7 specimens, the T4/T5 index (62.3–101.4, mean = 81.9; $n = 17$) falls within the range of *M. arvalis* or higher. Because the dentition of *M. arvalis* and *Microtus rossiameridionalis* is extremely similar [79], making their distinction impossible, we assign the CHO-7 material to *Microtus* sp. (cf. *M. arvalis*). Moreover, some *Microtus* molars show confluent T4 and T5 triangles (Figure 5m), and their dimensions are within the range of *Microtus* (*Terricola*) *subterraneus*; as such, they are ascribed to this species. Furthermore, the CHO-7 rodent assemblage includes three complete and five incomplete rootless m1s with three closed triangles (Figure 5j,k). The enamel of the mesial and distal sides of each triangle differs. The length of the complete specimens (mean = 2.71 mm; $n = 33$), the SDQ (Schmelzband–Differenzierung Quotient [80]) values (114.5–133.3, mean = 120.0; $n = 7$), and the shape of their occlusal surface permit their attribution to *Arvicola mosbachensis*. Finally, some murid molars that are similar to *Apodemus*, and some soricid ones, have not been studied yet.

The study of macro-mammals is pending. The material consists predominantly of bone fragments, which show clear evidence of rounding of their broken edges, and polishing and abrasive marks of their cortical surfaces, indicating substantial transport by water. Three cervid taxa are present at CHO-7 (following morphological and metrical data in [46,75], and comparison with corresponding specimens from other Megalopolis sites): a *Capreolus*-sized (molars, phalanges, and distal tibia), *Dama* sp. (upper molar, phalanges, and distal humerus), and *Cervus elaphus* (hemimandible, molars, and tibia; Figure 12i–k; Table S1). Several isolated, hypsodont molars are morphologically compatible (following [44]) with an attribution to *Bos* (Figure 12l–q).

3.11. Selected Isolated Finds

Two proboscidean large tusk fragments were discovered, one from the Marathousa mine (above LII) and the other from the Choremi mine (below LIII; Figure 13a,b). Both show obtuse Schreger angles in their cross-sections and, therefore, can be ascribed to *Palaeoloxodon antiquus*. Several other small tusk fragments collected from various findspots in the Megalopolis Basin also show the same Schreger pattern, also permitting the attribution to this species.

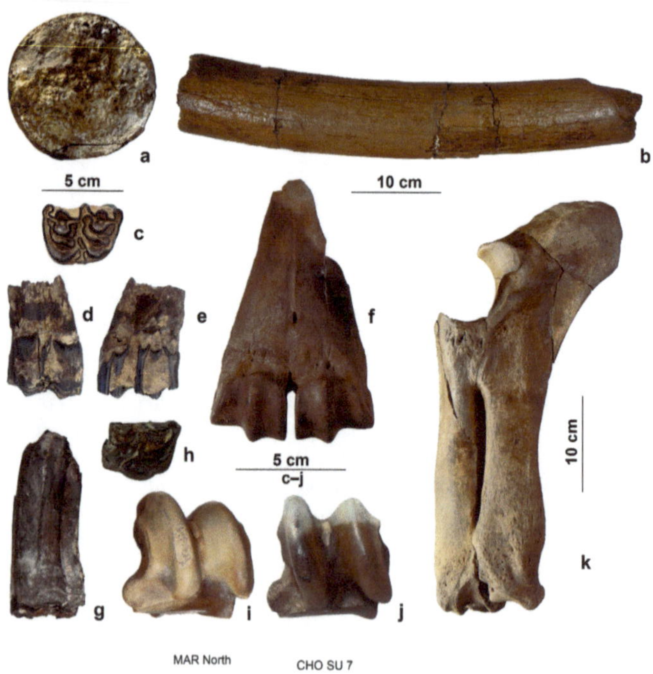

Figure 13. (**a**,**b**) Tusk fragment of *Palaeoloxodon antiquus* from CHO SU 11 in cross-sectional (**a**) and lateral (**b**) view. (**c**–**e**) Left upper molar of cf. *Bos* from Choremi 5 in occlusal (**c**), lingual (**d**), and buccal (**e**) view. (**f**) Distal metacarpal of *Bos* sp. from CHO SU 11 in dorsal view. (**g**,**h**) Right upper premolar (P3/P4) of *Equus* sp. from CHO SU 10, close to the site Choremi 6, in lingual (**g**) and occlusal (**h**) view. (**i**) Left astragalus of *Equus* sp. from MAR SU North in dorsal view. (**j**) Left astragalus of *Equus* sp. from CHO SU 7 in dorsal view. (**k**) Left radius and ulna of *Hippopotamus antiquus* from the Marathousa mine in lateral view.

Five isolated horse (*Equus*) specimens were discovered during the field surveys and are considered important, because horses constitute a rare faunal element in the Megalopolis Basin ([9,19,42] and a study in progress). An upper premolar (P3/P4) was identified close to CHO-6, at the same stratigraphic level as the site, thus preceding the Matuyama/Brunhes boundary (Figure 13g,h). Two more specimens were found in the southern–southwestern part of the Choremi mine within clastic sediments below LIIa, an astragalus, collected at CHO SU 7 (Figure 13j), and a proximal phalanx, collected during the sampling for magnetostratigraphy at the level of sample K7 [3] (Figure 4). Two more astragali were found within MAR SU 8 (below LII) and MAR SU North (surface; Figure 13i). Interestingly, all the newly collected horse material (including the few specimens from KYP-4) originated from the lower parts of the stratigraphic sequence.

Some more hippopotamus remains were collected during the field survey of 2013. These include a stratified M2 (MAR SU 1; Figure 9g) and some surface-collected finds (MAR SU 2) that comprise an M2 (Figure 9h), a navicular, and an astragalus. Both M2 specimens fall in the overlap region of *H. amphibius* and *H. antiquus* in their crown dimensions (Figure 9i). The dimensions of both the navicular and the astragalus are smaller than that of CHO-6 and very close to the MAR-1B hippopotamus skeleton (Figure 4). An additional surface-collected find is a complete radio-ulna (Figure 13k) from the lower parts of the sequence in the Marathousa mine, possibly originating from clastic sediments below LII.

Finally, some more Bovini remains were found in the Choremi mine. The oldest ones originated from the site Choremi 5 (CHO-5, Figure 1b), discovered during the field survey of 2018. The find horizon is located between two lignite seams, possibly LIIa and LIIb, and thus might correspond to MIS 14 (Figure 1d) [3]. From this layer, the collected Bovini remains include a lower premolar, an upper molar, a proximal phalanx, and a sesamoid. The morphology of the upper molar (Figure 13c–e; following [44]) fit better with an attribution to an aurochs (*Bos primigenius*) rather than to a bison. However, the material is very limited for a definite assignment and, therefore, the molar is attributed to cf. *Bos*. Additionally, a distal metacarpal fragment of a Bovini (Figure 13f) was collected within CHO SU 11 toward the upper the part of the Marathousa Mb sequence of the Choremi mine. The absence of marked supra-articular tubercles (so that the greatest distal width is measured across the trochleae [44]) allow for an attribution to *Bos* sp.

4. Discussion

4.1. Biostratigraphy–Biochronology

The updated faunal lists of the Megalopolis Basin sites and survey units are presented in Table S2. Most of the large mammals present in the Megalopolis Basin have a wide biostratigraphic/biochronological range within the Galerian and Aurelian European Land Mammal Ages, if not appearing even earlier, during the late Villafranchian–Epivillafranchian. As such, it is the study of the micromammals that offers the most important biochronological information that contributes to the age constraints of the sites and allows for the identification and dating of the currently known First and Last Greek Occurrence (FGO and LGO), as well as for the First and Last Megalopolis Occurrence (FMO and LMO) of large mammals.

The data from the level of CHO-6 are limited, yet based on the local stratigraphy, its dating at the end of the Lower Pleistocene renders CHO-6 as the oldest large mammal site of the Choremi Fm and marks the FMO of *Hippopotamus antiquus* and *Equus* (Figure 14).

The available data for KYP-4 are more informative. The presence of *Microtus* (*Microtus*) indicates that the fauna post-dates the *Allophaiomys*/*Microtus* transition that took place close to the Early/Middle Pleistocene transition (780 ka). The occurrence of *Mimomys savini* is also biochronologically important. The *Mimomys*/*Arvicola* transition was roughly synchronous throughout Europe and took place around MIS 15 [81–83]. Transitional populations showing primitive *Arvicola* traits were identified at Isernia (Italy, ca. MIS15/MIS14 [84]) and Mauer (Germany, MIS 15, ca. 600 ka [85]). Therefore, the presence of *Mimomys* at KYP-4 indicates an age older than 600–650 ka. Perhaps the most informative taxon at KYP-4 is the smaller-sized *Mimomys*, a taxon which is found only in a limited number of western Eurasian sites (identified as *Mimomys pusillus* or *Mimomys* sp.) dated to or predating MIS 16. These include the Choremiou Section of the Megalopolis Basin [2], Kärlich in Germany [86], Pakefield in the UK [87], and Chiu-Atasova in Russia (Southern Urals [88]). As a whole, the micromammal assemblage of KYP-4 indicates an early Middle Pleistocene age, after 780 ka and before MIS 16, in agreement with the stratigraphic placement of the site inside the lowermost lignite (LI).

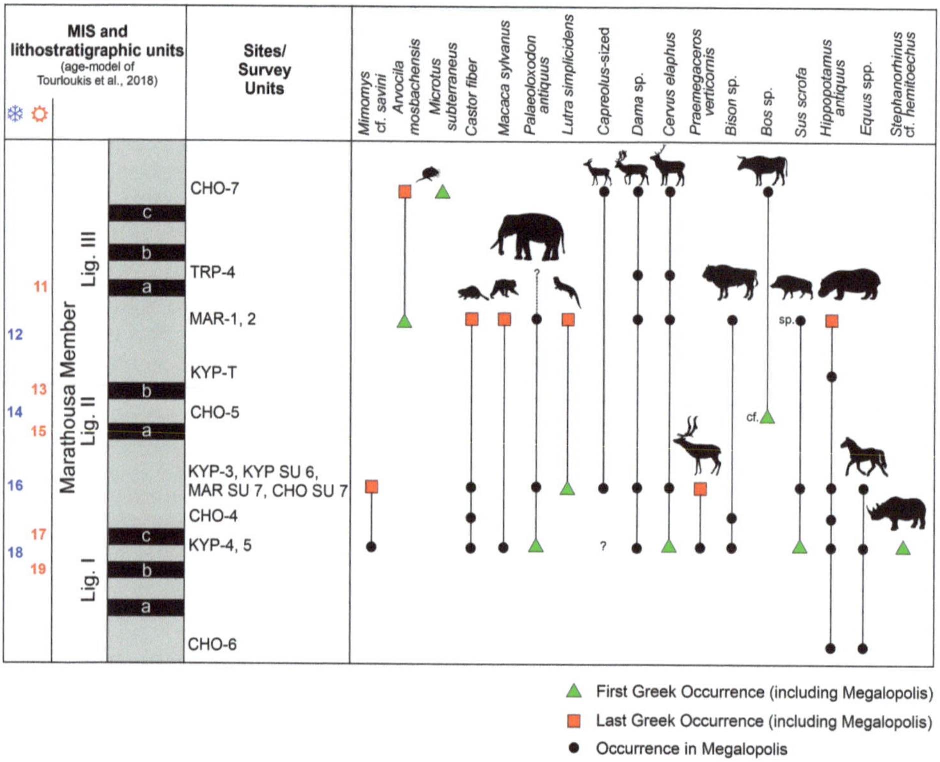

Figure 14. Simplified stratigraphic column of the Marathousa Member showing the lignite seams and the intercalating detrital intervals, their correlation to the Marine Isotope Stages (MIS) following the age model proposed by Tourloukis et al. [3], the stratigraphic position of the sites and survey units following Karkanas et al. [25,28], and the biostratigraphic range of selected micro- and macro-mammals from the Megalopolis Basin, indicating their First and Last Occurrences in the Greek fossil record. Silhouette images from PhyloPic, phylopic.org.

This age for KYP-4 marks the FMO and FGO for several of the large mammals (Figure 14). These include *Homo* (based on the presence of lithic artefacts), *Palaeoloxodon antiquus*, *Cervus elaphus*, *Sus scrofa*, and perhaps, *Stephanorhinus hemitoechus*, marking their currently known FGO in the early Middle Pleistocene, while depending on the followed taxonomy, this period may also mark the FGO of *Praemegaceros verticornis*.

The earliest palaeoloxodont elephants outside of Africa are documented at Gesher Benot Ya'aqov at ca. 780 ka (MIS 19, Israel) [89,90], while their arrival in Europe is documented with *P. antiquus* at Slivia (Slivia faunal unit) in Italy, estimated to be slightly older than the Early/Middle Pleistocene boundary based on micromammals [91]. *Palaeoloxodon* is present in Croatia in sediments dated between 940 and 470 ka [91,92]; therefore, the KYP-4 record is the secure first known occurrence of palaeoloxodont elephants is SE Europe.

The red deer remains are scarce and fragmentary at KYP-4, yet from the roughly same level of KYP-5 (slightly to the north of KYP-4; Figure 1b), the presence of *Cervus elaphus* is secure based on a large-sized antler, showing the characteristic adjacent disposition of brow and bez tines [42]. *Cervus elaphus* is already known from Epivillafranchian localities of Europe, dated between 1.0 and 0.8 Ma, but in Greece, the KYP-4, 5 red deer record is the earliest one identified so far.

The replacement of *Sus strozzii* by *Sus scrofa* took place at or soon after the Early/Middle Pleistocene transition [93], and the KYP-4 record is one of the earliest known occurrences of the species in Europe, biochronologically close perhaps to the record from Petralona Cave [94]. Accordingly, if *Stephanorhinus hemitoechus* is indeed present at KYP-4, it marks its FGO, again perhaps close to the Petralona occurrence [94].

The uncertain taxonomic status of the various *Praemegaceros* species/subspecies and the validity of several distinctive morphological traits makes the earliest occurrence of *Praemegaceros verticornis* in Europe problematic. Nonetheless, the rich KYP-4 material, including antlers with distal palmation, is more advanced than the Early Pleistocene *Praemegaceros* material from Greece, commonly attributed to *Praemegaceros pliotarantoindes* [19,42,95]. *Praemegaceros verticornis* is thought to have existed in Europe until ca. 400 ka [96], yet it is not present at MAR-1 (dated at ca. 450 ka). Its absence at MAR-1 and MAR-2, if not accidental, may be due to paleoecological reasons, or to earlier local extinction [19].

Additionally, *Castor fiber*, *Macaca sylvanus*, *Dama* sp., and *Bison* sp. document their FMO (Figure 14).

As in the case of the micromammals at KYP-4, the presence at the level of KYP-3/KYP SU 6 of both a larger *Mimomys* (*M.* cf. *savini*) and a smaller one (*Mimomys* sp.) indicates a correlation to or before MIS 16, and certainly older than the *Arvicola*-bearing MAR-1. The stratigraphic data suggest that KYP-3 overlies KYP-4, and following the age model of Tourloukis et al. [3], a correlation to MIS 16 is possible. This marks the LGO of *Mimomys* (Figure 14). As a whole, the level of KYP-3/KYP SU 6/MAR SU 7/CHO SU 7 marks the so far known LGO of *Praemegaceros*. The genus is present at Petralona [97], yet of unknown age. In addition, *Sus scrofa* and *Equus* record their LMO, at least a dated one, since both taxa were recorded by Melentis [9,10] and Sickenberg [17]. On the other hand, this level marks the FGO of *Lutra simplicidens* and of a *Capreolus*-sized cervid (Figure 14), the latter documented from unknown horizons [11,17,19].

At CHO-5 (likely MIS 14, based on the basin stratigraphy) the FGO of *Bos* is possibly recorded (Figure 14), a genus that is otherwise present at higher stratigraphic positions in the basin (e.g., CHO-7 and CHO SU 11) and was also documented by Melentis [13] but from unknown horizons.

The biostratigraphy/biochronology of the mammal fauna from MAR-1 and MAR-2 has been already discussed [24,45,52]. Of particular interest is *Arvicola mosbachensis* from MAR-1, whose molars' SDQ values and size suggest a correlation to MIS 12/11 [52]. This is also the so far known FGO of *Arvicola*, although its slightly advanced stage might indicate that an earlier occurrence of more primitive *Arvicola* in lower stratigraphic layers of the basin is possible. In terms of large mammals, the levels of MAR-1 and 2 mark the LGO of *Castor fiber*, *Macaca sylvanus*, *Lutra simplicidens*, and *Hippopotamus antiquus* (Figure 14). Particularly, the MAR-1 and 2 hippopotamus record corresponds to one of the last occurrences of *H. antiquus* in Europe at ca. 450 ka (see also [98]). The last occurrences of *L. simplicidens* in Europe are reported at Hundsheim (Austria) and Mosbach-2 (Germany), both correlated to MIS 15 or MIS 13 (ca. 600–500 ka), while the modern *Lutra lutra* is reported from Hoxne, England, at MIS 11 (ca. 400 ka) [99,100] (note, however, that the Hoxne lutrine is regarded as *Lutra* sp. in [101]). As such, the MAR-1 record during the MIS 12 may document one of the last occurrences of *L. simplicidens* in Europe.

The rich rodent assemblage from CHO-7 is important in terms of biochronology. The earliest occurrence of *Microtus* (*Terricola*) *subterraneus* in Central Europe is recorded at Maastricht-Belvédère 4 (MIS 7, The Netherlands) [102,103], although a slightly older age (MIS 8) for the earliest occurrence of the species is also possible [104] (Figure 2). This indicates that CHO-7 is significantly younger than MAR-1. The presence of *Microtus* sp. (cf. *M. arvalis*) and *Arvicola mosbachensis* is common between CHO-7 and MAR-1. The mean SDQ values of the *Arvicola* molars are also comparable. Yet, the available molars of *Arvicola* from CHO-7 are limited and, as such, the mean SDQ value should be considered only as indicative.

In a recent study, Konidaris and Kostopoulos [105] introduced the Late Pliocene to Middle Pleistocene large mammal faunal units (FUs) of Greece, which are defined by a set of First and Last Occurrences. In this biochronological unit scheme, they recognized the distinct Marathousa FU, dated between <800 and ca. 450 ka, which encompasses the Galerian faunal assemblages from the Megalopolis Basin and the Petralona Cave (the older faunal assemblage) in Chalkidiki Peninsula (e.g., [94,97]). Marathousa FU records the FGO of *Homo*, *Palaeoloxodon antiquus*, *Lutra simplicidens*, *Ursus deningeri*, *Panthera spelaea*, *Crocuta crocuta*, *"Hyaena" prisca"*, *Stephanorhinus hemitoechus*, *Bison*, *Sus scrofa*, the *Cervus elaphus* lineage, and possibly *Bos*, while additional large mammals may also include *Panthera pardus* (if present at KYP-3) and *Mammuthus trogontherii* from Loussika (Peloponnese) and other localities [71]. On the other hand, this FU marks the LGO of *Macaca sylvanus*, *Equus altidens*, *Hippopotamus antiquus*, *Praemegaceros*, and *Lutra simplicidens* [105]. In a wider European context, the Marathousa FU represents a unit within Isernia and Fontana Ranuccio FUs of Italy, and corresponds to Gran Dolina TD8, Sima de los Huesos (Spain), and L'Escale and Arago CM (France) faunal assemblages (e.g., [91], and references therein). Marathousa FU is part of the major large mammal reorganization that took place close to the Early/Middle Pleistocene transition and is associated with the "Mid-Pleistocene Revolution" that affected the terrestrial ecosystems (e.g., [91]). The faunal turnover documents the disappearance of several Villafranchian–Epivillafranchian taxa (whose last occurrences in Greece are recorded in the Apollonia FU) and the arrival of several Galerian immigrants (Marathousa FU). The Marathousa FU is followed by the Apidima FU, which is characterized by the establishment of modern large mammals from the end of the Middle Pleistocene onwards [105,106].

4.2. Hippopotamuses in the Megalopolis Basin

In the last few years, several studies dealing with the status of the Middle Pleistocene hippopotamuses from Europe were published (e.g., [24,67,107–110]). Most of these studies underline that in the absence of the diagnostic cranium and mandible from securely dated contexts, the taxonomy (largely focused on postcranial dimensions) and the time of replacement of the two or three commonly recognized species remain vague. Despite this, and some different opinions among the researchers, there is a broad consensus that *H. antiquus* or closely related forms classified under *H.* ex gr. *antiquus* were replaced by *H. amphibius* broadly between 500 and 400 ka [111].

The hippopotamus material from the Megalopolis Basin does not include adequately preserved cranial material, yet the quite abundant dental and postcranial specimens (in some cases belonging to partial skeletons), with secure chronological placement and from different stratigraphic levels covering the late Early–Middle Pleistocene, are important and contribute to the current investigations and discussions.

The study of Athanassiou et al. [20] on the hippopotamuses from the early Middle Pleistocene sites of the Kyparissia mine (KYP-1, KYP-3, KYP-4, and KYP-T) showed their intermediate position in terms of bone dimensions between *H. amphibius* and *H. antiquus*, and a better match with specimens classified as *H. tiberinus*. The present study confirms that the hippopotamuses from other Middle Pleistocene deposits of the basin (CHO-4, MAR SU 2, and MAR-1; a larger-sized third metacarpal from MAR-1 Area A [45] is of comparable size to a recently discovered third metacarpal from KYP-4) are also consistently placed at the lower range of *H. antiquus*, reflected also in the estimated body masses (Figure 4; Tables 1 and S1; note that the MAR-1 individual is skeletally immature). The skeletal elements (and the estimated body mass) of the CHO-6 hippopotamus indicate a large-sized individual, larger than the other hippopotamuses from Megalopolis and as large as those of "typical" *H. antiquus* [40], in agreement with the older (Lower Pleistocene) age of the find-bearing sediment.

Therefore, in terms of taxonomy, either the middle-sized *H. tiberinus* is a distinct species (according to [112] after 1.4–1.2 Ma; pending confirmation from cranial remains), implying a second dispersal of African hippopotamuses to Europe (related perhaps to *Hippopotamus gorgops* [33,112]), or the smaller dimensions of the Middle Pleistocene hippopotamuses correspond to a body size decrease of the single Early Pleistocene arrival of hippopotamuses in Europe (between ca. 2.1 and 1.6 Ma; *H. antiquus*) as a response to climatic conditions [43]. Considering that, at present, there exists no clear-cut morphological distinction between *H. tiberinus* and *H. antiquus* (e.g., [67,113]) that would support the first option, and the fact that the studied *Hippopotamus* remains from Megalopolis Basin originated from glacial (or stadial) sediments with supposedly harsher paleoenvironmental conditions (see also below), the second option is preferred for the moment (besides hippopotamuses, the potential size reduction of other large mammals in the Megalopolis Basin requires further investigation). This is also in agreement with the study on dental enamel hypoplasia (related to physiological stress) in hippopotamus incisors and canines from Early Pleistocene sites of Europe [114]. In this study, a high incidence of hypoplasia was observed at sites dated to ca. 900–860 ka when a long and severe glacial phase was recorded.

Therefore, the Middle Pleistocene hippopotamuses from the Megalopolis Basin are attributed to the smaller-sized representatives of *H. antiquus* (the *H.* ex gr. *H. antiquus* of [43]). The even smaller-sized canine from TRP-4 (younger than MAR-1 and 2) might hint at an attribution to *H. amphibius*, but the evidence is too fragmentary and, therefore, is ascribed to *Hippopotamus* sp.

4.3. Paleoenrvironmental Remarks

Palaeoloxodon antiquus had wide ecological adaptations, inhabiting a broad spectrum of environments and having flexible dietary preferences and foraging behavior (e.g., [115–117]). Despite this, in central and northern Europe, the species existed predominantly during the warmer interglacial stages and was generally absent from the intervening cold stages of open conditions, when its geographic range was restricted in refugia of peri-Mediterranean Europe (e.g., [118,119]). The isotopic study on the elephant individual from the Area A of MAR-1 (i.e., during the glacial MIS 12) corroborated the refugial status of Greece and the Megalopolis Basin in particular, by finding a restricted geographical range within a C_3-dominated open woodland environment under relatively stable climatic conditions [120]. This conclusion is further supported by the continuous presence of straight-tusked elephants in several stratigraphic levels of the Megalopolis Basin, corresponding to both interglacial and glacial, which further highlights the adaptability of the species to survive under diverse climatic conditions and to inhabit a variety of environments. The refugial status of the region is additionally attested from the also temperate-adapted fallow deer (present in several sites/findspots) and macaques (the latter present at both KYP-4 and MAR-1) that prefer settings with milder conditions (e.g., [121,122]), and especially the former is found in several stratigraphic layers of the Megalopolis Basin.

Although not exclusively linked with a warm/temperate environment, the presence during glacial stages of beavers, hippopotamuses, and otters, which are highly dependent on permanent freshwater bodies for their survival (e.g., [43]), indicates the capacity of the basin to retain freshwater bodies throughout the annual cycle, even under the harsher climatic conditions [54]. Such an environment is also supported from the presence of two freshwater turtles, the European pond turtle *Emys orbicularis* and the Balkan terrapin *Mauremys rivulata*, which further require high enough average summer temperatures that allow for sexual differentiation ([60]; see also [123] regarding the refugial status of Greece for *E. orbicularis*).

The presence of warmer-adapted floral and faunal taxa, for instance, Palmae and the darter (*Anhinga*; present at both MAR-1 and KYP-T), a bird that is not found today in Greece and is associated with warm/tropical regions, indicate, together with the other botanical material (aquatic, waterside, and damp ground vegetation, combined with willow, alder, elm, deciduous oak, and maple) and avian remains (ducks, geese, and swans), a warm-temperate environment [20,56,58]. On the other hand, cooler conditions than today's Greece are inferred for MAR-1 by the multiproxy paleoenvironmental study including ostracods that indicate mean summer temperatures of 10–15 °C [54], as well as for MAR-2 by the mollusk study [24]; yet, in both cases, temperatures allowed the existence of water bodies under non-freezing conditions throughout the year.

For KYP-4, most paleoenvironmental analyses are currently in progress. The large mammal fauna generally shows similarities with MAR-1 in the occurrence of elephants, hippopotamuses, macaques, deer, and beavers. However, the presence at KYP-4 of the giant deer *Praemegaceros verticornis* (abundant in the faunal assemblage), rhinoceroses, and horses [19,42] might indicate slightly more open conditions compared to MAR-1. This stands in agreement with the preliminary results from the mollusk fauna from KYP-4, which show a slight increase in the terrestrial/semi-terrestrial taxa [124] as well as with the presence of the (terrestrial) tortoise *Testudo marginata* in the wider Kyparissia region (e.g., at KYP-3), a species absent in the particularly abundant (freshwater) turtle assemblage of MAR-1 [20,60]. The ostracod analysis for KYP-4 indicates the presence of rather cold (constantly lower than 15 °C) carbonate springs and a relatively stable temperature throughout the year [125].

Overall, despite the evidence of colder conditions than present-day Greece, the presence/abundance of taxa adapted to milder climatic conditions (elephants, hippopotamuses, macaques, and fallow deer), and some even to warmer ones (Palmae and the darter), indicate that the prevailing climate at MAR-1, and possibly at the other Megalopolis sites (at least at MAR-2, KYP-3, and KYP-4), was temperate, even during cold—oxygen isotopic—stages, while both flora and fauna support the existence of a landscape with substantial woodland components and more open areas, close to permanent and large freshwater bodies. The presence of taxa generally indicative of milder conditions in different stratigraphic layers of the basin, including detrital deposits (corresponding to the cold, glacial/stadial stages), suggests their continuous presence and a refugium region for their populations.

5. Conclusions

The late Early to Middle Pleistocene localities of the Megalopolis Basin permit the identification and tracking of faunal and evolutionary changes within the stratigraphic succession of the same sedimentary basin, thus allowing the study of the taxonomy and evolution of several vertebrate taxa and mammal turnovers between ca. 900 and 300 ka. Additionally, they provide paleoenvironmental (e.g., the refugial status of the basin for fauna and flora during glacial stages, dietary preferences of mammals, and seasonal climatic variations) and taphonomic interpretations (e.g., carnivore and anthropogenic modifications of mammal bones and particularly megafauna, site formation processes) that offer valuable insights into the Middle Pleistocene terrestrial ecosystems of Europe and hominin adaptations in particular, thus comprising a unique record for SE Europe. Particularly for Greece, the Megalopolis Basin localities, together with the Lower Pleistocene ones, e.g., Dafnero and Mygdonia Basin, and the Middle–Upper Pleistocene caves Petralona and Apidima, all of which are characterized by long stratigraphic sequences, altogether represent a rather continuous faunal sequence that captures the entire Pleistocene and allows for the detailed study of the faunal changes and the paleoenvironmental record of Greece.

Ongoing dating (e.g., electron spin resonance (ESR) and post-infrared infrared stimulated luminescence (pIRIR)), paleontological, paleoenvironmental (e.g., pollen, ostracods, mollusks, and stable isotope biogeochemistry, and dental micro- and meso-wear analyses in the mammal remains), and archaeological research in the Megalopolis Basin will refine the age constraints of the localities and establish a secure paleoecological and cultural framework, leading to more accurate interpretations for the first occurrence of hominins and other taxa in the southern-most tip of the Balkans.

Supplementary Materials: The following supporting information can be downloaded at: https://www.mdpi.com/article/10.3390/quat7040041/s1. Table S1: Dental and postcranial measurements (in mm) of new large mammal fossils from the Megalopolis Basin. Table S2: Updated faunal lists of the Megalopolis Basin sites and survey units. Data from [19,20,24,41,42,45,52,57] and the present study.

Author Contributions: Conceptualization, G.E.K.; methodology, G.E.K., A.A. and T.v.K.; validation, G.E.K., A.A., V.T., K.C., T.v.K., D.G., N.T., G.T., E.R., E.P., P.K. and K.H.; formal analysis, G.E.K., A.A., K.C. and T.v.K.; data curation, G.E.K. and A.A.; writing—original draft preparation, G.E.K. and K.C.; writing—review and editing, G.E.K., A.A., V.T., K.C., T.v.K., D.G., N.T., G.T., E.R., E.P., P.K. and K.H.; visualization, G.E.K. and K.C.; project administration, K.H., T.v.K. and E.P.; funding acquisition, K.H. All authors have read and agreed to the published version of the manuscript.

Funding: Excavation at Marathousa 1 was supported by the ERC-StG-283503 ("PaGE") and the ERC-CoG-724703 ("CROSSROADS"), both awarded to K.H. The 2018–2022 field survey in the Megalopolis Basin was supported by the ERC-CoG-724703 ("CROSSROADS"). G.E.K., V.T., and K.H. were supported by the Deutsche Forschungsgemeinschaft (DFG Project No. 463225251; "MEGALOPOLIS"). K.C., E.R., and K.H. were supported by the ERC-AdG-101019659 ("FIRSTSTEPS").

Data Availability Statement: Data are contained within the article and the Supplementary Materials.

Acknowledgments: Excavation at Marathousa 1 was conducted under a permit granted to the Ephorate of Paleoanthropology–Speleology, Hellenic Ministry of Culture. The 2018–2022 field survey in the Megalopolis Basin was conducted in the framework of the Megalopolis Paleoenvironmental Project (MegaPal) under a permit granted to the Ephorate of Paleoanthropology–Speleology, Hellenic Ministry of Culture, and the American School of Classical Studies at Athens. Many thanks to all field team members for their contribution in the fieldwork, the preparation of material, and the sieving and sorting of micromammals, particularly, A. Varis, C. Siali, M. Psyrri, G. Boni, N. Nikolakopoulou, and K. Kouveli. We greatly thank D. Kostopoulos for his valuable help with the bovine material from Megalopolis. Many thanks to the academic editors and the four reviewers for providing constructive comments and suggestions that improved the manuscript.

Conflicts of Interest: The authors declare no conflicts of interest.

References

1. Vinken, R. Stratigraphie und Tektonik des Beckens von Megalopolis (Peloponnes, Griechenland). *Geol. Jahrb.* **1965**, *83*, 97–148.
2. van Vugt, N. Orbital Forcing in Late Neogene Lacustrine Basins from the Mediterranean. A Magnetostratigraphic and Cyclostratigraphic Study. Ph.D. Thesis, Utrecht University, Utrecht, The Netherland, 2000.
3. Tourloukis, V.; Muttoni, G.; Karkanas, P.; Monesi, E.; Scardia, G.; Panagopoulou, E.; Harvati, K. Magnetostratigraphic and chronostratigraphic constraints on the Marathousa 1 Lower Palaeolithic site and the Middle Pleistocene deposits of the Megalopolis basin, Greece. *Quat. Int.* **2018**, *497*, 154–169. [CrossRef]
4. Melentis, J.K. Die Dentition der pleistozänen Proboscidier des Beckens von Megalopolis im Peloponnes (Griechenland). *Ann. Géologiques Pays Helléniques* **1961**, *12*, 153–262.
5. Darlas, A. Palaeolithic finds from the Megalopolis Basin: Their relation to the fossils of the same area. In *Argonautis: Honorary Issue to C.G. Doumas*; Vlachopoulos, A., Birtaha, K., Eds.; Antiquity: Kathimerini, Athens, 2003; pp. 27–37. (In Greek with English Summary)
6. Bürchner, L. Wichtige Funde fossiler Knochen in Arkadien. *Berichte Naturwissenschaftlichen Vereines Regensbg.* **1903**, *9*, 119–123.
7. Skouphos, T. Über die paläontologischen Ausgrabungen in Griechenland in Beziehung auf das Vorhandensein des Menschen. In Proceedings of the Comptes Rendus du Congrès International d'Archéologie, 1re Session, Athens, Greece, 7–13 April 1905; pp. 231–236.
8. Melentis, J.K. Die Osteologie der pleistozänen Proboscidier des Beckens von Megalopolis im Peloponnes (Griechenland). *Ann. Géologiques Pays Helléniques* **1963**, *14*, 1–107.
9. Melentis, J.K. Über *Equus abeli* aus dem Mittelpleistozän des Beckens von Megalopolis im Peloponnes (Griechenland). *Ann. Géologiques Pays Helléniques* **1963**, *16*, 507–519.
10. Melentis, J.K. *Sus scrofa* L. aus dem Jungpleistozän des Beckens von Megalopolis im Peloponnes (Griechenland). *Ann. Géologiques Pays Helléniques* **1965**, *16*, 436–445.
11. Melentis, J.K. Die pleistozänen Cerviden des Beckens von Megalopolis im Peloponnes (Griechenland). *Ann. Géologiques Pays Helléniques* **1965**, *16*, 1–92.
12. Melentis, J.K. Über *Hippopotamus antiquus* Desmarest aus dem Mittelpleistozän des Beckens von Megalopolis im Peloponnes (Griechenland). *Ann. Géologiques Pays Helléniques* **1965**, *16*, 403–435.
13. Melentis, J.K. Die Boviden des Jungpleistozäns des Beckens von Megalopolis im Peloponnes (Griechenland). *Ann. Géologiques Pays Helléniques* **1965**, *16*, 446–472.

14. Melentis, J.K. Die pleistozänen Nashörner des Beckens von Megalopolis im Peloponnes (Griechenland). *Ann. Géologiques Pays Helléniques* **1965**, *16*, 363–402.
15. Melentis, J.K. *Clemmys caspica* aus dem Pleistozän des Beckens von Megalopolis im Peleponnes (Griechenland). *Ann. Géologiques Pays Helléniques* **1966**, *17*, 169–181.
16. Melentis, J.K. Erster Fund von *Palaeoloxodon antiquus germanicus* in den jungpleistozänen Ablagerungen des Beckens von Megalopolis (Peloponnes). *Prakt. Akad. Athin.* **1965**, *40*, 197–207.
17. Sickenberg, O. Eine Säugetierfauna des tieferen Bihariums aus dem Becken von Megalopolis (Peloponnes, Griechenland). *Ann. Géologiques Pays Helléniques* **1976**, *27*, 25–73.
18. Theodorou, G. Megalopolis—112 years after the first excavation by National and Kapodistrian University of Athens (NKUA)—And the Post Lignite era. In Proceedings of the VIth International Conference on Mammoths and their Relatives, Grevena-Siatista, Greece, 5–12 May 2014; pp. 195–196.
19. Athanassiou, A. Pleistocene vertebrates from the Kyparíssia lignite mine, Megalopolis Basin, S. Greece: Rodentia, Carnivora, Proboscidea, Perissodactyla, Ruminantia. *Quat. Int.* **2018**, *497*, 198–221. [CrossRef]
20. Athanassiou, A.; Michailidis, D.; Vlachos, E.; Tourloukis, V.; Thompson, N.; Harvati, K. Pleistocene vertebrates from the Kyparíssia lignite mine, Megalopolis Basin, S. Greece: Testudines, Aves, Suiformes. *Quat. Int.* **2018**, *497*, 178–197. [CrossRef]
21. Röding, C.; Zastrow, J.; Scherf, H.; Doukas, C.; Harvati, K. Crown outline analysis of the hominin upper third molar from the Megalopolis Basin, Peloponnese, Greece. In *Ancient Connections in Eurasia*; Reyes-Centeno, H., Harvati, K., Eds.; Kerns Verlag: Tübingen, Germany, 2021; pp. 13–36.
22. Thompson, N.; Tourloukis, V.; Panagopoulou, E.; Harvati, K. In search of Pleistocene remains at the Gates of Europe: Directed surface survey of the Megalopolis Basin (Greece). *Quat. Int.* **2018**, *497*, 22–32. [CrossRef]
23. Panagopoulou, E.; Tourloukis, V.; Thompson, N.; Konidaris, G.; Athanassiou, A.; Giusti, D.; Tsartsidou, G.; Karkanas, P.; Harvati, K. The Lower Palaeolithic site of Marathousa 1, Megalopolis, Greece: Overview of the evidence. *Quat. Int.* **2018**, *497*, 33–46. [CrossRef]
24. Konidaris, G.E.; Tourloukis, V.; Boni, G.; Athanassiou, A.; Giusti, D.; Thompson, N.; Syrides, G.; Panagopoulou, E.; Karkanas, T.; Harvati, K. Marathousa 2: A new Middle Pleistocene locality in the Megalopolis Basin (Greece) with evidence of hominin exploitation of megafauna (*Hippopotamus*). *PaleoAnthropology* **2023**, *2023*, 34–55.
25. Karkanas, P.; Tourloukis, V.; Thompson, N.; Giusti, D.; Tsartsidou, G.; Athanassiou, A.; Konidaris, G.; Roditi, E.; Panagopoulou, E.; Harvati, K. The Megalopolis Paleoenvironmental Project (MegaPal). In *Human Evolution at the CROSSROADS: Research in Greece and Beyond*; Harvati, K., Ioannidou, M., Eds.; Tübingen University Press: Tübingen, Germany, 2024.
26. Nickel, B.; Riegel, W.; Schönherr, T.; Velitzelos, E. Environments of coal formation in the Pleistocene lignite at Megalopolis, Peloponnesus (Greece)—Reconstructions from palynological and petrological investigations. *Neues Jahrb. Geol. Paläontologie—Abh.* **1996**, *200*, 201–220. [CrossRef]
27. Okuda, M.; van Vugt, N.; Nakagawa, T.; Ikeya, M.; Hayashida, A.; Yasuda, Y.; Setoguchi, T. Palynological evidence for the astronomical origin of lignite–detritus sequence in the Middle Pleistocene Marathousa Member, Megalopolis, SW Greece. *Earth Planet. Sci. Lett.* **2002**, *201*, 143–157. [CrossRef]
28. Karkanas, P.; Tourloukis, V.; Thompson, N.; Giusti, D.; Panagopoulou, E.; Harvati, K. Sedimentology and micromorphology of the Lower Palaeolithic lakeshore site Marathousa 1, Megalopolis basin, Greece. *Quat. Int.* **2018**, *497*, 123–136. [CrossRef]
29. Jacobs, Z.; Li, B.; O'Gorman, K.; Tourloukis, V.; Panagopoulou, E.; Karkanas, P.; Harvati, K. Single-grain optical dating of sediment samples from Marathousa 1, southern Greece. In *Human Evolution at the CROSSROADS: Research in Greece and Beyond*; Harvati, K., Ioannidou, M., Eds.; Tübingen University Press: Tübingen, Germany, 2024.
30. van Kolfschoten, T.; Konidaris, G.E.; Doukas, C.; Athanassiou, A.; Tourloukis, V.; Panagopoulou, E.; Karkanas, P.; Harvati, K. Voles (Rodentia, Mammalia) as a proxy to date the site Kyparissia 4 (Megalopolis Basin, Greece). In *Human Evolution at the CROSSROADS: Research in Greece and Beyond*; Harvati, K., Ioannidou, M., Eds.; Tübingen University Press: Tübingen, Germany, 2024.
31. van der Meulen, A.J. Middle Pleistocene smaller mammals from the Monte Peglia (Orvieto, Italy) with special reference to the phylogeny of *Microtus* (Arvicolidae, Rodentia). *Quaternaria* **1973**, *17*, 1–144.
32. Luzi, E.; López-García, J.M. Patterns of variation in *Microtus arvalis* and *Microtus agrestis* populations from Middle to Late Pleistocene in southwestern Europe. *Hist. Biol.* **2019**, *31*, 535–543. [CrossRef]
33. Mazza, P. New evidence on the Pleistocene hippopotamuses of Western Europe. *Geol. Romana* **1995**, *31*, 61–241.
34. Kahlke, R.D. Die *Hippopotamus*-Reste aus dem Unterpleistozän von Untermaßfeld. In *Das Pleistozän von Untermaßfeld bei Meiningen (Thüringen), Teil 1*; Kahlke, R.D., Ed.; Monographien des Römisch-Germanischen Zentralmuseums Mainz, Dr. Rudolf Habelt: Bonn, Germany, 1997; Volume 40, pp. 277–374.
35. Galobart, À.; Ros, X.; Maroro, J.; Vila, B. Descripción del material de hipopótamo (*Hippopotamus antiquus* Desmarest, 1822) de los yacimientos del Pleistoceno inferior de Incarcal (Girona, NE de la Península Ibérica). *Paleontol. Evol.* **2003**, *34*, 153–173.
36. Tsoukala, E.; Chatzopoulou, K. A new Early Pleistocene (latest Villafranchian) site with mammals in Kalamotó (Mygdonia Basin, Macedonia, Greece)—Preliminary report. *Mitteilungen Komm. Quartärforschung Osterr. Akad. Wiss.* **2005**, *14*, 213–233.
37. Georgitsis, M.K.; Liakopoulou, D.E.; Theodorou, G.E.; Tsiolakis, E. Functional morphology of the hindlimb of fossilized pygmy hippopotamus from Ayia Napa (Cyprus). *J. Morphol.* **2022**, *283*, 1048–1079. [CrossRef]
38. Faure, M. Les hippopotames quaternaires non-insulaires d'Europe occidentale. *Nouv. Arch. Muséum D'histoire Nat. Lyon* **1985**, *23*, 13–79. [CrossRef]

39. Scott, K.M. Postcranial dimensions of ungulates as predictors of body mass. In *Body Size in Mammalian Paleobiology. Estimation and Biological Implications*; Damuth, J., MacFadden, B.J., Eds.; Cambridge University Press: Cambridge, UK, 1990; pp. 301–335.
40. Palmqvist, P.; Rodríguez-Gómez, G.; Figueirido, B.; García-Aguilar, J.M.; Pérez-Claros, J.A. On the ecological scenario of the first hominin dispersal out of Africa. *L'Anthropologie* **2022**, *126*, 102998. [CrossRef]
41. Konidaris, G.E.; Athanassiou, A.; Panagopoulou, E.; Karkanas, P.; Harvati, K. Fossil macaques (Cercopithecidae, Primates) from the Middle Pleistocene of the Megalopolis Basin (Greece) with description of a new specimen from Kyparissia 4. In *Human Evolution at the CROSSROADS: Research in Greece and Beyond*; Harvati, K., Ioannidou, M., Eds.; Tübingen University Press: Tübingen, Germany, 2024.
42. Athanassiou, A.; Konidaris, G.E.; Tourloukis, V.; Thompson, N.; Giusti, D.; Panagopoulou, E.; Karkanas, P.; Harvati, K. The Middle Pleistocene large mammal fauna from Kyparissia (Peloponnese, S. Greece): New collected material. In *Human Evolution at the CROSSROADS: Research in Greece and Beyond*; Harvati, K., Ioannidou, M., Eds.; Tübingen University Press: Tübingen, Germany, 2024.
43. Mazza, P.P.A.; Bertini, A. Were Pleistocene hippopotamuses exposed to climate-driven body size changes? *Boreas* **2013**, *42*, 194–209. [CrossRef]
44. Sala, B. *Bison schoetensacki* Freud. from Isernia la Pineta (early Mid-Pleistocene-Italy) and revision of the European species of bison. *Palaeontogr. Ital.* **1986**, *74*, 113–170.
45. Konidaris, G.E.; Athanassiou, A.; Tourloukis, V.; Thompson, N.; Giusti, D.; Panagopoulou, E.; Harvati, K. The skeleton of a straight-tusked elephant (*Palaeoloxodon antiquus*) and other large mammals from the Middle Pleistocene butchering locality Marathousa 1 (Megalopolis Basin, Greece): Preliminary results. *Quat. Int.* **2018**, *497*, 65–84. [CrossRef]
46. Stefaniak, K. *Neogene and Quaternary Cervidae from Poland*; Institute of Systematics and Evolution of Animals of the Polish Academy of Sciences: Kraków, Poland, 2015; p. 260.
47. Cherin, M. New material of *Lutra simplicidens* (Carnivora, Mustelidae, Lutrinae), a key taxon for understanding the evolution of European otters. *Riv. Ital. Paleontol. Stratigr.* **2017**, *123*, 433–441.
48. Löhnert, E.; Nowak, H. Die Braunkohlenlagerstätte von Khoremi im Becken von Megalopolis/Peloponnes. *Geol. Jahrb.* **1965**, *82*, 847–868.
49. Panagopoulou, E.; Tourloukis, V.; Thompson, N.; Athanassiou, A.; Tsartsidou, G.; Konidaris, G.E.; Giusti, D.; Karkanas, P.; Harvati, K. Marathousa 1: A new Middle Pleistocene archaeological site from Greece. *Antiquity* **2015**, *343*, 1–8. Available online: https://antiquity.ac.uk/projgall/panagopoulou343 (accessed on 28 May 2024).
50. Harvati, K.; Konidaris, G.; Tourloukis, V. Paleoanthropology at the Gates of Europe: Recent research in Greece in the frame of the PaGE ERC Starting Grant project. *Quat. Int.* **2018**, *497*, 1–3. [CrossRef]
51. Blackwell, B.A.B.; Sakhrani, N.; Singh, I.K.; Gopalkrishna, K.K.; Tourloukis, V.; Panagopoulou, E.; Karkanas, P.; Blickstein, J.I.B.; Skinner, A.R.; Florentin, J.A.; et al. ESR dating ungulate teeth and molluscs from the Paleolithic site Marathousa 1, Megalopolis Basin, Greece. *Quaternary* **2018**, *1*, 22. [CrossRef]
52. Doukas, C.; van Kolfschoten, T.; Papayianni, K.; Panagopoulou, E.; Harvati, K. The small mammal fauna from the palaeolithic site Marathousa 1 (Greece). *Quat. Int.* **2018**, *497*, 95–107. [CrossRef]
53. Jacobs, Z.; Li, B.; Karkanas, P.; Tourloukis, V.; Thompson, N.; Panagopoulou, E.; Harvati, K. Optical dating of K-feldspar grains from Middle Pleistocene lacustrine sediment at Marathousa 1 (Greece). *Quat. Int.* **2018**, *497*, 170–177. [CrossRef]
54. Bludau, I.J.E.; Papadopoulou, P.; Iliopoulos, G.; Weiss, M.; Schnabel, E.; Thompson, N.; Tourloukis, V.; Zachow, C.; Kyrikou, S.; Konidaris, G.E.; et al. Lake-level changes and their paleoclimatic implications at the MIS12 Lower Paleolithic (Middle Pleistocene) site Marathousa 1, Greece. *Front. Earth Sci.* **2021**, *9*, 668445. [CrossRef]
55. Boni, G. Paleoenvironmental reconstruction based on mollusc macrofaunal analysis: A case study from the Middle Pleistocene of Megalopolis basin (Peloponnesus, Greece). Master's Thesis, Aristotle University of Thessaloniki, Thessaloniki, Greece, 2022.
56. Field, M.H.; Ntinou, M.; Tsartsidou, G.; van Berge Henegouwen, D.; Risberg, J.; Tourloukis, V.; Thompson, N.; Karkanas, P.; Panagopoulou, E.; Harvati, K. A palaeoenvironmental reconstruction (based on palaeobotanical data and diatoms) of the Middle Pleistocene elephant (*Palaeoloxodon antiquus*) butchery site at Marathousa, Megalopolis, Greece. *Quat. Int.* **2018**, *497*, 108–122. [CrossRef]
57. Konidaris, G.E.; Athanassiou, A.; Panagopoulou, E.; Harvati, K. First record of *Macaca* (Cercopithecidae, Primates) in the Middle Pleistocene of Greece. *J. Hum. Evol.* **2022**, *162*, 103104. [CrossRef] [PubMed]
58. Michailidis, D.; Konidaris, G.E.; Athanassiou, A.; Panagopoulou, E.; Harvati, K. The ornithological remains from Marathousa 1 (Middle Pleistocene; Megalopolis Basin, Greece). *Quat. Int.* **2018**, *497*, 85–94. [CrossRef]
59. Tourloukis, V.; Thompson, N.; Panagopoulou, E.; Giusti, D.; Konidaris, G.E.; Karkanas, P.; Harvati, K. Lithic artifacts and bone tools from the Lower Palaeolithic site Marathousa 1, Megalopolis, Greece: Preliminary results. *Quat. Int.* **2018**, *497*, 47–64. [CrossRef]
60. Vlachos, E.; Georgalis, G.L.; Konidaris, G.E.; Athanassiou, A.; Tourloukis, V.; Thompson, N.; Panagopoulou, E.; Harvati, K. Preliminary results on the reptiles from the Middle Pleistocene of Marathousa 1, Megalopolis Basin (Greece). In *Human Evolution at the CROSSROADS: Research in Greece and Beyond*; Harvati, K., Ioannidou, M., Eds.; Tübingen University Press: Tübingen, Germany, 2024.
61. Guibert-Cardin, J.; Tourloukis, V.; Thompson, N.; Panagopoulou, E.; Harvati, K.; Nicoud, E.; Beyries, S. The function of small tools in Europe during the Middle Pleistocene: The case of Marathousa 1 (Megalopolis, Greece). *J. Lithic Stud.* **2022**, *9*, 1–29. [CrossRef]

62. Laws, R.M. Dentition and ageing of the hippopotamus. *East Afr. Wildl. J.* **1968**, *6*, 19–52. [CrossRef]
63. Hammer, Ø.; Harper, D.A.T.; Ryan, P.D. PAST: Paleontological statistics software package for education and data analysis. *Palaeontol. Electron.* **2001**, *4*, 1–9.
64. Hooijer, D.A. The fossil Hippopotamidae of Asia, with notes on the recent species. *Zool. Verh.* **1950**, *8*, 1–123.
65. Alberdi, M.T.; Ruiz Bustos, A. Descripción y significado biostratigráfico y climático del *Equus* e *Hippopotamus*, en el yacimiento de Venta Micena (Granada). *Estud. Geológicos* **1985**, *41*, 251–261. [CrossRef]
66. Kahlke, R.D. Schädelreste von *Hippopotamus* aus dem Unterpleistozän von Untermaßfeld. In *Das Pleistozän von Untermaßfeld bei Meiningen (Thüringen), Teil 2*; Kahlke, R.D., Ed.; Monographien des Römisch-Germanischen Zentralmuseums Mainz, Dr. Rudolf Habelt: Bonn, Germany, 2001; Volume 40, pp. 483–500.
67. Fidalgo, D.; Rosas, A.; Madurell-Malapeira, J.; Pineda, A.; Huguet, R.; García-Tabernero, A.; Cáceres, I.; Ollé, A.; Vallverdú, J.; Saladie, P. A review on the Pleistocene occurrences and palaeobiology of *Hippopotamus antiquus* based on the record from the Barranc de la Boella Section (Francolí Basin, NE Iberia). *Quat. Sci. Rev.* **2023**, *307*, 108034. [CrossRef]
68. Martínez-Navarro, B.; Palmqvist, P.; Madurell-Malapeira, J.; Ros-Montoya, S.; Espigares, M.P.; Torregrosa, V.; Pérez-Claros, J.A. La fauna de grandes mamíferos de Fuente Nueva-3 y Barranco León-5: Estado de la cuestión. In *Ocupaciones Humanas en el Pleistoceno Inferior y Medio de la Cuenca de Guadix-Baza*; Toro, I., Martínez-Navarro, B., Agustí, J., Eds.; Consejería de Cultura, Junta de Andalucía, Arqueología Monografías: Sevilla, Spain, 2010; pp. 197–236.
69. Inuzuka, N.; Hasegawa, Y.; Nogariya, H.; Kamei, T. On the stylohyoid bone of Naumann's elephant (*Elephas naumanni* Makiyama) from Lake Nojiri. *Mem. Fac. Sci. Kyoto Univ. Ser. Geol. Mineral.* **1975**, *41*, 49–65.
70. Palombo, M.R. Intra-specific variation of stylohyoid bones in *Palaeoloxodon*: A case study of Neumark-Nord 1 (Geiseltal, Sachsen-Anhalt, Germany) straight-tusked elephants. *Quat. Int.* **2012**, *276–277*, 77–92. [CrossRef]
71. Athanassiou, A. A skeleton of *Mammuthus trogontherii* (Proboscidea, Elephantidae) from NW Peloponnese, Greece. *Quat. Int.* **2012**, *255*, 9–28. [CrossRef]
72. Shoshani, J.; Ferretti, M.P.; Lister, A.M.; Agenbroad, L.D.; Saegusa, H.; Mol, D.; Takahashi, K. Relationships within the Elephantinae using hyoid characters. *Quat. Int.* **2007**, *169–170*, 174–185. [CrossRef]
73. Guenther, E.W. Die Biber (*Trogontherium cuvieri* Fisch. und *Castor fiber* L.) der altpleistozänen Fundstelle von Voigtstedt in Thüringen. *Paläontologische Abh. A* **1965**, *2*, 566–582.
74. Fischer, K. Postkraniale Skelettreste von Bibern (*Castor* L., *Trogontherium* Fischer, Castoridae, Rodentia, Mamm.) aus dem Mittelpleistozän von Bilzingsleben. In *Bilzingsleben IV. Homo erectus—Seine Kultur und Seine Umwelt*; Fischer, K., Guenther, E.W., Heinrich, W.D., Mania, D., Musil, R., Nötzold, T., Eds.; Veröffentlichungen des Landesmuseums für Vorgeschichte in Halle: Berlin, Germany, 1991; pp. 63–70.
75. Lister, A.M. The morphological distinction between bones and teeth of fallow deer (*Dama dama*) and red deer (*Cervus elaphus*). *Int. J. Osteoarchaeol.* **1996**, *6*, 119–143. [CrossRef]
76. Pfeiffer, T. Die Stellung von *Dama* (Cervidae, Mammalia) im System plesiometacarpaler Hirsche des Pleistozäns. *Cour. Forschungsinstitut Senckenberg* **1999**, *211*, 1–218.
77. Heintz, E. Les cervidés Villafranchiens de France et d'Espagne. Volume II: Figures et tableaux. *Mem. Mus. Natl. D'histoire Nat. Série C* **1970**, *22*, 1–206.
78. Zimmerman, K. Zur Evolution der Molarenstruktur der Erdmaus, *Microtus agrestis* (L.). *Zool. Jahrbücher Abt. Syst. Okol. Und Geogr. Tiere* **1956**, *84*, 270–274.
79. Markova, E.; Malygin, V.; Montuire, S.; Nadachowski, A.; Quéré, J.P.; Ochman, K. Dental variation in sibling species *Microtus arvalis* and *M. rossiaemeridionalis* (Arvicolinae, Rodentia): Between-species comparisons and geography of morphotype dental patterns. *J. Mamm. Evol.* **2010**, *17*, 121–139. [CrossRef]
80. Heinrich, W.D. Zur Evolution und Biostratigraphie von *Arvicola* (Rodentia, Mammalia) im Pleistozän Europas. *Z. Geol. Wiss.* **1982**, *10*, 683–735.
81. Mahmoudi, A.; Maul, L.C.; Khoshyar, M.; Darvish, J.; Aliabadian, M.; Kryštufek, B. Evolutionary history of water voles revisited: Confronting a new phylogenetic model from molecular data with the fossil record. *Mammalia* **2020**, *84*, 171–184. [CrossRef]
82. Masini, F.; Maul, L.C.; Abbazzi, L.; Petruso, D.; Savorelli, A. Independent water vole (*Mimomys savini*, *Arvicola*: Rodentia, Mammalia) lineages in Italy and Central Europe. *Foss. Impr.* **2020**, *76*, 59–83. [CrossRef]
83. Maul, L.C. Die Kleinsäugerfauna aus der Forschungsbohrung Ummendorf 1/2012. *Mitteilungen Geol. Bergwesen Sachs. -Anhalt* **2019**, *20*, 99–108.
84. Peretto, C.; Arnaud, J.; Moggi-Cecchi, J.; Manzi, G.; Nomade, S.; Pereira, A.; Falguères, C.; Bahain, J.J.; Grimaud-Hervé, D.; Berto, C.; et al. A human deciduous tooth and new $^{40}Ar/^{39}Ar$ dating results from the Middle Pleistocene archaeological site of Isernia La Pineta, southern Italy. *PLoS ONE* **2015**, *10*, e0140091. [CrossRef]
85. Wagner, G.A.; Maul, L.C.; Löscher, M.; Schreiber, H.D. Mauer—The type site of *Homo heidelbergensis*: Palaeoenvironment and age. *Quat. Sci. Rev.* **2011**, *30*, 1464–1473. [CrossRef]
86. van Kolfschoten, T.; Turner, E. Early Middle Pleistocene mammalian faunas from Kärlich and Miesenheim I and their biostratigraphical implications. In Proceedings of the Early Middle Pleistocene in Europe, Proceedings of the SEQS Cromer Symposium, Norwich, UK, 3–7 September 1996; pp. 227–253.
87. Parfitt, S.A.; Barendregt, R.W.; Breda, M.; Candy, I.; Collins, M.J.; Coope, G.R.; Durbidge, P.; Field, M.H.; Lee, J.R.; Lister, A.M.; et al. The earliest record of human activity in northern Europe. *Nature* **2005**, *438*, 1008. [CrossRef]

88. Danukalova, G.; Yakovlev, A.; Osipova, E.; Kurmanov, R.; van Kolfschoten, T. Biostratigraphy of the early Middle Pleistocene of the Southern Fore-Urals. *Quat. Int.* **2016**, *420*, 115–135. [CrossRef]
89. Goren-Inbar, N.; Lister, A.; Werker, E.; Chech, M. A butchered elephant skull and associated artifacts from the Acheulian site of Gesher Benot Ya'aqov, Israel. *Paléorient* **1994**, *20*, 99–112. [CrossRef]
90. Goren-Inbar, N.; Feibel, C.S.; Verosub, K.L.; Melamed, Y.; Kislev, M.E.; Tchernov, E.; Saragusti, I. Pleistocene milestones on the out-of-Africa corridor at Gesher Benot Ya'aqov, Israel. *Science* **2000**, *289*, 944–947. [CrossRef] [PubMed]
91. Palombo, M.R. Deconstructing mammal dispersals and faunal dynamics in SW Europe during the Quaternary. *Quat. Sci. Rev.* **2014**, *96*, 50–71. [CrossRef]
92. Malez, M.; Lenardić-Fabić, J. New subspecies of the southern elephant (*Mammuthus meridionalis adriacus* n. ssp.) from the bottom of the Adriatic Sea (Croatia, Yugoslavia). *Palaeontol. Jugosl.* **1988**, *37*, 1–36.
93. Cherin, M.; Alba, D.M.; Crotti, M.; Menconero, S.; Moullé, P.É.; Sorbelli, L.; Madurell-Malapeira, J. The post-Jaramillo persistence of *Sus strozzii* (Suidae, Mammalia) in Europe: New evidence from the Vallparadís Section (NE Iberian Peninsula) and other coeval sites. *Quat. Sci. Rev.* **2020**, *233*, 106234. [CrossRef]
94. Tsoukala, E.; Guérin, C. The Rhinocerotidae and Suidae of the Middle Pleistocene from Petralona Cave (Macedonia, Greece). *Acta Zool. Bulg.* **2016**, *68*, 243–264.
95. Athanassiou, A. The fossil record of continental fossil deer (Mammalia: Artiodactyla: Cervidae) in Greece. In *Fossil Vertebrates of Greece Vol. 2: Laurasiatherians, Artiodactyles, Perissodactyles, Carnivorans, and Island Endemics*; Vlachos, E., Ed.; Springer: Cham, Switzerland, 2022; pp. 205–247.
96. Pfeiffer, T. The first complete skeleton of *Megaloceros verticornis* (Dawkins, 1868) Cervidae, Mammalia, from Bilshausen (Lower Saxony, Germany): Description and phylogenetic implications *Mitteilungen Mus. Naturkunde Berl. Geowiss. Reihe* **2002**, *5*, 289–308. [CrossRef]
97. Tsoukala, E. Contribution to the Study of the Pleistocene Fauna of Large Mammals (Carnivora, Perissodactyla, Artiodactyla) from Petralona Cave, Chalkidiki (N. Greece). Ph.D. Thesis, Aristotle University of Thessaloniki, Thessaloniki, Greece, 1989.
98. Martino, R.; Marra, F.; Ríos, M.; Pandofli, L. The Middle Pleistocene hippopotamus from Malagrotta (Latium, Italy): New data and future perspectives. *Quaternary* **2024**, *7*, 13. [CrossRef]
99. Stuart, A.J.; Wolff, R.G.; Lister, A.M.; Singer, R.; Egginton, J.M. Fossil vertebrates. In *The Lower Paleolithic Site at Hoxne, England*; Singer, R., Gladfelter, B.G., Wymer, J.J., Eds.; University of Chicago Press: Chicago, IL, USA, 1993; pp. 163–206.
100. Kahlke, R.D.; García, N.; Kostopoulos, D.S.; Lacombat, F.; Lister, A.M.; Mazza, P.P.A.; Spassov, N.; Titov, V.V. Western Palaearctic palaeoenvironmental conditions during the Early and early Middle Pleistocene inferred from large mammal communities, and implications for hominin dispersal in Europe. *Quat. Sci. Rev.* **2011**, *30*, 1368–1395. [CrossRef]
101. Willemsen, G.F. A revision of the Pliocene and Quaternary Lutrinae from Europe. *Scr. Geol.* **1992**, *101*, 1–115.
102. van Kolfschoten, T. The evolution of the mammal fauna in the Netherlands and the middle Rhine Area (Western Germany) during the late Middle Pleistocene. *Meded. Rijks Geol. Dienst* **1990**, *43*, 1–69.
103. van Kolfschoten, T. The mammal fauna from the interglacial deposits at Maastricht-Belvédère. *Meded. Rijks Geol. Dienst* **1993**, *47*, 51–60.
104. Maul, L.C.; Markova, A.K. Similarity and regional differences in Quaternary arvicolid evolution in Central and Eastern Europe. *Quat. Int.* **2007**, *160*, 81–99. [CrossRef]
105. Konidaris, G.E.; Kostopoulos, D.S. The Late Pliocene–Middle Pleistocene Large Mammal Faunal Units of Greece. *Quaternary* **2024**, *7*, 27. [CrossRef]
106. Harvati, K.; Röding, C.; Bosman, A.M.; Karakostis, F.A.; Grün, R.; Stringer, C.; Karkanas, P.; Thompson, N.C.; Koutoulidis, V.; Moulopoulos, L.A.; et al. Apidima Cave fossils provide earliest evidence of *Homo sapiens* in Eurasia. *Nature* **2019**, *571*, 500–504. [CrossRef]
107. Martino, R.; Pandolfi, L. The Quaternary *Hippopotamus* records from Italy. *Hist. Biol.* **2022**, *34*, 1146–1156. [CrossRef]
108. Martino, R.; Ríos, M.I.; Mateus, O.; Pandolfi, L. Taxonomy, chronology, and dispersal patterns of Western European Quaternary hippopotamuses: New insight from Portuguese fossil material. *Quat. Int.* **2023**, *674–675*, 121–137. [CrossRef]
109. Fidalgo, D.; Madurell-Malapeira, J.; Martino, R.; Pandolfi, L.; Rosas, A. An updated review of the Quaternary hippopotamus fossil records from the Iberian Peninsula. *Quaternary* **2024**, *7*, 4. [CrossRef]
110. Mecozzi, B.; Iannucci, A.; Mancini, M.; Tentori, D.; Cavasinni, C.; Conti, J.; Messina, M.Y.; Sarra, A.; Sardella, R. Reinforcing the idea of an early dispersal of *Hippopotamus amphibius* in Europe: Restoration and multidisciplinary study of the skull from the Middle Pleistocene of Cava Montanari (Rome, central Italy). *PLoS ONE* **2023**, *18*, e0293405. [CrossRef]
111. Pandolfi, L.; Petronio, C. A brief review of the occurrences of Pleistocene *Hippopotamus* (Mammalia, Hippopotamidae) in Italy. *Geol. Croat.* **2015**, *68*, 313–319. [CrossRef]
112. Made, J.v.d.; Sahnouni, M.; Boufaghraief, K. *Hippopotamus gorgops* from El Kherba (Aigeria) and the context of its biogeography. In Proceedings of the II Meeting of African Prehistory, Burgos, Spain, 15–16 April 2017; pp. 135–169.
113. Petronio, C. Note on the taxonomy of Pleistocene hippopotamuses. *Ibex* **1995**, *3*, 53–55.
114. Fidalgo, D.; Rosas, A.; Bartolini-Lucenti, S.; Boisserie, J.R.; Pandolfi, L.; Martínez-Navarro, B.; Palmqvist, P.; Rook, L.; Madurell-Malapeira, J. Increase on environmental seasonality through the European Early Pleistocene inferred from dental enamel hypoplasia. *Sci. Rep.* **2023**, *13*, 16941. [CrossRef] [PubMed]

115. Palombo, M.R.; Albayrak, E.; Marano, F. The straight-tusked elephants from Neumark-Nord. A glance into a lost world. In *Elefantenreich. Eine Fossilwelt in Europa*; Meller, H., Ed.; Landesamt für Denkmalpflege und Archäologie Sachsen-Anhalt (Halle): Halle, Germany, 2010; pp. 219–247.
116. Saarinen, J.; Lister, A.M. Dental mesowear reflects local vegetation and niche separation in Pleistocene proboscideans from Britain. *J. Quat. Sci.* **2016**, *31*, 799–808. [CrossRef]
117. Rivals, F.; Semprebon, G.M.; Lister, A.M. Feeding traits and dietary variation in Pleistocene proboscideans: A tooth microwear review. *Quat. Sci. Rev.* **2019**, *219*, 145–153. [CrossRef]
118. Tsoukala, E.; Lister, A. Remains of straight-tusked elephant, *Elephas (Palaeoloxodon) antiquus* Falc. & Caut. (1847) ESR-dated to oxygen isotope Stage 6 from Grevena (W. Macedonia, Greece). *Boll. Soc. Paleontol. Ital.* **1998**, *37*, 117–139.
119. Lister, A.M. Ecological interactions of elephantids in Pleistocene Eurasia: *Palaeoloxodon* and *Mammuthus*. In *Human paleoecology in the Levantine Corridor*; Goren-Inbar, N., Speth, J.D., Eds.; Oxbow Books: Oxford, UK, 2004; pp. 53–60.
120. Roditi, E.; Bocherens, H.; Konidaris, G.E.; Athanassiou, A.; Tourloukis, V.; Karkanas, P.; Panagopoulou, E.; Harvati, K. Life-history of *Palaeoloxodon antiquus* reveals Middle Pleistocene glacial refugium in the Megalopolis basin, Greece. *Sci. Rep.* **2024**, *14*, 1390. [CrossRef]
121. Lister, A.M. Evolutionary and ecological origins of British deer. *Proc. R. Soc. Edinb.* **1984**, *82B*, 205–229. [CrossRef]
122. Elton, S.; O'Regan, H.J. Macaques at the margins: The biogeography and extinction of *Macaca sylvanus* in Europe. *Quat. Sci. Rev.* **2014**, *96*, 117–130. [CrossRef]
123. Vlachos, E.; Delfino, M. Food for thought: Sub-fossil and fossil chelonian remains from Franchthi Cave and Megalopolis confirm a glacial refuge for *Emys orbicularis* in Peloponnesus (S. Greece). *Quat. Sci. Rev.* **2016**, *150*, 158–171. [CrossRef]
124. Boni, G.; Syrides, G.; Konidaris, G.E.; Athanassiou, A.; Tourloukis, V.; Koukousioura, O.; Panagopoulou, E.; Karkanas, P.; Harvati, K. Preliminary results on the taxonomy and palaeoenvironmental analysis of the mollusc fauna from Marathousa 1, Marathousa 2 and Kyparissia 4 (Middle Pleistocene, Megalopolis Basin, Greece). In *Human Evolution at the CROSSROADS: Research in Greece and Beyond*; Harvati, K., Ioannidou, M., Eds.; Tübingen University Press: Tübingen, Germany, 2024.
125. Papadopoulou, P.; Tsoni, M.; Konidaris, G.E.; Tourloukis, V.; Panagopoulou, E.; Karkanas, P.; Harvati, K.; Iliopoulos, G. Ostracod contribution to the palaeoenvironmental study of Kyparissia 4 (Megalopolis Basin, Greece). In *Human Evolution at the CROSSROADS: Research in Greece and Beyond*; Harvati, K., Ioannidou, M., Eds.; Tübingen University Press: Tübingen, Germany, 2024.

quaternary

Article

Middle Pleistocene Hippopotamuses from the Italian Peninsula: An Overview

Beniamino Mecozzi [1,*], Alessio Iannucci [2], Marta Arzarello [3], Marco Carpentieri [3], Marie-Hélène Moncel [4], Carlo Peretto [3], Benedetto Sala [3] and Raffaele Sardella [3]

1 Department of Earth Sciences (PaleoFactory), Sapienza University of Rome, I-00185 Rome, Italy
2 Department of Geosciences, Section of Terrestrial Palaeoclimatology, Eberhard-Karls-University Tübingen, D-72074 Tübingen, Germany; alessio.iannucci@mnf.uni-tuebingen.de
3 Department of Humanities, University of Ferrara, I-44121 Ferrara, Italy; rzrmrt@unife.it (M.A.); mark.carpentieri@gmail.com (M.C.); carlo.peretto@unife.it (C.P.); benedetto.sala@unife.it (B.S.); raffaele.sardella@uniroma1.it (R.S.)
4 UMR 7194 Natural History of Prehistoric Humans (MNHN-CNRS-UPVD), Department of Human and Environment, National Museum of Natural History, F-66720 Paris, France; marie-helene.moncel@mnhn.fr
* Correspondence: beniamino.mecozzi@uniroma1.it

Abstract: Our work presents an updated overview of the Italian Middle Pleistocene records of hippopotamuses, including the two species *Hippopotamus antiquus* and *Hippopotamus amphibius*. In addition to reviewing several well-known fossils in the literature, a large number of samples are described herein for the first time. Following the recent results published in the literature, where the skull from the Middle Pleistocene of Cava Montanari (ca. MIS 13) was confirmed to belong to *H. amphibius*, one of the aims of this work was to investigate the *H. antiquus*–*H. amphibius* transition. A morphological analysis applied to a large sample confirmed the validity of the arrangement of the enamel ridges of the external surfaces of the lower canines as a diagnostic character for specific identifications. Finally, biometric analyses allowed us to test the size variability during the Middle Pleistocene, which confirmed that *H. antiquus* was generally larger than *H. amphibius*. Nevertheless, the remains of *H. antiquus* dated to ca. 600 ka show a reduced size when compared to older fossils of the same taxon, probably as a response to severe glacial conditions that occurred during MIS 16.

Keywords: Quaternary; Mediterranean Europe; *Hippopotamus*; taxonomy; biochronology; large mammals; Epivillafranchian; Galerian

Citation: Mecozzi, B.; Iannucci, A.; Arzarello, M.; Carpentieri, M.; Moncel, M.-H.; Peretto, C.; Sala, B.; Sardella, R. Middle Pleistocene Hippopotamuses from the Italian Peninsula: An Overview. *Quaternary* **2024**, 7, 20. https://doi.org/10.3390/quat7020020

Academic Editors: Juan Rofes and Jef Vandenberghe

Received: 19 January 2024
Revised: 19 March 2024
Accepted: 21 March 2024
Published: 22 April 2024

1. Introduction

The Middle Pleistocene was an important period for the evolution of European terrestrial ecosystems, which experienced a progressive transformation toward modern biodiversity [1,2]. This evolution was driven by substantial changes in the Earth's climate system, with a progressive yet not gradual increase in the amplitude of climate oscillations (from 41 ka to approximately 85–125 ka; [3,4]). These climatic changes started during the late Early Pleistocene and continued during the Middle Pleistocene [5–7]. This phase is known in the literature as the Middle Pleistocene Revolution (MPR) or the Early–Middle Pleistocene Transition (EMPT, 1.2–0.4 Ma) [8–10]. The end of the EMPT coincides with the Mid-Brunhes Event (MBE, ca. 424 ka; MIS 12–11 transition), an event that, from a climatic perspective, marks the consolidation of the glacial cycles ruled by a 100 kyr periodicity, the longest cyclicity recognized during the Quaternary.

These intense climatic oscillations caused a renewal of large mammal faunas, reflected in the identification of two European Land Mammal Ages (ELMAs), the Epivillafranchian for the late Early Pleistocene (ca. 1.2–0.8 Ma) and the Galerian for the latest Early to early Middle Pleistocene (ca. 0.8–0.4 Ma), witnessing multiple mammal dispersals, including those of species that still live in Europe (e.g., *Cervus elaphus*, *Sus scrofa*) or that disappeared only during the early Late Pleistocene (*Palaeoloxodon antiquus*), but also the survival of late Villafranchian holdovers [2,10–18]. The Acheulean technocultural complex also spread into Europe during the EMPT (e.g., [13,15,18]). Finally, multiple key bioevents occurred in Europe shortly before or coinciding with the MBE (the end of the EMPT), among which was the diffusion of *Bos primigenius*, *Equus hydruntinus*, *Dama clactoniana*, and *Canis lupus* [19–22].

The hippopotamuses occupied Europe for a long time, with their first dispersal recognized during the Early Pleistocene (ca. 2.2 Ma, "Hippo Event" *sensu* [23] and references therein). During the Villafranchian, hippopotamuses were dispersed across the continent, but their fossil record is limited to a few European localities [24–29]. Late Early Pleistocene hippopotamuses were often extremely abundant and widely spread, as documented, for instance, by the impressive record of Untermassfeld [30,31], on the one hand, or the fossil from Westbury Cave [29], which testifies to the diffusion of these animals into Britain, on the other.

The evolution of European Quaternary hippopotamuses has been long disputed, with diverging taxonomic opinions expressed by different authors [24–27,32–35]. Excluding the Mediterranean islands, the model proposed by [36] was adopted in many works (e.g., [13,22,29,31,37–40], but see [35]). This model involves only two species: *H. antiquus* (=*H. major* = *H. amphibius antiquus* = *H. tiberinus* = *H.* ex gr. *antiquus*) and *H. amphibius* (=*H. incognitus* = *H. amphibius incognitus*).

These two forms, however, morphologically differ in cranial and mandibular features [24,26,40,41], while a distinction based on postcranial bones was proposed by [26] but followed only by a few authors ([40] and references therein).

Dental remains display a large intraspecific variability in the development cusps, cingula, and valleys, which are, therefore, hardly useful for taxonomic assignments [24,26]. One of the features proposed for the distinction between *H. antiquus* and *H. amphibius* was the configuration of the enamel ridges and grooves on the external surfaces of the lower canines [24,42]. According to [42], the canine enamel ridges of *H. antiquus* are prominent and convergent, whereas in *H. amphibius*, they are characterized by parallel development. Later, [26] argued that the two distinct morphotypes are observed with the same frequency in *H. antiquus*. However, [26] did not provide specific details on the specimens used to score this character, and since he attributed all Middle Pleistocene European hippopotamuses to *H. antiquus* (=*H. tiberinus*), it cannot be ruled out that this interpretation is biased by the inclusion in his sample of specimens actually belonging to *H. amphibius*. Indeed, [22] recently reviewed the chronostratigraphic setting of Cava Montanari and confirmed the attribution of a skull known from the site to *H. amphibius*. With an age of ca. 500 ka, the Cava Montanari skull implies that the appearance of *H. amphibius* in the European Middle Pleistocene record is another bioevent that occurred during the end of the EMPT, confirming the hypothesis previously proposed by several authors [24,34,36,39,41].

This work offers an updated overview of the Italian Middle Pleistocene and, to give perspective on the subject, selected late Early Pleistocene occurrences of hippopotamuses, comprising both *H. antiquus* and *H. amphibius*. Fossils from a number of localities are presented herein for the first time. Finally, size variability during the Middle Pleistocene is investigated, testing the existence of differences between species or evolutionary trends across time.

2. An Overview of the Late Early to Late Middle Pleistocene Hippopotamus Record

The Italian hippopotamus record from the late Early Pleistocene is quite sparse and known from few localities (Figure 1, Table 1).

Of great importance are the remains recovered from Colle Curti (or Collecurti), where 402 specimens were found (90.7% of the total mammal fossils collected at the site; [43]). The sample includes three skulls, a real rarity in the European record that, in addition to other remains, allowed researchers to explore and to define the variability of the Epivillafranchian hippopotamuses [26,43]. This locality is definitely a reference point for *H. antiquus* of Europe.

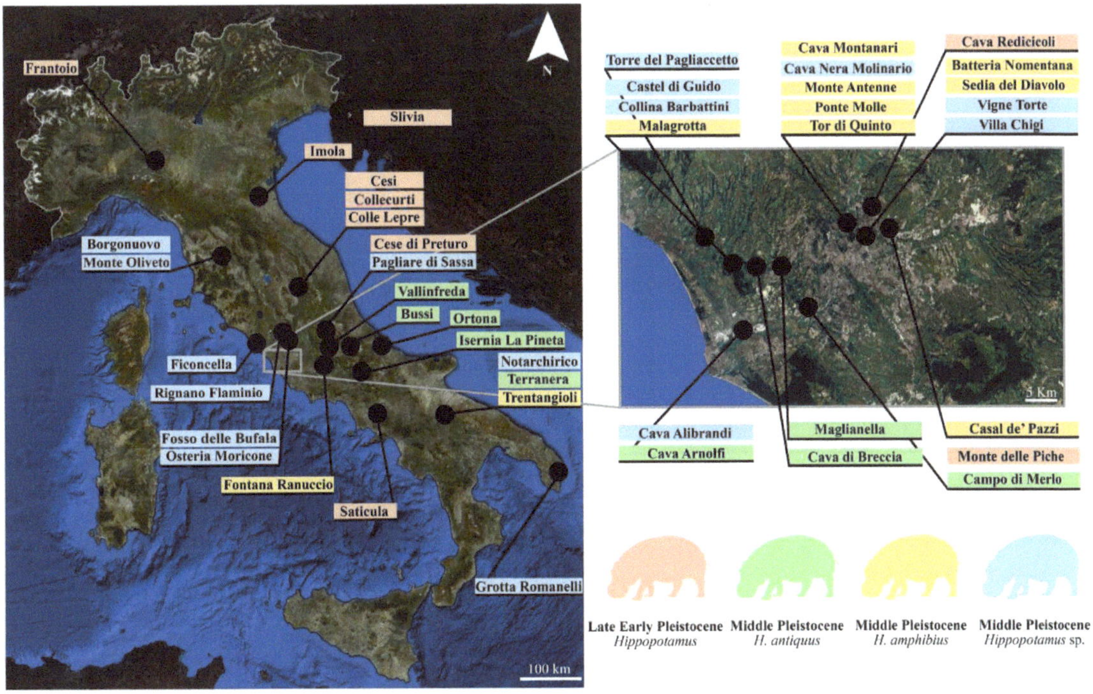

Figure 1. Geographic positions of the main late Early to late Middle Pleistocene sites in the Italian Peninsula yielding *Hippopotamus* remains.

Another important area for hippopotamuses is the late Early Pleistocene basin of Scoppito-Madonna della Strada, where several fossil-bearing sites were discovered. The first local fauna was collected from Cese di Preturo, where hippopotamuses were initially described by [44]. These fossils were examined by [26], who confirmed the presence of *H. antiquus* (the author also attributed several specimens to *H. tiberinus*, today a synonym of *H. antiquus*). The sample was also studied by [45], confirming again its attribution to *H. antiquus*. Other fossils recovered from the Scoppito-Madonna della Strada basin were collected in the Cava Santarelli succession, and they were referred to *H. antiquus* by [45]. Additional remains were found at Civitatomassa, Coppito, and Genzano di Sassa, listed by [45] and ascribed to *H. antiquus*. In the Scoppito-Madonna della Strada basin, a rich mammal sample was recovered from Pagliare di Sassa, including some isolated hippopotamus remains [46]. These fossils were considered of limited taxonomic value and attributed to *Hippopotamus* sp. cf. *H. antiquus* by [46]. Later, the sample from Pagliare di Sassa was attributed to *Hippopotamus* ex gr. *H. antiquus* by [45] and *H. antiquus* by [47].

In sum, at Scoppito-Madonna della Strada basin, hippopotamuses were documented from three late Early Pleistocene deposits, Cese di Preturo, Cava Santarelli, and Genzano di Sassa (between ca. 1.3 and 1.1 Ma), and from three early Middle Pleistocene sites, Civitatomassa, Coppito, and Pagliare di Sassa.

Early Pleistocene mammals were discovered at Saticula, a lacustrine deposit located in the intramontane basin in the north-western part of the Caudina valley [48]. The sample included a partially articulated skeleton and other isolated fossils attributed to *H. antiquus* [48].

Table 1. Hippopotamuses from the late Early Pleistocene to the Middle Pleistocene sites in the Italian Peninsula, with chronological attribution, previous and present taxonomical attribution, references, and present location of the material. Light pink—Epivillafranchian specimens of *H. antiquus;* green—Middle Pleistocene specimens of *H. antiquus;* light orange—Middle Pleistocene specimens of *H. amphibius;* blue—Middle Pleistocene specimens with open nomenclature. MIS = Marine Isotopic Stage.

Sites	Previous Attribution	References for Previous Attribution	Geochronology	Age (Ma)	MIS	References for Chronological Assigment	Analyzed in This Work	Current Attribution
Cese di Preturo	*Hippopotamus antiquus*	[26]	late Early Pleistocene	1.3–1.1		[26,45]		
Cava Santarelli	*Hippopotamus antiquus*	[45]	late Early Pleistocene	1.3–1.1		[26,45]		
Colle Curti	*Hippopotamus antiquus*	[26,43]	late Early Pleistocene	1.1		[26,43]		
Saticula	*Hippopotamus antiquus*	[48]	late Early Pleistocene	1.1		[48]		
Cesi	*Hippopotamus* sp.	[49]	late Early Pleistocene	ca 1.0		[49]		
Colle Lepre	*Hippopotamus* ex gr. *H. antiquus*	[34]	late Early Pleistocene	1.0		[34]		
Frantoio	*Hippopotamus* sp.	[50]	late Early Pleistocene	0.99		[50]		
Slivia	*Hippopotamus* sp.	[51,52]	late Early Pleistocene	ca. 0.9		[51,52]		
Imola	*Hippopotamus amphibius (?)*	[53]	late Early Pleistocene	0.85–0.8	21/20	[54]		
Cava Redicicoli	*Hippopotamus* cf. *antiquus*	[40]	late Early Pleistocene	ca. 0.8	20/19	[40,55]		
Monte delle Piche	*Hippopotamus* sp.	[56]	late Early Pleistocene	ca. 0.8	20/19	[56]		
Bussi	*Hippopotamus antiquus*	[57]	early Middle Pleistocene	0.78–0.6	19–15	[57]	X	*Hippopotamus antiquus*
Ortona	*Hippopotamus antiquus*	[34,58]	early Middle Pleistocene	0.78–0.7	19–17	[34]		
Vallinfreda	*Hippopotamus antiquus*	[24,59]	early Middle Pleistocene	0.78–0.7	19–17	[59]	X	*Hippopotamus antiquus*
Campo di Merlo	*Hippopotamus antiquus*	[36]	early Middle Pleistocene	ca. 0.7	18–17	[60]	X	*Hippopotamus antiquus*
Cava Alibrandi	*Hippopotamus* sp.	[61]	early Middle Pleistocene	ca. 0.7	18–17	[55]	X	*Hippopotamus* sp.
Cava Arnolfi	*Hippopotamus antiquus*	[36]	early Middle Pleistocene	ca. 0.7	18–17	[55]	X	*Hippopotamus antiquus*
Monte Oliveto	*Hippopotamus* sp. cf. *H. antiquus*	[62]	early Middle Pleistocene	0.8–0.6	19–15	[62]		
Notarchirico	*Hippopotamus antiquus*	[15]	early Middle Pleistocene	0.7–0.65	17	[15]	X	*Hippopotamus* sp.
Pagliare di Sassa	*Hippopotamus* sp.	[46]	early Middle Pleistocene	ca. 0.7–0.6	18–17	[45,46]		
Borgonuovo	*Hippopotamus* sp.	[63]	early Middle Pleistocene	ca. 600	15	[64]		

Table 1. *Cont.*

Sites	Previous Attribution	References for Previous Attribution	Geochronology	Age (Ma)	MIS	References for Chronological Assigment	Analyzed in This Work	Current Attribution
Cava di Breccia—Casal Selce 2	*Hippopotamus antiquus*	[2]	early Middle Pleistocene	0.615–0.611	15	[2,55]	X	*Hippopotamus antiquus*
Isernia La Pineta	*Hippopotamus antiquus*	[26]	early Middle Pleistocene	ca. 600	15	[65]	X	*Hippopotamus antiquus*
Maglianella	*Hippopotamus antiquus*	[36]	early Middle Pleistocene	ca. 600	15	[55]		
Terranera			early Middle Pleistocene	ca. 600	15	[66]	X	*Hippopotamus antiquus*
Cava Montanari	*Hippopotamus amphibius*	[22]	Mid Middle Pleistocene	ca. 540–460	13	[22]		
Cava Nera Molinario	*Hippopotamus* sp.	[55]	mid Middle Pleistocene	ca. 500	13	[55]	X	*Hippopotamus* sp.
Ficoncella	*Hippopotamus* sp.	[67]	mid Middle Pleistocene	499–441	13	[67]		
Fosse della Bufala	*Hippopotamus* cf. *antiquus*	[68]	mid Middle Pleistocene	ca. 600–500	15	[68]		
Monte Antenne	*Hippopotamus amphibius*	[69]	Mid Middle Pleistocene	ca. 500	13	[55]	X	*Hippopotamus amphibius*
Osteria Moricone	*Hippopotamus* cf. *antiquus*	[68]	mid Middle Pleistocene	ca. 600–500	15	[68]		
Ponte Molle	*Hippopotamus* ex gr. *antiquus*	[70]	Mid Middle Pleistocene	ca. 540–460	13	[70]	X	*Hippopotamus* cf. *amphibius*
Rignano Flaminio	*Hippopotamus* cf. *amphibius*	[71]	mid Middle Pleistocene	ca. 500	13	[71]	X	*Hippopotamus* sp.
Tor di Quinto	*Hippopotamus* sp. and *Hippooptamus antiquus*	[72]	mid Middle Pleistocene	ca. 500	13	[41]	X	*Hippopotamus amphibius*
Trentangioli			early Middle Pleistocene	ca. 500	13	This work	X	*Hippopotamus amphibius*
Castel di Guido	*Hippopotamus* sp.	[73]	mid Middle Pleistocene	ca. 400	11	[73–75]		
Collina Barbattini	*Hippopotamus* sp. and *Hippooptamus antiquus*	[60,76,77]	mid Middle Pleistocene	ca. 400	11	[60,75,76]	X	*Hippopotamus* sp.
Fontana Ranuccio	*Hippopotamus amphibius*	[78]	mid Middle Pleistocene	ca. 400	11	[79]		
Malagrotta	*Hippopotamus* sp.	[80]	mid Middle Pleistocene	ca. 400	11	[60,75,80]		
Serbaro di Romagnano	*Hippopotamus amphibius*	[81]	mid Middle Pleistocene	ca. 400	11	[82,83]	X	*Hippopotamus* sp.
Batteria Nomentana	*Hippopotamus (Tetraprotodon) major*	[84]	late Middle Pleistocene	0.370–0.290	9–8	[41]	X	*Hippopotamus amphibius*
Sedia del Diavolo	*Hippopotamus* cf. *amphibius*	[24]	late Middle Pleistocene	0.370–0.290	9–8	[41]	X	*Hippopotamus amphibius*

Table 1. *Cont.*

Sites	Previous Attribution	References for Previous Attribution	Geochronology	Age (Ma)	MIS	References for Chronological Assigment	Analyzed in This Work	Current Attribution
Vigne Torte	*Hippopotamus* sp.	[72]	Late Middle Pleistocene	0.370–0.290	9–8	[41]	X	*Hippopotamus* sp.
Villa Chigi	*Hippopotamus amphibius*	[72]	Late Middle Pleistocene	0.370–0.290	9–8	This work	X	*Hippopotamus* sp.
Grotta Romanelli—ISU1-2	*Hippopotamus antiquus*	[26]	late Middle Pleistocene	0.320–0.200	9–7	[85]	X	*Hippopotamus* sp.
Casal de' Pazzi	*Hippopotamus* cf. *amphibius*	[86]	late Middle Pleistocene	ca. 240	7	[86]		
Torre del Pagliaccetto—upper level	*Hippopotamus* sp.	[26,87]	late Middle Pleistocene	0.270–0.240	7	[88]		

Hippopotamus remains were also found at Cesi, a site located in the Colfiorito basin, near the aforementioned locality of Collecurti [49]. The state of preservation of the limited number of dental and postcranial bones prevents their specific identification [49]. In the same area, other hippopotamus remains were accidentally discovered at Colle Lepre, thanks to agricultural activities [34]. The sample, including cranial fragments, teeth, and postcranial bones, was ascribed to *H.* gr. *H. antiquus* [34].

An interesting mammal record was collected from the Frantoio deposit, located in the proximity of the Arda river [50]. Two hippopotamus femurs, belonging to juvenile individuals, were identified, but considering the limited taxonomic value of the remains, they were attributed to *Hippopotamus* sp. [50].

A key site for the Italian Land Mammal Biochronological scheme is Slivia, which gave the name of the last Epivillafranchian faunal unit ([11], Galerian in [1]). The deposit consists of an ossiferous breccia developed in a karst cavity, which was exposed by quarry activities during the XX century [51]. Hippopotamus remains collected during illegal excavations [51] were presented by [52]. The authors ascribed them to *H. amphibius*, but later, they were considered *Hippopotamus* sp. by [51].

Four remains were found at Imola: a fragment of a mandible, two fragments of a lower canine, and a pelvis [53]. The authors attributed, with uncertainty, the sample to *H. amphibius* (?).

The rich sample of Cava Redicicoli was recently examined by [40], who attributed the remains to *H.* cf. *antiquus*. The sample included several complete postcranial bones, in addition to a number of dental remains, which allowed the expansion of the knowledge on Epivillafranchian hippopotamus variability.

A fragmentary femur was found at Monte delle Piche, a deposit located in an area of Rome discovered during the XIX century ([56] and references therein). Due to its state of preservation, the femur was attributed to *Hippopotamus* sp. [56].

In the Middle Pleistocene, the hippopotamuses were widely dispersed in the Italian Peninsula, especially in the Rome basin (Figure 1; Table 1).

Five postcranial bones recovered at Monte Oliveto were attributed to *Hippopotamus antiquus* (*H. major* considered synonym of *H. antiquus*) [62].

A complete mandible and other isolated remains were accidentally found at Ortona during construction works [58]. The fossils were ascribed to *H. antiquus* by [34].

An interesting record comes from the lacustrine level of Maglianella, a locality in the area of Rome described by [89]. The deposit was located approximately along the Maglianella Road at the level of the intersection with the Aurelia Statal Road (ca. 11 km of the Aurelia Statal Road). One articulated skeleton with its skull and other remains, including two crania, were found at Maglianella. The first skull was considered by [25] to be the holotype of *Hippopotamus tiberinus*, and the other two crania were subsequently described to expand the known variability of this form [26]. As aforementioned, *H. tiberinus* is now considered a synonym of *H. antiquus* (see [36] for discussion). Nonetheless, the Maglianella record offers an important insight into the morphological variability of Middle Pleistocene hippopotamuses, and more generally, the site yielded one of the latest occurrences of *H. antiquus* in the Italian Peninsula.

Another important skull attributed to *H. amphibius* was found at Tor di Quinto (urban area of Rome), but for a long time, remained of uncertain geographical provenance and stratigraphical context and hence was not properly considered in reconstructing the evolution of Quaternary hippopotamuses. Fabiani & Maxia [90] first described the hippopotamus skull and indicated it was recovered from Cava Montanari ("Cava" means quarry in Italian). Subsequently, other authors doubted the provenance of the skull and suggested it could have been collected from another quarry (always opening in the Tor di Quinto area), Cava Nera Molinario, ascribed to the Middle Pleistocene (MIS 13, [24,37,41]). The skull from Tor di Quinto, nowadays part of the historical collection of the MUST, was recently restudied after the end of the restoration work carried out in 2021 [22]. The results of this work confirmed its taxonomic attribution as *H. amphibius* and supported the hypothesis published by [90], where Cava Montanari was indicated as the toponym of its provenance. This, integrated with sedimentological analysis and the revision of historical and updated geological and topographic maps, allowed the specimen to be chronologically attributed to the Middle Pleistocene, precisely between 560 ka and 460 ka [22]. According to these recent results, the Cava Montanari skull represents the earliest attested occurrence of *H. amphibius* in Europe.

Hippopotamus remains were also collected from the Cretone basin, where several deposits were identified [68]. Fossils were found at Fosso delle Bufala and Osteria Moricone (one specimen at each site) and attributed to *Hippopotamus* cf. *antiquus* [68]. The authors identified a first phalange but did not specify which site the specimen came from (Fosso della Bufala or Osteria Moricone). A second phalange was also recovered from this area, but no data were reported on this specimen.

Aureli and coauthors [67] reported the presence of *Hippopotamus* sp. from La Ficoncella, but without information about the findings.

Along the Via Aurelia, in the area of Rome, several sites dated to the mid-late Middle Pleistocene have been discovered, and several of these also yielded hippopotamus remains (Figure 1; Table 1). The first is Castel di Guido, the most important site in this area for both the quality of the fossil record and the human presence documented based on an isolated fossil and a considerable number of artifacts [91]. The mammal sample was described by [73], who attributed a fragment of an incisor and a distal epiphysis of the fourth metacarpal to *Hippopotamus* sp.

The second locality is Malagrotta, where vertebrate remains were described by [80]. Similarly to other Middle Pleistocene deposits, the two remains, a fragment of a vertebra and a fragment of a metacarpal, were ascribed to *Hippopotamus* sp.

The third locality is Torre del Pagliacetto (or Torre in Pietra), where hippopotamus material was collected from the upper level. Four dental fragments were attributed to *H. amphibius* [87]. According to [26], these specimens possess no diagnostic features for specific attribution.

Fossil remains from other sites located along the Via Aurelia are described in our work (see Section 4).

Hippopotamus material was found at Fontana Ranuccio, one of the most important localities for human evolution in Quaternary Europe (e.g., [79,92]). Initially, a single canine fragment was ascribed to *Hippopotamus* sp. by [93]. Several fieldwork activities were carried out later in the work of [93], and a rich vertebrate sample was recovered (a total of 1360 mammal remains [78]). According to [78], 11 isolated specimens were attributed to *H. amphibius*.

The latter site of the urban area of Rome is Casal de' Pazzi, excavated in the 1980s, and nowadays, part of the deposit is still preserved in a museum aptly named Casal de' Pazzi Museum (see [94] for discussion). The hippopotamus sample, consisting of a fragment of an upper canine, a fragment of a lower canine, a lower deciduous tooth, and two lower third molars, was ascribed to *H. amphibius* [86].

Three other Middle Pleistocene records are excluded from this work since the fossils were not fully described, and/or their chronostratigraphic setting needs to be revised.

The first is Pignataro Interamna, initially described by [95]. The findings included a nearly complete mandible of a hippopotamus, in addition to other postcranial remains. The sample, now stored at the American Museum of Natural History of New York, was attributed to *H. amphibius* (*H. antiquus* for [26]). Additional dental and postcranial remains, also ascribed to common hippopotamuses, were collected during the 1930s and are currently stored at the Paleontological Museum of the Naples University Federico II [96]. These specimens need to be re-examined, as does their chronostratigraphic assessment.

The second is the Mercure basin, where hippopotamus remains were collected in different localities [97]. For instance, several authors [96,98,99] reported the presence of hippopotamus teeth from the lignite bed of Fornaci, attributed to *H. amphibius*. At Calorie, isolated teeth were ascribed to *H. antiquus* [97]. These fossils are dispersed among different repositories, including the Civic Museum of Natural History of Milan, the Paleontological Museum of the Naples University Federico II, and the Natural Museum of Pollino. Fossils collected in the Mercure basin were generally dated to the early Middle Pleistocene (0.7–0.5 Ma; [100] and references therein). Nevertheless, a review of the fossil materials and a better definition of their chronostratigraphic contexts is needed.

The last important sample excluded from this work is the skeleton recovered from Sant'Oreste (MPUR/V 1950). The material of Sant'Oreste is preserved at the Earth Science University Museum (MUST), and the skeleton MPUR/V 1950 is displayed in the vertebrate hall (now closed for restoration work). According to [25], the skeleton represented the paratype of *H. tiberinus*. By contrast, Petronio [36] stated that the skull is slightly deformed due to taphonomic processes, and several parts were reconstructed during restoration work carried out during the second part of the XX century. In 2021, a new restoration project started on mammal skeletons displayed at the MUST, including MPUR/V 1950 from Sant'Oreste. This work is still in progress, and therefore, the material is excluded from our review.

3. Materials and Methods

The fossil remains described in this work were recovered from a number of Italian localities and are preserved in different museums and/or institutions (Figure 1, Table 1): Museo Universitario di Scienze della Terra, Dipartimento di Scienze della Terra, Sapienza Università di Roma (MUST, including the former Museo di Paleontologia di Roma, MPUR): Batteria Nomentana, Campo di Merlo, Cava Alibrandi, Cava Arnolfi, Collina Barbattini, Monte Antenne, Ponte Molle, Rignano Flaminio, Sedia del Diavolo, Tor di Quinto Vallinfreda, Vigne Torte and Villa Chigi; Laboratorio PaleoFactory, Sapienza Università di Roma (PF): Bussi, Cava di Breccia di Casal Selce, Grotta Romanelli; Museo Archeologia Nazionale "Mario Torelli" (MANV; Venosa): Notarchirico, Terranera and Trentangioli; Museo di Paleontologia e Preistoria "Piero Leonardi"(MPPL): Isernia La Pineta; Istituto Italiano di Paleontologia Umana (IsIPU, Anagni): Cava Nera Molinario, Collina Barbattini, Grotta Romanelli; Museo delle Civiltà (MUCIV, Roma): Grotta Romanelli; Museo di Storia Naturale di Verona: Serbaro di Romagnano; and Collegio San Giuseppe—Istituto De Merode: Ponte Molle. Other comparative materials were also considered: Cava Redicicoli (MUST, IsIPU), Monte delle Piche (MUST), and Saticula (PF). Extant material of *H. amphibius* was also studied: one complete mounted skeleton and one complete skull displayed at the Museo di Anatomia comparata "Battista Grassi", Sapienza Università di Roma, and the skull from PF. In addition to the material directly examined, we considered literature data on Italian Quaternary hippopotamuses and extant *H. amphibius* (see Table 1). We also included the material of *H. amphibius* from Barrington (Britain) as representative of this species for size comparison.

The taxonomic attribution of the hippo specimens was carried out by morphological and biometric comparisons of the studied materials with corresponding specimens of other fossils and extant hippopotamuses [24,26,33].

Following [26], six dental variables were considered: length (L) and breadth (B) for upper and lower incisors and canines, and outer length (OL), inner length (IL), anterior breadth (AB), and posterior breadth (PB) for the upper and lower premolars and molars.

The measurements reported by [33] on extant hippopotamuses were excluded since they were taken differently from [26] and later studies.

Size variations in fossil and extant hippopotamuses were investigated by comparing the maximum length of the upper first (M^1), second (M^2), and third molars (M^3), and the lower first (M_1), second (M_2), and third molars (M_3). We grouped the Italian fossils of *H. antiquus* from the Early Pleistocene in the Villafranchian and Epivillafranchian samples (following the Mammal Ages identified in the Italian biochronological scheme). We considered the sample from the Britain locality of Barrington (the only fossil group recovered from a site not located in Italy) as representative of fossil *H. amphibius*. Finally, extant specimens of *H. amphibius* were also included.

A standard bivariate plot of the M_2 length vs. M_3 length of fossil and extant hippopotamuses was also generated in order to investigate possible size differences among groups.

4. Newly Studied Hippopotamus Material in Italy

4.1. Vallinfreda

The material was accidentally discovered in yellowish sands near Vallinfreda village by a farmer during the XIX century [59]. The hippopotamus sample included several specimens, among which was a cranium that was probably broken at the time of discovery [101]. The authors of [24] only studied the complete mandible, attributing it to *Hippopotamus* cf. *antiquus*. The authors of [59] carried out a systematic review of the fossils found in the Oricola–Carsoli intramontane basin and their chronostratigraphic setting. The authors attributed the fossils to the early Middle Pleistocene (Ponte Galeria FU). The study of all remains from Vallinfrenda confirms that in the MUST collection are preserved specimens belonging to the same damaged cranium mentioned by [101], the right and left maxillaries, and a fragment of the left premaxillary and maxillary (including I^2 and C^X; Figures 2 and 3). It cannot be excluded that the mandible was associated with the cranium, since the teeth of these four specimens possess a similar degree of wear. The sample shows features that are generally observed in *H. antiquus*: a concave profile of the mandibular corpus in lateral view and parallel grooves on the external surfaces of the canines [22,24,42].

4.2. Venosa Basin

The Venosa Basin is characterized by the preservation of long archeological sequences in volcano–sedimentary complexes linked to the eruptive activity of the Vulture stratovolcano, mainly dated to the Middle Pleistocene [102]. Notarchirico, discovered in 1979, is one of the most important and best-known sites of the area, yielding a 7 m thick sequence of fluvial sediments including eleven archeological and paleontological levels [103]. Recent $^{40}Ar/^{39}Ar$ ages and ESR dates have revised the chronology of the sedimentary sequence excavated by Marcello Piperno [103] and constrained archeological levels α to F to between ca. 610 and 675 ka (MIS 16, [104]). According to the new chronostratigraphic setting, the hominin femur fragment found in level α is the oldest human finding in the Italian Peninsula. New excavations coordinated by one of the authors (MHM), authorized by the Soprintendenza Archeologia Belle Arti e Paesaggio of the Basilicata Region, began in 2016, with the aim to better investigate the bottom of the sequence [15,70,105]. Five sedimentary units have been identified, including five archeo- and paleosurfaces (G, H, I1, I2, and J). These levels were dated by $^{40}Ar/^{39}Ar$ to between 690.3 $\pm$ 5.8 ka and 695.2 $\pm$ 6.2 ka, which corresponds to the end of MIS 17. As such, Notarchirico is one of the key sites for better understanding the age of the earliest Acheulean occupations and the impact of climatic changes on mammal paleocommunities between 700 and 600 ka. During 2017-2018 fieldwork activities, three hippopotamus remains were found in levels G and I1, representing an addition to the species previously known at Notarchirico. Although the specimens recovered are of limited taxonomic value, they testified to the presence of hippopotamuses in southern Italy during the end of MIS 17.

Terranera is another locality in the Venosa Basin, though it is lesser known than Notarchirico and Loreto. The toponym Terranera probably indicates an area where several deposits (e.g., Cicoria, Mandorleto, Siniscalchi) were identified, but also systematically excavated ([106] and references therein). Two fieldwork activities were carried out during the first decades of the XX century, coordinated by Quintino Quagliati in 1909 and Ugo Rellini in 1914 ([106] and references therein). The archeological and paleontological materials of Terranera have been dispersed among several Italian institutions since the beginning of the research in the area (National Archeological Museum of Taranto, Museum of Civilizations [Rome], National Archeological Museum of Venosa, Italian Institute of Human Paleontology [Anagni], Museum of Paleontology, Naples University Federico II, Museum of origins, Sapienza University of Rome).

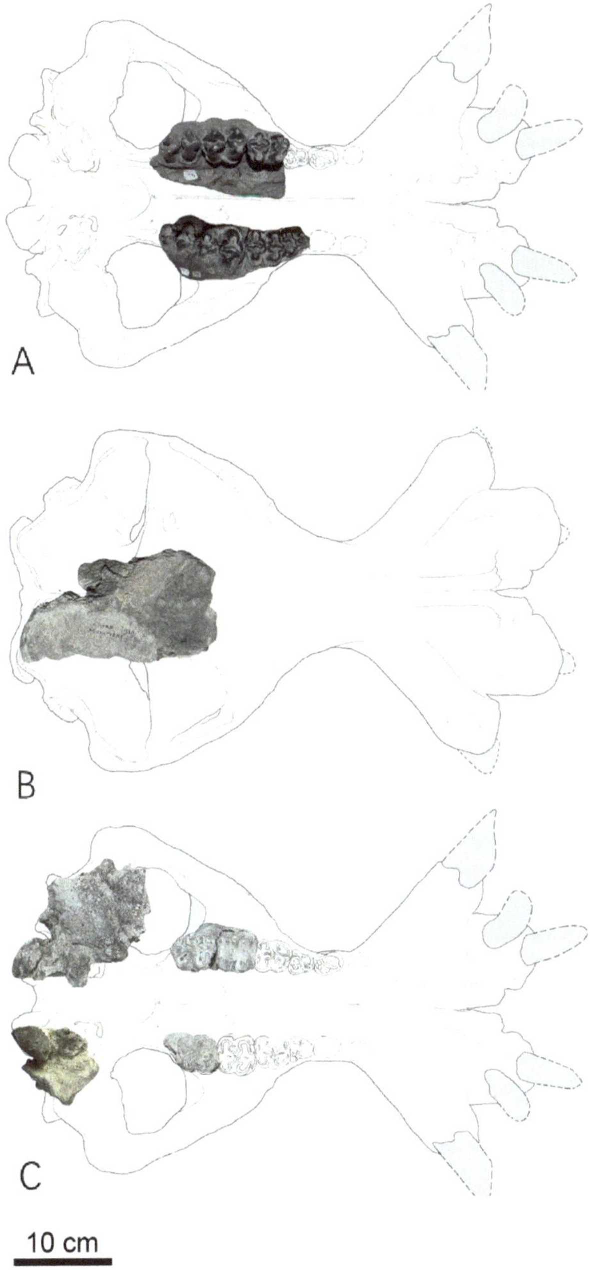

Figure 2. Cranial remains of *H. antiquus* from Vallinfreda (52a and 52; (**A**)) and *H. amphibius* from Cava Nera Molinario (IsIPU-CNM10, IsIPU-CNM11, IsIPU-CNM14, IsIPU-CNM15, IsIPU-CNM43; **B**,**C**). Fossil in dorsal (**B**) and occlusal (**A**,**C**) views.

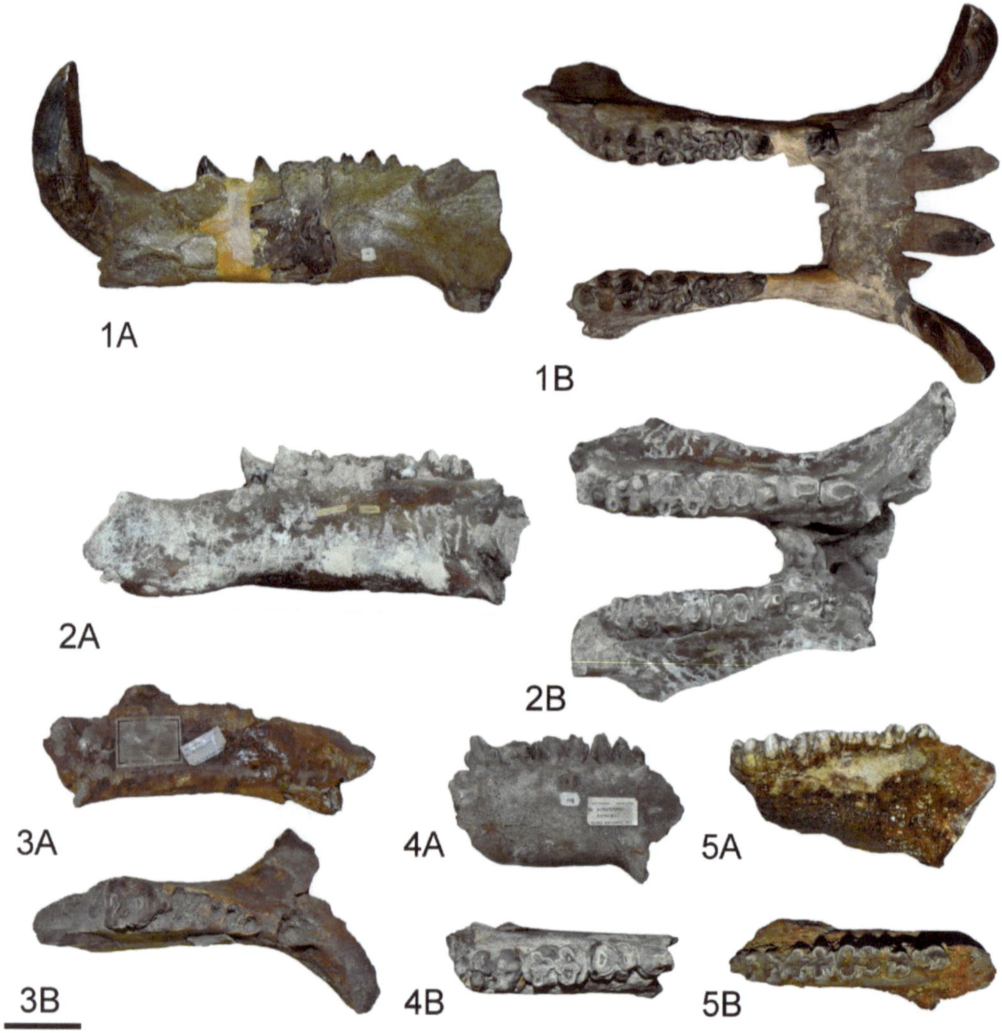

Figure 3. Hemimandibles remains from the Italian Peninsula: *H. antiquus* from Vallinfreda (SN180FS; (**1A**,**B**)), Terranera (SN 1936; (**2A**,**B**)), and Campo di Merlo (24; (**3A**,**B**)); *H. amphibius* from Monte Antenne (128; (**4A**,**B**)) and Tor di Quinto (SN16; (**5A**,**B**)). Fossils in labial (**A**) and occlusal (**B**) views. Scale bar 10 cm.

Following the modern geological mapping of the Venosa Basin, three units are recognized in the area of Terranera: Monte San Marco Formation (Early Pleistocene), Foggianello Synthem (early Middle Pleistocene, ca. between 740 and 687 ka), and Barile Synthem (early Middle Pleistocene, ca. between 670 and 610 ka) [107]. The first formation, however, is related to marine deposition [107]. According to [108], fossils and artifacts were collected from a sand bed with the presence of volcanic material (number 3 in the deposit, ca. 1.2–1.5 m thick) in a succession characterized by the alternation of fluvio-lacustrine and volcanic levels [108], which, more generally, suggests continental deposition.

The fauna from Terranera was never described, except for a fragment of an antler attributed to *Megaceros* (=*Praemegaceros*) *solilhacus* by [66]. This species is also indicative of an early Middle Pleistocene age, being its earliest occurrence in Isernia La Pineta (ca. 600 ka; [72]).

Among the fossils collected between 1928 and 1940 belonging to the Bríscese collection and stored at Museo Archeologico Nazionale "Mario Torelli", a nearly complete mandible and other remains of hippopotamuses are preserved (Figure 3). The mandible is long and slender and displays a concave basal profile of the mandibular corpus in lateral view. The isolated lower canines (not preserved in the mandible) have parallel grooves on the external surface. These features align with the morphology of *H. antiquus* [22,24,26].

Trentangioli (or Trentangeli) is another site located in the Venosa Basin, where mammal remains were collected during geo-paleontological surveys between 1928 and 1940 ([84] and references therein). The fossil material is currently stored at the National Archeological Museum "Mario Torelli" of Venosa, part of the Briscese collection. In the other deposits of the Venosa Basin, fossils were found in fluvio-lacustrine levels with abundant volcanic material and, therefore, attributed to the Middle Pleistocene. The sample consists of several dental remains, among which are two fragmentary lower canines. These display convergent grooves on the external surfaces, a feature shared with *H. amphibius* [22,24,26].

4.3. Isernia La Pineta

The open-air archeological and paleontological site of Isernia La Pineta is considered a reference for studies of human evolution in Europe ([109] and references therein). The site was discovered during work for the Napoli–Vasto highway in 1978, in the proximity of the Isernia–Roma railway line. Decades of systematic excavations have been carried out, coordinated by one of the authors (CP) with the permission of Soprintendenza Archeologia Belle Arti e Paesaggio of the Molise Region, the "Direzione Regionale Musei Molise", and the support of Ferrara University. The site yielded hundreds of fossil remains and artifacts, among which was a deciduous tooth attributed to *Homo* cf. *heidelbergensis* ([109] and references therein).

The age of archeological and paleontological levels of Isernia La Pineta was initially considered close to 700 ka and later dated at ca. 610 ka with the $^{40}Ar/^{39}Ar$ method [65]. In 2015, the age was revised again, attributing Unit 4 (a tephra level) to 586 ± 1 ka [109]. The authors also dated reworked volcanic material from levels 3coll, 3s10, and 3s6 (just above the main archeosurface, 3a), yielding an age of 586 ± 2 ka, 583 ± 3 ka, and 587 ± 2 ka, respectively. The upper part of Unit 1 (U1) was dated with the $^{40}Ar/^{39}Ar$ method at 499 ± 13 ka [65].

The mammal assemblage is considered characteristic of the Galerian Mammal Age [73,110,111], giving the name to a Faunal Unit of the Italian biochronological scheme (see [1]). Hippopotamus remains were ascribed to *Hippopotamus* cf. *antiquus* [26,73]. The sample includes several dental remains, among which is a lower canine that displays parallel grooves on the external surface, as typically observed in *H. antiquus* [22,24,42].

4.4. Bussi

The paleontological deposit of Bussi was never fully described but only mentioned in several geological works [57]. The locality is part of the Tirino valley, an intermontane basin of the Abruzzi Apennines [57]. *Hippopotamus antiquus* remains, associated with other fossils of *Elephas (=Palaeoloxodon) antiquus*, were found in a whitish silt level resulting from lacustrine deposition, identified as the Convento dei Cappuccini Unit by [57]. The mentioned unit was chronologically attributed to the early part of the Middle Pleistocene [57]. The hippopotamus remains from Bussi consist of several isolated dental remains and a fragment of a hemimandible. These specimens probably belong to the same skull and were likely disarticulated during the excavation. The lower canine, the only element useful for taxonomical purposes, displays parallel grooves on the external surface, a trait commonly observed in *H. antiquus* [22,24,42].

4.5. Rignano Flaminio

A fluvial–lacustrine deposit exposed at Rignano Flaminio (north of Rome) was recently described by [71]. The base of the succession was represented by the Tufo Rosso a Scorie Nere, a deposit dated at approximately 449 ± 2 ka [112]. The fossil-bearing level was attributed to MIS 11 (between 430 and 405 ka; [71]). The analysis of a radius/ulna attributed to *Hippopotamus* cf. *amphibius* revealed that this bone was intentionally fractured and retouched by humans [71]. The sample from Rignano Flaminio is preserved at the MUST. After examining the material, we suggest an open attribution to *Hippopotamus* sp. due to the lack of morphological features diagnostic at the species level.

4.6. Serbaro di Romagnano

This locality is a karst-infilling deposit near the village of Romagnano. The vertebrate assemblages found in the cavities of the area were biochronologically dated based on the micromammal content [113]. However, no microvertebrate remains were identified at Serbaro di Romagnano [113]. According to [82,83], the mammal assemblage can be attributed to the Fontana Ranuccio FU (sites whose faunas are associated with this FU date to ca. 500-400 ka). Pasa [81] attributed two dental remains to *H. amphibius*: an upper third molar and a lower third molar. The latter tooth is in an advanced stage of wear, group XVIII, according to the scheme of [14], which should correspond to an age of 38 years. The measurements of this tooth are excluded from the biometric comparison. The limited taxonomical value of these remains prevents a specific attribution.

4.7. Grotta Romanelli

Grotta Romanelli was immediately considered a key site for paleontological and archeological research on the Mediterranean Quaternary [114–117]. Its importance in the European context originates from the coupling of the extraordinary number of findings and the robust chronostratigraphic setting described by [118,119]. The author interpreted the basal conglomerate level as a Thyrrenian beach deposit (early Late Pleistocene), constraining the whole succession to approximately the last 128 ka. This schematic model was corroborated by radiometric datings performed at the end of the 1960s [120–122].

After more than 40 years of inactivity in the cave, a new project was started in Grotta Romanelli in 2015, coordinated by the Sapienza University of Rome under the direction of one of us (RS) and with the authorization of the Soprintendenza ai Beni Archeologici della Puglia (now Soprintendenza Archeologia Belle Arti e Paesaggio di Brindisi e Lecce). One of the main results of the new project was the review of the chronostratigraphic setting of the cave [85]. New geomorphological and geological data, in addition to new radiometric datings, revealed that the basal levels of the cave have to be ascribed to the late Middle Pleistocene (levels K and I *sensu* [118,119]; ISU1 and ISU2 *sensu* [85]).

Fossils from the lower levels stored at IsIPU were studied by [26] and attributed to *H. antiquus*. Other remains of hippopotamuses from the lower levels of Grotta Romanelli are preserved at MUCIV. The entire sample consists of isolate.ed postcranial bones, generally of juvenile individuals (long bones show disarticulated epiphyses), and two deciduous teeth. These specimens are of limited taxonomic value, limiting the attribution to the genus *Hippopotamus*.

4.8. Rome Area

The territory of Rome, including the urban area of the city and its surroundings, is generally known in the literature as Campagna Romana (Roman Campaign). This area has yielded an impressive archeological and paleontological record, collected in dozens of sites. The historical deposits of Rome were generally destroyed by the urban growth of the XX century or otherwise buried by roads, places, houses, train stations, etc., and are located in what today is the urban area of the city [123,124]. Since the Quaternary, the area of Rome was shaped by the action of the Tiber and Aniene Rivers, combined with the volcanic activity of the Alban Hills and the Sabatini Mounts. This could be schematized as follows: the rivers hollowed out the fluvial valleys and coastal plain incisions during a glacial stage, filled with deposits, which often included pyroclastic products, during the following phases (e.g., [124–126]). The presence of volcanic materials (pumice, tephra, as primary deposition or reworked) in alluvial–deltaic sediments allows a reliable age to be obtained for many of these localities (obtained by radiometric dating or through correlations with previously dated deposits).

The Rome basin, however, can be divided into several areas where fossiliferous deposits show, more or less, a similar age.

4.9. Cava Alibrandi, Cava Arnolfi, Campo di Merlo, and Cava di Breccia di Casal Selce

These four fossiliferous sites are located in the Ponte Galeria area, in the surroundings of Rome (at ca. 15 km from the center of the city). In this area, several quarries have been opened for the extraction of gravels used for construction works since the 1960s, exposing long geological successions studied by many authors ([55,127] and references therein). Fossils in this area, collected from the gravel and sand levels from many quarries, are commonly dated to the early Middle Pleistocene. The richness and completeness of the fossils collected in this area led paleontologists to identify Ponte Galeria as the first Faunal Unit of the Middle Pleistocene (ca. 700 ka).

Bonadonna [128] reported the presence of mammal remains at Cava Arnolfi (11 Km of Portuense Road), identifying a lower canine as *H. amphibius* (later considered *H. antiquus* by [36,129]). The lower canine displays parallel grooves on the external surfaces of the canines, features commonly detected in *H. antiquus* [22,24,42].

Capasso Barbato and Petronio [61] described remains collected at Cava Alibrandi (13.5 Km of Aurelia Statal Road). A rib was ascribed to *Hippopotamus* sp. by the authors, an attribution confirmed in this work.

Petronio [129] described a mandibular fragment found at Campo di Merlo (a quarry located in the proximity of the crossroad between the Magliana Road and Portuense Road; Figure 3). The fragment was attributed to *H. antiquus* based on a concave profile of the mandibular corpus in lateral view [129]. However, the material from Campo di Merlo, currently stored at the MUST, was probably recovered during the XX century and acquired by the museum during the early 1900s. In addition to the mandibular fragment, at Campa di Merlo, a lower canine was found, which displays parallel grooves on the external surfaces of the canines, a morphology shared with *H. antiquus* [22,24,42]. Following this, the attribution to *H. antiquus* proposed by [129] can be confirmed.

Additional mammal remains from Ponte Galeria were listed in [130], but the author mentioned a number of localities without providing information about the exact toponym of the provenance of the hippopotamus material. As aforementioned, the ages of these fossiliferous deposits are quite close, all attributed to the Ponte Galeria FU. Three dental remains, two fragments of a canine and a deciduous tooth, and other postcranial bones preserved at the MUST show only a general indication of Ponte Galeria. The canine, in particular, shows parallel grooves on the external surfaces of the canines, a trait generally observed in *H. antiquus* [22,24,42].

Cava di Breccia of Casal Selce is located along the Castel di Guido Road, near the Aurelia Statal Road in the proximity of the highway exit of Castel di Guido. The deposit preserves a long sedimentary succession, with two different fossiliferous levels (e.g., [2,55,72]). Abundant mammal remains were collected from the salmon sand level, which was correlated to the Santa Cecilia Formation [131], while [55] dated the pumice falls intercalating with this deposit at 611 $\pm$ 6 ka. The hippopotamus sample includes several fragments of dental remains and a fragment of a femur attributed to *H. antiquus* in previous work (e.g., [2,55]). For taxonomical purposes, a fragmentary lower canine displays fairly parallel enamel ridges and grooves along the external surface. This feature is generally observed in *H. antiquus* [22,24,42].

4.10. Cava Nera Molinario, Monte Antenne, Ponte Molle, and Tor di Quinto

The area of Tor di Quinto is located on the right bank of the Tiber River. Several quarries were opened for the extraction of gravels for building during the XIX and early XX centuries ([70] and references therein). The names of most of these deposits commonly correspond to those of the quarries (Cava means quarry in Italian), for example, Cava Nera Molinario or Cava Montanari. In the Tor di Quinto area, the fossil remains were collected by gravel and sand deposits quarried for the extraction of building materials [40,125]. The most important toponyms are Cava Montanari (described in Section 2), Cava Nera Molinario, Monte Antenne, Ponte Molle, and Tor di Quinto; all of these yielded hippopotamus remains.

The sedimentary succession of Cava Nera Molinario was carefully described by [132]. This is one of the few still-existing deposits in the urban area of Rome described during the XX century, which was revised by [133]. The authors recognized the presence of three volcanic deposits: Tufo del Palatino (533 $\pm$ 2 ka), Tufo Giallo di Prima Porta (516 $\pm$ 1 ka), and Tufo Rosso a Scorie Nere (452 $\pm$ 2–447 $\pm$ 7 ka). Fossils have been attributed to the Valle Giulia Formation (MIS 13, [55,133]). Material from Cava Nera Molinario was attributed to *Hippopotamus* sp. [55], but without a formal description. Five specimens, the left and right fragmentary maxillaries, the right and left fragmentary occipital–parietal portions, and a fragment of parietal–frontal bone, could belong to the same cranium (Figure 2). The studied sample, however, is of poor taxonomic value and, therefore, is ascribed herein to *Hippopotamus* sp.

In general, the Tor di Quinto area was first described by [134], who explored several quarries opened at that time in this district but also observed several successions exposed during the construction of some main roads, for example, the Cassia and Flaminia roads. One of the sites described by [134] was Tor di Quinto (or Torretta di Quinto). The geological setting of this area was recently revised by [70], where a number of boreholes were integrated with the data published during the XX century and collected in still-existing deposits in this district. The results demonstrated that the majority of the fluvial deposits with gravels and sands and rich in volcanic material can be ascribed to the Valle Giulia Formation [70]. A similar age was also proposed by [56], who attributed the fossils from Tor di Quinto to the Valle Giulia Formation.

The hippopotamus material from Tor di Quinto includes two fragments of a hemimandible, isolated teeth, a fragmentary tibia, and a vertebra (Figure 3). Unfortunately, one of the two hemimandibles preserved only the portion of the first and second molars, while the second consists of the portion between the third premolar and the third molar, but the mandibular corpus is damaged, and the basal profile cannot be observed. Two lower canines show convergent grooves on the external surface, a trait commonly observed in *H. amphibius* [22,24,42].

The stratigraphical succession of Monte Antenne was described by [69], who reported the presence of sands at the base of the deposit with several volcanic levels above.

The remains of hippopotamuses were collected from the sandy levels, where a complete skull was identified during the first geo-paleontological survey. The author stated that, after its discovery, the skull was left in its original position in the outcrop, to be recovered the next day. Unfortunately, the skull was significantly damaged, and only a fragment of the left hemimandible and several isolated teeth were found [69]. These were ascribed to *H. amphibius* [69]. The sedimentary succession of Monte Antenne was attributed to the Valle Giulia Formation by [55]. According to [69], the hippopotamus material from Monte Antenne belonged to a single individual. The hemimandible consists of a fragment of the portion between the fourth premolar and the third molar. Considering its fragmentary status, the morphology of the lateral profile (concave or convex) cannot be recognized. The two lower canines show convergent grooves on the external surface. This feature would confirm the attribution proposed by [69] to *H. amphibius*.

Ponte Molle is another historical locality of Tor di Quinto, well known in the literature thanks to the richness and completeness of the fossil specimens [70]. The chronostratigraphic setting of the site was recently redefined, and the whole deposit was attributed to the Middle Pleistocene [70]. In particular, the fossiliferous level, gravel with a sandy matrix and volcanic materials, was constrained to between 540 ka and 460 ka [70]. The faunal assemblage identified at the site suggested an interglacial period (MIS 13). The authors of [70] discussed the problematic taxonomical attribution of Quaternary hippopotamuses. The authors ascribed the fossils from Ponte Molle to *H.* ex gr. *antiquus*, exclusively based on the medium size of the remains. This attribution can be revised following the work of [22].

The sample from Ponte Molle includes two fragments of mandibles, several isolated teeth, and postcranial bones. The morphological analysis revealed that the corpora of the two hemimandibles were damaged, and the lateral profile could not be observed. Three lower canines from Ponte Molle show convergent grooves on the external surface, whereas only one possesses a parallel arrangement. Ponte Molle yielded one of the most impressive vertebrate records of Quaternary mammals, with thousands of remains dispersed among several Italian institutions since the early stages of research. Fossils were mainly found at the end of 1800 and in the early decades of 1900 in several quarries opened in the Tor di Quinto area or in long, extensive sedimentary succession exposed by construction works on the two main roads of this district (Flaminia and Cassia roads; [70]). As discussed by [70], the putative presence of fossils recovered from older deposits outcropping in the area of Tor di Quinto cannot be excluded. Another possibility is that some fossils could have been removed from older deposits due to the activity of the Tiber River and been deposited in the fluvial levels of Ponte Molle, resulting in reworked elements.

The overall features of the fossils from Ponte Molle fall within the variability of extant hippopotamuses, but considering the variation in lower canine morphologies, the sample is attributed to *H.* cf. *amphibius* [22,24,42].

4.11. Collina Barbattini—Via Aurelia Km 18.0–19.3

A number of Middle Pleistocene localities were discovered during the XX century along the Aurelia Statal Road. Several of these take their name from the kilometer of the Statal Road in the proximity of the deposits.

Two main areas can be identified: the first is the hill located between 18.0 km and 19.3 km of the Aurelia Statal Road toward the city of Rome, called Collina Barbattini; the second is at about 18.9 Km of the Aurelia Statal Road toward the city of Civitavecchia [76,77].

Paleontological findings in this area were initially collected by Ernesto Longo and later donated to the MUST. This sample, in addition to other mammal fossils found in the same area, was described by [77]. Additional remains were found between 18 and 20 km during work on the motorway junction of Castel Guido [76]. These sites are geographically really close to each other, and fossils were recovered from a fluvial–lacustrine level correlated to the San Paolo Formation (MIS 11) by [75].

At Collina Barbattini, [77] reported the presence of four dental remains of juvenile individuals (Longo collection) attributed to *Hippopotamus* sp. and three other fossils, including a fragment of a canine, a tibia, and a second metacarpal, ascribed to *Hippopotamus* cfr. *antiquus*. For the tibia and metacarpal, we previously mentioned that the putative postcranial characters proposed by [26] for specific attribution need to be confirmed. At the moment, a taxonomic attribution based on size repartition is avoided. The canine has only its apical portion, and it would seem to belong to a juvenile individual (as for the other teeth of the Longo collection). Even these three fossils are ascribed to *Hippopotamus* sp. here.

The fossils listed by [76] from Collina Barbattini were also attributed to *Hippopotamus* sp. The sample consists of a IV metatarsal, a fragment of a humerus, and four fragmentary dental remains. These fossils are of limited taxonomic value, and therefore, the attribution proposed by [76] is confirmed here.

4.12. Batteria Nomentana, Sedia del Diavolo, Vigne Torte, and Villa Chigi

These deposits are located in the proximity of the Aniene River, just outside the old Roman wall that encloses the center of Rome. A number of quarries were opened for the extraction of building materials and commonly show the "Tufo litoide" as the basal level (e.g., [84,135,136]). As aforementioned, the volcanic deposit historically called "Tufo litoide" is today known as Tufo Lionato, a pyroclastic-flow deposit erupted from the Colli Albani volcanic district dated to approximately 367 ka, representing an important chronostratigraphic marker in the area of Rome [137].

Sedia del Diavolo is undoubtedly one of the most famous sites in this part of Rome for its paleontological content and the great number of papers that described the site and its stratigraphic succession. Important Italian scientists wrote about Sedia del Diavolo, such as Alberto Carlo Blanc, Alessandro Portis, Enrico Clerici, Giuseppe Ponzi, Gugielmo Terrigi, and Romollo Meli ([138] and references therein). The deposit can be divided into three main units, listed from bottom to top, as follows: the Tufo Lionato deposit, the fluvial gravel levels, and the pumiceous ash flow deposit. The third-listed deposit was dated at approximately 284 ka by [139]. Caloi and coauthors [138] attributed the fragmentary dental remains to *Hippopotamus* cf. *amphibius*, since the lower canines show convergent grooves on the external surface. These fossils are stored at the MUST, and our review confirms this morphology, generally observed in extant *H. amphibius* [22,24,42]. However, a nearly complete mandible is currently displayed in the exhibition called "Animals, Plants, Rocks and Minerals: The ISPRA Collections" in the Museum of Civilization. Considering its important taxonomic value, the study of this fossil can help in the classification of hippopotamuses from Sedia del Diavolo.

The stratigraphic succession of Batteria Nomentana, located near the crossroad between the homonymous road (Batteria Nomentana) and Nomentana road, was discovered by [84]. When Meli described the deposit, he highlighted a strong resemblance to that exposed in the well-known site of Sedia del Diavolo (only 500 m from Batteria Nomentana). The fossiliferous levels of Batteria Nomentana were constrained from 367 to 287 ka by [56]. Three fragmentary lower canines and a second lower incisor were collected at the site. The canines possess convergent grooves on the external surface, a typical trait of extant *H. amphibius* [22,24,42].

Fossils from Vigne Torte were initially mentioned by [136] when the author described the bovine cranial material preserved in the "Museo geologico universitario di Roma" (most of the collections today are part of the MUST). The name Vigne Torte indicated a number of quarries opened along the Salaria Road for the extraction of gravels from 1894 to 1896 [136]. As for the other sites in this district, the lower deposit of these quarries was the Tufo Lionato [136]. The deposition of the fossils collected from the fluvial gravel levels of Vigne Torte was constrained to approximately between 367 and 287 ka [56]. Remains of hippopotamuses were attributed to *Hippopotamus* sp. by [72] but were never described. The sample consists of several isolated dental remains and one astragalus. These specimens are of poor taxonomic value and are generically attributed to *Hippopotamus*.

The last deposit in this area is Villa Chigi, which was never described in the literature. The authors of [72] only listed the mammal remains found at Villa Chigi and housed in the MUST. By analyzing the original specimen labels, it is possible to note that the fossils were collected from the gravel levels of the Aniene River from a quarry opened near the Villa Chigi ("Nelle ghiaie dell'Aniene (quaternarie) alla cava di breccia presso la Villa Chigi fuori porta Salaria"). The deposit is not far from the Vigne Torte quarries always opening for the extraction of gravels for building materials. Also, considering the geological context of this area, where a number of fossils were found in fluvial–lacustrine deposits attributed to the Aurelia Formation, it could be assumed that the age of the fluvial gravel deposit of Villa Chigi is similar to that of other localities known in this geographic area (Monte delle Gioie, Prati Fiscali, and others described here). One fragment of a maxillary and three isolated teeth were found at Villa Chigi. Considering the limited taxonomic value of these remains, the sample is attributed to *Hippopotamus* sp.

5. Discussion

5.1. Taxonomic Attribution

The taxonomic attribution of Pleistocene hippopotamuses is mainly based on cranial morphologies [22,24,26]. Diagnostic features in the postcranial bones were proposed by [26], but these were adopted in only a few works. When skulls are unavailable, taxonomic identification is often based on chronological grounds or limited to the genus level (Table 1). Several examples come from the Italian sites of Slivia [51], Monte delle Piche [56], Pagliare di Sassa [46], and Borgonuovo [63], whose remains are attributed to *Hippopotamus* sp.

One operational strategy for the identification of postcranial bones considered the dimensions of the remains, with large-sized specimens attributed to *Hippopotamus antiquus* [24,36]. Conversely, the large sample from Cava Redicicoli highlighted that Pleistocene hippopotamuses display huge variations and, more generally, that the size of the bones can also be affected by sex and age [22]. The oversimplified model involving the attribution of large-sized bones to *H. antiquus* and medium- and small-sized bones to *H. amphibius* cannot be confirmed. The differences presented by [26] should be tested on other Pleistocene samples so that they could virtually represent another tool for the specific identification of fossil hippopotamuses. Keeping these considerations in mind, it is understandable why this work is based on cranial and dental materials.

Mandibular remains were found in several of the studied sites, but only those from Terrenera and Vallinfreda display useful features for specific identification. The two fossils possess a concave profile of the mandibular corpus in lateral view, a feature typically observed in *H. antiquus* [22,24,26]. For the identification of other fossil samples, the only character that can be analyzed is the arrangement of the enamel ridges and grooves on the external tooth surfaces of the lower canines. Initially, this feature was considered diagnostic by [42], who analyzed the sample from Valdarno Superiore (Upper Valdarno). The authors stated that parallel enamel ridges and grooves on the external tooth surfaces of lower canines were observed in 15 out of 16 remains. Subsequently, [24] confirmed the validity of this feature for specific identification, but only if concurrently observed together with other diagnostic characters on the same specimens. The author of [26] argued that convergent enamel ridges are rather frequent in *H. amphibius* (extant and fossils), while both morphotypes are observed in *H. antiquus*. It should be considered that [26] set the dispersal of *H. amphibius* only during the early Late Pleistocene, attributing the entire Middle Pleistocene to *H. antiquus* (=*H. tiberinus*). Another aspect is that the author reported no detailed information on the distribution of these morphotypes in the *H. antiquus* record. Recent results published by [40] reinforced the view that *H. amphibius* dispersed in Europe during the Middle Pleistocene (ca. 500 ka), dismantling the hypothesis of [26]. The parallel arrangement of the enamel ridges and grooves on the external surfaces of the lower canines, in addition to the Valdarno Superiore, was identified in specimens of *H. antiquus* from the Italian sites of Cava Redicicoli [22], Cava Santarelli [45], Chiusi [41], Colle Curti [26,43], Ortona [34], and Saticula [48]. Other European remains of *H. antiquus* show this feature, as, for example, documented at Incarcal [27] and Untermassfeld [30]. A parallel arrangement is also found in several samples analyzed in this work, such as Terranera and Vallinfreda, attributed to *H. antiquus* (Table 2). A convergent arrangement of the enamel ridges and grooves on the external surfaces of the lower canines was observed in the skull of Cava Montanari [40] and in the sample from Barrington [26], attributed to *H. amphibius*. In the studied material, a convergent arrangement was observed in several samples attributed to the mid-to-late Middle Pleistocene (Table 2).

Table 2. Morphology of hippopotamus canines. Light pink—Epivillafranchian specimens of *H. antiquus;* green—Middle Pleistocene specimens of *H. antiquus;* light orange—Middle Pleistocene specimens of *H. amphibius.* Canines drawn from [26]: *H. antiquus* from Valdarno Superiore—National History Museum of Paris (3866); *H. amphibius* from Barrington—Sedgwick Museum of Cambridge (D 3980); extant *H. amphibius* from Zoo—Comparative Anatomy Laboratory of Paris (1917–249).

			Grooves on the external surface of canines		
			Parallel	Convergent	
Sites	Species	References	 Fossil *H. antiquus*	 Fossil *H. amphibius*	 Extant *H. amphibius*
Cava Santarelli	*Hippopotamus antiquus*	[45]	X		
Colle Curti	*Hippopotamus antiquus*	[26,43]	X		
Saticula	*Hippopotamus antiquus*	[48], this work	X		
Cava Redicicoli	*Hippopotamus* cf. *antiquus*	[40], this work	X		
Bussi	*Hippopotamus antiquus*	This work	1 specimen		
Ortona	*Hippopotamus antiquus*	[34,58]	X		
Vallinfreda	*Hippopotamus antiquus*	This work	1 specimen		
Cava Arnolfi	*Hippopotamus antiquus*	This work	1 specimen		
Campo di Merlo	*Hippopotamus antiquus*	This work	1 specimen		
Cava di Breccia—Casal Selce 2	*Hippopotamus antiquus*	This work	1 specimen		
Isernia La Pineta	*Hippopotamus antiquus*	This work	3 specimens		
Terranera	*Hippopotamus antiquus*	This work	5 specimens		
Cava Montanari	*Hippopotamus amphibius*	[22]		1 specimen	
Monte Antenne	*Hippopotamus amphibius*	This work		2 specimens	
Ponte Molle	*Hippopotamus* cf. *amphibius*	This work	1 specimen	3 specimens	
Tor di Quinto	*Hippopotamus amphibius*	This work		3 specimens	
Trentangioli	*Hippopotamus amphibius*	This work		2 specimens	
Batteria Nomentana	*Hippopotamus amphibius*	This work		3 specimens	
Sedia del Diavolo	*Hippopotamus amphibius*	This work		6 specimens	

Considering the lack of diagnostic features in the teeth, which represent the majority of hippopotamus remains known from the Quaternary in Europe, we reiterate the importance of the morphology of the lower canines for the specific distinction between *H. antiquus* and *H. amphibius*. This feature might be somewhat variable, as exemplified by the presence of 1 specimen (out of 16) from the Valdarno Superiore that does not show a parallel arrangement of the canines [42]. Nonetheless, there is an almost complete chronological separation between canines with parallel (prior to ca. 500 ka) and convergent (since ca. 500 ka) arrangements from the Middle Pleistocene in Italy, which coincides with the earliest appearance of *H. amphibius* in the European fossil record. The only exception to this sharp separation is a "parallel" canine from Ponte Molle, whereas the other three specimens from the site show a "convergent" morphology. This might indicate intraspecific variability, as in the case of the Valdarno Superiore; it could also indicate that both *H. antiquus* and *H. amphibius* were present at Ponte Molle, and indeed, a short chronological overlap between the two species in Europe would be supported by the last occurrence of *H. antiquus* (ca. 450 ka) in the Iberian Peninsula [140] or that the "parallel" canine is one of the few spurious older remains collected at Ponte Molle (see [70] for discussion).

In any case, our review confirms that the *H. antiquus*–*H. amphibius* transition occurred during the Middle Pleistocene (MIS 13, Figure 4).

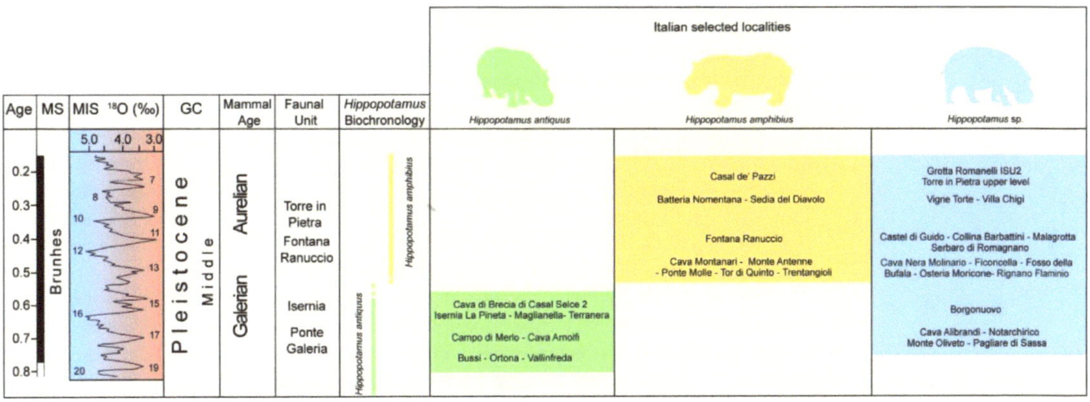

Figure 4. Biochronological scheme on quaternary time scale with selected Italian localities mentioned in the text. Abbreviations: MS—Magnetostratigraphy; GC—Geochronology; MIS—Marine Isotope Stage(s).

5.2. Size Variability and Climatic Impact on Hippopotamuses

Size variation in fossil mammals and its relationship with climate oscillations have been investigated in several species, for example, *Crocuta crocuta* [141], *Bos primigenius* [142,143], *Canis lupus* [144,145], and *Sus scrofa* [146]. In the majority of these studies, the size variation was regarded as linked to climatic changes, in agreement with the ecogeographical rule proposed by [147] that predicts larger sizes in colder climates. The opposite situation was observed in an *S. scrofa* fossil from the Late Pleistocene of Apulia, with smaller individuals occurring during glacial stages, a pattern interpreted as resulting from the decrease in available trophic resources [146]. Hippopotamuses have a peculiar ecology, which similarly suggests a complex response to fluctuations in climate and resources. Indeed, they are strongly dependent on the presence of water, to the extent that they are considered indicators of humid conditions and mild winters [29,34,48,148]. Although it is generally believed that *H. antiquus* was larger than *H. amphibius*, size variations in Pleistocene fossil hippopotamuses of Europe have been poorly investigated [24,26,149–151]. Mazza and Bertini [34] suggested that size fluctuations observed in Quaternary hippopotamuses reflect changes in resources, which, in turn, are dependent on temperature and precipitation, silt levels also allowing for the recognition of a chronological subdivision: large-sized specimens during the Early Pleistocene; small-sized specimens during the Middle Pleistocene (with sharper climatic oscillations); and large-sized specimens during the early Late Pleistocene (in warm and humid environments). Recently, [22] studied the large sample from Cava Redicicoli, highlighting two important aspects: the large dimensional variability of postcranial bones (also affected by sex and age) not connected with climatic changes and the general size similarity between Middle and Late Pleistocene specimens and extant *H. amphibius*, confirming the large size for Early Pleistocene remains of *H. antiquus*.

In this work, a considerable number of fossils were considered by integrating the data present in the literature.

Investigating the tooth length in upper and lower molars reveals a similar pattern for most tooth positions, with Early Pleistocene (Villafranchian and Epivillafranchian) *H. antiquus* being larger (on average) than extant *H. amphibius*, but with a wide overlap (Figures 5 and 6). The range of extant *H. amphibius* encompasses that of most fossil hippopotamuses, both *H. antiquus* and *H. amphibius*. There is, in general, no clear-cut dimensional separation between *H. antiquus* and *H. amphibius* from the Middle Pleistocene in Italy. Rather, a decrease in size seems to have occurred within *H. antiquus* after or during MIS 16, with smaller specimens documented, for instance, at Isernia La Pineta and Terranera. Marked size differences are mainly observed in M_3 L (Figure 6C). Therefore, in order to better explore the dimensional variability of M_3, two bivariate plots were generated (Figures 7 and 8). The first (Figure 7), depicting the length vs. breadth of the tooth, shows two distinct groups (Villafranchian and Epivillafranchian *H. antiquus*, and fossil *H. amphibius* from Barrington and extant specimens). For Middle Pleistocene specimens, Terranera and Maglianella display a small size compared to other *H. antiquus*, whereas Cava Montanari and Ponte Molle possess a larger size than the *H. amphibus* variability. The second plot (Figure 8), the M_2 length vs. the M_3 length, shows clear differences between groups, with *H. antiquus* being larger than *H. amphibius*. The only exception is the sample from Maglianella, which falls in the range of extant *H. amphibius*. This would seem to confirm that *H. antiquus* has a larger size if compared with *H. amphibius*, but these differences are more evident when more biometric measurements are considered.

Our results suggest that multiple factors should be considered when discussing size fluctuations in Quaternary hippopotamuses, in agreement with previous studies [34]. *Hippopotamus antiquus* was larger, on average, than extant *H. amphibius*, but the huge size variation in the latter indicates the need to exercise caution in the attribution of isolated remains. This is especially true for the Middle Pleistocene around 500 ka, where both species might be present, although no dimensional sorting is clear. A decrease in size occurred instead during the early Middle Pleistocene and within *H. antiquus*, probably triggered by the severe glacial conditions known to have characterized MIS 16, which indeed allows an effective separation between large- (Bussi, Ortona, Vallinfreda) and small-sized (Isernia La Pineta, Maglianella, Terranera) samples. A few specimens do not adhere to this pattern, perhaps indicating peculiar local conditions or merely due to individual variability. Indeed, the availability of fossil material is still rather limited to consider this interpretation conclusive, but our overview of the Italian Middle Pleistocene provides important new data.

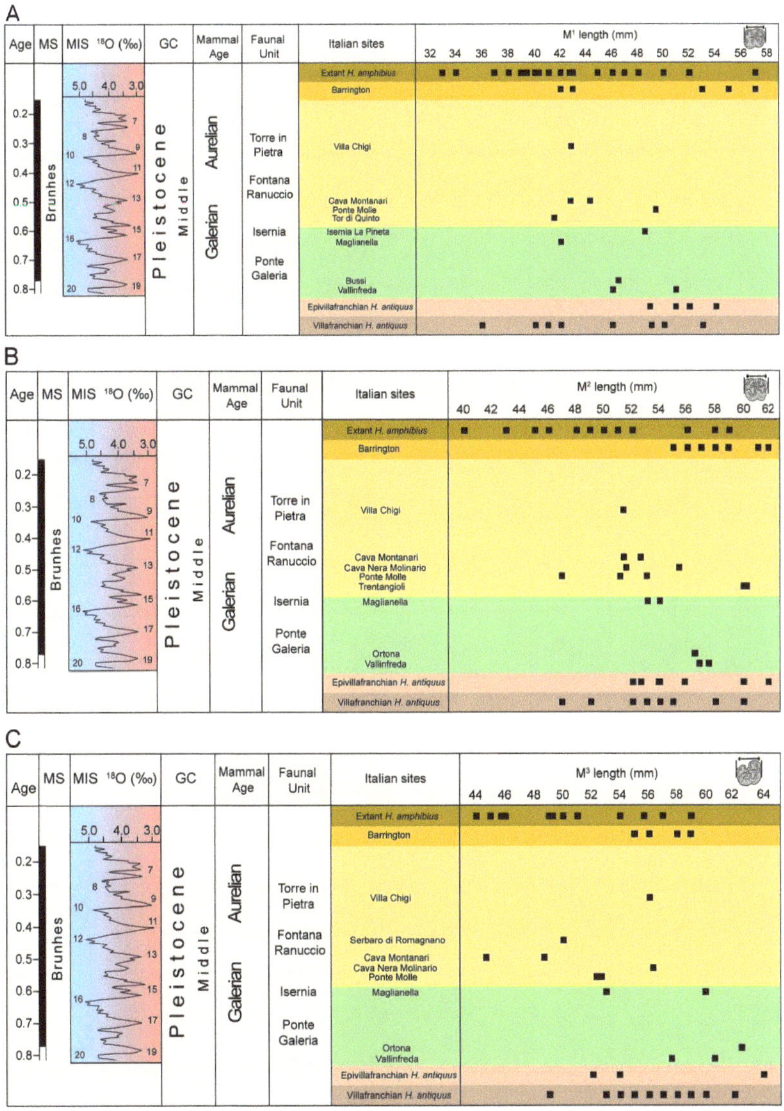

Figure 5. The variation in the size of *Hippopotamus*, as indicated by the length of the upper first (M^1, (**A**)), second (M^2, (**B**)), and third (M^3, (**C**)) molars. Age in millions of years (vertical axis) and oxygen isotope curves (horizontal axis; after [152]) are on the left, followed by the fossiliferous sites and their suggested ages. Dark pink—Villafranchian specimens of *H. antiquus*; light pink—Epivillafranchian specimens of *H. antiquus;* green—chronological range of Middle Pleistocene *H. antiquus;* light orange—chronological range of Middle Pleistocene *H. amphibius;* dark orange—fossil specimens of *H. amphibius* from Barrington; brown—extant specimens of *H. amphibius*.

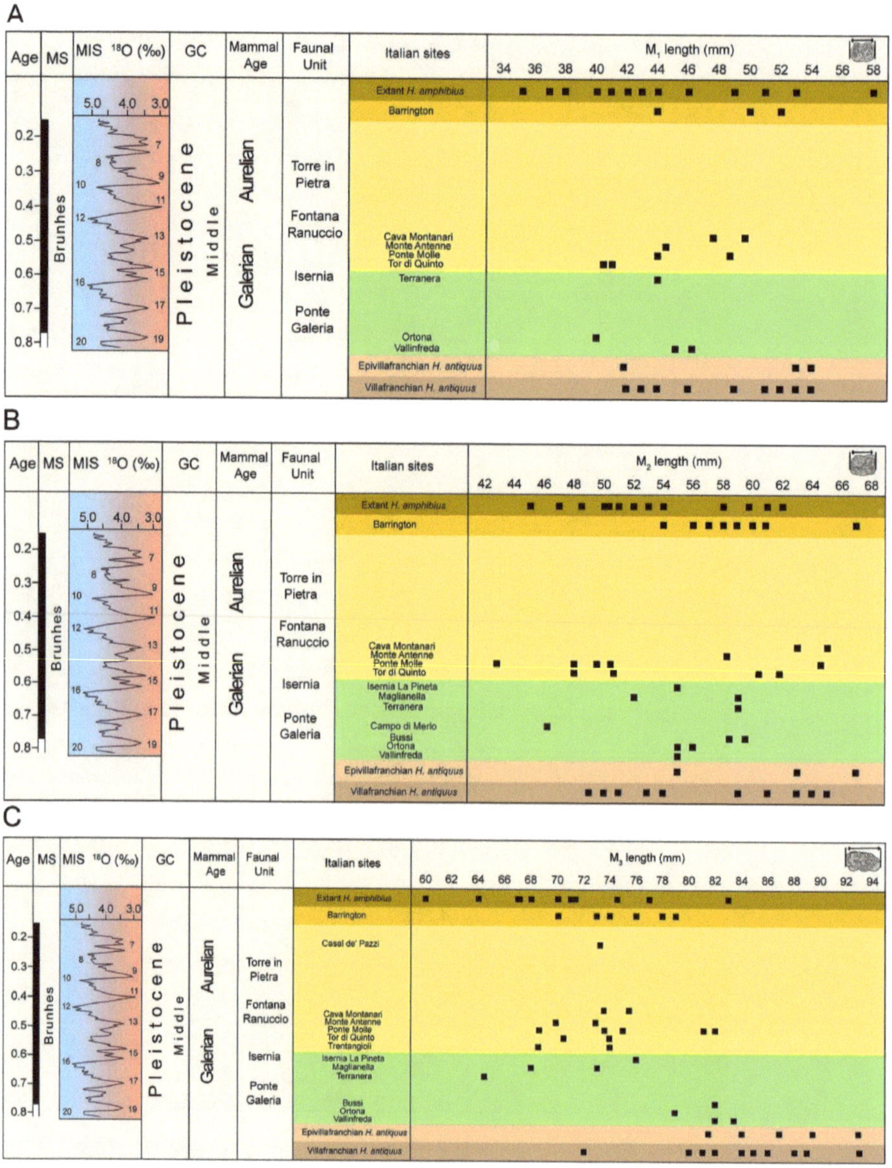

Figure 6. The variation in the size of *Hippopotamus*, as indicated by the length of the lower first (M_1, (**A**)), second (M_2, (**B**)), and third (M_3, (**C**)) molars. Age in millions of years (vertical axis) and oxygen isotope curves (horizontal axis; after [152]) are on the left, followed by the fossiliferous sites and their suggested ages. Dark pink—Villafranchian specimens of *H. antiquus;* light pink—Epivillafranchian specimens of *H. antiquus;* green—chronological range of Middle Pleistocene *H. antiquus;* light orange—chronological range of Middle Pleistocene *H. amphibius;* dark orange—fossil specimens of *H. amphibius* from Barrington; brown—extant specimens of *H. amphibius*.

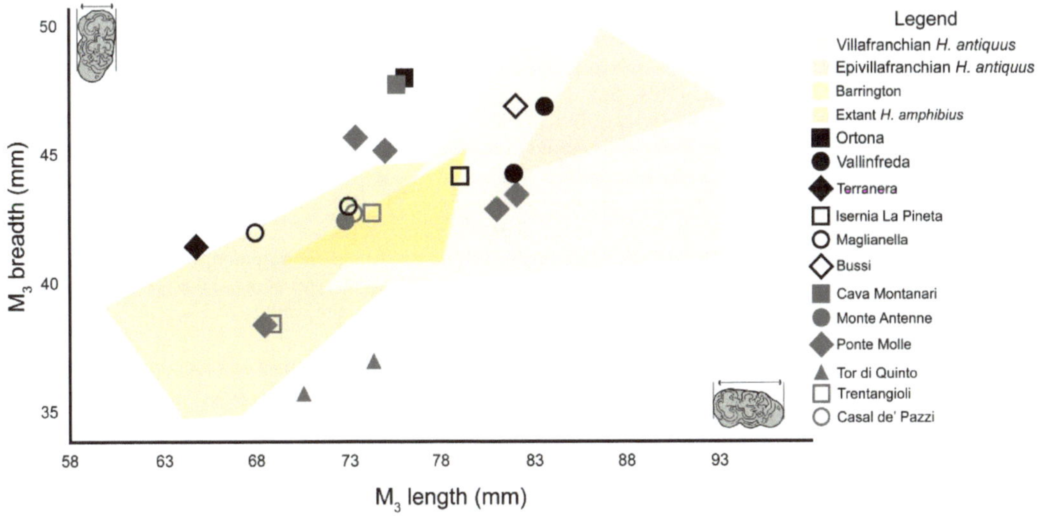

Figure 7. Standard bivariate plots of the breadth vs. length of the lower third molars (M3) of hippopotamuses. Dark pink—Villafranchian specimens of *H. antiquus;* light pink—Epivillafranchian specimens of *H. antiquus;* orange—fossil specimens of *H. amphibius* from Barrington; brown—extant specimens of *H. amphibius;* black—Middle Pleistocene specimens of *H. antiquus;* gray—Middle Pleistocene specimens of *H. amphibius*.

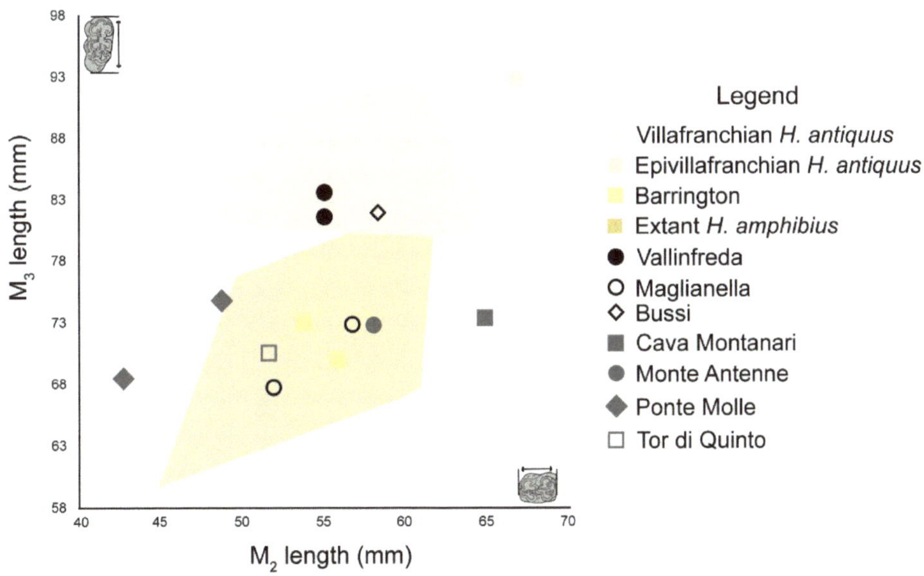

Figure 8. Standard bivariate plots of length of third molar (M3) vs. length of lower second molar (M2) of hippopotamuses. Dark pink—Villafranchian specimens of *H. antiquus;* light pink—Epivillafranchian specimens of *H. antiquus;* orange—fossil specimens of *H. amphibius* from Barrington; brown—extant specimens of *H. amphibius; black*—Middle Pleistocene specimens of *H. antiquus;* gray—Middle Pleistocene specimens of *H. amphibius*.

6. Conclusions

This work offers an important overview of the Middle Pleistocene hippopotamuses of the Italian Peninsula, a territory that, for a long time, has been considered key for the evolution of this group in Europe.

Our results confirm that the *Hippopotamus antiquus–Hippopotamus amphibius* transition occurred during the Middle Pleistocene, probably during MIS 13. The analysis of a large sample also reinforced the idea that, in the absence of cranial remains, taxonomic attributions of Quaternary hippopotamuses are quite difficult. The only feature useful for specific identification is the arrangement of the enamel ridges on the external surfaces of the lower canines. Our results demonstrated that a parallel arrangement was observed in remains dated prior to ca. 500 ka, whereas a convergent arrangement was generally detected in fossils dated after ca. 500 ka. This would confirm the validity of this feature for taxonomical purposes.

Finally, biometric comparisons show that *H. antiquus* was larger, on average, than extant *H. amphibius*, in agreement with previous studies. Nevertheless, *H. antiquus* specimens dated to ca. 600 ka display a smaller size when compared to older samples of the same species, suggesting that severe glacial conditions recognized to have occurred during MIS 16 could have triggered a decrease in size.

Author Contributions: Conceptualization, B.M.; methodology, B.M.; formal analysis, B.M.; investigation, B.M.; writing—original draft preparation, B.M.; writing—review and editing, B.M., A.I., B.S., C.P., M.A., M.C., M.-H.M. and R.S; visualization, B.M.; supervision, R.S.; funding acquisition, B.M., M.-H.M. and R.S. All authors have read and agreed to the published version of the manuscript.

Funding: This work was supported by the Grandi Scavi 2019 (Grant no.: SA11916B513E7C4B, Sapienza University of Rome); Grandi Scavi 2020 (Grant no.: SA120172B2C05E68, Sapienza University of Rome), Grandi Scavi 2021 (Grant no.: SA12117A87BC3F0A), and Grandi Scavi 2022 (Grant no.: SA1221816893E2AB) to RS; and Sapienza, University of Rome "Progetti per Avvio alla Ricerca—Tipo 2 anno 2022" (Grant no.: AR222181333C1B88) and the "Regione Lazio—Contributi premiali per i ricercatori e assegnisti di ricerca per rafforzarne la condizione professionale e potenziare il sistema della ricerca del Lazio" call for proposals of the Lazio Region (DE G05411 del 05/05/2022) to BM. Fieldwork activities at Notarchirico were carried out with the financial and scientific support of the Leakey Foundation ("Early Evidence of Acheulean bifacial technology in Europe" grant, 2015–2016 and 2019–2021), the National Museum of Natural History, Paris, France (ATM Action Transversale du Muséum, 2016–2018), and the ERC-Adv. LATEUROPE (nÅã101052653), supervised by M-HM.

Data Availability Statement: The material studied in this work is curated and available for study in public institutions. Other data are available in the cited references.

Acknowledgments: First of all, We thank the organizer of the session "Mammals Biochronology and Paleoecology of the Euro-Mediterranean Quaternary" during the XXIINQUA conference (Rome, Italy, 14–20 July 2023), where a preliminary data on Quatgernary hippopotamuses was presented and discussed with other participants. B.M. thanks all the staff of the museums and institutions for access to the fossil collections, their kindness, availability, and support: Linda Riti and Michele Macrì (Museo Universitario di Scienze della Terra, MUST, Sapienza Università di Roma, including the former Museo di Paleontologia); Roberto Zorzin and Irene Tomelleri (Museo di Storia Naturale di Verona); Luciano Bruni, Stefano Grimaldi and Barbara Saracino (Istituto Italiano di Paleontologia Umana, IsIPU, Anagni); Edoardo Di Russo and Riccardo Castiglia (Museo di Anatomia comparata "Battista Grassi", Sapienza Università di Roma); Francesca Alhaique (Museum of Civilizations, including the National Prehistoric Ethnographic Museum "Luigi Pigorini"); Ursula Thun Hohenstein (Museo di Paleontologia e Preistoria "Piero Leonardi"(MPPL); Alessandro Cacciotti, Marina Pescarmona and Lorenzo Zaffagnini (Collegio San Giuseppe—Istituto De Merode). RS is thankful to the Soprintendenza Archeologia, Belle Arti e Paesaggio delle province di Brindisi, Lecce e Taranto (Maria Piccarreta, Laura Masiello, and Serena Strafella) for authorizing the research and field activities (2015–2017 and 2018–2020 and extension in 2021 due to COVID-19 to 2022-2024, resp. R. Sardella). M-HM thanks the Soprintendenza Archeologia, Belle Arti e Paesaggio della Basilicata (Teresa Elena Cinquantaquattro, Francesco Canestrini, Romina Pirraglia, and Sabrina Mutino), for authorizing the research and field activities and technical and scientific support. We also thank Rosanna Calabrese for access to the collections preserved at Museo Archeologico Nazionale "Mario Torelli" di Venosa. Finally, the Two anonymous reviewers provided valuable comments and suggestions, for which I am grateful.

Conflicts of Interest: The authors declare no conflict of interest.

References

1. Gliozzi, E.; Abbazzi, L.; Argenti, P.; Azzaroli, A.; Caloi, L.; Capasso Barbato, L.; Di Stefano, G.; Esu, D.; Ficcarelli, G.; Girotti, O.; et al. Biochronology of selected mammals, molluscs and ostracods from the Middle Pliocene to the Late Pleistocene in Italy. The state of the art. *Riv. Ital. Paleontol. Stratig.* **1997**, *103*, 369–388.
2. Strani, F.; Bellucci, L.; Iannucci, A.; Iurino, D.A.; Mecozzi, B.; Sardella, R. Palaeoenvironments of the MIS 15 site of Cava di Breccia-Casal Selce 2 (central Italian Peninsula) and niche occupation of fossil ungulates during Middle Pleistocene interglacials. *Hist. Biol.* **2022**, *34*, 555–565. [CrossRef]
3. Lisiecki, L.E.; Raymo, M.E. A Pliocene–Pleistocene stack of 57 globally distributed benthic δ18O records. *Paleoceanogr. Paleoclimatol.* **2005**, *20*, PA1003.
4. Lisiecki, L.E.; Raymo, M.E. Plio-Pleistocene climate evolution: Trends and transition in glacial cycle dynamics. *Quat. Sci. Rev.* **2007**, *26*, 56–69. [CrossRef]
5. Candy, I.; Rose, J.; Lee, J. A seasonally 'dry' interglacial climate in eastern England during the early Middle Pleistocene: Palaeopedological and stable isotopic evidence from Pakefield, UK. *Boreas* **2006**, *35*, 255–265. [CrossRef]
6. Candy, I.; Coope, G.R.; Lee, J.R.; Parfitt, S.A.; Preece, R.C.; Rose, J.; Schreve, D.C. Pronounced warmth during early Middle Pleistocene interglacials: Investigating the Mid-Brunhes Event in the British terrestrial sequence. *Earth-Sci. Rev.* **2010**, *103*, 183–196. [CrossRef]
7. Clark, P.; Archer, D.; Pollard, D.; Blum, J.D.; Rial, J.A.; Brovkin, V.; Mix, A.C.; Pisias, N.G.; Roy, M. The Middle Pleistocene transition: Characteristics, mechanisms, and implications for long-term changes in atmospheri pCO_2. *Quat. Sci. Rev.* **2006**, *25*, 3150–3184. [CrossRef]
8. Head, M.J.; Gibbard, P.L. Early-Middle Pleistocene transitions: Linking terrestrial and marine realms. *Quat. Int.* **2015**, *389*, 7–46. [CrossRef]
9. Maslin, M.A.; Brierley, C.M. The role of orbital forcing in the early middle Pleistocene transition. *Quat. Int.* **2015**, *389*, 47–55. [CrossRef]
10. Kahlke, R.-D. Late Early Pleistocene European large mammals and the concept of an Epivillafranchian biochron. In *Late Neogene and Quaternary Biodiversity and Evolution: Regional Developments and Interregional Correlations, Volume II*; Kahlke, R.-D., Maul, L.C., Mazza, P.P.A., Eds.; Cour. Forschungsinst. Senckenberg 259; Schweizerbart'sche Verlagsbuchhandlung: Stuttgart, Germany, 2007; pp. 265–278.
11. Bellucci, L.; Sardella, R.; Rook, L. Large mammal biochronology framework in Europe at Jaramillo: The Epivillafranchian as a formal biochron. *Quat. Int.* **2015**, *389*, 84–89. [CrossRef]
12. van der Made, J.; Rosell, J.; Blasco, R. Faunas from Atapuerca at the Early-Middle Pleistocene limit: The ungulates from level TD8 in the context of climatic change. *Quat. Int.* **2017**, *433*, 296–346. [CrossRef]
13. Konidaris, G.E.; Kostopoulos, D.S.; Koufos, G.D. *Mammuthus meridionalis* (Nesti, 1825) from Apollonia-1 (Mygdonia Basin, Northern Greece) and its importance within the Early Pleistocene mammoth evolution in Europe. *Geodiversitas* **2020**, *42*, 69–91. [CrossRef]
14. Laws, R.M. Dentition and ageing of the Hippopotamus. *East Afr. Wildl. J.* **1968**, *6*, 19–52. [CrossRef]
15. Moncel, M.-H.; Santagata, C.; Pereira, A.; Nomade, S.; Voinchet, P.; Bahain, J.-J.; Daujeard, C.; Curci, A.; Lemorini, C.; Hardy, B.; et al. The origin of early Acheulean expansion in Europe 700 ka ago: New findings at Notarchirico (Italy). *Sci. Rep.* **2020**, *10*, 1–16. [CrossRef] [PubMed]
16. Kahlke, R.-D. The Pleistocene of Untermassfeld—A synopsis of site origin, palaeobiodiversity, taphonomic characteristics, palaeoanvironment, chronostratigraphy, and significance in western Palaearctic faunal history. In *The Pleistocene of Untermassfeld near Meiningen (Thu ¨ringen, Germany). Part 5*; Kahlke, R.-D., Ed.; Monographien des Romisch-Germanischen Zentralmuseums: Köln, Germany, 2022; pp. 1635–1734.
17. Kahlke, R.-D.; García, N.; Kostopoulos, D.S.; Lacombat, F.; Lister, A.M.; Mazza, P.P.A.; Spassov, N.; Titov, V.V. Western Palaearctic palaeoenvironmental conditions during the Early and early Middle Pleistocene inferred from large mammal communities, and implications for hominin dispersal in Europe. *Quat. Sci. Rev.* **2011**, *30*, 1368–1395. [CrossRef]
18. Ollé, A.; Lombao, D.; Asryan, L.; Garcìa-Medrano, P.; Arroyo, A.; Fernàndez-Marchena, J.L.; Yeşilova, G.C.; Cáceres, I.; Huguet, R.; López-Polín, L.; et al. The earliest European Acheulean: New insights into the large shaped tools from the late Early Pleistocene site of Barranc de la Boella (Tarragona, Spain). *Front. Earth Sci.* **2023**, *11*, 1188663. [CrossRef]
19. Breda, M.; Lister, A.M. *Dama roberti*, a new species of deer from the early Middle Pleistocene of Europe, and the origins of modern fallow deer. *Quat. Sci. Rev.* **2013**, *69*, 155–167. [CrossRef]
20. Boulbes, N.; Van Asperen, E.N. Biostratigraphy and palaeoecology of European *Equus*. *Front. Ecol. Evol.* **2019**, *7*, 301. [CrossRef]
21. Iannucci, A.; Mecozzi, B.; Sardella, R. Large mammals from the Middle Pleistocene (MIS 11) site of Fontignano 2 (Rome, central Italy), with an overview of "San Cosimato" assemblages. *Alp. Mediterr. Quat.* **2021**, *34*, 155–164.

22. Mecozzi, B.; Iannucci, A.; Mancini, M.; Tentori, D.; Cavasinni, C.; Conti, J.; Messina, M.Y.; Sarra, A.; Sardella, R. Reinforcing the idea of an early dispersal of *Hippopotamus amphibius* in Europe: Restoration and multidisciplinary study of the skull from the Middle Pleistocene of Cava Montanari (Rome, central Italy). *PLoS ONE* **2023**, *18*, e0293405. [CrossRef]
23. Iannucci, A.; Mecozzi, B.; Sardella, R. Beware of the "Wolf Event"—Remarks on large mammal dispersals in Europe and the late Villafranchian faunal turnover. *Alp. Mediterr. Quat.* **2023**, *36*, 1–16.
24. Caloi, L.; Palombo, M.R.; Petronio, C. Resti cranici di *Hippopotamus antiquus* (=*H. major*) e *Hippopotamus amphibius* conservati nel Museo di Paleontologia dell'Università di Roma. *Geol. Rom.* **1980**, *19*, 91–119.
25. Mazza, P. Interrelations between Pleistocene hippopotami of Europe and Africa. *Boll. Soc. Paleont. Ital.* **1991**, *30*, 153–186.
26. Mazza, P. New evidence on the Pleistocene hippopotami of western Europe. *Geol. Rom.* **1995**, *31*, 24–61.
27. Galobart, A.; Ros, X.; Maroto, J.; Vila, B. Descripción del material de hipopótamo (*Hippopotamus antiquus* Desmarest, 1822) de los yacimientos del Pleistoceno inferior de Incarcal (Girona, NE de la Península Ibérica). *Paleontol. Evol.* **2003**, *34*, 153–173.
28. Fidalgo, D.; Galli, E.; Madurell-Malapeira, J.; Rosas, A. Earliest Pleistocene European hippos: A review. *Comun. Geológicas* **2021**, *108*, 65–69.
29. Adams, N.F.; Candy, I.; Schreve, D.C. An Early Pleistocene hippopotamus from Westbury Cave, Somerset, England: Support for a previously unrecognized temperate interval in the British Quaternary record. *J. Quat. Sci.* **2022**, *37*, 28–41. [CrossRef]
30. Kahlke, R.-D. Schädelreste von *Hippopotamus* aus dem Unterpleistozän von Untermassfeld. In *Das Pleistozän von Untermassfeld bei Meiningen (Thüringen): Teil 2*; Kahlke, R.-D., Ed.; Dr Rudolf Habelt GmbH: Bonn, Germany, 2001; pp. 483–500.
31. Kierdorf, U.; Kahlke, R.-D. Pathological findings on remains of hippopotamids from the Early Pleistocene site of Untermassfeld. In *The Pleistocene of Untermassfeld Near Meiningen (Thüringen, Germany): Part 4.*; Kahlke, R.-D., Ed.; Verlag des Römisch-Germanischen Zentralmuseums: Mainz, Germany, 2020; pp. 1251–1272.
32. Faure, M. *Hippopotamus incognitus* nov. sp., un hippopotame (Mammalia, Artiodactyla) du Pléistocène d'Europe occidentale. *Geobios* **1984**, *17*, 427–437. [CrossRef]
33. Faure, M. Les hippopotames quaternaires non-insulaires d'Europe occidentale. *Nouv. Arch. Mus. Hist. Nat. Lyon* **1985**, *23*, 13–79. [CrossRef]
34. Mazza, P.P.A.; Bertini, A. Were Pleistocene hippopotamuses exposed to climate-driven body size changes? *Boreas* **2013**, *42*, 194–209. [CrossRef]
35. van der Made, J.; Sahnouni, M.; Boulaghraief, K. Hippopotamus gorgops from El Kherba (Algeria) and the context of its biogeography. In Proceedings of the II Meeting of African Prehistory, Burgos, Spain, 15–16 April 2015; Sahnouni, M., Semaw, S., Rios Garaizar, J., Eds.; CENIEH: Burgos, Spain, 2017; pp. 135–169.
36. Petronio, C. Note on the taxonomy of Pleistocene hippopotamuses. *Ibex* **1995**, *3*, 53–55.
37. Martino, R.; Pandolfi, L. The Quaternary *Hippopotamus* records from Italy. *Hist. Biol.* **2022**, *34*, 1146–1156. [CrossRef]
38. Martino, R.; Rìoz, M.I.; Mateus, O.; Pandolfi, L. Taxonomy chronology and dispersal pattern of Western European Quaternary hippopotamuses Portuguese. *Quat. Int.* **2023**, *674-675*, 121–137. [CrossRef]
39. Fidalgo, D.; Rosas, A.; Madurell-Malapeira, J.; Pineda, A.; Huguet, R.; García-Tabernero, A.; Cáceres, I.; Ollé, A.; Vallverdú, J.; Saladie, P. A review on the Pleistocene occurrences and palaeobiology of *Hippopotamus antiquus* based on the record from the Barranc de la Boella Section (Francoli Basin NE Iberia). *Quat. Sci. Rev.* **2023**, *307*, 108304. [CrossRef]
40. Mecozzi, B. The *Hippopotamus* remains from the latest Early Pleistocene site of Cava Redicicoli (Rome, central Italy). *Boll. Soc. Paleont. Ital.* **2023**, *62*, 263–279.
41. Pandolfi, L.; Petronio, C. A brief review of the occurrences of Pleistocene *Hippopotamus* (Mammalia, Hippopotamidae) in Italy. *Geol. Croat.* **2015**, *68*, 313–319. [CrossRef]
42. Blandamura, F.; Azzaroli, A. *L'Ippopotamo Maggiore di Filippo Nesti*; serie 8; Atti dell'Accademia Nazionale dei Lincei: Rome, Italy, 1977; Volume 14, pp. 169–188.
43. Mazza, P.P.A.; Ventra, D. Pleistocene debris-flow deposition of the hippopotamus-bearing Collecurti bonebed (Macerata, Central Italy): Taphonomic and paleoenvironmental analysis. *Palaeogeogr. Palaeoclimatol. Palaeoecol.* **2011**, *310*, 296–314. [CrossRef]
44. Marini, G. *Il Lago Pleistocenico Della Conca de L'Aquila*; CET: Lanciano, Italy, 1967; 64p.
45. Mancini, M.; Cavuoto, G.; Pandolfi, L.; Petronio, C.; Salari, L.; Sardella, R. Coupling basin infill history and mammal biochronology in a Pleistocene intramontane basin: The case of western L'Aquila Basin (central Apennines, Italy). *Quat. Int.* **2012**, *267*, 62–67. [CrossRef]
46. Palombo, M.R.; Mussi, M.; Agostini, S.; Barbieri, M.; Di Canzio, E.; Di Rita, F.; Fiore, I.; Iacumin, P.; Magri, D.; Speranza, F.; et al. Human peopling of Italian intramontane basins: The early Middle Pleistocene site of Pagliare di Sassa (L'Aquila, central Italy). *Quat. Int.* **2010**, *223*, 170–178. [CrossRef]
47. Agostini, S.; Rossi, M.A.; Tallini, M. Geologia e paleontologia del Quaternario nel territorio aquilano. In *Atti del Convegno "Nei Dintorni di L'Aquila. Ricerche Archeologiche nel Territorio dei Vestini Cismontani Prima e Dopo il Terremoto del 6 Aprile 2009", Roma, 12–13 Febbraio 2010*; Collection de l'Ecole Francaise de Rome: Rome, Italy, 2014; Volume 494, pp. 7–19.
48. Russo Ermolli, E.; Sardella, R.; Di Maio, G.; Petronio, C.; Santangelo, N. Pollen and mammals from the late Early Pleistocene site of Saticula (Sant'Agata de' Goti, Benevento, Italy). *Quat. Int.* **2010**, *225*, 128–137. [CrossRef]
49. Ficcarelli, G.; Abbazzi, L.; Albianelli, A.; Bertini, A.; Coltorti, M.; Magnatti, M.; Masini, F.; Mazza, P.; Mezzabotta, C.; Napoleone, G.; et al. Cesi, and early Middle Pleistocene site in Colfiorito Basin (Umbro-Marchean Apennine), central Italy. *J. Quat. Sci.* **1997**, *12*, 507–518. [CrossRef]

50. Bona, F.; Sala, B. Villafranchian-Galerian mammal faunas transition in South-western Europe. The case of the late early Pleistocene mammal fauna of the Frantoio locality, Arda River (Castell'Arquato, Piacenza, northern Italy). *Geobios* **2016**, *49*, 329–347. [CrossRef]
51. Bon, M.; Piccoli, G.; Sala, B. *La Fauna Pleistocenica Della Breccia di Slivia (Carso triestino) Nella Collezione del Museo Civico di Storia Naturale di Trieste*; Atti del Museo Civico Storia Naturale: Trieste, Italy, 1992; Volume 44, pp. 33–51.
52. Benussi, B.; Melato, M. Considerazioni preliminari su reperti di una fauna fossile a pachidermi in una breccia ossifera a Slivia-Visogliano. *Att. E Mem. Della Comm. Grotte E. Boegan* **1970**, *10*, 77–95.
53. Azzaroli, A.; Berzi, A. On an upper Villafranchian fauna at Imola, northern Italy, and its correlation with the marine Pleistocene sequence of the Po plain. *Paleontogr. Ital.* **1970**, *66*, 1–12.
54. Muttoni, G.; Scardia, G.; Kent, D.V.; Morsiani, E.; Tremolada, F.; Cremaschi, M.; Peretto, C. First dated human occupation of Italy at~ 0.85 Ma during the late Early Pleistocene climate transition. *Earth Planet. Sci. Lett.* **2011**, *307*, 241–252. [CrossRef]
55. Marra, F.; Pandolfi, L.; Petronio, C.; Di Stefano, G.; Gaeta, M.; Salari, L. Reassessing the sedimentary deposits and vertebrate assemblages from Ponte Galeria area (Rome, central Italy): An archive for the Middle Pleistocene faunas of Europe. *Earth-Sci. Rev.* **2014**, *139*, 104–122. [CrossRef]
56. Pandolfi, L.; Marra, F. Rhinocerotidae (Mammalia, Perissodactyla) from the chrono-stratigraphically constrained Pleistocene deposits of the urban area of Rome (Central Italy). *Geobios* **2015**, *48*, 147–167. [CrossRef]
57. Bosi, C.; Galadini, F.; Giacico, B.; Messina, P.; Sposato, A. Plio-Quaternary continental deposits in the Latium-Abruzzi apennines: The correlation of geological events across different intermontane basins. *Il Quat. Ital. J. Quat. Sci.* **2003**, *16*, 55–76.
58. Agostini, S.; Bertini, A.; Caramiello, S.; De Flaviis, A.G.; Mazza, P.; Rossi, M.A.; Satolli, S. A new mammalian bone bed from the lower Middle Pleistocene of Ortona (Chieti, Abruzzo, central Italy). In *Proceedings of the Giornate di Paleontologia*; Coccioni, R., Marsili, A., Eds.; Grzybowski Foundation: Kraków, Poland, 2007; Volume 12, pp. 1–5.
59. Fabbi, S.; Romano, M.; Strani, F.; Sardella, R.; Bellucci, L. The Pleistocene vertebrate fauna of the Oricola-Carsoli intermontane Basin (Latium-Abruzzi, Italy): State of the art and historical review. *Boll. Soc. Paleont. Ital.* **2021**, *60*, 255–268.
60. Ceruleo, P.; Rolfo, M.F.; Petronio, C.; Salari, L. Review of Middle Pleistocene archaeological and biochronological data in Malagrotta—Ponte Galeria area (Rome, Central Italy). *Atti Della Soc. Toscana Di Sci. Nat. Mem. Ser. A* **2021**, *128*, 85–103.
61. Capasso Barbato, L.; Petronio, C. Nuovi resti di mammiferi del Pleistocene medio-inferiore di Ponte Galeria (Roma). *Boll. Serv. Geol. d'Ital.* **1983**, *104*, 157–176.
62. Berzi, A. An early Middle Pleistocene fauna at Monte Oliveto (S. Giaminiano, Siena, Italy). *Palaeontogr. Ital.* **1972**, *68*, 28–33.
63. Azzaroli, A. A skull of the Giant deer *Megaceros verticornis* from Eastern Tuscany. *Att. Della Accad. Naz. Dei Lincei. Cl. Sci. Fis. Mat. E Naturali. Rend. Ser. 8* **1976**, *61*, 485–487.
64. Palombo, M.R.; Azanza, B.; Alberdi, M.T. Italian Mammal Biochronology from the latest Miocene to the Middle Pleistocene: A multivariate approach. *Geol. Romana* **2003**, *36*, 355–368.
65. Coltorti, M.; Feraud, G.; Marzoli, A.; Peretto, C.; Ton-That, Y.; Voinchet, P.; Bahain, J.J.; Minelli, A.; Thun Hohenstein, U. New 40Ar/39 Ar, stratigraphic and palaeoclimatic data on the Isernia La Pineta Lower Palaeolithic site, Molise, Italy. *Quat. Int.* **2005**, *131*, 11–22. [CrossRef]
66. Caloi, L.; Palombo, M.R. *Megaceros solilhacus* Robert, da Terranera (Bacino di Venosa, Potenza). *Quaternaria* **1979**, *21*, 129–2138.
67. Aureli, D.; Rocca, R.; Lemorini, C.; Modesti, V.; Scaramucci, S.; Milli, S.; Giaccio, B.; Marano, F.; Palombo, M.R.; Contardi, A. Mode 1 or mode 2? "Small tools" in the technical variability of the European Lower Palaeolithic: The site of Ficoncella (Tarquinia, Lazio, central Italy). *Quat. Int.* **2016**, *393*, 169–184. [CrossRef]
68. Marra, F.; Ceruleo, P.; Jichac, B.; Pandolfi, L.; Petronio, C.; Salari, L.; Giaccio, B.; Sottili, G. Glacio-eustatic and tectonic forcing on the lacustrine succession of the Cretone basin: Chronostratigraphic constraints to Acheulian industry and Middle Pleistocene faunal assemblages of Latium (central Italy). *J. Quat. Sci.* **2016**, *31*, 641–658. [CrossRef]
69. Clerici, E. Resti di ippopotamo al Monte Antene. *Boll. Soc. Geol. Ital.* **1926**, *45*, 118–122.
70. Mecozzi, B.; Iannucci, A.; Mancini, M.; Sardella, R. Redefining Ponte Molle (Rome, central Italy): An important locality for Middle Pleistocene mammal assemblages of Europe. *Alp. Mediterr. Quat.* **2021**, *34*, 131–154.
71. Petronio, C.; Di Stefano, G.; Kotsakis, T.; Salari, L.; Marra, F.; Jicha, B.R. Biochronological framework for the late Galerian and early-middle Aurelian Mammal Ages o peninsular Italy. *Geobios* **2019**, *53*, 35–50. [CrossRef]
72. Di Stefano, G.; Petronio, C.; Sardella, R. Biochronology of Pleistocene mammal faunas from Rome urban area. *Il Quat.* **1998**, *11*, 191–199.
73. Sala, B.; Barbi, G. Descrizione della fauna. In *Gli Scavi a Castel di Guido. Il Più Antico Giacimento di Cacciatori Nell'agro Romano*; Radmilli, A.M., Boschian, G., Eds.; ETS: Pisa, Italy, 1996; pp. 55–91.
74. Villa, P.; Boschian, G.; Pollarolo, L.; Saccà, D.; Marra, F.; Nomade, S.; Pereira, A. Elephant bones for the Middle Pleistocene toolmaker. *PloS ONE* **2021**, *16*(8), e0256090. [CrossRef] [PubMed]
75. Marra, F.; Pereira, A.; Boschian, G.; Nomade, S. MIS 13 and MIS 11 aggradational successions of the Paleo-Tiber delta: Geochronological constraints to sea-level fluctuations and to the Acheulean sites of Castel di Guido and Malagrotta (Rome, Italy). *Quat. Int.* **2022**, *616*, 1–11. [CrossRef]
76. Anzidei, A.P.; Caloi, L.; Giacopini, L.; Mantero, D.; Palombo, M.R.; Sebastiani, R.; Segre, A.G. Saggi di scavo nei depositi pleistocenici del Km 18,900 della Via Aurelia e di Collina Barbattini (Castel di Guido-Roma). *Quad. Di Archeol. Etrusco-Ital.* **1993**, *21*, 81–90.

77. Capasso Barbato, L.; Petronio, C. La mammalofauna pleistocenica di Castel di Guido (Roma). *Boll. Serv. Geol. d'Ital.* **1981**, *102*, 95–108.
78. Strani, F.; Demiguel, D.; Bona, F.; Sardella, R.; Biddittu, I.; Bruni, L.; De Castro, A.; Guadagnoli, F.; Bellucci, L. Ungulate dietary adaptations and palaeoecology of the Middle Pleistocene site of Fontana Ranuccio (Anagni, Central Italy). *Palaeogeogr. Palaeoclimatol. Palaeoecol.* **2018**, *496*, 238–247. [CrossRef]
79. Grimaldi, S.; Santaniello, F.; Angelucci, D.E.; Bruni, L.; Parenti, F. A Techno-Functional Interpretation of the Lithic Assemblage from Fontana Ranuccio (Anagni, Central Italy): An Insight into a MIS 11 Human Behaviour. *J. Paleolit. Archaeol.* **2020**, *3*, 944–966. [CrossRef]
80. Caloi, L.; Palombo, P. Resti di mammiferi del Pleistocene Medio di Malagrotta (Roma). *Boll. Ser. Geol. d'Ital.* **1979**, *100*, 141–188.
81. Pasa, A. I mammiferi di alcune antiche brecce veronesi. *Mem. Museo Civ. St. Nat. Verona* **1947**, *1*, 1–111.
82. Caloi, L.; Palombo, M.R. Osservazioni sugli equidi italiani del Pleistocene medio inferiore. *Geol. Rom.* **1990**, *26*, 187–221.
83. Caloi, L.; Palombo, M.R. Gli equidi del Pleistocene medio-inferiore di Soave il Castello, Monte Tenda e Romagnano il Serbaro (Verona, Italia). *Boll. Mus. Civ. St. Nat. Verona* **1991**, *15*, 447–468.
84. Meli, R. Sopra alcune ossa fossili rinvenute nelle ghiaie allivionali presso la via Nomentana, al 3° chilom. da Roma. *Boll. Reg. C. Geol. d'Ital.* **1886**, *17*, 265–280.
85. Pieruccini, P.; Forti, L.; Mecozzi, B.; Iannucci, A.; Yu, T.L.; Shen, C.-C.; Bona, F.; Lembo, G.; Muttillo, B.; Sardella, R.; et al. Stratigraphic reassessment of Grotta Romanelli sheds light on Middle-Late Pleistocene palaeoenvironments and human settling in the Mediterranean. *Sci. Rep.* **2022**, *12*, 13530. [CrossRef] [PubMed]
86. Pandolfi, L.; Martino, R.; Palombo, M.R. New insights on the fossil mammals from Casal de'Pazzi (Rome). *J. Mediterr. Earth Sci.* **2023**, *15*. [CrossRef]
87. Caloi, L.; Palombo, M.R. Anfibi, rettili e mammiferi di Torre del Pagliaccetto (Torre in Pietra, Roma). *Quaternaria* **1978**, *20*, 315–428.
88. Villa, P.; Soriano, S.; Grün, R.; Marra, F.; Nomade, S.; Pereira, A.; Boschian, G.; Pollarolo, L.; Fang, F.; Bahain, J.-J. The Acheulian and Early Middle Paleolithic in Latium (Italy): Stability and Innovation. *PLoS ONE* **2016**, *11*, e0160516. [CrossRef] [PubMed]
89. Blanc, G.A.; Tongiorgi, E.; Trevisan, L. Giacimento di fauna ad Ippopotamo nel quaternario antico della Maglianella presso Roma. *Att. Della Soc. Ital. Per Il Prog. Delle Sci.* **1951**, *42*, 505–506.
90. Fabiani, R.; Maxia, C. *L'Istituto e i Musei di Geologia e Paleontologia*; Pubblicazioni Università degli Studi di Roma Istituto di Geologia e Paleontologia; Istituto grafico tiberino: Roma, Italy, 1953; Volume 9, pp. 1–31.
91. Radmilli, A.M.; Boschian, G. *Gli Scavi a Castel di Guido. Il più Antico Giacimento di Cacciatori Nell'agro Romano*; ETS: Pisa, Italy, 1996.
92. Rubini, M.; Cerroni, V.; Festa, G.; Sardella, R.; Zaio, P. A revision of hominin fossil teeth from Fontana Ranuccio (middle Pleistocene, Anagni, Frosinone, Italy). *J. Hum. Evol.* **2014**, *77*, 204–216. [CrossRef] [PubMed]
93. Biddittu, I.; Cassoli, P.F.; Radicati di Brozolo, F.; Segre, A.G.; Segre Naldini, E.; Villa, I. Anagni a K/Ar dated Lower Middle Pleistocene site, Central Italy: Preliminary report. *Quaternaria* **1979**, *21*, 53–71.
94. Gioia, P.; Silvestri, L.; Zanzi, G.L. Introduction: Casal de' Pazzi: The Pleistocene on the outskirts between past, present, and future. The reasons for a conference. *J. Mediterr. Earth Sci.* **2023**, *15*.
95. Osborn, H.F. *Palaeoloxodon antiquus italicus* sp. nov., final stage in the '*Elephas antiquus*' phylum. *Am. Mus. Novit.* **1931**, *460*, 1–24.
96. De Lorenzo, G.; D'Erasmo, G.D. Avanzi di ippopotamo nell'Italia meridionale. *Att. R. Acc. Sci. Fis. Mat. Napoli* **1935**, *1*, 1–11.
97. Cavinato, G.P.; Petronio, C.; Sardella, R. The mercure river basin (southern Italy): Quaternary stratigraphy and large mammal biochronology. *World Elephants Proc. 1st Int. Congr. Rome* **2001**, 187–190.
98. Airaghi, C. L'ippopotamo (*Hippopotamus amphibius* L.) dell'antico lago di Mércure (Calabria). *Att. Soc. Ital. Sci. Nat. E Mus. Civ. Stor. Nat. Milano* **1922**, *9*, 408–418.
99. Leonardi, P. Nuovi resti di Ippopotamo nelle ligniti del Mercure. *Studi Trentini Sci. Nat.* **1932**, *14*, 3–6.
100. Petrosino, P.; Russo Ermolli, E.; Donato, P.; Jicha, B.; Robustelli, G.; Sardella, R. Using Tephrochronology and palynology to date the MIS 13 lacustrine sediments of the Mercure Basin (Southern Apennines–Italy). *Ital. J. Geosci.* **2014**, *133*, 169–186. [CrossRef]
101. Ponzi, G. *Le Ossa Fossili Subappennine dei Dintorni di Roma*; Serie III; Atti della Reale Accademia dei Lincei: Roma, Italy, 1878; Volume 2, pp. 1–30.
102. Lefévre, D.; Raynal, J.-P.; Vernet, G.; Kieffer, G.; Piperno, M. Tephrostratigraphy and the age of ancient Southern Italian Acheulean settlements: The sites of Loreto and Notarchirico (Venosa, Basilicata, Italy). *Quat. Int.* **2010**, *223-224*, 360–368. [CrossRef]
103. Piperno, M. *Notarchirico, Un sito del Pleistocene Medio Iniziale nel Bacino di Venosa*; Edizioni Osanna: Venosa, Italy, 1999.
104. Pereira, A.; Nomade, S.; Voinchet, P.; Bahain, J.J.; Falguères, C.; Garon, H.; Lefévre, D.; Raynal, J.P.; Scao, V.; Piperno, M. The earliest securely dated hominin fossil in Italy and evidence of Acheulian occupation during glacial MIS 16 at Notarchirico (Venosa, Basilicata, Italy). *J. Quat. Sci.* **2015**, *30*, 639–650. [CrossRef]
105. Moncel, M.H.; Lemorini, C.; Eramo, G.; Fioretti, G.; Daujeard, C.; Curci, A.; Berto, C.; Hardy, B.; Pineda, A.; Rineau, V.; et al. A taphonomic and spatial distribution study of the new levels of the middle Pleistocene site of Notarchirico (670–695 ka, Venosa, Basilicata, Italy). *Archaeol. Anthropol. Sci.* **2023**, *15*, 106. [CrossRef]
106. Ferrara, F.; Piperno, M. Località di interesse preistorico nel bacino di Venosa. Collezioni, ritrovamenti isolati e scavi. In *Notarchirico, Un Sito del Pleistocene Medio Iniziale nel Bacino di Venosa*; Piperno, M., Ed.; Edizioni Osanna: Venosa, Italy, 1999; pp. 41–66.

107. A cura di Colella, A.; Schiattarella, M.; Giannandrea, P.; Lazzari, M.; Longhitano, S.G.; Vitale, G. APAT-SERVIZIO GEOLOGICO D'ITALIA-Carta Geologica d'Italia Alla Scala 1:50.000 e Relative Note Illustrative. Foglio 452 Rionero in Vulture. *Serv. Geol. d'Ital.* **2018**, 1–95. The Documents Are Available for Consultation on the Website. Available online: www.apat.it (accessed on 19 February 2024).
108. Rellini, U. *Sulle Stazioni Quaternarie di Tipo "Chelléen" Del1'agro Venosino*; Tipografia della R. Accademia dei Lincei address: Roma, Italy, 1915; Volume 2, pp. 183–210.
109. Peretto, C.; Arnaud, J.; Moggi-Cecchi, J.; Manzi, G.; Nomade, S.; Pereira, A.; Falguéres, C.; Bahain, J.J.; Grimaud-Hervé, D.; Berto, C.; et al. A Human Deciduous Tooth and New 40Ar/39Ar Dating Results from the Middle Pleistocene Archaeological Site of Isernia La Pineta, Southern Italy. *PLoS ONE* **2015**, *10*, e0140091. [CrossRef]
110. Sala, B. Le nuove specie rinvenute a La Pinet. In *Preistoria in Molise, gli Insediamenti del Territorio di Isernia*; Peretto, C., Ed.; Centro Europeo di Ricerche Preistoriche, Collana Ricerche 3; Aracne editrice: Pomezia, Italy, 2006; pp. 36–38.
111. Breda, M.; Thun-Hohenstein, U.; Peretto, C. The deer from the early Middle Pleistocene site of Isernia La Pineta (Molise, Italy): Revised identifications and new remains from the last 10 years of excavation. *Geol. J.* **2015**, *50*, 290–305. [CrossRef]
112. Karner, D.B.; Marra, F.; Renne, P.R. The history of the Monti Sabatini and Alban Hills volcanoes: Groundwork for assessing volcani-tectonic hazards for Rome. *J. Volcanol. Geotherm. Res.* **2001**, *107*, 185–219. [CrossRef]
113. Bon, M.; Piccoli, G.; Sala, B. I giacimenti Quaternari di vertebrati fossili nell'Italia Nord-Orientale. *Mem. Di Sci. Geol.* **1991**, *18*, 185–231.
114. Sardella, R.; Mazzini, I.; Giustini, F.; Mecozzi, B.; Brilli, M.; Iurino, D.A.; Lembo, G.; Muttillo, B.; Massussi, M.; Sigari, D.; et al. Grotta Romanelli (Southern Italy, Apulia): Legacies and issues in excavating a key site for the Pleistocene of the Mediterranean. *Riv. Ital. Paleontol. Stratigr.* **2018**, *124*, 247–264.
115. Sardella, R.; Iurino, D.A.; Mecozzi, B.; Sigari, D.; Bona, F.; Bellucci, L.; Coltorti, M.; Conti, J.; Lembo, G.; Muttillo, B.; et al. Grotta Romanelli (Lecce, Southern Italy) Between Past and Future: New Studies and Perspectives for an Archaeo-geosite Symbol of the Palaeolithic in Europe. *Geoheritage* **2019**, *11*, 1413–1432. [CrossRef]
116. Mecozzi, B.; Iannucci, A.; Bona, F.; Mazzini, I.; Pieruccini, P.; Sardella, R. Rediscovering *Lutra lutra* from Grotta Romanelli (southern Italy) in the framework of the puzzling evolutionary history of Eurasian otter. *PalZ* **2022**, *96*, 161–174. [CrossRef]
117. Mecozzi, B.; Buzi, C.; Iannucci, A.; Micarelli, I.; Bona, F.; Forti, L.; Lembo, G.; Manzi, G.; Mazzini, I.; Muttillo, B.; et al. New human fossil from the latest Pleistocene Pieruccini of Grotta Romanelli (Apulia, southern Italy). *Archaeol. Anthropol. Sci.* **2022**, *14*, 27. [CrossRef]
118. Blanc, G.A.; Grotta Romanelli, I. Stratigrafia dei depositi e natura e origine di essi. *Arch. Per L'antropologia E La Etnol.* **1920**, *50*, 1–39.
119. Blanc, G.A. Grotta Romanelli II. Dati ecologici e paletnologici. *Arch. Per L'antropologia E La Etnol.* **1928**, *58*, 1–49.
120. Fornaca-Rinaldi, G. Il metodo 230Th/238U per la datazione di stalattiti e stalagmite. *Boll. Geofis. Teor. Appl.* **1968**, *10*, 3–14.
121. Fornaca-Rinaldi, G. 230Th/234Th dating of cave concretions. *Earth Planet. Sci. Lett.* **1968**, *5*, 120–122. [CrossRef]
122. Fornaca-Rinaldi, G.; Radmilli, A.M. Datazione con il metodo 230Th/238U di stalagmiti contenute in depositi musteriani. *Att. Della Soc. Toscana Sci. Nat.* **1968**, *75*, 639–646.
123. Romano, M.; Mecozzi, B.; Sardella, R. The Quaternary paleontological research in the Campagna Romana (Central Italy) at the 19th-20th century transition: Historical overview. *Alp. Mediterr. Quat.* **2021**, *34*, 1–21.
124. Iannucci, A.; Conti, J.; Curcio, F.; Iurino, D.A.; Mancini, M.; Mecozzi, B.; Strani, F.; Sardella, R. Middle Pleistocene mammal faunas of the area of Rome: Recent results and ongoing work on the MUST collection. *J. Mediterr. Earth Sci.* **2023**, *15*, 117–135.
125. Mecozzi, B.; Iannucci, A.; Sardella, R.; Curci, A.; Daujeard, C.; Moncel, M.-H. *Macaca* ulna from new excavations at the Notarchirico Acheulean site (Middle Pleistocene, Venosa, southern Italy). *J. Hum. Evol.* **2021**, *153*, 102946. [CrossRef] [PubMed]
126. Mancini, M.; Di Salvo, C.; Giallini, S.; Marini, M.; Simionato, M.; Caciolli, M.C.; Cavinato, G.P.; Moscatelli, M.M.; Polpetta, F.; Sirianni, P.; et al. The subsoil of the Colosseum and the detection of the ancient Tiber river Paleovalley (MIS 12-11) in Rome. *J. Mediterr. Earth Sci.* **2023**, *15*. [CrossRef]
127. Milli, S.; Palombo, M.R. The high-resolution sequence stratigraphy and the mammal fossil record: A test in the Middle-Upper Pleistocene deposits of the Roman Basin (Latium, Italy). *Quat. Int.* **2005**, *126/128*, 251–270. [CrossRef]
128. Bonadonna, P. Resti di *Hippopotamus amphibius* Linn. nei sedimenti del Pleistocene medio-inferiore della via Portuense, Roma. *Boll. Soc. Geol. Ital.* **1965**, *84*, 29–39.
129. Petronio, C. Nuovi resti di Ippopotami del Pleistocene medio inferiore dei dintorni di Roma e problemi di tassonomia e filogenesi del gruppo. *Geol. Rom.* **1986**, *25*, 63–73.
130. Ambrosetti, P. Cromerian fauna of the Rome area. *Quaternaria* **1967**, *9*, 267–283.
131. Karner, D.B.; Marra, F. Correlation of fluviodeltaic aggradational sections with 1 glacial climate history: A revision of the classical Pleistocene stratigraphy of Rome. *Geol. Soc. Am. Bull.* **1998**, *110*, 748–758. [CrossRef]
132. Blanc, A.C.; Lona, F.; Settepassi, F. Una torba ad *Abies*, malacofauna montana e criosedimenti nel Pleistocene inferiore di Roma—Il periodo glaciale Cassio. Ricerche sul Quaternario Laziale 1. *Quaternaria* **1955**, *2*, 151–158.
133. Marra, F.; Nomade, S.; Pereira, A.; Petronio, C.; Salari, L.; Sottili, G.; Bahain, J.-J.; Boschian, G.; Di Stefano, G.; Falguéres, C.; et al. A review of the geologic sections and the faunal assemblages of Aurelian Mammal Age of Latium (Italy) in the light of a new chronostratigraphic framework. *Quat. Sci. Rev.* **2018**, *181*, 173–199. [CrossRef]

134. Portis, A. *Contribuzioni Alla Storia Fisica del Bacino di Roma e Studii Sopra L'estensione da Darsi al Pliocene Superiore*; L. Roux e C: Roma, Italy, 1893; Volume I.
135. Terrigi, G. Le formazioni vulcaniche del bacino Romano considerate nella loro fisica costituzione e giacitura. *Att. Della Regia Accad. Dei Lincei* **1881**, *10*, 389–419.
136. Portis, A. Di alcuni avanzi fossili di grandi ruminanti principalmente della provincia di Roma. *Palaeontologr. Ital.* **1907**, *13*, 141–198.
137. Marra, F.; Karner, D.B.; Freda, C.; Gaeta, M.; Renne, P.R. Large mafic eruptions at the Alban Hills Volcanic District (Central Italy): Chronostratigraphy, petrography and eruptive behavior. *J. Volcanol. Geotherm. Res.* **2009**, *179*, 217–232. [CrossRef]
138. Caloi, L.; Palombo, M.R.; Petronio, C. La fauna quaternaria di Sedia del Diavolo (Roma). *Quaternaria* **1980**, *22*, 177–209.
139. Palombo, M.R.; Milli, S.; Rosa, C. Remarks on the biochronology of the late Middle Pleistocene mammalian faunal complexes of the Campagna Romana (Latium, Italy). *Geol. Rom.* **2004**, *37*, 135–143.
140. Fidalgo, D.; Madurell-Malapeira, J.; Martino, R.; Pandolfi, L.; Rosas, A. An Updated Review of The Quaternary *Hippopotamus* Fossil Records from the Iberian Peninsula. *Quaternary* **2024**, *7*, 4. [CrossRef]
141. Klein, R.G.; Scott, K. Glacial/Interglacial Size Variation in Fossil Spotted Hyenas (*Crocuta crocuta*) from Britain. *Quat. Res.* **1989**, *32*, 88–95. [CrossRef]
142. Cerilli, E.; Petronio, C. Biometrical variations of *Bos primigenius* Bojanus. In Proceedings of the International Symposium "Ongules/Ungulates 91", Toulouse, France, 2–6 September 1991; pp. 37–42.
143. Wright, E. The History of the European Aurochs (*Bos primigenius*) from the Middle Pleistocene to its Extinction: An Archaeological Investigation of Its Evolution, Morphological Variability and Response to Human Exploitation. Ph.D. Thesis, University of Sheffield, Sheffield, UK, 2013.
144. Mecozzi, B.; Bartolini Lucenti, S. The Late Pleistocene *Canis lupus* (Canidae, Mammalia) from Avetrana (Apulia, Italy): Reappraisal and new insights on the European glacial wolves. *Ital. J. Geosci.* **2018**, *137*, 138–150. [CrossRef]
145. Sansalone, G.; Bertè, D.F.; Maiorino, L.; Pandolfi, L. Evolutionary trends and stasis in carnassial teeth of European Pleistocene wolf *Canis lupus* (Mammalia, Canidae). *Quat. Sci. Rev.* **2015**, *110*, 36–48. [CrossRef]
146. Iannucci, A.; Sardella, R.; Strani, F.; Mecozzi, B. Size shifts in late Middle Pleistocene to Early Holocene *Sus scrofa* (Suidae, Mammalia) from Apulia (southern Italy): Ecomorphological adaptations? *Hystrix* **2020**, *31*, 10–20.
147. Bergmann, C. Über die Verhältnisse der Wärmeökonomie der Thiere zu ihrer Grösse. Güttinger Studien. *Gott. Stud.* **1847**, *3*, 595–708.
148. Bellucci, L.; Mazzini, I.; Scardia, G.; Bruni, L.; Parenti, F.; Segre, A.G.; Segre Naldini, E.; Sardella, R. The site of Coste San Giacomo (Early Pleistocene, central Italy): Palaeoenvironmental analysis and biochronological overview. *Quat. Int.* **2012**, *267*, 30–39. [CrossRef]
149. Palmqvist, P.; Gröcke, D.R.; Arribas, A.; Fariña, R.A. Paleoecological reconstruction of a lower Pleistocene large mammal community using biogeochemical (δ13C, δ15N, δ18O, Sr: Zn) and ecomorphological approaches. *Paleobiology* **2003**, *29*, 205–229. [CrossRef]
150. Palmqvist, P.; Pérez-Claros, J.A.; Janis, C.M.; Figueirido, B.; Torregrosa, V.; Grocke, D.R. Biogeochemical and ecomorphological inferences on prey selection and resource partitioning among mammalian carnivores in an early Pleistocene community. *Palaios* **2008**, *23*, 724–737. [CrossRef]
151. Palmqvist, P.; Rodríguez-Gómez, G.; Figueirido, B.; García-Aguilar, J.M.; Pérez-Claros, J.A. On the ecological scenario of the first hominin dispersal out of Africa. *L'Anthropologie* **2022**, *126*, 102998. [CrossRef]
152. Shackleton, N.J. New data on the evolution of pliocene climatic variability. In *Paleoclimate and Evolution, with Emphasis on Human Origins*; Vrba, E.S., Denton, G.H., Partidge, T.C., Burckle, L.H., Eds.; Yale University Press: New Haven, CT, USA, 1995; pp. 242–248.

quaternary

MDPI

Article

The Early Pleistocene Carnivoran of Coste San Giacomo (Anagni, Central Italy): Biochronological Implications

Luca Bellucci [1,2,*], Fabio Bona [3], Jacopo Conti [4], Beniamino Mecozzi [5], Flavia Strani [6] and Raffaele Sardella [5]

1 Museo di Geologia e Paleontologia, Sistema Museale di Ateneo, Università di Firenze, via Giorgio La Pira 4, I-50121 Firenze, Italy
2 Istituto Italiano di Paleontologia Umana, Piazza Ruggero Bonghi 2, I-03012 Anagni, Italy
3 Museo Civico dei Fossili di Besano, via Prestini 5, I-21050 Besano, Italy; fabgeo@libero.it
4 Polo Museale, Sapienza, Università di Roma, p.le Aldo Moro 5, I-00185 Roma, Italy; jacopo.conti@uniroma1.it
5 Dipartimento di Scienze della Terra, Sapienza, Università di Roma, p.le Aldo Moro 5, I-00185 Roma, Italy; beniamino.mecozzi@uniroma1.it (B.M.); raffaele.sardella@uniroma1.it (R.S.)
6 Departamento de Ciencias de La Tierra, Instituto Universitario de Investigación en Ciencias Ambientales de Aragón (IUCA), Universidad de Zaragoza, ES-50009 Zaragoza, Spain; flavia.strani@unizar.es
* Correspondence: luca.bellucci@unifi.it

Abstract: Coste San Giacomo (CSG) represents a significant paleontological site to investigate the faunal and environmental changes that occurred in Mediterranean Europe during the Early Pleistocene. In this work, we described for the first time the Carnivoran assemblage. We ascribed the fossil remains to the following taxa: *Ursus* sp., *Homotherium latidens*, *Canis etruscus*, *Pliocrocuta perrieri*, *Martellictis ardea* and *Vulpes alopecoides*. Considering the value of the carnivoran taxa here identified, we discuss their particular biochronological significance, since the CSG site records the last occurrence of *P. perrieri* and the first occurrences of *H. latidens*, *C. etruscus*, *M. ardea* and *V. alopecoides* for the Italian Peninsula. These results will allow us to improve the data of the biochronological scheme of the Villafranchian European Land Mammal Age, recognizing the earliest dispersals and latest occurrences across Europe.

Keywords: Carnivora; Pleistocene; Villafranchian; *Homotherium*; *Pliocrocuta*; *Canis*; *Vulpes*; *Pannonictis*

Citation: Bellucci, L.; Bona, F.; Conti, J.; Mecozzi, B.; Strani, F.; Sardella, R. The Early Pleistocene Carnivoran of Coste San Giacomo (Anagni, Central Italy): Biochronological Implications. *Quaternary* **2024**, *7*, 57. https://doi.org/10.3390/quat7040057

Academic Editor: Miriam Belmaker

Received: 31 May 2024
Revised: 16 October 2024
Accepted: 20 October 2024
Published: 12 December 2024

1. Introduction

The Coste San Giacomo site (hereinafter CSG), recently dated around 2.2 Ma [1], represents a key palaeontological site to investigate the environmental and faunal changes that occurred in the European Mediterranean region during the Gelasian (Early Pleistocene) as a result of major climatic changes at a global scale [2,3]. Forest diversity decreased because of a progressive decline in, and loss of, subtropical taxa during the Early and Middle Pleistocene [4].

The CSG site was discovered by Italo Biddittu, archaeologist of the Istituto Italiano di Paleontologia Umana (hereinafter IsIPU) during a survey carried out in September 1978. Intense fieldwork and scientific activities were performed by IsIPU researchers and, in the last fifteen years, in collaboration with the Dipartimento di Scienze della Terra of the Sapienza Università di Roma ([5–11] and references therein).

The CSG site is located 70 km southeast of Rome (central Italy), in the Anagni basin (Figure 1). This deeply faulted and extensional depression covers an area of around 20 km^2 and was produced during the initial phases of the neotectonic evolution of the Apennines Chain. This basin developed largely between the Late Pliocene and the early part of the Middle Pleistocene [1,12] and, in addition to CSG, other palaeontological sites such as Fontana Acetosa (FA), Colle Marino (CM) and Fontana Ranuccio (FR) have yielded important mammal records spanning from the earliest (CSG and FA) to the Middle (CM and FR) Pleistocene [5].

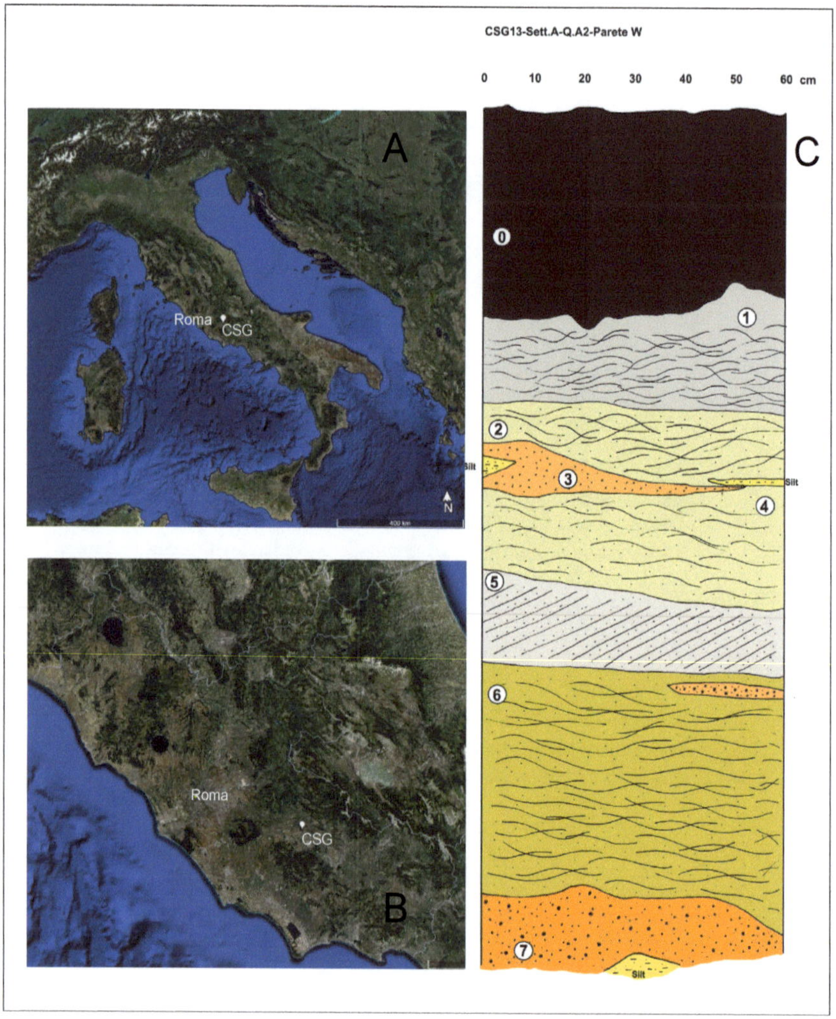

Figure 1. (**A**) CSG geographical position within the Italian Peninsula. (**B**) CSG geographical position within the Latium Region. (**C**) Stratigraphical scheme of the CSG section excavated in 2013: (0) The first 50 cm below the surface have been subject to the action of a plough. (1) Compact silt partly cemented by calcium carbonate recognised by the presence of whitish lineations shows an oblique trend in relationship to the sedimentation. The precipitation of the mineral is probably catalysed by the presence of the root systems of plants. (2, 3, 4) Fine festooned sand with scant traces of iron and manganese oxides. (5) Laminated sands with strong inclination of approximately 28°. (6, 7) Coarse sands with gravelly lenses; the gravels are mostly composed of travertine fragments. The upper portion is characterised by slightly undulating sands and gravelly lenses (20–30 cm high). Towards the bottom, the gravelly component increases, with abundant presence of iron and manganese oxides probably due to a braided river course.

Owing to the extensive large and small mammal sample, CSG has been regarded as representative of the Middle Villafranchian Large Mammal Age and of the Late Villanyan Small Mammal Age, respectively (Coste San Giacomo Faunal Unit) [13,14]. It is worth recalling the following significant biochronological occurrences recorded in the CSG site: the coexistence of both the European mastodon *Anancus arvernensis* and the newcomer southern mammoth *Mammuthus meridionalis*; the occurrence of *Hippopotamus*, providing clear evidence of early dispersal events of African taxa prior to the early *Homo* radiation into Europe (Hippo event *sensu* Iannucci et al. [15]); the occurrence of the vole *Mimomys pliocaenicus*—the CSG specimens represent the largest collection in Europe—and the occurrence of the water mole *Galemys kormosi*, the first Desmaninae reported in the Italian Peninsula.

In this paper, the carnivoran palaeoguild from CSG is studied for the first time, providing the first formal description of the fossil specimens. Our results will improve the data of the biochronological scheme of the European Land Mammal Age, considering the value of the carnivoran taxa identified here for recognizing earliest dispersals and latest occurrences across the continent.

2. Geological and Paleontological Setting

The site of Coste San Giacomo was discovered by Italo Biddittu during a survey carried out in September 1978 and is located near Anagni (Frosinone), at approximately 50 km southeast of Rome (Latium, central Italy) (41°45′21.7″ N; 13°05′49.4″ E).

After its discovery, trenches were excavated in 1985, 1989 and 1990, identifying traceable levels with fossil vertebrates in the yellow sands [6]. In September 2009, a 46 m deep core was drilled, which allowed researchers to obtain new magnetostratigraphical, pollen and small mammal data and confirmed the possible age of the mammal assemblage to be around 2.1 Ma, in a reversed phase before the base of the Olduvai chron [7]. In order to better investigate the chronostratigraphic setting of the site, four pits were excavated in 2011, and three of these yielded vertebrate remains [7]. The last fieldwork activities were performed in 2013, coordinated by Fabio Parenti (IsIPU) and Raffaele Sardella (Sapienza Università di Roma) and authorised by the Soprintendenza per i Beni Archeologici del Lazio. During these excavations, vertebrate bones were collected in a single fossiliferous level.

The age of the faunal assemblage of CSG was recently refined by Florindo et al. [1] at 2.233 ± 0.032 Ma.

The stratigraphic series of CSG here described is based on the 2013 field excavation. This is characterised by the presence of five main sedimentary bodies (Figure 1):

- Starting from the surface, about 50 cm of deposit have been subjected to the action of the plough (0).
- Compact silt partly cemented by calcium carbonate, which can be recognised by the presence of whitish lineations that have an oblique trend with respect to the sedimentation, and the precipitation of the mineral is probably catalysed by the presence of the root systems of plants (1).
- Fine festooned sand with scant traces of iron and manganese oxides (2, 3, 4).
- Sand laminate with strong inclination (approximately 28°) (5).
- Coarse sand with gravelly lenses (the gravels are mostly composed of fragments of travertine). The upper portion of this last sedimentary body is characterised by slightly undulating sands and gravelly lenses (20–30 cm of development); towards the bottom, the gravelly component increases, with an abundant presence of iron and manganese oxides (probable braided river course) (6, 7).

Most of the fossil finds, and the best-preserved ones, come from the gravelly basal bodies (units 6–7; Figure 1). The accumulation of bone was possibly related to the presence of river channels in a fluviolacustrine environment.

The updated fauna list of the CSG large-mammal assemblage, excluding the carnivorans that are the focus of this work, is *Anancus arvernensis*, *Mammuthus meridionalis*, *Stephanorhinus* sp., *E. senezensis* aff. *E. sen. stehlini*, *Eucladoceros* sp., *Axis* cf. *lyra*, *Croizetoceros* cf. *ramosus*, *Leptobos* sp., *Gallogoral meneghini*, *Gazellospira torticornis*, *Gazella borbonica*, *Sus strozzii*, *Hippopotamus* sp. and *Macaca sylvanus*. Small mammals were described by Bona et al. [9]: *Mimomys pliocaenicus*, *Mimomys* gr. *tigliensis/tornensis*, *Apodemus* sp., *Sciurus* cf.

S. warthae, *Castor fiber*, *Hystrix refossa*, *Prolagus italicus*, *Soricinae indet.*, *Beremendia fissidens*, *Sorex* cf. *S. minutus*, *Talpa minor*, *Talpa* sp. and *Galemys kormosi*.

Finally, Villa et al. [16]) identified Amphibia (Bufonidae indet., *Pelophylax* sp., *Rana* sp., Ranidae indet., Anura indet.) and Reptilia remains (Lacertidae indet., *Pseudopus* sp., *Natrix* sp., Serpentes indet., Reptilia indet.). This assemblage comprises taxa that are common in the Italian Quaternary, but also presents the first occurrence of the large anguid *Pseudopus* from the Apennine Peninsula in this period.

The dietary analyses of the ungulate community revealed the presence of a mosaic of habitats, from woodlands to open landscapes [8,11,17]. However, the presence of humid environments was probably limited to the surroundings of the watercourse, as documented by fossils of *Hippopotamus*, *Castor fiber*, *Mimomys pliocaenicus* and *Galemys kormosi* and by the herpetofauna.

3. Materials and Methods

The carnivoran specimens from Coste San Giacomo were discovered in 1985, 1989 and 1990 in the yellow sands that were exposed when trenches were excavated at the base of a terrace during the IsIPU field activities. Others were collected from the ground surface where the fossil-bearing yellow sands are ploughed for agriculture. Finally, some specimens were excavated during the 2011 and 2013 campaigns, and, consequently, labelled "CSG 11-. . ." or "CSG 13-. . .".

These specimens are currently housed at the IsIPU depository and in the Museo Archeologico Ernico, both in Anagni (Frosinone, Italy). The determination of the palaeontological specimens is based on the anatomical feature descriptions and morphometric comparative analyses. Morphological and morphometric data used for comparison are from the cited literature. All measurements are taken with a digital calliper to the nearest 0.1 mm.

4. Systematic Palaeontology

4.1. *Class Mammalia LINNAEUS, 1758*

Order Carnivora BOWDICH, 1821
Family Ursidae FISCHER DE WALDHEIM, 1817
Genus *Ursus* LINNAEUS, 1758

4.1.1. Referred Material

CSG 981147: upper canine; CSG 82-5: lower canine; CSG 981149: lower fourth premolar; CSG 13-272: lower fourth premolar; CSG sd-76: lower third molar (Figure 2, *Ursus*).

Figure 2. CSG *Ursus* specimens. CSG 981149: fourth premolar, A1 labial view, A2 buccal view; CSG sd-76: right lower third molar, B occlusal view. Scale bar 10 mm.

4.1.2. Description

The upper left canine CSG 981147 shows an elongated and narrow profile mainly consisting of only the root; the crown is broken almost at the height of the collar, probably due to post-mortem damage. The lower left canine CSG 82-5 is poorly preserved; the tip is missing and it lacks the root. The enamel profile of the crown is jagged rather than worn. The right lower fourth premolars CSG 981149 and CSG 13-272 show a narrow occlusal surface with an ellipsoid profile; a thin crest joins a barely developed metaconid and paraconid, passing through a pronounced protoconid that occupies a large part of the chewing surface. There are no tubercles or accessory cusps; the root is double and divided, even if CSG 981149 lacks the anterior branch and CSG 13-272 lacks the posterior one. There are no signs of wear on the occlusal surface; thus, they probably belonged to young individuals. The third right lower molar CSG sd-76 shows an ellipsoid morphology with a rounded proximal margin. On the chewing surface, there are several ridges and accessory cusps, but it is still possible to recognise the mesoconid, the entoconid and the hypoconid, which is less visible, whereas the protoconid is the only cuspid well developed. This tooth does not present any sign of wear on the occlusal surface; juvenile status is also confirmed by the absence of the root and any sign of its fusion with the crown.

4.1.3. Discussion

The *Ursus* material from CSG is composed of only isolated teeth that do not allow a specific determination, since an extremely high morphological variability of the teeth is observed in different Plio-Pleistocene bear species [18]. The CSG specimens have been morphometrically compared with fossil remains reported in Europe during the Early Pleistocene belonging to the *U.* cf. ex. group *etruscus* and *U. thibetanus-minimus*. These two species are mainly distinguished by the size and morphology of the carnassial teeth, the upper fourth premolar and the first lower molar, respectively, that represent the most taxonomically significant teeth according to Wagner et al. [18] and by the general latero-medial compression of the teeth in *U. minimus* [19]. The lower fourth premolar and the third molar teeth were compared with other specimens from different Italian and European localities (Figures 3 and 4) (Table S1). These morphometric analyses identify the main size groups, the larger *U. etruscus* and the smaller *U. minimus*. The CSG teeth occupy the central part of the graphs, lacking distinct identification with either of these groups. Considering the morphological variability observed in Pleistocene bears and the intermediate size of the CSG specimens, the material is here ascribed to *Ursus* sp.

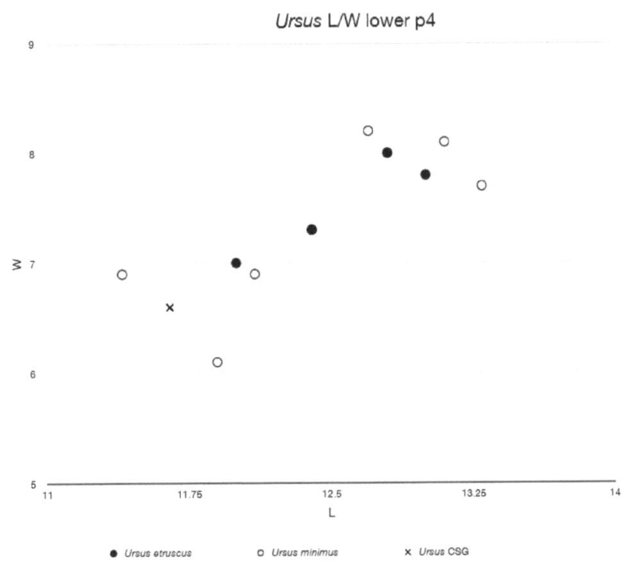

Figure 3. Scatter plot of the length/width of the *Ursus* lower p4.

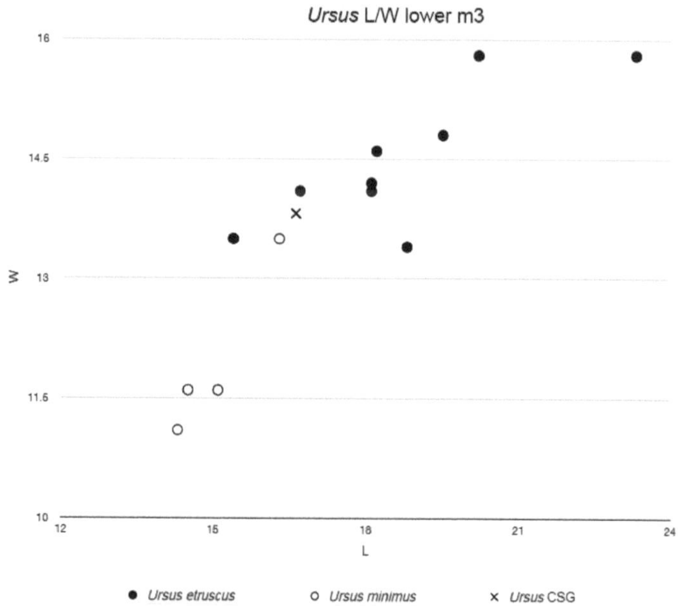

Figure 4. Scatter plot of the length/width of the *Ursus* lower m3.

4.2. Family Felidae GRAY, 1821

Genus *Homotherium* FABRINI, 1890
Homotherium latidens OWEN, 1846

4.2.1. Referred Material

CSG 11-172: a fragment of an upper canine; CSG 13-464: an upper fourth premolar; CSG 13-469: a lower canine; CSG 90-13: a lower third premolar; CSG sd-61 a fourth premolar (Figure 5).

4.2.2. Description

The scimitar-toothed cat *Homotherium* is represented by only isolated teeth. The small portion (28 mm long) of the upper canine is serrate (22 in 10 mm). In the anterior edge of the upper fourth premolar, the presence of a preparastyle is not observed. The morphology of the parastyle displays a vertical anterior side. The wear degree suggests an aged specimen. The lower third premolar is quite reduced in comparison with the fourth premolar; it is a single-rooted and unicuspid tooth with crenulations in both the distal and mesial margins. A marked cingulum is also present. The lower fourth premolar is not well preserved; it displays a high main cuspid and two mesial and distal cuspids, the last one buccolingually wider. These cuspids are mesially oriented.

4.2.3. Discussion

The length/width ratio of the upper fourth premolar from CSG has been compared with specimens from selected European Early to Middle Pleistocene localities, as listed in Table S2. The morphometric analysis confirms the attribution of the CSG specimen to *Homotherium latidens* (Figure 6). As many studies have stated, the Eurasian homotheres show a high degree of variation in the overall size and craniodental morphology and proportions. In this context, the extensive sample collected at Incarcal (Spain) shows a high morphological polymorphism, embracing nearly all the variation in the *Homotherium* record from Eurasia [20]. Following this, Turner and Antón [20] proposed attributing all the

Plio-Pleistocene Eurasian fossils to *Homotherium latidens*. In a splitter perspective, the latest Pliocene to earliest Pleistocene Eurasian homotheres, which have relatively longer upper canines and a more robust and convex mental area in the mandibula, could be referred, respectively, to *Homotherium nestianus* (western Europe) and to *Homotherium davitasvili* (Caucasus region) [21]. Although the isolated teeth from CSG cannot be useful clues to solve such a taxonomical dilemma, the hypothesis to consider only one polymorphic species, *Homotherium latidens*, for the Plio-Pleistocene Eurasian material seems to be the most conservative approach.

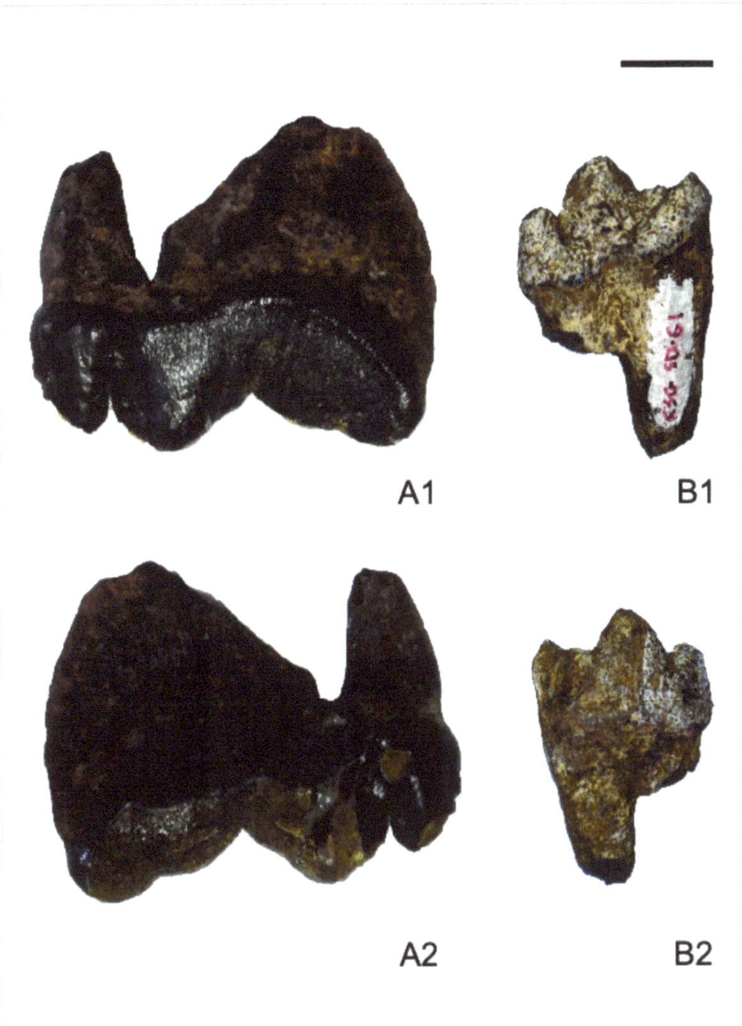

Figure 5. CSG *Homotherium* specimens. CSG 13-464: right upper fourth premolar, A1 buccal view, A2 lingual view; CSG sd-61: right lower fourth premolar, B1 buccal view, B2 lingual view. Scale bar 10 mm.

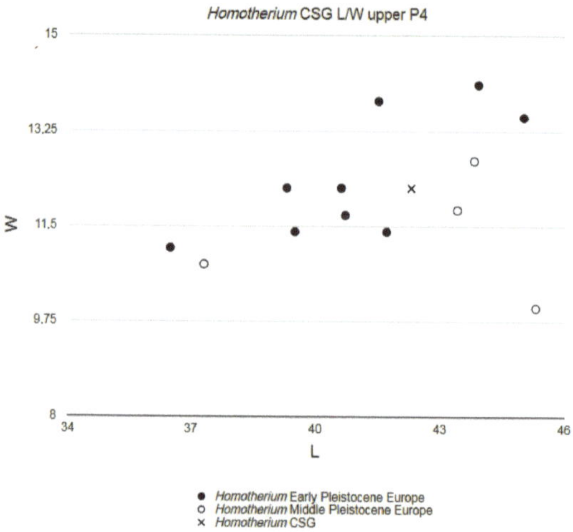

Figure 6. Scatter plot of the length/width of the *Homotherium* upper p4.

4.3. *Family Hyaenidae GRAY, 1821*

Genus *Pliocrocuta* Kretzoi, 1938

Pliocrocuta perrieri (Croizet and Jobert, 1828)

4.3.1. Referred Material

CSG 11-48: an upper third incisor; CSG 11-99: an upper juvenile third deciduous tooth; CSG sd-59: an incomplete upper fourth premolar; CSG 11-87: a lower third premolar; CSG 981146 and CSG 11-80: two lower broken fourth premolars; CSG 56700: three coprolites (another 54 without any inventory number) (Figure 7).

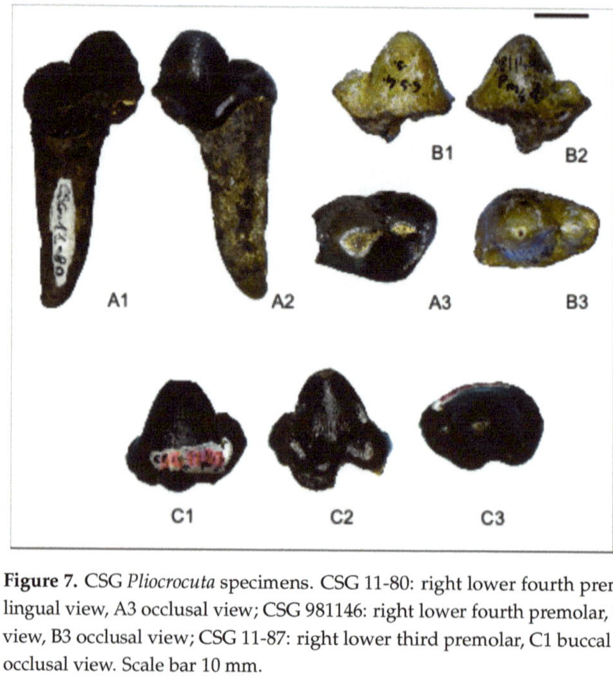

Figure 7. CSG *Pliocrocuta* specimens. CSG 11-80: right lower fourth premolar, A1 buccal view, A2 lingual view, A3 occlusal view; CSG 981146: right lower fourth premolar, B1 buccal view, B2 lingual view, B3 occlusal view; CSG 11-87: right lower third premolar, C1 buccal view, C2 lingual view, C3 occlusal view. Scale bar 10 mm.

4.3.2. Description

The upper fourth premolar CSG sd-59 is broken, and only the parastyle is preserved. The lower premolars display a rectangular occlusal outline. The lower third premolar CSG 11-87 is formed by a well-developed protoconid and a defined distal accessory cuspid and is lacking the mesial accessory cuspid. The upper deciduous carnassial (CSG 11-99) has a sub-rectangular outline; the parastyle is well-defined and labially set and the paracone and the metastyle blade are aligned. Both the lower fourth premolars, CSG 981146 and CSG 11-80, are broken, and the mesial cuspids are not preserved. The protoconid is well-developed, as in the living bone-cracking hyaenas. The distal accessory cuspid is also well-defined and is labially placed behind the protocone.

4.3.3. Discussion

The morphology of these teeth strongly resembles the specimens described by Viret ([22], pp. 46–52, Planches 5: Figures 1 and 2; Planches 6: Figures 1–8; Planches 7: Figures 1–5; Planches 8: Figures 1 and 2) as "*Crocuta* (*Plesiocrocuta*) *perrieri* subgen. nov.", in particular in the reduction stage of the accessory cuspids in the lower premolars [23]. Only the lower third premolar CSG 11-87 was morphometrically compared with *Pachycrocuta brevirostris* and *Pliocrocuta perrieri* specimens from different selected sites (Table S3), since the other teeth are incomplete. As shown in Figure 8, the CSG specimen fits in the *Pliocrocuta perrieri* variability. For these reasons, the CSG remains are here ascribed to the bone-cracking *Pliocrocuta perrieri*.

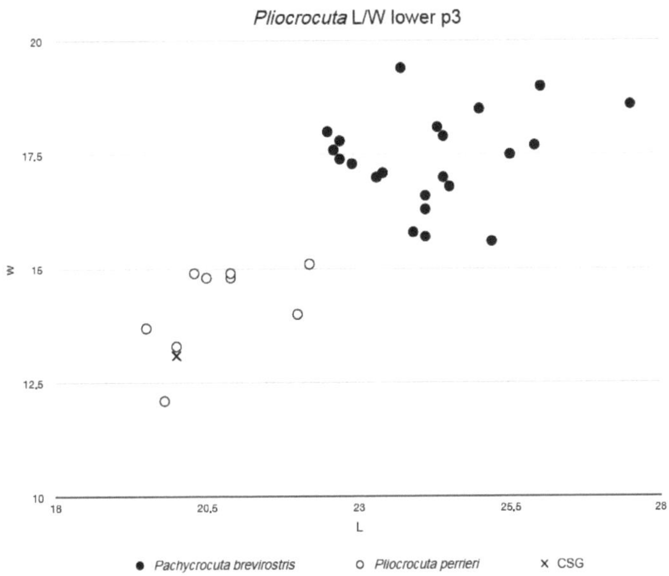

Figure 8. Scatter plot of the length/width of the *Pliocrocuta* lower p3.

4.4. *Family Canidae GRAY, 1821*

Genus *Canis* LINNAEUS, 1758
Canis etruscus FORSYTH MAJOR, 1877

4.4.1. Referred Material

CSG sd-81: upper canine; CSG sd-74: portion of an upper canine; CSG sd-63: portion of a lower canine; CSG sd-75: portion of a canine; CSG 78-40: portion of a canine; CSG 56699, CSG 581152: lower fourth premolars; CSG sd-62: lower fourth premolar; CSG 961159: fragmented fourth premolar; CSG 56690, CSG 13-233: two lower first molars; CSG 13-439: lower second molar; CSG sd-60: coprolite (Figure 9).

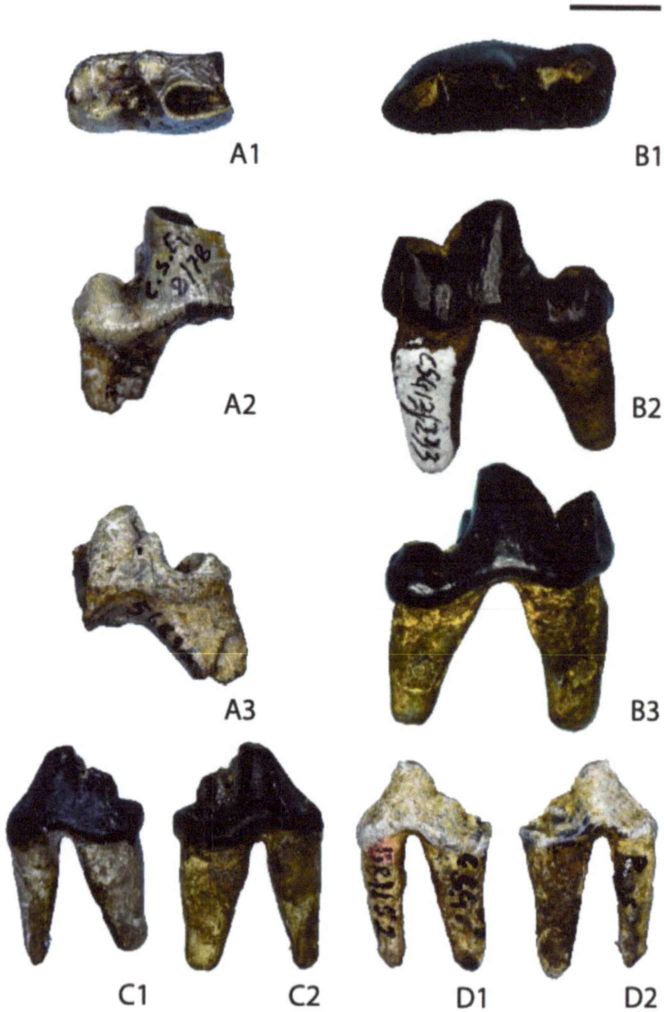

Figure 9. CSG *Canis* specimens. CSG 56690: left lower first molar, A1 occlusal view, A2 lingual view, A3 buccal view; CSG 13-233: right lower first molar, B1 occlusal view, B2 lingual view, B3 buccal view; CSG 56699: left lower fourth premolar, C1 buccal view, C2 lingual view; CSG 581152: left lower fourth premolar, D1 buccal view, D2 lingual view. Scale bar 10 mm.

4.4.2. Description

This species is the most represented amongst the CSG carnivoran specimens. The canines, both upper and lower, have curved crowns and are mesiodistally elongated. The fourth lower premolars are also mesiodistally elongated and show a symmetric protoconid and two distal accessory cuspids; the first is well developed, whereas the second is smaller and set on the distal cingulum. Only one lower first molar is complete (CSG 13-233). Its paraconid is less developed than the protoconid, which represents the main cusp of the tooth. The metaconid is well developed and is distinguished from the protoconid. The talonid basin is deep and surrounded by three cuspids. The hypoconid, labially set, is the largest cuspid of the talonid, whereas the entoconid, lingually placed, is less developed. The other cuspids generally present in the Pleistocene canids are reduced (mesoconid) or absent (entoconulid). The lower second molar is mesiodistally elongated, appearing rectangular in the occlusal view. It displays three main cuspids; the protoconid and hypoconid are lingually set, whereas the metaconid is labial. The protoconid is the largest, and is mesially located, whereas the hypoconid is quite reduced and is set on the distal cingulum. The metaconid is quite large and is developed along the mesiolabial cingulum. The entoconid is present and is distally set in relationship to the metaconid.

4.4.3. Discussion

The morphological characteristics useful to distinguish the two *Canis* species recorded in the Italian Peninsula during the Early Pleistocene, *Canis etruscus* and *Canis arnensis*, are that the Etruscan wolf possesses some wolf-like dental features which cannot be found in *C. arnensis*. These are detected, for example, in the lower molars, with the first molar hypoconid being stronger than the entoconid, and a larger protoconid compared with the metaconid on the second molar [24]. These features are observed in the CSG first and second molars. Moreover, the lower M_1 CSG 13-233 has been morphometrically compared with selected *Canis etruscus* and *C. arnensis* specimens from the Italian Peninsula (Table S4), strengthening the attribution of the CSG canids to the Etruscan wolf (Figure 10). In fact, the Etruscan wolf is considerably larger in general terms compared with *C. arnensis* [24].

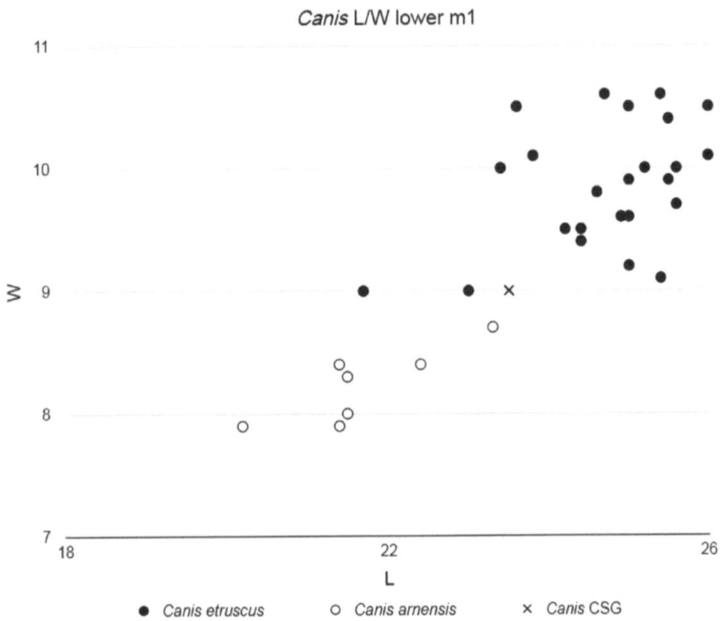

Figure 10. Scatter plot of the length/width of the *Canis* lower m1.

4.5. *Genus Vulpes FRISCH, 1775*

Vulpes alopecoides (DEL CAMPANA, 1913) pro parte

4.5.1. Referred Material

CSG 981154: upper fourth premolars; CSG sd-80: lower fourth premolar; CSG 11-173, CSG sd-79: lower first molars (Figure 11).

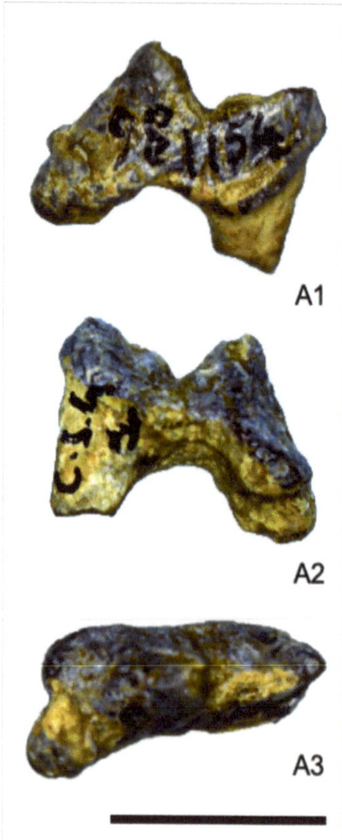

Figure 11. CSG *Vulpes* specimen. CSG 981154: left upper fourth premolar, A1 lingual view, A2 buccal view, A3 occlusal view. Scale bar 10 mm.

4.5.2. Description

The upper fourth premolar is mesiodistally elongated and shows a well-developed protocone, which is anteriorly projected in the occlusal view. The mesiolabial cingulum is quite marked, as is the distal cingulum. The parastyle is absent. The paracone is the largest and highest cuspid. The metastyle is quite elongated, with an incipient cuspid on the distal part which is connected to the distolabial cingulum. The lower fourth premolar preserves only the distal portion, with a marked distal cuspid and cingulum. The two lower first molars are both fragmentary. The lower carnassial is relatively mesiodistally elongated; the paraconid is large, but it is, however, less developed and lower than the protoconid, being the largest cuspids of the tooth. The metaconid is set on the lingual margin, just distally to the protoconid, and is reduced compared with the other trigonid cuspids, but is larger than the other talonid cuspids. Four cuspids are on the talonid, which is delimited by a round and deep talonid basin. On the labial margin, two cuspids occur; the larger entoconid is distally located, and the incipient entoconulid is set between the metaconid and the entoconid. The cristid-like cingulid is distally placed in relationship to the entoconid and to the hypoconid, just along the distal margin of the tooth. The hypoconid is stout and as high as the entoconid, and it is well separated from the protoconid. The mesoconid is absent.

4.5.3. Discussion

The CSG *Vulpes* teeth specimens can be ascribed to *V. alopecoides*, according to the diagnosis reported in Bartolini Lucenti and Madurell-Malapeira [25]. In particular, the upper fourth premolar is mesiodistally elongated, with a large paracone and a pointed and large protocone, while the parastyle is absent.

The presence of fox fossils in the Plio-Pleistocene Eurasian record is quite scarce, leading to the identification of different species and several hypotheses of their relative taxonomic relationships. A large review of *Vulpes* fossil material was published by Bartolini Lucenti and Madurell-Malapeira [25] that attributed all Early Pleistocene European fossils to the single, polymorphic species *Vulpes alopecoides*. According to the chronological setting of the site, the CSG record represents one of the oldest attested occurrences in western Europe.

4.6. Family Mustelidae FISCHER, 1817

Genus *Martellictis* BARTOLINI LUCENTI, 2018
Martellictis ardea (GERVAIS, 1848-1852)
Referred material
CSG sd-78: hemimandible

4.6.1. Description

This specimen (Figure 12) consists of an incomplete hemimandible, which preserves only the lower first molar. The corpus is quite gracile and low; the opening of the masseteric fossa is distally placed close to the margin of the lower first molar. The lower carnassial is mesiodistally elongated and is quite slender. The paraconid is well extended, and it is less high than the protoconid. Even though the tooth shows advanced wear of the cuspids, the metaconid seems to be located posteriorly to the protoconid. In addition, the metaconid is as high as the paraconid. The talonid is short; only the hypoconid is recognizable. Considering the wear stage, the presence of other cuspids cannot be excluded.

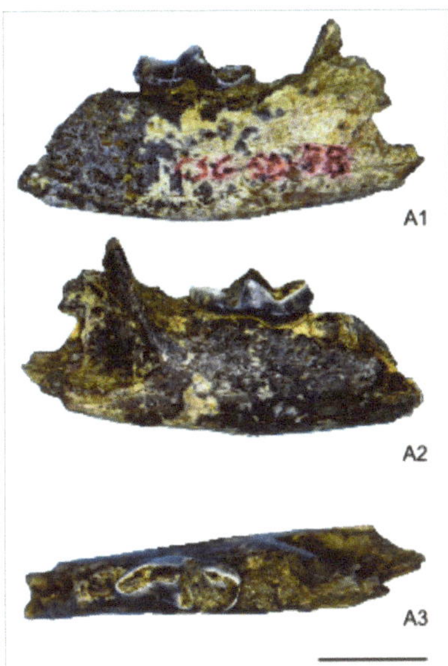

Figure 12. CSG sd-78: *Martellictis ardea* hemimandible, A1 buccal view, A2 lingual view, A3 occlusal view. Scale bar 10 mm.

4.6.2. Discussion

Lyncodontini remains are relatively scarce in the European fossil record, complicating the reconstruction of the evolutionary history of this group. Recently, Bartolini Lucenti [26] introduced a new genus, *Martellictis*, based on the samples from the Late Pliocene to the Early Pleistocene of Europe, Olivola (Italy), Perrier-Les Étouaires (France) and Saint Vallier (France), recognizing only the species *Martellictis ardea*.

Morphologically, there is no clear separation between *Pannonictis* and *Martellictis* based on the lower first molar. The only feature observable in the CSG fossil is the development of the mandible corpus, which appears to be relatively gracile (even if incomplete), resembling *Martellictis* [26]. The height (11.6 mm) and the breadth (5.6 mm) of the CSG hemimandible at m1 falls in the *Martellictis* variability, according to the data reported in Ros-Montoya et al. [27], strengthening the attribution to this taxon.

5. Discussion

5.1. Biochronological Implications

The CSG carnivoran assemblage includes *Ursus* spp.; the scimitar-toothed cat, *Homotherium latidens*; the bone-cracking hyena, *Pliocrocuta perrieri*; the Etruscan wolf, *Canis etruscus*; *Vulpes alopecoides* and the mustelid *Martellictis ardea*. These remains represent a key sample to investigate carnivoran diversity and, consequently, its biochronological and palaeogeographical implications during the Gelasian (Early Pleistocene) in Mediterranean Europe.

A "historic" major faunal turnover that affected the Carnivoran palaeoguild in western Europe during the Gelasian was the so-called "Wolf event", initially dated to ca. 2.0–1.7 Ma [28,29], best represented by the First Appearance Datum of *Canis* spp. However, findings of *C. etruscus* and other species traditionally considered to be involved in the "Wolf event" are today known from several European sites [15]. These fossil remains, in fact, backdated the "Wolf event" as defined by Azzaroli [28], pointing out also its diachronous nature [30], and thus, questioning its biochronological significance [31]. A recent, extensive review of this significant biochronological event is discussed in Iannucci et al. [15], where it is reported that the arrival of wolf-like canids in Europe is at least as early as the Coste San Giacomo FU (middle Villafranchian), even if putative evidence could suggest an even older age. Moreover, there has been a conceptual shift from considering the "Wolf event" from the earliest appearance of *Canis* in the fossil record to an abundance increase, i.e., the late Villafranchian "massive expansion", as it has been called by Azzaroli et al. [29], and is generally agreed in subsequent research [15]. At the present time, *Canis etruscus* remains from CSG are the clear, earliest evidence of wolf-like canid occurrence from the Early Pleistocene in the Italian Peninsula, concurrently with those of Pantalla, now also dated to ca. 2.2 Ma [32], and maybe in Europe as a whole, pending the taxonomical attribution and the chronology of other Gelasian specimens. In Italy, for instance, several findings of *Canis* sp. are known from localities placed within the CSG Faunal Unit, namely, Fontana Acetosa [33], Montagnola Senese [34], Quercia [35], Torre Picchio [36] and Vigna Nuova [37]. The CSG specimens slightly pre-date the Italian findings from the Olivola (Val di Magra, Tuscany, ca. 2 Ma; Rook and Martínez-Navarro [38]) and Poggio Rosso (Upper Valdarno, Tuscany; ca. 1.9–1.8 Ma; Napoleone et al., 2001 [39,40]; Bartolini Lucenti and Rook [24]) sites, where three different species of canids, the wolf-like *C. arnensis* and *C. etruscus* and the Lycaon-like *Xenocyon falconeri*, occur [28,41,42].

Another important biochronological signal recorded in the CSG large-mammal assemblage is the last occurrence in the Italian Peninsula of the medium-sized hyaenid *Pliocrocuta perrieri*. This taxon also occurred in the Montopoli site (Tuscany; ~2,6 Ma, Gelasian, Montopoli FU) [43], thus, well characterising the Italian middle Villafranchian assemblages. *Pliocrocuta perrieri* (= *Pachycrocuta perrieri*) was also reported in the Valle Catenaccio site (Latium; Gelasian, Coste San Giacomo FU?) by Cassoli and Segre Naldini [33]; Segre Naldini and Valli [44] verbatim indicated "*Pachicrocuta* sp.". This hyaenid tooth (V.C. 981247), housed at the IsIPU depository (Anagni, Frosinone), is a worn lower premolar and does not provide here any specific determination. Recently, Iannucci et al. [45] redescribed and revised the taxonomic attribution of a hyena hemimandible recovered from Paciano (Umbria, Italy), originally reported in the early 1900s and attributed to *Hyaena striata* (= *Hyaena hyaena*) and subsequently listed as a record of the giant hyena *Pachycrocuta brevirostris*, but now assigned to *Pliocrocuta perrieri*. Unfortunately, the exact provenance of the hyena from Paciano is unknown, as is its dating [45]. The last occurrences in Mediterranean Europe of *P. perrieri* are recorded in the Iberian Peninsula, where it is documented at the site of Fonelas P-1 and dated around ~2.1–1.9 Ma [46], and in Greece, at the site of Gerakarou 1 (= *Pliohyaena perrieri*; Koufos [47]), biochronologically referred to the late Villafranchian, and in particular, to the Olivola Faunal Unit [48]. In Georgia, it is recorded at Dmanisi and dated around 1.8 Ma [49]. It is worth noting that in these last sites, *P. perrieri* is recorded together with *Pachycrocuta brevirostris* [46]. The short-faced giant hyaena *Pachycrocuta brevirostris*, sister-taxon of *Pliocrocuta* [50,51] or even its direct descendant [23,50–52], in fact gradually replaced *Pliocrocuta*, becoming one of the most widespread carnivoran species in Eurasia during the second half of the Early Pleistocene. Moreover, the dispersal across Eurasia of *Pachycrocuta brevirostris* has been regarded as a major event during the Early Pleistocene (e.g., the "large canids and *Pachycrocuta*" events in Sardella and Palombo [31]; the "*Pachycrocuta brevirostris* event" in Martínez-Navarro et al. [53]. According to Vinuesa et al. [54], *Pliocrocuta* probably displayed a less-complex social behaviour than extant bone-cracking hyenas, especially compared with the spotted hyaena, *Crocuta*. These scholars, on the basis on the relative size of the anterior cerebrum, have hypothesised that *Pliocrocuta* is more comparable in its home range and forelimb usage to the living genera *Hyaena* and *Parahyaena* and is probably also lacking the complex pack-hunting and territorial behaviours of *Crocuta*. The spotted hyaena is, in fact, an active pack hunter [55,56], whereas both *Hyaena* and *Parahyaena* are basically scavengers, although they can also chase small prey alone [55,57,58]. In conclusion, *P. perrieri* went extinct in Mediterranean Europe around 2–1.8 Ma, suffering from ecological competition with both the larger, bone-cracking hyaena *P. brevirostris* in accessing carcasses, and with the several other carnivoran species present at the time, especially the hypercarnivorous and ambush-hunting felids, which limited its possibilities to deviate towards a different feeding behaviour [28,59–61]. Finally, several coprolites have been found in the CSG site, thus questioning the active role in the massing of carcass parts played by the hyaenas as in the Italian site of Poggio Rosso [62]. Taphonomic studies could clarify this hypothesis.

The lion-sized sabre-toothed cat *Homotherium latidens* was the dominant predator in the CSG large-mammal assemblage. *Homotherium* spread into Europe around 3 Ma, probably when the climate became cooler and drier and vegetation cover decreased, becoming an important faunal taxon in the Plio-Pleistocene Eurasian carnivore guild. According to [63], the First Appearance Datum of *Homotherium* in Europe is recorded in the fauna of the Odessa Catacombs (end of MN15). Nowadays, *H. latidens* is recognised as a single variable species in the Plio-Pleistocene of Eurasia (see, e.g., Sardella and Iurino [21]; Antón et al. [64] and references therein). The first occurrence of this taxon in the Italian Peninsula is not clear. Two upper canines of a Machairodontinae specimen were collected at the site of "Fornaci al Ponte di Castiglione" (= "Fornaci di Pievefosciana" or "Pievefosciana" or "Pieve Fosciana") in the Garfagnana area (Tuscany) by De Stefani in 1882. These fossils, IGF 53V and IGF 112V, were originally labelled as "*Meganthereon Nestianus*" (IGF 53V) and *Meganthereon (Machairodus) Nestianus* (IGF 112V) and are housed at the Museo di Geologia e Paleontologia of the Università di Firenze. Azzaroli [65] reported from the site of Pieve Fosciana the following taxa: *Anancus arvernensis*, *Tapirus arvernensis*, *Sus minor*, *Megantereon megantereon* and *Cervus* sp. Later, Azzaroli [66] referred, with question, this fauna to the Triversa FU (Early Villafranchian). Subsequently, Ficcarelli [67] ascribed these two upper canines to *Homotherium crenatidens*. Finally, Rustioni [68] revised the Pievefosciana mammal association indicating the occurrence of these taxa: *Anancus arvernensis*, *Stephanorhinus* sp., *Tapirus arvernensis*, *Cervus* sp., *Sus minor* (now recognised as a junior synonym of *Sus arvernensis*) [69], *Lynx issiodorensis*, *Megantereon cultridens*, *Vulpes alopecoides* and an undetermined Carnivora without any evidence of *Homotherium* remains. Since it is not possible at the present time to provide for the remains of "Fornaci al Ponte di Castiglione", any accurate taxonomical attribution of the *Homotherium* remains from CSG currently represents the First (clear) Appearance Datum in the Italian Peninsula.

Although different extinct fox species and several hypotheses of their relative taxonomic relationships have been proposed in the literature, the Early Pleistocene European sample was recently referred to as a single and polymorphic taxon, *Vulpes alopecoides* [25]. In western Europe, the oldest fossils of this taxon come from the Iberian Peninsula, in particular from Fonelas P-1 (dated around ~2.1–1.9 Ma; Arribas et al. [70]), La Puebla de Valverde (calibrated between the Reunion and Olduvai subchrons at ~2.12–1.92 Ma; Sinusía et al. [71] and Cuccu et al. [72]) and Villaroya (calibrated to the Reunion chron at ~2.14–2.11 Ma; Pueyo et al. [73]). The CSG record thus represents one of the oldest attested occurrences of *Vulpes alopecoides* in western Europe.

Finally, the *Martellictis ardea* at CSG is one of a few occurrences of this species in Europe during the middle Villafranchian. Ros-Montoya et al. [27] recently reported the presence of this small Lyncodontini in the Spanish localities of Barranco León and Fuente Nueva-3, dated approximately between 1.4 and 1.3 Ma, suggesting its survival during the late Villafranchian.

5.2. Paleoenvironmental Implications

The CSG paleoenvironment was mainly a braided stream surrounded by a watercourse, with the presence of mainly grasslands, as testified by the abundant presence of large arvicolids such as *Mimomys pliocaenicus* [7]. The very poorly represented Eulipotyphla and Muridae could be a clue to reduced and unstructured covered areas. The humid environment was limited to the surroundings of the watercourse, and it was the habitat of hippopotamuses, beavers, arvicolids and the water mole *Galemys kormosi* [9].

This paleoenvironmental reconstruction is also strengthened by the coexistence of two proboscideans, which clearly testifies to the gradual replacement of the European mastodons by early species of *Mammuthus*, since the arrival of the Southern Mammoth in the Italian Peninsula corresponds with an opening up of the vegetation during the Early Pleistocene. Forest diversity decreased, in fact, as a result of a progressive decline in and loss of subtropical taxa during the Early and Middle Pleistocene [4].

Studies carried out by Strani et al. [8,11,17] on the dietary adaptations of the CSG fossil herbivorous ungulates allowed them to obtain information on the palaeoenvironmental context that characterised the area during the Early Pleistocene. A wide range of feeding behaviours has been recorded, suggesting that herbivores had access to a variety of plant resources in a mosaic of biomes that spanned wetlands to woodlands to grasslands. The abundant remains of the stenonid equid *E. senezensis* aff. *E. s. stehlini* (one of the best-represented taxa at CSG), which display mesowear patterns and an isotopic signal compatible with a grazing behaviour in open habitats [11], indicate that grasslands were widespread in the area, which is in accordance with vegetation data reported from other localities on the Italian Peninsula [74].

In this palaeoenvironmental scenario, the Etruscan wolf operated as a pack-hunting animal able to harry rodents and lagomorphs, which are assumed be the main meat resources. However, sporadic ungulate hunting in the 10–45 kg size category is also considered feasible according to Rodríguez et al. (2012), as, for example, the cervids *Axis* cf. *lyra* and *Croizetoceros* cf. *ramosus* and the bovid *Gazella borbonica*. Carrion would be eaten when available.

The killing behaviour and prey preferences of *Homotherium* is a controversial topic (see Rodríguez et al. [75] and references therein). Anatomical and morphofunctional analyses suggest that *Homotherium* killed large prey, the size of a horse or a large bovid; probably, its postcranial skeleton was better adapted to long-distance travel than that of the pantherines, but their jumping abilities were less developed [76]. These characteristics suggest an adaptation to open environments, as reconstructed in the CSG site, and a large home range. According to Palmqvist et al. [77], juvenile *Mammuthus* were an important part of the diet of *Homotherium* at Venta Micena (Spain), together with *Bison* sp. (52%) and *Equus altidens*

(38%) [78]. *Homotherium* was a top predator and likely a pack hunter, with a preferred prey size in the 90–360 kg range; it could also kill prey in the 10 to 1000 kg range, depending on their availability. Juveniles of species weighing more than one ton are also considered vulnerable to *Homotherium* attack according to Rodríguez et al. [75]. However, *Homotherium* was itself vulnerable to hyaenid *Pliocrocuta perrieri* ambush. *Pliocrocuta perrieri* was, in fact, a medium-sized hyaenid that Turner et al. [51] classifies in their Ecomorph Group 6, "fully developed bone crackers".

In consideration of all the above, the CSG assemblages described testify to a high Carnivoran richness, apporting new data for studies that will address the predator/prey relationships at the regional and/or local community fauna scales during the Early Pleistocene in Mediterranean Europe.

Supplementary Materials: The following supporting information can be downloaded at: https://www.mdpi.com/article/10.3390/quat7040057/s1, Table S1: *Ursus* teeth measurements [18,19,79,80]; Table S2: *Homotherium* teeth measurements [21,67,81,82]; Table S3: *Pliocrocuta* teeth measurements [45,48,70,79,83–93]; Table S4: *Canis* teeth measurements [94].

Author Contributions: Conceptualization, L.B.; methodology, L.B.; formal analysis, L.B., J.C., B.M. and R.S.; geological analysis, F.B.; writing—original draft preparation, L.B., F.B., J.C., B.M. and R.S.; visualization, L.B. and F.B.; supervision, R.S.; funding acquisition, L.B. and R.S.; writing—review and editing, L.B., F.B., J.C., B.M., R.S. and F.S. All authors have read and agreed to the published version of the manuscript.

Funding: Research was financially supported by (i) Italian Ministry of University and Research Grants, "Ricerca di Ateneo federato di Scienze e Tecnologia AST Sapienza 2007" (Project: "I mammiferi del Pleistocene Inferiore italiano: evoluzione, migrazioni, biocronologia"; Resp.: Raffaele Sardella); (ii) Sapienza Progetto Università 2011 (prot. C26A11SNA3; Resp.: Raffaele Sardella); (iii) Sapienza Progetto Università 2012 (prot. C26A12PZA2; Resp.: Raffaele Sardella); (iv) Sapienza Progetto Università 2012 (prot. C26A12NW4X; Resp.: Donatella Magri). Fieldwork was financially supported by (i) BancAnagni and (ii) IsIPU. Thanks to the Italian Ministry of University and Research, Project "Valorizzazione Fisica e Virtuale dei siti preistorici del Bacino di Anagni" (legge 6/2000 per la diffusione della cultura scientifica).

Data Availability Statement: The considered material is curated and available for study in public institutions, and all the data analysed in this work are either directly available within this article in the Supplementary Tables or from the cited references.

Acknowledgments: Firstly: L.B. thanks the organiser of the session, "Mammals Biochronology and Paleoecology of the Euro-Mediterranean Quaternary" during the XXI INQUA conference (Rome, Italy, 14–20 July 2023). L.B. thanks the Istituto Italiano di Paleontologia Umana for access to the fossil collections, in particular, Stefano Grimaldi, Fabio Parenti, Barbara Saracino and Luciano Bruni. R.S. also thanks Annalisa Zarattini (Soprintendenza per i Beni Archeologici del Lazio), who authorised ongoing field activity at Coste San Giacomo. R.S., B.M., J.C., and F.S. also thank Daniele Natalia for supporting the opening of Museo Archeologico Ernico (MAE), where selected fossils of Coste San Giacomo are today displayed. The authors are also indebted to Chiara Delpino, Soprintendenza Archeologia, Belle Arti e Paesaggio per le province di Frosinone, Latina e Rieti, for assistance and valuable support during the construction of the MAE. The authors are thankful to Alessio Iannucci, Dawid Adam Iurino and Saverio Bartolini Lucenti for their useful comments and suggestions. This research received support from the SYNTHESYS+ Project http://www.synthesys.info/ (accessed on 30 May 2024), which is financed by the European Community Research Infrastructure Action under the H2020 Integrating Activities Programme (grants HU-TAF-Call4-Mecozzi at the Hungarian Natural History Museum, Budapest). B.M. thanks Mihaly Gasparik for access to the Hungarian Pleistocene collections. F.S. is supported by the Spanish Ministry of Science, Innovation, and Universities ("Juan de la Cierva—Formación", ref. FJC2020-042982-I).

Conflicts of Interest: The authors declare no conflicts of interest.

References

1. Florindo, F.; Marra, F.; Angelucci, D.E.; Biddittu, I.; Bruni, L.; Florindo, F.; Gaeta, M.; Guillou, H.; Jicha, B.; Macri, P.; et al. Environmental evolution, faunal and human occupation since 2 Ma in the Anagni basin, central Italy. *Sci. Rep.* **2021**, *11*, 7056. [CrossRef] [PubMed]
2. Lourens, L.J.; Antonarakou, A.; Hilgen, F.J.; Van Hoof, A.A.M.; Vergnaud-Grazzini, C.; Zachariasse, W.J. Evaluation of the Plio-Pleistocene astronomical timescale. *Paleoceanography* **1996**, *11*, 391–413. [CrossRef]
3. Lourens, L.J.; Hilgen, F.J.; Laskar, J.; Shackleton, N.J.; Wilson, D.S. The Neogene Period. In *A Geologic Time Scale 2004*; Gradstein, F.M., Ogg, J.G., Smith, A.G., Eds.; Cambridge University Press: Cambridge, UK, 2004; pp. 409–440.
4. Combourieu-Nebout, N.; Bertini, A.; Russo-Ermolli, E.; Peyron, O.; Klotz, S.; Montade, V.; Fauquette, S.; Allen, J.; Fusco, F.; Goring, S.; et al. Climate changes in the central Mediterranean and Italian vegetation dynamics since the Pliocene. *Rev. Palaeobot. Palyno.* **2015**, *218*, 127–147. [CrossRef]
5. Cassoli, P.F.; Segre Naldini, E. Le faune Villafranchiane: Costa san Giacomo e Fontana Acetosa. In *Dives Anagnia*; Ascenzi, A., Ed.; L'Erma di Bretscheider: Roma, Italy, 1993; pp. 31–33.
6. Bellucci, L.; Mazzini, I.; Scardia, G.; Bruni, L.; Parenti, F.; Segre, A.G.; Naldini, E.S.; Sardella, R. The site of Coste San Giacomo (Early Pleistocene, central Italy): Palaeoenvironmental analysis and biochronological overview. *Quat. Int.* **2012**, *267*, 30–39. [CrossRef]
7. Bellucci, L.; Bona, F.; Corrado, P.; Magri, D.; Mazzini, I.; Parenti, F.; Scardia, G.; Sardella, R. Evidence of late Gelasian dispersal of African fauna at Coste San Giacomo (Anagni Basin, central Italy): Early Pleistocene environments and the background of early human occupation in Europe. *Quat. Sci. Rev.* **2014**, *96*, 72–85. [CrossRef]
8. Strani, F.; DeMiguel, D.; Sardella, R.; Bellucci, L. Paleoenvironments and climatic changes in the Italian Peninsula during the Early Pleistocene: Evidence from dental wear patterns of the ungulate community of Coste San Giacomo. *Quat. Sci. Rev.* **2015**, *121*, 28–35. [CrossRef]
9. Bona, F.; Bellucci, L.; Sardella, R. The Gelasian (Late Villanyan-MN17) diversified micromammal assemblage with *Mimomys pliocaenicus* from Coste San Giacomo (Anagni basin, central Italy), taxonomy and comparison with selected European sites. *Hystrix* **2015**, *26*, 141–151. [CrossRef]
10. Palombo, M.R.; Alberdi, M.T.; Bellucci, L.; Sardella, R. An intriguing middle-sized horse from Coste San Giacomo (Anagni Basin, central Italy). *Quat. Res.* **2017**, *87*, 347–362. [CrossRef]
11. Strani, F.; Pushkina, D.; Bocherens, H.; Bellucci, L.; Sardella, R.; DeMiguel, D. Dietary adaptations of early and middle pleistocene equids from the Anagni basin (Frosinone, Central Italy). *Front. Ecol. Evol.* **2019**, *7*, 176. [CrossRef]
12. Galadini, F.; Messina, P. Early-Middle Pleistocene eastward migration of the Abruzzi Apennine (central Italy) extensional domain. *J. Geody.* **2004**, *37*, 57–81. [CrossRef]
13. Gliozzi, E.; Abbazzi, L.; Argenti, P.; Azzaroli, A.; Caloi, L.; Capasso Barbato, L.; Di Stefano, G.; Esu, D.; Ficcarelli, G.; Girotti, O.; et al. Biochronology of selected mammals, molluscs and ostracods from the Middle Pliocene to the Late Pleistocene in Italy. The state of the art. *Riv. Ital. Paleontol. Stratig.* **1997**, *103*, 369–388.
14. Sala, B.; Masini, F. Late Pliocene and Pleistocene small mammal chronology in the Italian peninsula. *Quat. Int.* **2007**, *160*, 4–16. [CrossRef]
15. Iannucci, A.; Mecozzi, B.; Sardella, R. Beware of the "Wolf event"–Remarks on large mammal dispersals in Europe and the late Villafranchian faunal turnover. *Alp. Mediterr. Quat.* **2023**, *36*, 75–90.
16. Villa, A.; Minnelli, E.; Bona, F.; Bellucci, L.; Sardella, R.; Delfino, M. The amphibians and reptiles from the Early Pleistocene of Coste San Giacomo (Anagni Basin, Italy). *Hist. Biol.* **2021**, *33*, 3075–3083. [CrossRef]
17. Strani, F.; DeMiguel, D.; Sardella, R.; Bellucci, L. Resource and niche differentiation mechanisms by sympatric Early Pleistocene ungulates: The case study of Coste San Giacomo. *Quat. Int.* **2018**, *481*, 157–163. [CrossRef]
18. Wagner, J.; Čermák, S.; Horáček, I. The presence of *Ursus* ex gr. *minimus-thibetanus in the Late Villányian and its position among the Pliocene and Pleistocene black bears in Europe. Quaternaire* **2011**, *4*, 39–58.
19. Mazza, P.; Rustioni, M. On the phylogeny of the Eurasian bears. *Palaeontogr. Abt A* **1994**, *230*, 1–38. [CrossRef]
20. Turner, A.; Antón, M. *The Big Cats and Their Fossil Relatives*; Columbia University Press: New York, NY, USA, 1997; 245p.
21. Sardella, R.; Iurino, D.A. The latest early pleistocene sabertoothed cat homotherium (felidae, Mammalia) from Monte Peglia (Umbria, central Italy). *Boll. Soc. Paleontol. It.* **2012**, *51*, 15–22.
22. Viret, J. Le Loess à bancs durcis de Saint-Vallier (Drôme) et sa faune de mammifères villafranchiens. *Nouv. Arch. Du Muséum D'histoire Nat. De Lyon* **1954**, *4*, 1–200.
23. Olive, F. Évolution des grands Carnivores au Plio Pléistocène en Afrique et en Europe occidentale. *L'Anthropologie* **2006**, *110*, 850–869. [CrossRef]
24. Bartolini Lucenti, S.; Rook, L. A review on the Late Villafranchian medium-sized canid *Canis arnensis* based on the evidence from Poggio Rosso (Tuscany, Italy). *Quat. Sci. Rev.* **2016**, *151*, 58–71. [CrossRef]
25. Bartolini Lucenti, S.; Madurell-Malapeira, J. Unraveling the fossil record of foxes: An updated review on the Plio-Pleistocene *Vulpes* spp. from Europe. *Quat. Sci. Rev.* **2020**, *236*, 106296. [CrossRef]
26. Bartolini Lucenti, S. Revising the species "*Mustela*" *ardea* Gervais, 1848–1852 (Mammalia, Mustelidae): *Martellictis* gen. nov. and the systematics of the fossil "Galictinae" of Eurasia. *Comptes Rendus Palevol* **2018**, *17*, 522–535. [CrossRef]

27. Ros-Montoya, S.; Bartolini-Lucenti, S.; Espigares, M.P.; Palmqvist, P.; Martínez-Navarro, B. First review of Lyncodontini material (Mustelidae, Carnivora, Mammalia) from the lower Pleistocene archaeo-palaeontological sites of Orce (southeastern Spain). *Riv. Ital. Paleontol. Stratigr.* **2021**, *127*, 33–47.
28. Azzaroli, A. Quaternary mammals and the "End-Villafranchian" dispersal event-a turning point in the history of Eurasia. *Palaeogeogr. Palaeoclimatol. Palaeoecol.* **1983**, *44*, 117–139. [CrossRef]
29. Azzaroli, A.; De Giuli, C.; Ficcarelli, G.; Torre, D. Late Pliocene to early mid-Pleistocene mammals in Eurasia: Faunal succession and dispersal events. *Palaeogeogr. Palaeoclimatol. Palaeoecol.* **1988**, *66*, 77–100. [CrossRef]
30. Sotnikova, M.; Rook, L. Dispersal of the Canini (Mammalia, Canidae: Caninae) across Eurasia during the Late Miocene to Early Pleistocene. *Quat. Int.* **2010**, *212*, 86–97. [CrossRef]
31. Sardella, R.; Palombo, M.R. The Pliocene-Pleistocene boundary: Which significance for the so called "wolf event"? Evidences from Western Europe. *Quaternaire* **2007**, *18*, 65–71. [CrossRef]
32. Cherin, M.; Basilici, G.; Duval, M.; Shao, Q.; Sier, M.J.; Parés, J.M.; Gliozzi, E.; Mazzini, I.; Magri, D.; Di Rita, F.; et al. The dawn of the Late Villafranchian: Paleoenvironment and age of the Pantalla paleontological site (Italy; Early Pleistocene). *Quat. Sci. Rev.* **2023**, *316*, 108279. [CrossRef]
33. Cassoli, P.F.; Segre Naldini, E. Nuovo contributo alla conoscenza delle faune villafranchiane e del Pleistocene medio del bacino di Anagni (Frosinone). Atti Della XXIV Riunione Scientifica dell'Istituto Italiano di Preistoria e Protostoria. 1984. Available online: https://www.amazon.com/contributo-conoscenza-villafranchiane-Pleistocene-Frosinone/dp/B01N52ST98 (accessed on 30 May 2024).
34. Fondi, R. Fauna cromeriana della Montagnola senese. *Palaeontogr. Ital.* **1972**, *68*, 1–27.
35. Iannucci, A. *Sus strozzii* (Suidae, Mammalia) from the historical locality of Quercia (Early Pleistocene, Italy). *Geobios* **2023**, *77*, 27–36. [CrossRef]
36. Girotti, O.; Capasso Barbato, L.; Esu, D.; Gliozzi, E.; Kotsakis, T.; Martinetto, E.; Petronio, C.; Sardella, R.; Squazzini, E. The section of Torre Picchio (Terni, Umbria, Central Italy): A Villafranchian site rich in mammals, molluscs, ostracods and plants. *Riv. It. Paleont. Stratigr.* **2003**, *109*, 77–98.
37. Azzarà, B.; Breda, M.; Cirilli, O.; Madurell-Malapeira, J.; Ruzza, F.; Sorbelli, L.; Tancredi, D.; Cherin, M. Vigna Nuova: The first Middle Villafranchian mammal assemblage from the Valdichiana Basin, Perugia (Italy). *Boll. Soc. Paleontol. It.* **2022**, *61*, 223–247.
38. Rook, L.; Martínez-Navarro, B. Villafranchian: The long story of a Plio-Pleistocene European large mammal biochronologic unit. *Quat. Int.* **2010**, *219*, 134–144. [CrossRef]
39. Napoleone, G.; Albianelli, A.; Azzaroli, A.; Mazzini, M. The Poggio Rosso locality calibrated to the end-Pliocene and its significance for dating the Late Villafranchian mammal faunas of the Upper Valdarno, Central Italy. *Riv. Ital. Paleontol. Stratigr.* **2001**, *107*, 287–296.
40. Napoleone, G.; Albianelli, A.; Azzaroli, A.; Bertini, A.; Magi, M.; Mazzini, M. Calibration of the Upper Valdarno basin to the Plio-Pleistocene for correlating the Apennine continental sequences. *Il Quat.* **2003**, *16*, 159–195.
41. Torre, D.; Ficcarelli, G.; Masini, F.; Rook, L.; Sala, B. Mammal dispersal events in the early Pleistocene of Western Europe. *Cour. Forschungsinstitut Senckenberg* **1992**, *153*, 51–58.
42. Rook, L. I Cani dell'Eurasia Dal Miocene Superiore Al Pleistocene Medio. Ph.D. Thesis, Modena-Bologna-Firenze and Roma "La Sapienza" Universities, Roma, Italy, 1993.
43. Palombo, M.R.; Azanza, B.; Alberdi, M.T. Italian mammal biochronology from Latest Miocene to Middle Pleistocene: A multivariate approach. *Geol. Romana* **2003**, *36*, 335–368.
44. Segre Naldini, E.; Valli, A.M.F. Villafranchian Cervides from central Italy. *Quat. Nova* **2004**, *7*, 159–204.
45. Iannucci, A.; Cherin, C.; Sardella, R. The lost hyena from Paciano (Umbria, Italy) reconsidered. *Alp. Mediterr. Quat.* **2022**, *35*, 105–117.
46. Iannucci, A.; Mecozzi, B.; Sardella, R.; Iurino, D.A. The extinction of the giant hyena *Pachycrocuta brevirostris* and a reappraisal of the Epivillafranchian and Galerian Hyaenidae in Europe: Faunal turnover during the Early–Middle Pleistocene Transition. *Quat. Sci. Rev.* **2021**, *272*, 107240. [CrossRef]
47. Koufos, G.D. The Villafranchian carnivoran guild of Greece: Implications for the fauna, biochronology and paleoecology. *Integr. Zool.* **2014**, *9*, 444–460. [CrossRef] [PubMed]
48. Koufos, G.D. The Pleistocene carnivores of the Mygdonia basin (Macedonia, Greece). *Ann. Paléontol.* **1992**, *78*, 205–257.
49. Vekua, A. Die Wirbeltierfauna des Villafranchium von Dmanisi und ihre biostratigraphische Bedeutung. *Jahrb. RGZ* **1995**, *42*, 77–180.
50. Werdelin, L.; Solounias, N. The Hyaenidae: Taxonomy, systematics and evolution. *Foss. Strat.* **1991**, *30*, 1–104.
51. Turner, A.; Antón, M.; Werdelin, L. Taxonomy and evolutionary patterns in the fossil Hyaenidae of Europe. *Geobios* **2008**, *41*, 677–687. [CrossRef]
52. Howell, F.C.; Petter, G. The *Pachycrocuta* and *Hyaena* lineages (Plio-Pleistocene and extant species of the Hyaenidae). Their relationships with Miocene ictitheres: *Palhyaena* and *Hyaenictitherium*. *Geobios* **1980**, *13*, 579–623. [CrossRef]
53. Martínez-Navarro, B.; Palmqvist, P.; Madurell-Malapeira, J.; Ros-Montoya, S.; Espigares, M.P.; Torregrosa, V.; Pérez-Claros, J.A. La fauna de grandes mamíferos de Fuente Nueva-3 y Barranco Léon-5. Estado de la cuestión. In *Ocupaciones Humanas en el Pleistoceno Inferior y Medio de la Cuenca de Guadix-Baza, Memoria Científica*; Toro, I., Martínez-Navarro, B., Agustí, J., Eds.; Junta de Andalucía, Consejería de Cultura: Sevilla, Spain, 2010; pp. 197–236.

54. Vinuesa, V.; Madurell-Malapeira, J.; Fortuny, J.; Alba, D.M. The endocranial morphology of the Plio-Pleistocene bone-cracking hyena *Pliocrocuta perrieri*: Behavioral implications. *J. Mammal. Evol.* **2015**, *22*, 421–434. [CrossRef]
55. Kruuk, H. *The Spotted Hyena: Study of Predation and Social Behavior*; Chicago University Press: Chicago, IL, USA, 1972; pp. 1–335.
56. Cooper, S.M.; Holekamp, K.E.; Smale, L. A seasonal feast: Long-term analysis of feeding behavior in the spotted hyaena *Crocuta Crocuta* (Erxbelen). *Afr. J. Ecol.* **1999**, *1*, 178–180. [CrossRef]
57. Mills, M.G.L. *Kalahari Hyaenas: The Comparative Behavioral Ecology of Two Species*; Unwin Hyman: London, UK, 1990; pp. 1–304.
58. Wagner, A.P. Behavioral ecology of the striped hyaena (*Hyaena hyaena*). Ph.D. Thesis, Montana State University, Bozeman, MT, USA, 2006.
59. Palombo, M.R.; Sardella, R.; Novelli, M. Carnivora dispersal in western Mediterranean during the last 2.6 Ma. *Quat. Int.* **2008**, *179*, 176–189. [CrossRef]
60. Konidaris, G.E. Guilds of large carnivorans during the Pleistocene of Europe: A community structure analysis based on foraging strategies. *Lethaia* **2022**, *55*, 1–18. [CrossRef]
61. Iannucci, A. Ecospace occupancy and disparity in Pleistocene large carnivorans of Europe and implications for hominin dispersal and ecological role. *Quat. Sci. Rev.* **2024**, *329*, 108562. [CrossRef]
62. Mazza, P.; Bertini, A.; Magi, M. The late Pliocene site of Poggio Rosso (central Italy): Taphonomy and paleoenvironment. *Palaios* **2004**, *19*, 227–248. [CrossRef]
63. Sotnikova, M.; Titov, V. Carnivora of the Tamanian faunal unit (the Azov Sea area). *Quat. Int.* **2009**, *201*, 43–52. [CrossRef]
64. Antón, M.; Salesa, M.J.; Galobart, A.; Tseng, Z.J. The Plio-Pleistocene scimitar-toothed felid genus *Homotherium* Fabrini, 1890 (Machairodontinae, Homotherini): Diversity, palaeogeography and taxonomic implications. *Quat. Sci. Rev.* **2014**, *96*, 259–268. [CrossRef]
65. Azzaroli, A. Pliocene. In *Geologia dell'Italia*; Desio, A., Ed.; UTET: Torino, Italy, 1973; pp. 669–676.
66. Azzaroli, A. The Villafranchian stage in Italy and the Plio–Pleistocene boundary. *G. Di Geol.* **1977**, *41*, 61–79.
67. Ficcarelli, G. The Villafranchian Machairodonts of Tuscany. *Palaeontogr. Ital.* **1979**, *71*, 17–26.
68. Rustioni, M. On Pliocene tapirs from France and Italy. *Boll. Soc. Paleontol. It.* **1992**, *31*, 269–294.
69. Iannucci, A.; Bellucci, L.; Conti, J.; Mazzini, I.; Mecozzi, B.; Sardella, R.; Iurino, D.A. Neurocranial anatomy of *Sus arvernensis* (Suidae, Mammalia) from Collepardo (Early Villafranchian; central Italy): Taxonomic and biochronological implications. *Hist. Biol.* **2022**, *34*, 108–120. [CrossRef]
70. Arribas, A.; Garrido, G. Hiénidos [Pachycrocuta brevirostris (Aymard, 1846) y Hyaena brunnea Thunberg, 1820] del yacimiento de Fonelas P-1 (cuenca de Guadix, Granada). In *Vertebrados del Plioceno Superior Terminal en el Suroeste de Europa: Fonelas P-1 y el Proyecto Fonelas*; Arribas, A., Ed.; Instituto Geológico y Minero de España, Serie Cuadernos del Museo Geominero 10: Madrid, Spain, 2008; pp. 201–230.
71. Sinusía, C.; Pueyo, E.L.; Azanza, B.; Pocoví, A. Datación magnetoestratigráfica del yacimiento paleontológico de la Puebla de Valverde (Teruel). *Geo-Temas* **2004**, *6*, 339–342.
72. Cuccu, A.; Valenciano, A.; Azanza, B.; DeMiguel, D. A new lynx mandible from the Early Pleistocene of Spain (La Puebla de Valverde, Teruel) and a taxonomical multivariate approach of medium-sized felids. *Hist. Biol.* **2023**, *35*, 127–138. [CrossRef]
73. Pueyo, E.L.; Muñoz, A.; Laplana, C.; Parés, J.P. The last appearance datum of *Hipparion* in Western Europe: Magnetostratigraphy along the Pliocene–Pleistocene boundary in the Villarroya basin (Northern Spain). *Int. J. Earth Sci.* **2023**, *105*, 2203–2220. [CrossRef]
74. Bertini, A. Pliocene to Pleistocene palynoflora and vegetation in Italy: State of the art. *Quat. Int.* **2010**, *225*, 5–24. [CrossRef]
75. Rodríguez, J.; Rodríguez-Gómez, G.; Martín-González, J.A.; Goikoetxea, I.; Mateos, A. Predator–prey relationships and the role of Homo in Early Pleistocene food webs in Southern Europe. *Palaeogeogr. Palaeoclimatol. Palaeoecol.* **2012**, *365*, 99–114. [CrossRef]
76. Antón, M.; Galobart, A.; Turner, A. Co-existence of scimitar-toothed cats, lions and hominins in the European Pleistocene. Implications of the post-cranial anatomy of *Homotherium latidens* (Owen) for comparative palaeoecology. *Quat. Sci. Rev.* **2005**, *24*, 1287–1301. [CrossRef]
77. Palmqvist, P.; Grôcke, D.R.; Arribas, A.; Fariña, R.A. Paleoecological reconstruction of a Lower Pleistocene large mammal community using biogeochemical ($\delta^{13}C$, $\delta^{15}N$, $\delta^{18}O$, Sr: Zn) and ecomorphological approaches. *Paleobiology* **2003**, *29*, 205–229. [CrossRef]
78. Palmqvist, P.; Pérez-Claros, J.A.; Janis, C.M.; Gröcke, D.R. Tracing the ecophysiology of ungulates and predator–prey relationships in an early Pleistocene large mammal community. *Palaeogeogr. Palaeoclimatol. Palaeoecol.* **2008**, *266*, 95–111. [CrossRef]
79. Medin, T.; Martínez-Navarro, B.; Rivals, F.; Madurell-Malapeira, J.; Ros-Montoya, S.; Espigares, M.-P.; Figueirido, B.; Rook, L.; Palmqvist, P. Late Villafranchian *Ursus etruscus* and other large carnivorans from the Orce sites (Guadix-Baza basin, Andalusia, southern Spain): Taxonomy, biochronology, paleobiology, and ecogeographical context. *Quat. Int.* **2017**, *431*, 20–41. [CrossRef]
80. Baryshnikov, G.F.; Lavrov, A.V. Pliocene bear *Ursus minimus* Devèze de Chabriol et Bouillet, 1827 (Carnivora, Ursidae) in Russia and Kazakhstan. *Russ. J. Theriol.* **2013**, *12*, 107–118. [CrossRef]
81. De Bonis, L. Un Félidé à longues canines de la Colline de Perrier (Puy-de-Dome): Ses rapports avec les Félinés machairodontes. *Ann. Paleont.* **1976**, *62*, 159–198.
82. Galobart, A.; Antón, M.; Maroto, J. Hiénidos y cánidos de los yacimientos de Incarcal (Girona, NE Peninsula Ibérica). Una aproximación a la paleobiologia del Pleistoceno inferior. *Paleontol. I Evol.* **2003**, *34*, 69–88.
83. Sotnikova, M.V.; Baigusheva, V.S.; Titov, V.V. Carnivores of the Khapry faunal assemblage and their stratigraphic implications. *Stratigr. Geol. Correl.* **2002**, *10*, 375–390.

84. Liu, J.; Liu, J.; Zhang, H.; Wagner, J.; Jiangzuo, Q.; Song, Y.; Liu, S.; Wang, Y.; Jin, C. The giant short-faced hyena *Pachycrocuta brevirostris* (Mammalia, Carnivora, Hyaenidae) from Northeast Asia: A reinterpretation of subspecies differentiation and intercontinental dispersal. *Quat. Int.* **2021**, *577*, 29–51. [CrossRef]
85. Martínez-Navarro, B. Revisión Sisteméatica y Estudio Cuantitativo de la Fauna de Macromamíferos del Yacimiento de Venta Micena (Orce, Granada). Ph.D. Thesis, Universidad Autónoma de Barcelona, Barcelona, Spain, 1991.
86. Turner, A.; Antón, M. The giant hyaena *Pachycrocuta brevirostris* (Mammalia, Carnivora, Hyaenidae). *Geobios* **1996**, *29*, 455–468. [CrossRef]
87. Fourvel, J.B.; Lateur, N. Note sur deux mandibules de *Pachycrocuta brevirostris* (Aymard, 1846) du Pléistocène ancien du sud-est de la France: La grotte de la Grosse Marguerite (Aiguèze) et l'aven des Trois Pigeons (Nîmes). *Paláeo* **2015**, *26*, 187–200. [CrossRef]
88. Kurtén, B.; Garevski, R. Giant hyena, *Hyaena brevirostris* Aymard (Mammalia, Carnivora), from the middle Pleistocene of Manastirec, Yugoslavia. *Ann. Zool. Fenn.* **1989**, *26*, 145–147.
89. Boule, M. Description de l'*Hyaena brevirostris* du Pliocèene de Sainzelles près le Puy (Haute-Loire). *Ann. Sci. Nat.* **1893**, *15*, 85–97.
90. Kurtén, B. Fossil Hyaenidae from the excavations at Stránská Skála. *Anthropos* **1972**, *20*, 113–120.
91. Turner, A. Remains of *Pachycrocuta brevirostris* (Carnivora, Hyaenidae) from the lower Pleistocene site of Untermaßfeld. In *Das Pleistozän von Untermaßfeld bei Meinigen (Thüringen), Teil 2*; Kahlke, R.-D., Ed.; Habelt Verlag: Bonn, Germany, 2001; pp. 673–690.
92. Qiu, Z. Die Hyaeniden aus dem Ruscinium und Villafranchium Chinas. Munchner Geowissen. *Abh. A Geol. Paläontol.* **1987**, *9*, 1–108.
93. Weithofer, K.A. Die fossilen hyanen des Arnothales in Toskana. *Denkschriften der kaiserliche Akademie der Wissenschaften. Matemathisch-Naturwissenschaftliche;* Classe 55. Kaiserlich-Königlichen Hof und Staatsdruckerel: Wien, Austria, 1889; pp. 337–360. Available online: https://www.biodiversitylibrary.org/item/31432#page/9/mode/1up (accessed on 30 May 2024).
94. Berté, D. L'evoluzione del Genere *Canis* (Carnivora, Canidae, Caninae) in Italia dal Wolf-Event a Oggi: Implicazioni Biocronologiche, Paleoecologiche e Paleoambientali. Ph.D. Thesis, Sapienza Università di Roma, Roma, Italy, 2013.

quaternary

MDPI

Article

The Occurrence of Suids in the Post-Olduvai to Pre-Jaramillo Pleistocene of Europe and Implications for Late Villafranchian Biochronology and Faunal Dynamics

Alessio Iannucci

Department of Geosciences, Section of Terrestrial Palaeoclimatology, Eberhard-Karls-University Tübingen, Sigwartstrasse 10, 72076 Tübingen, Germany; alessio.iannucci@mnf.uni-tuebingen.de

Abstract: It has been proposed that suids were absent from Europe during the post-Olduvai to pre-Jaramillo Early Pleistocene (from less than 1.8 to more than 1.2 Ma) and that their "re-appearance" in the late Early Pleistocene would mark the end of the late Villafranchian and the beginning of the Epivillafranchian. Arguments enumerated in favor of this "suid gap" are the lack of suid remains from extensively sampled fossil localities of this age and the high reproductive potential (r-strategy) of suids, which would translate in a high commonness of their remains in the fossil record. However, here it shown that while suids' reproductive potential is certainly exceptional within artiodactyls, there is no direct relationship between the reproductive strategy and preservation rate of a taxon in the fossil record. In Early Pleistocene localities of Europe and adjoining areas, where suids are present in a fossil assemblage, they are always rare. In terms of number of occurrences (frequency), suids range from being moderately common (~2.0–1.8 Ma) to moderately rare (~1.1–1.0 Ma). Suid material is also described herein from Peyrolles (Issoire, France; reference locality for MNQ 19), a site dated at 1.47 Ma, providing direct evidence for the presence of suids within the purported "suid gap". The case of suids underlines an important source of caveat in inferring faunal dynamics of the late Early Pleistocene of western Europe—including the dispersal of hominins—i.e., the unequal geographical distribution of the paleontological sites of post-Olduvai to pre-Jaramillo age. Indeed, Peyrolles is the only large mammal site in western Europe located outside the Iberian and Italian Peninsulas reliably dated around 1.5 Ma. In the post-Olduvai to pre-Jaramillo period, there is a paucity of radiometric estimates (or they have too coarse a resolution) and of paleomagnetic excursions detectable in continental deposits. Basically, for this time span, there is a high dependence on biochronological correlations, although, at the same time, these correlations are less reliable—because these are based on a few sites not covering the entire spectrum of environments present in Europe and the sites are not independently dated with methods that outperform biochronology—than those for other periods.

Keywords: bioevent; dispersal; Epivillafranchian; faunal turnover; large mammals; Mammalia; Mediterranean; pigs; reproductive strategy; Suidae

check for updates

Citation: Iannucci, A. The Occurrence of Suids in the Post-Olduvai to Pre-Jaramillo Pleistocene of Europe and Implications for Late Villafranchian Biochronology and Faunal Dynamics. *Quaternary* **2024**, *7*, 11. https://doi.org/10.3390/quat7010011

Academic Editor: Juan Manuel López García

Received: 6 November 2023
Revised: 24 January 2024
Accepted: 4 February 2024
Published: 21 February 2024

1. Introduction

The Pleistocene witnessed substantial oceanographic, meteorological, and environmental changes in a broad sense, including fluctuations in ice cover and sea level at a global scale. These phenomena, mediated by ecological interactions, shaped the evolution of the biota. In turn, the fossil record is key for reconstructing these changes, understanding their pace and magnitude, and comparing and correlating them between different areas [1–5].

In this context, one of the most intensively investigated and debated topics of research concerns the first hominin dispersal into Europe and the reason why it postdates by a considerable amount of time the earliest evidence Out of Africa, especially considering the extensive sample of Dmanisi at ~1.8 Ma [6,7]. Essentially, the time span bracketed between the earliest hominin presence in the Caucasus and their dispersal in western Europe can be considered the core chronology to examine to gain a better understanding of the hominin spread in the region. The latter point engenders vibrant debate, and while many authors favor a pre-Jaramillo age (e.g., [2,3]), the large uncertainties associated with the available dating methods should push to exercise caution [8,9]. In general, and focusing on the ~1.8–1.2 Ma crucial period, investigating the large mammal fossil record plays a fundamental role, not only owing to the paucity of hominin evidence, but also by recognizing that hominins were part of a wider faunal renewal featuring the appearance in Europe of species of African and Asian origin [2,3].

The perceived biochronological and paleoecological significance of the Early Pleistocene suids of Europe in the context of Quaternary faunal turnovers and in relation to hominins has long been quite low, being mainly limited to the inclusion of the appearance of *S. strozzii* in the bioevents denoting the late Villafranchian faunal turnover, the "Wolf event" [10–12]—although the species is now documented in several middle Villafranchian localities [13]—and to the emphasis placed on the replacement between *S. strozzii* Forsyth Major, 1881 [14] and *S. scrofa* Linnaeus, 1758 [15] in terms of the different ecological requirements of the two species [16,17]. In this framework, an important impetus was given by the work of Martínez-Navarro et al. [18], in which it is argued that suids were absent from Europe during the 1.8–1.2 Ma time span and that a species phylogenetically related to the extant wild boar (*Sus* gr. *scrofa*) appears at 1.2 Ma, approximately at the same time of other bioevents correlated with the beginning of the Epivillafranchian [19–22]. Martínez-Navarro et al. [18] also tentatively interpreted the absence of suids in terms of competitive displacement with hominins.

The attribution of the Epivillafranchian suids as *Sus* gr. *scrofa* proposed by Martínez-Navarro et al. [18] was challenged shortly after, when Bona and Sala [23] described an almost complete mandible of *S. strozzii* from the locality of Frantoio, placed at ~1.0 Ma based on paleomagnetism. Van der Made et al. [24] already assigned the latest Early Pleistocene suids to *Sus* sp. and did not recognize any gap in the European suid fossil record (this work, although issued later, was published online roughly at the same time as that of Martínez-Navarro et al. [18]). Cherin et al. [25] recognized that all Epivillafranchian samples that preserve lower male canines display a "verrucosic" cross-section, which differs from the "scrofic" morphology of *S. scrofa*, and consequently should be attributed to *S. strozzii*. Despite this, Cherin et al. [25] did not reject the existence of the "suid gap".

Prior to the work of Martínez-Navarro et al. [18], suids were either regarded as continuously present in Europe from the Early to the Middle Pleistocene or indeed experiencing a gap between the last record of *S. strozzii* and the earliest of *S. scrofa*, but shorter than that subsequently proposed and potentially considered an artefact of the low numbers of fossil localities of appropriate age [16,24,26–29].

Even though the absence of a taxon in a fossil record should only be taken as evidence of absence with great caution, Martínez-Navarro et al.'s [18] "suid gap" hypothesis, or at least part of the associated arguments, was followed by many authors (e.g., [25,30–33]). Indeed, the purported link between the reproductive potential of suids and their abundance in the fossil record was especially influential, and it is worth stressing that Martínez-Navarro et al. [18] were arguably the firsts to discuss the peculiar reproductive strategy of suids in relation to their fossil record at this level of detail. In brief, their reasoning can be summarized as follows: suids have a higher reproductive potential than related species of comparable size; consequently, when present, they are abundant in the fossil record; and therefore, their absence from extensively sampled localities dated between 1.8 and 1.2 Ma should be reliable. However, the link between the first and second points was assumed rather than tested, as was the abundance of suid remains in the European fossil record in the first place.

This work is articulated in several parts: (1) I describe the suid material from the site of Peyrolles (France), which is, at the moment, the only radiometrically dated (at 1.47 Ma) suid-bearing locality within the "suid gap"; (2) I investigate the relationship between body mass and reproductive strategy in artiodactyls to understand whether and how suids differ from other species of comparable size; and (3) I evaluate whether suids where truly abundant in the European fossil record before and after their putative absence. Finally, the case of suids is discussed in relation to the paucity and heterogeneous geographical distribution of European paleontological sites with mammal remains falling between ~1.8 and 1.2 Ma, and the implications for our comprehension of the faunal dynamics of this time span, including hominin dispersal.

2. Peyrolles

Peyrolles is located in the region of Issoire (Puy-de-Dôme, France), an area known since the 18th century for its extensive fossil record of Pliocene and Pleistocene of large mammals [34] (Figure 1). The paleontological site of Peyrolles was discovered and investigated mainly in the first half of the 19th century, and after a long period of oblivion, rediscovered in 1995 [34]. In the years elapsed between the first studies and the rediscovery of the site, and in those following, the area has been extensively investigated from a geochronological perspective, especially based on the products of the Mont-Dore stratovolcano, located ~30 Km west of Peyrolles (e.g., [35–38]). In particular, Nomade et al. [39] dated with the $^{40}Ar/^{39}Ar$ method the trachytic pumiceous alluvium interstratified within the Creux de Peyrolles fossiliferous beds, obtaining an age of 1.47 ± 0.01 Ma. Associated uncertainty aside, an equivalent age (1.42 ± 0.10 Ma, also $^{40}Ar/^{39}Ar$ dated) was reported in an unpublished doctoral thesis [40].

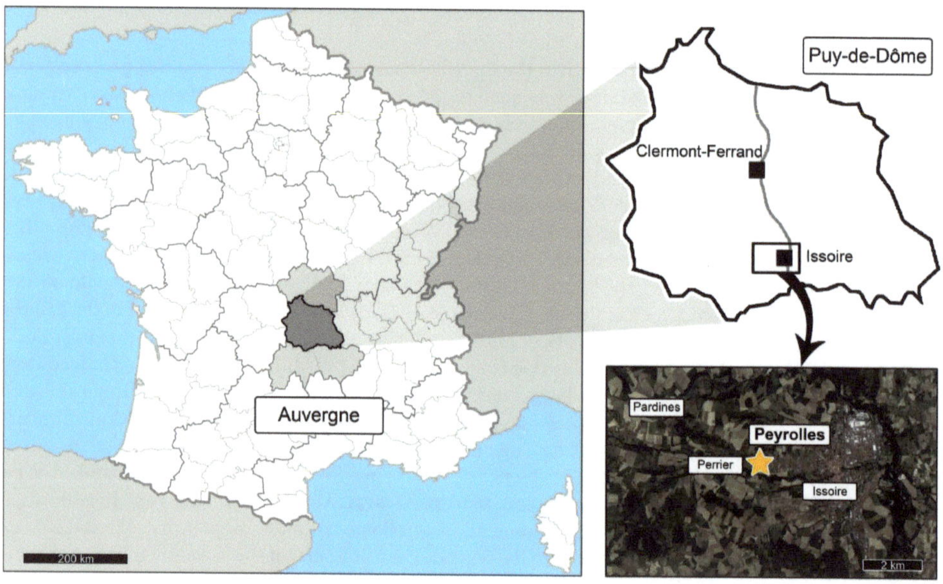

Figure 1. Outline of France with emphasis on the department of Puy-de-Dôme (squares indicate main cities) and geographical location of Peyrolles (star) and nearby localities in the area of Issoire.

The fauna of Peyrolles assumed an important biochronological role for the Villafranchian (Figure 2), being considered representative of MNQ 19 (French Biozones), and hence younger than Senèze (MNQ 18) and potentially close to the Tasso FU (Italian Faunal Units), although the presence of a derived form of *Eucladoceros* was already remarked as indicative of an age later than Tasso [11]. Few French faunas have been consistently referred to MNQ 19, namely Peyrolles (type fauna), Blassac-La Girondie, and La Sartanette [41]. La Sartanette is only constrained by the inverse polarity of the sediments to a likely post-Olduvai but pre-Jaramillo age [42]. On the other hand, Paquette et al. [43] obtained an age of 1.946 ± 0.028 Ma for Blassac-La Girondie, based on U/Pb dating on volcanic zircons. These authors reasoned that the combined evidence derived from the recent dating of Peyrolles

(1.47 ± 0.01 Ma) and Blassac (1.946 ± 0.028 Ma) would suggest a rather long duration for the MNQ 19 French biozone, corresponding to multiple Italian FUs (i.e., Olivola, Tasso, Farneta, and perhaps Pirro Nord). It is worth noting that Peyrolles has been taken as representative of MNQ 19 mainly for historical reasons, that is, having long been the only known fauna filling that chronology [41,44]. However, the fauna is neither abundant nor diverse in terms of number of species [34]. Considering the paucity of the late Villafranchian deposits in France already remarked in previous studies [41,45], the possibility that Peyrolles represents a unique faunal complex with respect to the others present in the country cannot be ruled out. Moreover, as will be stressed in this work, biochronological correlations for the 1.8–1.2 Ma time span in Europe are biased by the heterogeneous distribution of fossiliferous deposits. This is especially relevant when considering Peyrolles, as the biochronological placement of the site has been based on cervids, especially *"Cervus" perolensis* [34,44], for which different lineages in France and Italy are customarily recognized against the background of a myriad of divergent taxonomic proposals (e.g., [46]).

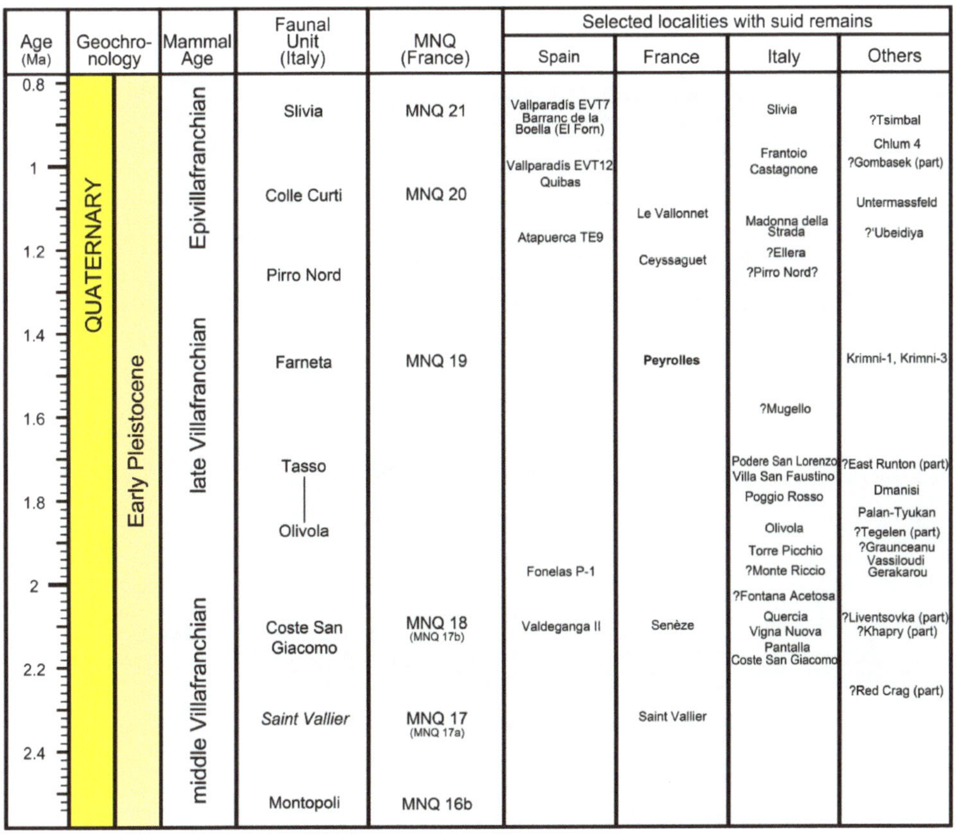

Figure 2. Biochronological scheme for the Early Pleistocene, indicating selected localities with suid remains from Europe and adjoining areas. Question marks indicate uncertain chronology (if placed before a locality) or uncertain occurrence (after).

3. Materials and Methods

3.1. Fossil Material

The fossil suid material from Peyrolles is housed in the Natural History Museum, London (NHMUK). It consists only of one fragmented metapodial, catalogued as NHMUK

PV OR 27621. The anatomical identification and comparative description of the specimen were mainly based on the extant wild boar, *S. scrofa* and, as far as possible considering the paucity of metapodial remains known for the latter species, *S. strozzii* (details are provided in the relevant section below). Material of the extant wild boar is available in several collections. Here, I especially used an individual part of the comparative osteological collection of the Hungarian Natural History Museum, Budapest (HNHM, specimen 56. 25. 3.). Measurements were taken with a digital caliper to the nearest 0.1 mm.

3.2. Reproductive Strategies

To understand whether and how much Suidae differ from other species of comparable size in terms of reproductive strategy, data for living Artiodacyla were downloaded from COMBINE [47], considering terrestrial non-volant species. Body mass (BM) and 9 other variables were selected, considering relatedness to reproduction and including only variables for which more than 95% of species have data: longevity (L), age at female sexual maturity (ASM), age at first reproduction (AFR), gestation length (GSL), litter size (LS), litters per year (LY), interbirth interval (II), weaning age (WA), and generation length (GNL). Details are provided in Table 1. In total, 247 species of artiodactyls were included. A non-parametric Spearman's rank-order correlation coefficient was calculated between the natural-logged variables to test for linear relationships, with and without the inclusion of suids. The correlation coefficient ranges from −1 (negative relationship) to +1 (positive relationship), while a score of 0 indicates no relationship. A principal component analysis (PCA) was performed on the correlation matrix of the same variables. The results of these analyses and comparisons were used to explore and discuss the differences in reproductive strategy between suids and other artiodactyls. The software PAST was used for the analysis [48].

Table 1. Variables considered in the analysis on reproductive strategies in Artiodactyla. After [47].

Variable	Abbreviation	Description
Body mass	BM	Body mass of an adult individual in grams
Longevity	L	Maximum reported age at death for the species in days
Age at female sexual maturity	ASM	The amount of time needed for a female to reach sexual maturity in days
Age at first reproduction	AFR	Age at which females give birth to their first litter or their young attach to teats in days
Gestation length	GSL	Length of time of fetal growth in days
Litter size	LS	Number of offspring born per litter per female
Litters per year	LY	Number of litters per female per year
Interbirth interval	II	Time between reproduction events in days
Weaning age	WA	Age at which primary nutritional dependency on the mother ends and independent foraging begins in days
Generation length	GNL	Average age of parents of the current cohort in days

3.3. Abundance and Frequency of Suids in the Fossil Record

To assess whether Suidae were abundant or not in the European fossil record during the Early Pleistocene, two approaches were applied. First, to understand the abundance of suids with respect to other artiodactyls in localities where their remains have been recovered. Second, to evaluate the frequency of suid occurrences with respect to well-known fossil localities from selected time spans.

In the first case, I collected data on the minimum number of individuals (MNI) or the number of identified specimens (NISP) for localities dated from ~2.0 to 0.9 Ma with suid remains for which this information is available. In chronological order, this comprises Poggio Rosso [49], 'Ubeidiya II-23 [50], Untermassfeld [51], and Vallparadís EVT7 [52]. The inclusion of 'Ubeidiya, although the site is not located in Europe, is justified by the presence of several faunal elements of European affinity [53–55].

For the second approach, I used data from Palombo [3], who provided an authoritative compilation of faunal lists for the Pleistocene of southwestern Europe (from Iberian, French, and Italian sites). This part of the analysis is focused on western Europe, as the presence of suids in the eastern Mediterranean was never denied [18,25] but it might be considered compatible with the "suid gap" hypothesis, if interpreted as a refuge with respect to western Europe [56]. Faunas from two Faunal Complexes (as referred to by Palombo [3]) were considered, V4 (~2.0–1.8 Ma, including faunas customarily referred to Olivola and Tasso FUs or MNQ 18, with the addition of Blassac, see below) and G1 (~<1.2–1.0 Ma, including faunas referred to the Colle Curti FU or MNQ 20). The two subtended time spans denote the periods before and after the purported absence of suids from Europe (Farneta and Pirro FUs, or MNQ 19), when the group has been stressed to be widely represented in the fossil record [18].

Some emendations to the faunal lists provided by Palombo [3] were made as detailed as follows, either regarding presence/absence data, chronology, or taxonomy (Supplementary Material). For the first time span, Palombo [3] listed *S. strozzii* from Casa Frata but the species is not present in this locality [57,58] and therefore I considered it absent. The local fauna referred to as "Valdarno 2" by Palombo [3] is not included, as it is a composite faunal list of historical findings of different provenance within the Upper Valdarno, which therefore does not satisfy the criterion of being "a list of the species identified from the same stratigraphical horizon at a given fossiliferous site" defined by the author, and besides is redundant with respect to Casa Frata, Matassino, and Poggio Rosso (which are also from the Upper Valdarno, but deriving from relatively recent collections). On the other hand, Blassac-La Girondie is included, in agreement with the dating of 1.946 $\pm$ 0.028 Ma obtained by Paquette et al. [43].

For the second time span, the fauna of Monte Peglia is not considered, as only two artiodactyl species were listed by Palombo [3] and both records were tentative. I also excluded Redicicoli and the Early Pleistocene layers of Vallparadís, which could be correlated with the following Slivia FU [46,59]. Likewise, I omitted all the faunas from the different layers of Gran Dolina, which were argued to be likely younger than previously assumed, considering the occurrence of Galerian newcomers [45], as eventually confirmed by new paleomagnetic investigations [60,61]. On the other hand, I added the fauna of Frantoio (Arda River; [23]).

Leptobos etruscus and *L. vallisarni* are treated together as they are commonly included in the same group or lineage (e.g., [62,63]), as opposite to another group consisting of *L. stenometopon*, *L. elatus*, *L. merlai*, and *L. furtivus*.

Dubious occurrences and taxa undetermined at the species level where not considered when they might represent a taxon already recorded, while tentative attributions at the species level (e.g., '*Pseudodama*' cf. *vallonnetensis*) were accepted as valid. The following three reasonable exceptions to this rule were made: (1) the *Praemegaceros* material of the ~1.1–1.0 time span (which is altogether referred to *P. verticornsis*); (2) *Hippopotamus* (all pragmatically referred to as *H. antiquus*); (3) suids, which, for the purpose of this analysis, are necessarily lumped into *S. strozzii*, although some localities yielded remains that are left in open taxonomy (e.g., Castagnone; [64]). Some dubious cases were checked critically by evaluating the primary literature. For instance, the presence of *Cervus elaphus acoronatus* at Saint Prest was not accepted, considering the paucity of the material described by Guérin et al. [65], who indeed originally assigned it to Cervidae cf. *Cervus elaphus*.

A simple quartile subdivision of the observed frequencies is used to objectively refer to a taxon as common (100–75% of occurrences), moderately common (74.9–50%), moderately rare (49.9–25%), or rare (less than 25%).

4. Systematic Paleontology

Class Mammalia Linnaeus, 1758 [15].
Order Artiodactyla Owen, 1848 [66].
Family Suidae Gray, 1821 [67].

Subfamily Suinae Gray, 1821 [67].
Genus *Sus* Linnaeus, 1758 [15].
Sus sp. (Figure 3).

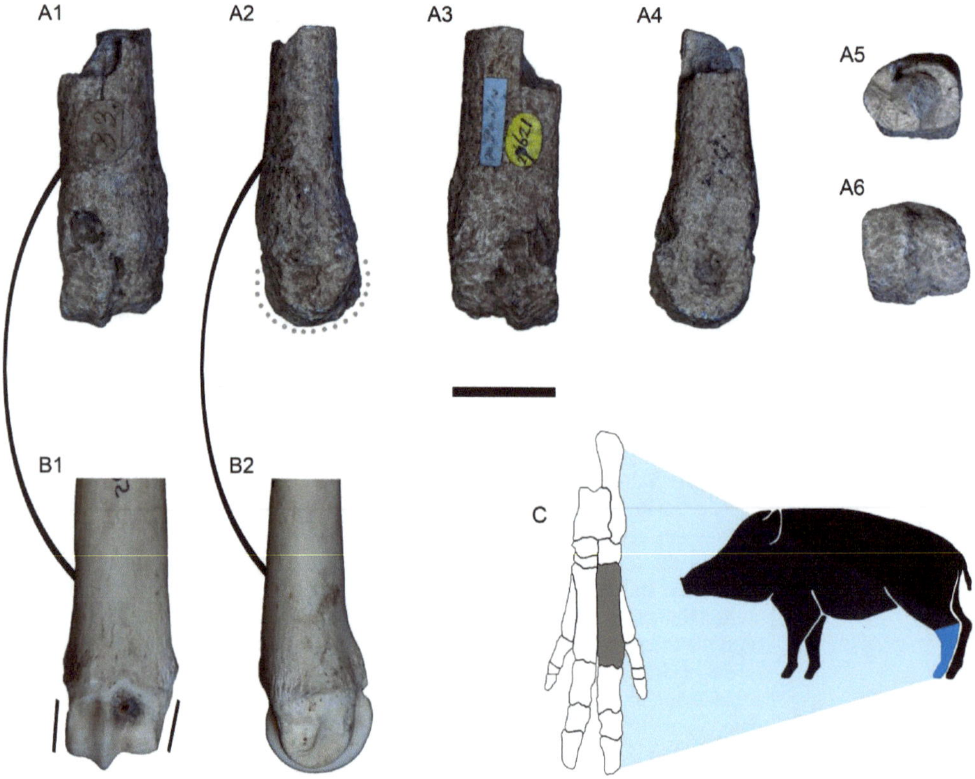

Figure 3. (**A**) The fragmentary left fourth metatarsal of *Sus* sp. from Peyrolles (NHMUK PV OR 27621) in anterior (**A1**), internal (**A2**), posterior (**A3**), external (**A4**), proximal (**A5**), and distal (**A6**) views; (**B**) the homologous element of *Sus scrofa* for comparison, in anterior (**B1**) and internal (**B2**) views; (**C**) anatomical sketch indicating the position of the bone. The dotted line in A2 approximately indicates the missing portion of the trochlea. The scale bar corresponds to 20 mm.

Referred material from Peyrolles. NHMUK PV OR 27621, a distal fragment of a left fourth metatarsal (henceforth, MT IV).

Remarks. Suids have a four-toe foot, with two small (second and fifth) and two large (third and fourth) digits and metapodials. The metapodial from Peyrolles is clearly large and hence represents the third or fourth. The proximal portion of the bone, which is the most diagnostic for anatomical identification, is not preserved in our specimen, but the section of the shaft and the distal end provide useful information as well. In the fourth metatarsals, the external side of the distal end is more strongly inclined toward the internal side of the bone than in third metatarsals. In the third and fourth metapodials, the internal side of the diaphysis is approximately straight antero-posteriorly and the anterior and posterior sides converge externally. In the metacarpals, this convergence is more pronounced than in the metatarsals; so, the section of the shaft is more rounded in the latter. As the metapodial from Peyrolles has a strongly inclined lateral side of the distal portion and a relatively rounded section of the diaphysis, it is identified as an MT IV.

NHMUK PV OR 27621 is, therefore, a distal fragment of a left MT IV. Just under half of the bone is preserved, measuring approximately 56 mm proximo-distally. The internal-external (transversal) diameter of the distal epiphysis measures 20 mm, while the perpendicular antero-posterior diameter cannot be taken reliably as the trochlea is eroded.

Comparing the Peyrolles MT IV with the homologous element in *S. scrofa* reveals a clear morphological resemblance, but I assign it to *Sus* sp., considering the limited knowledge on the postcranial anatomy of *S. strozzii* and the poor preservation of our specimen. In detail, I could not find an isolated and well-preserved MT IV of *S. strozzii* among the main collections that preserve remains of the species (see the institutions listed by Iannucci [13,68]). The MT IV is present in a mounted subadult skeleton from Senèze preserved in the Naturhistorisches Museum Basel (NMB) [69], which is the main source of information on the postcranial anatomy of *S. strozzii* and associated paleoecological inference [70]. Based on this specimen, Azzaroli [70] underlined that in *S. strozzii*, the metapodials are relatively shorter and hence stouter than in *S. scrofa*. Unfortunately, the incomplete preservation of the MT IV from Peyrolles precludes an assessment of its proportions.

5. Reproductive Strategies

In artiodactyls, most reproduction-related variables are significantly correlated with each other, with or without the inclusion of suids in the estimates (Figure 4A). The only exception is LS, which only has a weak negative relationship with GSL. Nonetheless, excluding suids from the computation always increases the correlation (positive or negative) between each variable, apart from relationships involving LS. The only other exception is the relationship between AFR and GSL.

Focusing on BM reveals a strong (correlation coefficient > 0.60) positive relationship with L, ASM, AFR, GSL, II, WA, and GNL, and a weaker negative relationship with LY (0.55) in all artiodactyls (Figure 4B). Including suids in the computation always diminishes the correlations between variables, and in the case of LS even has the effect of changing the relationship from positive to negative (although this relationship is not significant, Figure 4A).

In the PCA, the first two components jointly account for 67.5% of the explained variance (C1 = 53.9%, C2 = 13.6%) (Table 2). In the ecospace projection onto the plane described by the first two components, most artiodactyls are scattered along the first axis, which is positively influenced by all variables apart from LS and LY (Figure 5). This first axis can be mainly interpreted as the BM-related component of reproductive strategy in artiodactyls. Along the second component, the sample is mainly distributed according to the opposite contribution of LS and GSL, providing a partial separation of suids (complete for *Sus*) and, to a lesser extent, tayassuids, from other artiodactyls. This axis captures the rapidity of reproduction, with species plotting toward more positive values (like suids) producing a high number of offspring in a short time.

Table 2. Eigenvalues, percentage of explained variance, and loadings of PCA. Components explaining more than 5% of variance are reported.

PCA	PC1	PC2	PC3	PC4	PC5
% Variance	5.38938	1.35783	0.75328	0.677878	0.506962
BM	0.35411	0.17025	0.2838	−0.16842	0.37325
L	0.30613	0.090422	0.45011	0.40485	−0.43725
ASM	0.34086	0.065646	−0.42337	0.091518	0.39679
AFR	0.344	−0.04563	−0.204	0.38143	0.35725
GSL	0.28391	−0.50972	0.040859	−0.12475	−0.06277
LS	−0.078169	0.80313	0.057934	−0.074333	0.044958
LY	−0.33322	−0.089843	0.382	0.29656	0.54834
II	0.37192	0.17733	−0.28008	−0.25519	−0.18045
WA	0.30359	−0.053162	0.49521	−0.52393	0.17881
GNL	0.3437	0.096293	0.1462	0.45299	−0.11515

BM (as a proxy for most other reproduction-related variables), LS, LY, and GSL are considered in further detail given the aforementioned results. Violin and box plots are used to visualize these variables in different families of artiodactyls and separately for *Sus* (Figure 6).

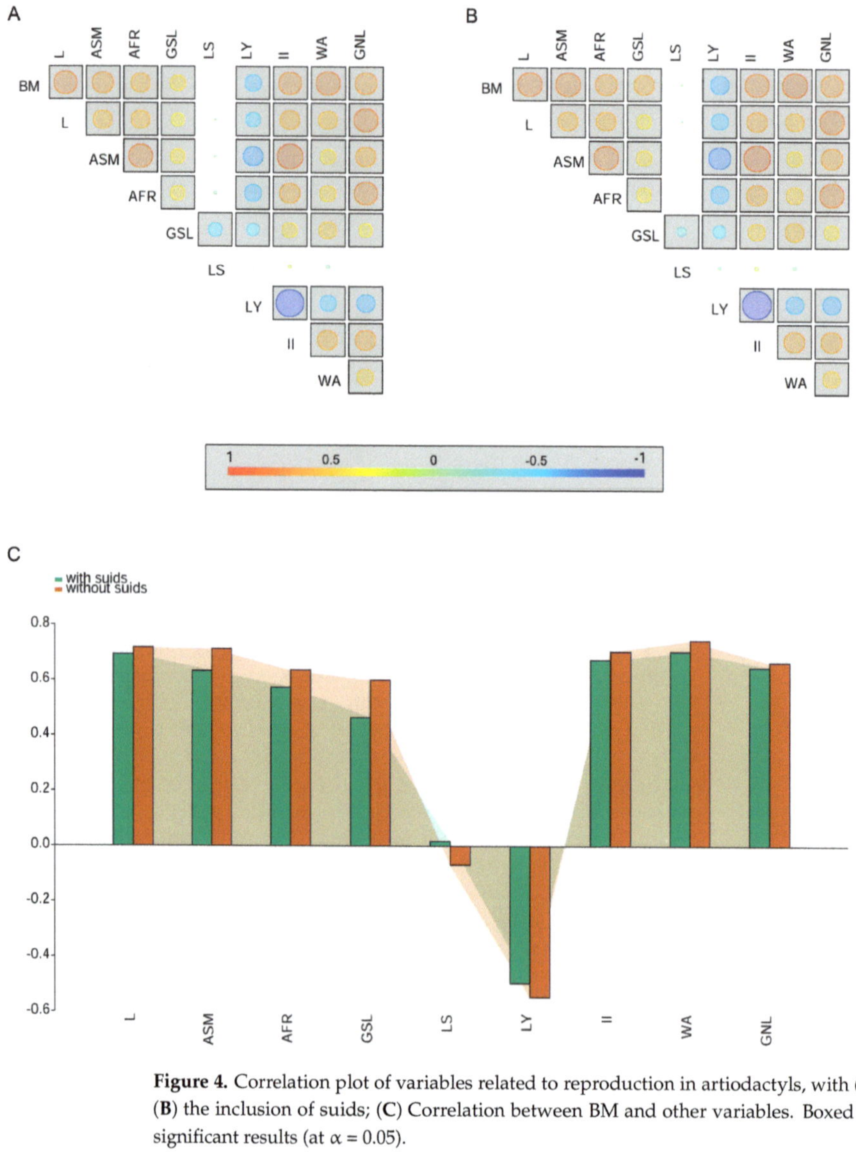

Figure 4. Correlation plot of variables related to reproduction in artiodactyls, with (**A**) or without (**B**) the inclusion of suids; (**C**) Correlation between BM and other variables. Boxed areas indicate significant results (at α = 0.05).

As regards BM, artiodactyls exhibit large variability, from ~2 to over 2500 kg (Figure 6A). Suids can be roughly described as middle-sized, falling within the variability of groups represented by most species (Bovidae and Cervidae) and outside that of clearly small- (Moschidae and Tragulidae) or large-sized (Giraffidae and Hippopotamidae) families. The smallest BM value reported for suids is 8 kg (*Porcula salvania*) and the largest 200 kg (*Hylochoerus meinertzhageni*). *Sus* does not exhibit such extreme values.

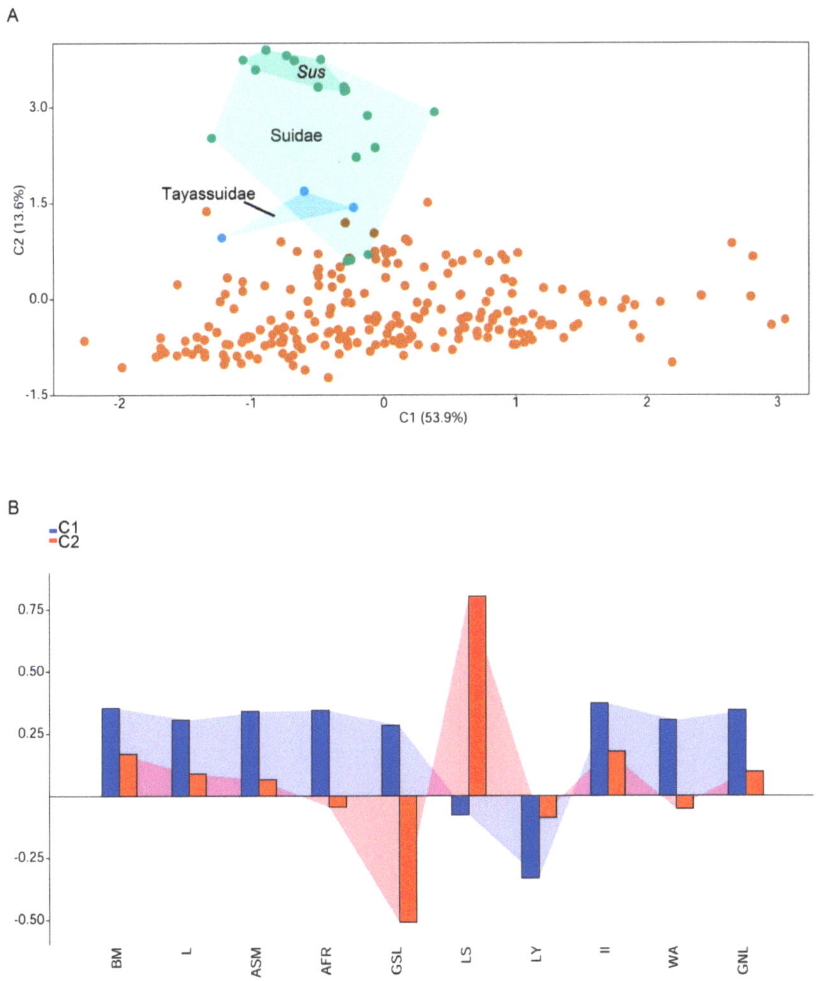

Figure 5. Ecospace projection of the PCA carried out on variables related to reproduction in artiodactyls onto the plane described by the first two components (**A**), and their respective loadings (**B**). In (**A**) Suidae are in green (darker for *Sus*), Tayassuidae in blue, and all other artiodactyls in orange.

In most artiodactyls, the LS is either one or ranges from one to two (Figure 6B). All suids and tayassuids have values greater than one (only in babirusas is it less than two) and in *Sus*, the minimum number of offspring is four. Bovids and cervids have some "outliers" (observations plotting outside the box plot of more than 1.5 times the box height), respectively, 14 (9.8%) and 6 (11%), with the most extreme case being the Chinese water deer (*Hydropotes inermis*) with a LS value of 3 [71]. Mean values for tragulids (1.27) and moschids (1.4) are also above one (but always below two).

Several artiodactyls, including suids and tayassuids, have generally one litter per year (LY) or more, except larger species (Giraffidae and Hippopotamidae) and camelids, which do not have a litter every year (Figure 6C). Bovids are highly variable, with 17 species (11.9%) having less than one LY.

As regards GSL, suids have low values, ranging from ~110 days in *P. salvania* to less than ~170 days in *Phacochoerus* spp. (Figure 6D). There is only little overlap between suids and the lower range of bovids and cervids, and tragulids.

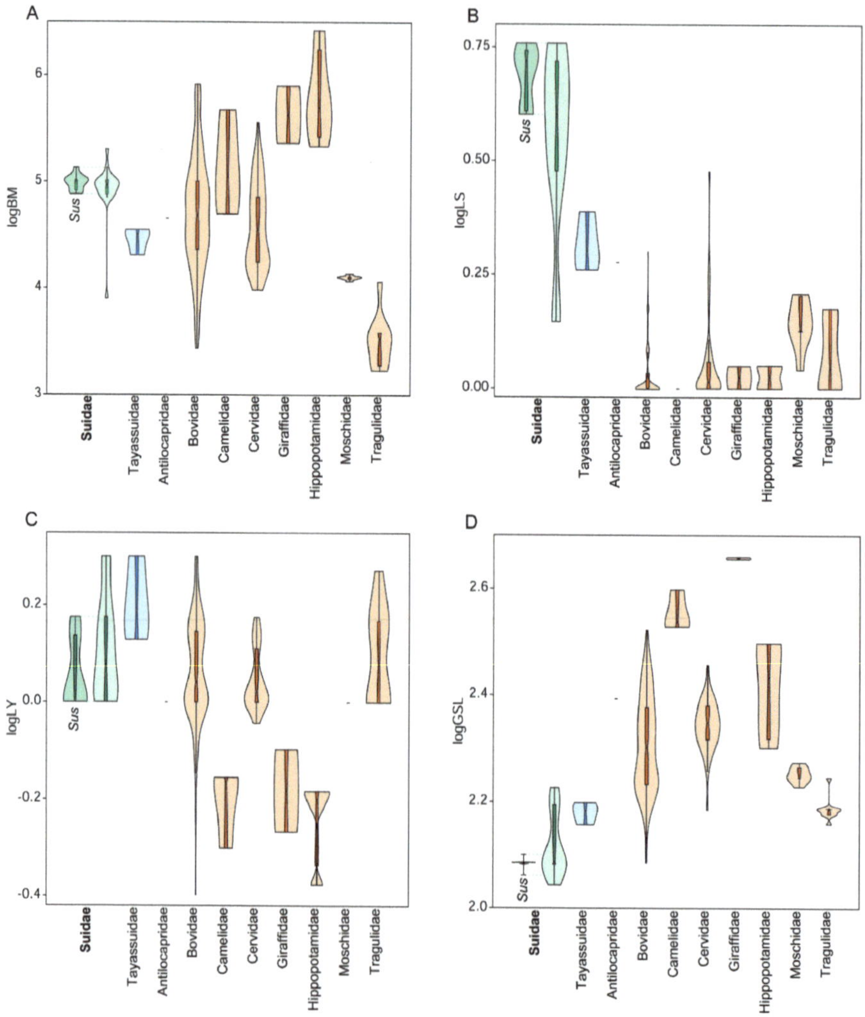

Figure 6. Violin and box plots of selected variables related to reproduction in artiodactyls. (**A**) Body mass (BM); (**B**) litter size (LS); (**C**) litters per year (LY); (**D**) gestation length (GSL). Suidae are in green (darker for *Sus*), Tayassuidae in blue, and all other artiodactyls in orange.

6. Abundance and Frequency in the Fossil Record

Mazza et al. [49] provided information on the number of identified specimens (NISP) and the minimum number of individuals (MNI) from Poggio Rosso. In both cases, around half of the mammal fauna corresponds to the large-sized deer referred to as *Eucladoceros dicranios*, whereas the percentages in terms of NISP and MNI represented by *S. strozzii* are 4% and 7%, respectively (Figure 7A).

The layer II-23 of 'Ubeidiya [50] is dominated by remains assigned to *Hippopotamus* cf. *behemoth* (22.8% of NISP) and to the middle-sized deer *'Pseudodama' nestii* (17.4%) (Figure 7B). Despite being represented by two species, suids are the rarest artiodactyl family in the sample, with few remains being referred to as *S. strozzii* (0.4%) and even fewer to *Kolpochoerus oludvaiensis* (0.1%).

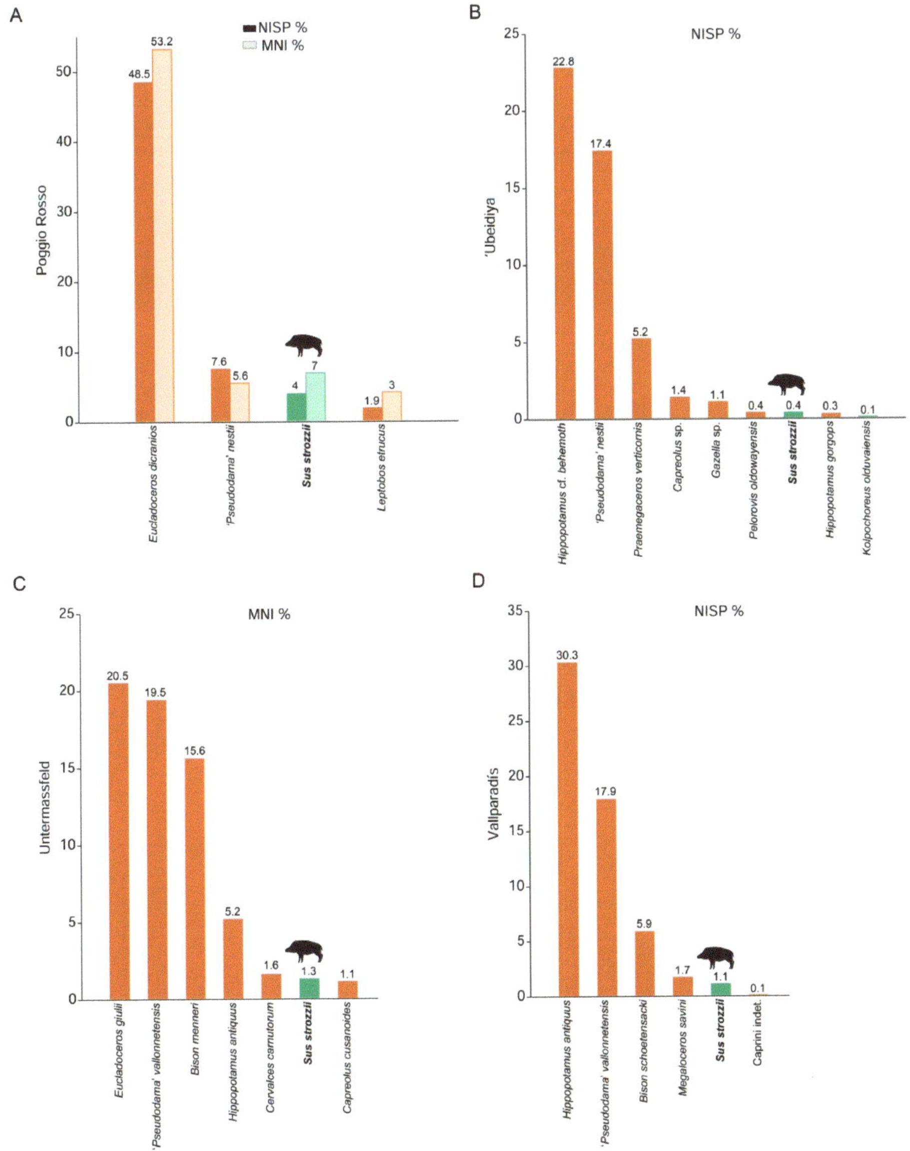

Figure 7. Abundance of suids (in green) and other artiodactyls (in orange) in terms of number of identified specimens (NISP) or minimum number of individuals (MNI) at: (**A**) Poggio Rosso [49]; (**B**) 'Ubeidiya layer II-23 [50]; (**C**) Untermassfeld [51]; (**D**) Vallparadís layer EVT7 [52]. Percentages refer to all large mammals but only artiodactyls are plotted.

At Untermassfeld [51], suids are represented by a MNI of six, which accounts for 1.3% of all large mammals (Figure 7C). The only artiodactyl species rarer than suids is *Capreoulus cusanoides*.

The artiodactyls recovered from the layer EVT7 of Vallparadís [52] are predominantly represented (in terms of NISP) by remains of hippopotamuses and middle-sized deer, referred to as *Hippopotamus antiquus* and *'Pseudodama' vallonnetensis*, respectively (Figure 7D).

Suids accounts for only 1.1% of all large mammals, with the only rarer taxon being an indeterminate representative of Caprini.

Considering the frequency of suid occurrences in comparison to those of other artiodactyls, in late Villafranchian faunas dated between ~2.0 and 1.8 Ma, suids are moderately common, being recorded in more than half (57.1%) of the paleontological localities (Figure 8A), while in the Epivillafranchian, ~1.1–1.0 Ma, they are moderately rare (37.5%) (Figure 8B).

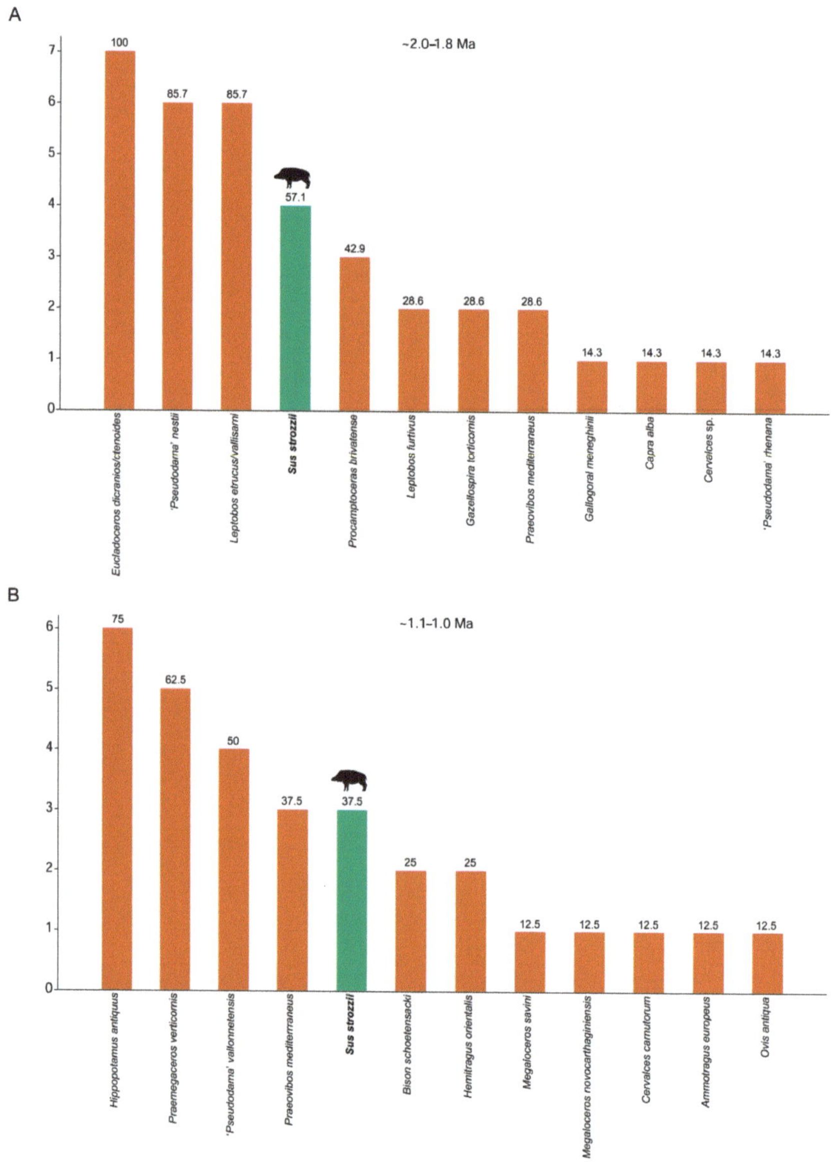

Figure 8. Frequency of occurrence of suids (in green) and other artiodactyl species (in orange) in southwestern Europe ~2.0–1.8 Ma (**A**) and ~1.1–1.0 Ma (**B**). The vertical axis indicates the number of localities. The value above each bar represents the percentage of occurrence over the total of localities. Data are from Palombo [3], with emendations detailed in Section 3.3.

7. Discussion

7.1. The Reproductive Strategy of Suids

Investigating the reproductive strategies of artiodactyls emphasized the uniqueness of suids, especially manifested in the large number of offspring per litter and short gestation length. These results agree with previous studies that underlined that extant suids are relatively r-selected in comparison to other ungulates, especially considering the wild boar, *S. scrofa* [18,32,72]. The r/K selection theory is a generalization of the ecological continuum of reproduction strategies, according to which r-selected species are those that maximize the quantity of offspring, while K-selected species focus on the "quality" (i.e., fewer offspring but with a greater investment) [73,74]. Many shades exist between the hypothetical endmembers of the spectrum and the use of the r/K selection theory as a predictive model in a strict sense is now surpassed (e.g., [75,76]), but considering species adaptations in this conceptual framework and its relative deviations toward one of the other ends—with respect to other related species—is often useful for understanding its ecology and evolution. This is because many biological features related to reproduction are also related to each other and to other life history traits (Figure 4). Indeed, the r/K terminology derives from the theory of island biogeography of MacArthur and Wilson [77,78], in which the different strategies are linked to different aspects of colonization and evolution on islands. From the above digression on reproduction strategies, the exceptionality of suids clearly emerges, as when the conditions are favorable, suids can reproduce at an extremely fast rate despite not being unlike other artiodactyls in the body-mass-related component of reproduction strategy (as exemplified by the PCA carried out herein, Figure 5).

Some aspects of the peculiar reproductive strategy of suids are, to some extent, shared with tayassuids (peccaries)—which are suids' closest living relatives—suggesting that related extinct species were also relatively r-selected. Indeed, tayassuids have a higher LS, LY, and shorter GSL than most other artiodactyls (Figure 6). At the same time, some suids, especially those belonging to *Sus*, specialized even further in the direction of the rapidity of reproduction, with the most emblematic example being the wild boar, *S. scrofa*.

The peculiar reproductive strategy of suids certainly render them capable to cope with environmental disturbance better than many other ungulates, favored the impressive demographic increase in wild boar populations observed over the last decades, and facilitated their natural or human-mediated introduction in many regions of the world [79–84]. The difference in reproduction strategies has also been argued to be a key factor to explain *S. scrofa*'s ability to cope with environmental changes occurred during the Late Pleistocene of Apulia (southern Italy) through shifts in body size, whereas other large mammals reacted by modifying their range or went extinct [32]. Moreover, there might be a relation between the reproductive potential of suids and the morphological changes observed along some lineages in the fossil record, in turn promoting their employment as a biochronological tool or paleoecological proxy, especially in the African Neogene-Quaternary (e.g., [85–90]).

7.2. Abundance and Frequency of Suids in the Early Pleistocene of Europe

Considering the above section, during the Pleistocene of Europe, it would have been reasonable to observe a high density of suid populations, which belonged to species either directly known (*S. scrofa*) or indirectly expected (*S. strozzii*) to reproduce extremely fast. The question is, does this translate in an abundant presence of their remains in the fossil record? Judging from the available evidence, the answer is negative. Rather, it appears that suids were rare. Remains assigned to *S. strozzii* are already known from some localities older than 2 Ma, but none of them include more than a handful of specimens [13]—though Senèze yielded an almost complete skeleton [70]. To the list of early occurrences of *S. strozzii* discussed by Iannucci [13] (Saint Vallier, Valdeganga II, Coste San Giacomo, Quercia, and Vigna Nuova) should be added Pantalla, which has recently been recognized as older than previously assumed, at ~2.2 Ma [91]. Even in the case of Pantalla, *S. strozzii* is represented only by a hemimandible [31]. The time span bracketed between 2 and 1.8 Ma can be considered the acme of abundance of *S. strozzii*, with several localities correlated with the Olivola and Tasso FUs documenting the presence of the species, especially in central Italy [13,31,68,70] (Figure 8A). The only site of this period for which detailed information on NISP and MNI is available is Poggio Rosso (Upper Valdarno), where suid remains account for a mere 4% of the total [49]. Despite this percentage being quite modest, the abundance of suid remains in the rest of the Early Pleistocene is even lower (Figure 7)—and, for completeness, it is worth mentioning that the situation does not change substantially in the early Middle Pleistocene, when *S. scrofa* is present [13,92].

Considering the frequency of occurrences of artiodactyl species at ~2.0–1.8 Ma and ~1.1–1.0 Ma, suids appear moderately common to moderately rare, being recorded in 57.1% and 37.5% of the localities, respectively (Figure 8). The general pattern indicates the presence of a few common species and a larger number of rare taxa, a skewed distribution that is frequently observed in ecology [93]. However, exploring commonness and rarity in the fossil record is biased by several factors that make some species seemingly rarer than they actually were. First, the sampled interval often intercepts only a portion of the temporal distribution of a taxon. For instance, *Bison schoetensacki* and *Megaloceros savini* are widely represented in the European fossil record [24,94], but since the chosen time span of ~1.1–1.0 Ma includes their earliest occurrences, they were obviously not recorded in all localities. The same applies to *Gazellospira torticornis* or *Hippopotamus* in the time span of ~2.0–1.8 Ma, which, respectively, intercepts only late (for *G. torticornis*) or early (for *Hippopotamus*) occurrences of the species [12,95].

Second, divergent taxonomic opinions might inflate the number of recognized species—and hence their perceived rarity—and here I necessarily followed a splitter approach, being a revision of taxa other than suids outside the scope of this work. For instance, this might be relevant for Caprini [96].

Of course, the geographical distribution of a species also affects its perceived rarity. For example, this case applies here to *Cervalces* sp. and *Cervalces carnutorum*, both of which are present in only one of the considered faunas, but merely because they had a mainly northern distribution [97].

As this study treated suids altogether and focused on time spans and a geographical area where they were deemed to be abundant, the aforementioned factors play a role in making suids apparently more common than they were. Critically evaluating the results of this investigation, it could be stressed that suids were certainly more common only than Caprini, but even this consideration should be taken with caution, considering that the apparent high diversity and rarity of Caprini might also be partly a result of their debated taxonomy [96].

In general, there is no clear relationship between reproductive strategy and commonness of a species in the fossil record, as exemplified by the frequency of occurrences and abundance of hippopotamuses remains (which are far from being fast-reproducing mammals).

In sum, suids are generally rare in the Pleistocene record of Europe, being only moderately common during the ~2.0–1.8 Ma time span, and only in terms of number of occurrences.

The reason why suids are rare is probably linked to multiple factors that act at different scales, although the paucity of sites with suid remains precludes to adequately test for causal relationships. Since suids are intelligent, robust, and more aggressive than many other artiodactyls [84], it is likely that predators usually preferred preys other than suids. This is supported by the low percentage of suid remains in sites whose faunal remains were mainly or partly collected by carnivores (Figure 7). For instance, the bone accumulation of Poggio Rosso has been interpreted as primarily resulting from the activity of the giant hyena *Pachycrocuta brevirostris* [49,98], a species which was also directly involved at Vallparadís [52], and generally regarded as a major taphonomic agent during the Early Pleistocene of Europe [45]. The extent of the influence of the first hominins dispersing into Europe on the environment and on specific sites is debated, and arguably it was less relevant than in later chronologies, although growing with the spread of the Acheulean since the very late Early Pleistocene and during the Middle Pleistocene [99,100]. In any case, hominins also probably preferred preys other than suids, as, for instance, testified by the several Middle Pleistocene archaeological sites in which suid remains are scanty or absent [92].

The social organization of suids is another aspect of their biology that likely contribute to their rarity in the fossil record. Indeed, although suids are social animals, they did not form large herds [84]. In several suid species, the most common social unit consists of a female and her litter, while adult males are often solitary [84,101]. Larger groups are of sporadic nature, only occasionally documented, and mainly observed in *S. scrofa* in relation to exceptional concentrations of resources that hardly have analogues in the Pleistocene of Europe (e.g., agricultural crops) [101]. The number of individuals that might die and potentially be preserved in the fossil record due to sporadic events like, e.g., a flood, is therefore usually rather limited.

Suids also do not display specific adaptations that would increase the chances of naturally died animals to be preserved as fossils, as it is likely the case for the semi-aquatic habits of hippopotamuses in open-air sites related to fluvio-lacustrine deposits.

In brief, suid remains in the Pleistocene of Europe are rare seemingly irrespective of type of site (e.g, open-air, karstic cavity) and type of accumulation (e.g., carnivore den, natural trap). In this situation, suids might easily be apparently absent in a time span during which few fossil sites and/or suboptimal environmental conditions occurred (see Section 7.4).

7.3. Fossil Localities and the "Suid Gap"

When discussing the potential suid occurrences falling in the time span of their supposed absence from Europe, Martínez-Navarro et al. [18] only listed and doubted, on biochronological grounds, the record from the Ellera Basin (Italy), placed by Pazzaglia et al. [102] at ~1.6–1.5 Ma. There are, however, several other localities that might fill the "suid gap" (Figure 2). Those listed by Van der Made et al. [24] included, in assumed chronological order from older to younger, Mugello, Selvella, Peyrolles, 'Ubeidiya, Pirro Nord, and Ceyssaguet. More recently, Kostopoulos et al. [56] reported the presence of suids from the sites of Krimni-1 and Krimni-3, both referred to as ~1.5 Ma. In the following, all these localities are briefly discussed.

The inferred age of Mugello and Selvella localities is based on their classical placement in Italian biochronological schemes (i.e., Farneta FU, e.g., [27]). Abbazzi et al. [103] provided a more detailed overview of the Mugello Basin faunas, referring them to the Tasso or Farneta FUs and reporting the occurrence of *S. strozzii* only from Pulicciano, together with *Hippopotamus* sp. and *Mammuthus meridionalis*. The latter species are both documented before, during, and after the "suid gap" [3,12,104], hence not providing robust biochronological constraints. On the other hand, the reference of the Selvella fauna to the Farneta FU has been accepted in several works (e.g., [61,105]). However, the material assigned by De Giuli [106] to a suid is limited to an isolated and fragmentary humerus, whose attribution seems doubtful [106] (p. 15).

The presence of suids at 'Ubeidiya is undeniable [107]; although, pending a revision of the sample, they should be referred to as *Sus* sp. [17]. The fauna contains several European elements, but, on the other hand, the site is not in Europe, hence offering arguments in favor or against the "suid gap" hypothesis. For instance, envisioning the Levant as a regium area or as an intermediate step in a new wave of dispersal of suids from Asia [25,56]. Moreover, the available chronological estimates for the site do not rule out an age younger than the "suid gap" end [53,54,108].

The presence of suids at Pirro Nord is mentioned in the reports of two independent collections, namely by Freudenthal [109] and De Giuli et al. [110], though unfortunately only in the form of a mere appearance in faunal lists. In any case, although the site of Pirro Nord has been extensively investigated over the years and especially after the discovery of lithic tools documenting one the earliest hominin evidence in Europe (e.g., [111–116]), the artiodactyl fauna gathered in the aforementioned collections has not been the object of a systematic revision, which would be necessary to confirm or debunk the presence of suids. In this regard, it is worth mentioning that the material collected by Freudenthal [109] was returned to the University of Bari, but it has been only recently "rediscovered", and it is currently under revision [117]. On the other hand, it is worth mentioning that, according to Falguères [9], unpublished radiometric estimates would point to an age close to 1 Ma for Pirro Nord, hence substantially younger than assumed so far.

The deposit of Ceyssaguet is placed above a lava flow dated at ~1.3 Ma [118] and hence it might fall within or be younger than the "suid gap" end, with prevailing opinions pointing to an age of ~1.2 Ma [119]. According to Van der Made et al. [24], Ceyssaguet would be the youngest locality with *S. strozzii*, although the sample is still unpublished.

The two sites with suid remains recently reported by Kostopoulos et al. [56] from the area of Krimni (Krimni-1 and Krimni-3) are, according to the authors, different spots likely belonging to the same fossiliferous level. The deposits are placed in the upper part of the Gerakarou Formation and the fauna has been biochronologically placed between those of Tsiotra Vryssi (~1.8–1.5 Ma) and Apollonia 1 (~1.2–1.0 Ma) [120,121].

In general, until now, accepting or not the absence of suids from Europe in the 1.8–1.2 Ma time span rested on the biochronological placement of some localities (e.g., Ceyssaguet, Ellera, Mugello), which is prone to subjectivity and biased by the paucity of late Villafranchian deposits in several areas (see next section), and on discounting or not occurrences merely based on faunal lists (especially Peyrolles and Pirro Nord).

Here, the presence of suids from the 1.47 Ma site of Peyrolles is confirmed, perfectly filling the purported "suid gap". The age of the site does not contrast with the traditional biochronological placement of the fauna [11,96] and, unlike other localities, is further supported by radiometric dating [39].

7.4. Implications for Late Villafranchian Biochronology and Faunal Dynamics

Alongside the different traits of suid biology discussed in Section 7.2, the environmental conditions also represent a factor affecting suid abundance, and hence potentially influencing their presence in the fossil record. *Sus strozzii* is traditionally considered a species adapted to humid and forested environments [16], especially after the considerations of comparative anatomy pointed out by Azzaroli [70]. Iannucci et al. [17] suggested a link between the ecological requirements of *S. strozzii* and its decrease in abundance from the late Villafranchian to the Epivillafranchian. Subsequently, Cherin et al. [25] pointed out the need to evaluate the ecology of *S. strozzii* with methods independent of morphology. Kostopoulos et al. [56] considered the "wet" ecological profile of *S. strozzii* an explanation for its absence from well-sampled but unsuitable localities such as Apollonia-1 and report preliminary microwear analysis that would indicate *S. strozzii* had indeed narrow trophic requirements ([122] not seen, *fide* [56]). Although further studies are needed to clarify the ecology of *S. strozzii*, Epivillafranchian suid remains are very scanty and a reduction in the frequency of occurrences with respect to late Villafranchian faunas dated at ~2.0–1.8 Ma is also detectable (Section 7.2). This indicates that a decrease in suid populations (or at least of their representation in the fossil record) took place, and a climatic or environmental drive would be the most reasonable explanation.

In Figure 9, the most important large mammal sites dated to or correlated between less than 1.8 and more than 1.2 Ma (placed within the Farneta and Pirro FUs; MNQ 19), that is, within the time span of the purported absence of suids from Europe [3,18,25], are plotted against a background representing the current Köppen-Geiger climate classification [123]. Of course, there is no one-to-one equivalence between current and past climate, but it is conceivable that similar relative differences existed between regions [3]. In any case, whatever the extent of the differences between regions truly was, the disproportion in the geographical distribution of the sites is striking. Peyrolles is the only locality in western Europe outside the Iberian and Italian Peninsulas reliably dated within this time span, and perhaps it is not a coincidence that the only site of this period where the presence of suids is verified is also the northernmost site.

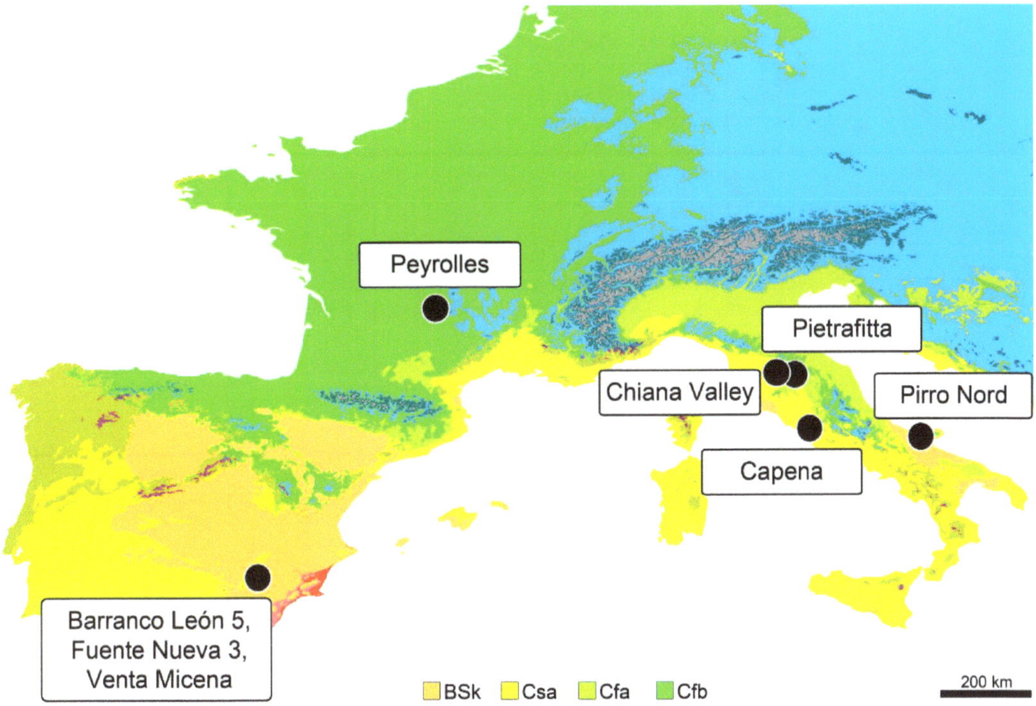

Figure 9. Large mammal paleontological sites in western Europe correlated with the Farneta and Pirro FUs or MNQ 19 (estimated to correspond to ~1.6–1.2 Ma) plotted on a map showing the current Köppen-Geiger climate classification. BSk: arid, steppe, cold; Csa: temperate, dry summer, hot summer (Mediterranean); Cfa: temperate, no dry season, hot summer; Cfb: temperate, no dry season, warm summer. Based on Beck et al. [123], where complete color scheme to climate information is available.

The Spanish localities of Orce (Venta Micena, Barranco Leon, and Fuente Nueva), whose climate is classified as temperate with dry and hot summers, and Pirro Nord in Italy (temperate with no dry season and hot summer) are also surrounded by a large belt of areas characterized by harsher conditions than those projected for the sites themselves. Considering the rarity of suids in the Pleistocene of Europe, only these localities have yielded the thousands of remains needed to be reasonably sure that their absence is not due to a taphonomic artefact (and they might be present at Pirro Nord; see Section 7.3). In this regard, the case of Dmanisi is emblematic. Indeed, although Vekua [124] described an isolated incisor of *Sus* sp. from the site, suids have long been considered absent [18]. Recently, Tappen et al. [7] clarified that 2 specimens of suids are recorded from B1, out of 2764 identified remains. This means that suids at Dmanisi are present, but account for less than 0.01% of the total of the findings.

The case of suids is representative of how this "geographical bias" seriously affects our comprehension of the faunal dynamics and correlations of the key time span comprised between the occurrence of hominins in the Caucasus and their spread into western Europe, but it is only an expression of a broader problem. For instance, even the giant hyena *Pachycrocuta brevirostris*, which is generally considered a ubiquitous predator in the Early Pleistocene of Europe since ~2.0–1.8 Ma of often-stressed importance for biochronology, taphonomy, etc., is not documented in France before its occurrence in some localities referred to as Epivillafranchian [45,119]. For *P. brevirostris*, it is evident that this absence is merely due to the lack of known deposits of the appropriate age, and it might easily pass unnoticed in the first place, being greatly counterbalanced by the abundant samples from the surrounding areas [45]. However, if *P. brevirostris* was evidently capable to maintain viable populations in most of Europe, the same was hardly the case of other species with more selective ecological requirements, which might have retreated to areas not yielding many fossiliferous deposits (or were simply better represented there than elsewhere). In this regard, the extreme paucity of localities in northern and central Europe, reasonably experiencing different environmental conditions than in the south, should be recognized as a substantial limitation in inferring the tempo and mode of the late Early Pleistocene faunal renewal—including the dispersal of hominins.

Another aspect of the thorny "suid gap" question offers room for discussing an additional general problem. As previously noted, while some authors recognized the existence of the "suid gap" [18,25], others did not [24]. Part of the reason rests on the divergent opinions on the biochronological placement of some localities (see previous section). In turn, this controversy is fueled by the low number of paleontological sites and exacerbated by their unbalanced geographical distribution, but also by the paucity of independent constraints. Indeed, in the post-Olduvai to pre-Jaramillo Early Pleistocene, there are few magnetic excursions of short duration [125] and few sites that are associated with robust radiometric ages [9]. Remarkably, the $^{40}Ar/^{39}Ar$ method, which can be considered the gold standard for dating Early (but also Middle) Pleistocene sites, has been successfully applied only on a fraction of them [9,39].

8. Conclusions

Suids have often been regarded as absent from Europe in the period bracketed between less than 1.8 and more than 1.2 Ma based on their great reproductive potential and the assumption than this would translate into an abundant fossil record [18]. Whilst the exceptionality of the fast reproductive strategy of suids is confirmed herein, the putative link between it and the commonness of pig remains in the fossil record is not. Suids are generally rare in the Early Pleistocene of Europe and adjoining areas, an emblematic case being Dmanisi, where they are present but account for less than 0.01% of the total of the remains [7].

Here, suid material is described from the French site of Peyrolles, radiometrically dated at 1.47 Ma, presenting direct evidence for the presence of suids within the alleged "suid gap". Peyrolles (reference for MNQ 19; correlative of Farneta or Pirro FUs) is the only large mammal late Villafranchian locality in western Europe located outside the Iberian and Italian Peninsulas that is reliably dated within this time span.

The case of the purported absence of suids from Europe is the most striking example of our still rather vague comprehension of the faunal dynamics of the ~1.8–1.2 Ma time span. This is the period between the spread of hominins in the Caucasus and the generally accepted age for their arrival in Europe. However, few sites of these age are known in Europe, even fewer are associated with robust radiometric constraints, their geographical distribution is disproportionally biased toward southern Europe, and the paleomagnetic signal of this period is rather monotonous (few excursions and seldom recorded). Collectively, these factors generate a high dependence on biochronological correlations while at the same time making these correlations less reliable—because these are based on a few sites not covering the entire spectrum of environments present in Europe and the sites are not dated independently with methods that outperform biochronology—than those available for other periods. The bias is evident, and it is predictable that several localities correlated within this time span will be reconsidered close to or even younger than the upper limits based on current biochronological estimates, when new independent dating or new findings will become available. In this regard, the possibility that, like suids, other groups survived or could be better documented in the late Villafranchian of scarcely sampled areas (e.g., France) is especially promising.

Supplementary Materials: The following supporting information can be downloaded at: https://www.mdpi.com/article/10.3390/quat7010011/s1, Table S1: data used in Figure 8.

Funding: This research received support from the SYNTHESYS+ Project http://www.synthesys.info/ which is financed by the European Community Research Infrastructure Action under the H2020 Integrating Activities Programme, Project number 823827 (grants GB-TAF-TA4-011 at the Natural History Museum London and HU-TAF-Call4-Iannucci at the Hungarian Natural History Museum). I am currently supported by the Alexander von Humboldt Foundation.

Data Availability Statement: The considered material is curated and available for study in public institutions and all the data analyzed in this work are either directly available within this article or from the cited references.

Acknowledgments: For access and support during the study of the material from Peyrolles and comparative samples, I am thankful to S. Pappa and N. Adams (Natural History Museum London) and to M. Gasparik (Hungarian Natural History Museum, Budapest). The organization of the session "Mammals Biochronology and Paleoecology of the Euro-Mediterranean Quaternary" during the XXI INQUA conference (Rome, Italy, 14–20 July 2023), where a preliminary version of this work has been presented, was a stimulating occasion of discussion. Two anonymous reviewers provided valuable comments and suggestions, for which I am grateful.

Conflicts of Interest: The author declares no conflicts of interest. The funders had no role in the design of the study; in the collection, analyses, or interpretation of data; in the writing of the manuscript; or in the decision to publish the results.

References

1. Shackleton, N.J. New data on the evolution of Pliocene climatic variability. In *Paleoclimate and Evolution, with Emphasis on Human Evolution*; Vrba, E.S., Denton, G.H., Patridge, T.C., Burckle, L.H., Eds.; Yale University Press: London, UK, 1995; pp. 242–248.
2. Kahlke, R.-D.; García, N.; Kostopoulos, D.S.; Lacombat, F.; Lister, A.M.; Mazza, P.P.A.; Spassov, N.; Titov, V.V. Western Palaearctic palaeoenvironmental conditions during the Early and early Middle Pleistocene inferred from large mammal communities, and implications for hominin dispersal in Europe. *Quat. Sci. Rev.* **2011**, *30*, 1368–1395. [CrossRef]
3. Palombo, M.R. Deconstructing mammal dispersals and faunal dynamics in SW Europe during the Quaternary. *Quat. Sci. Rev.* **2014**, *96*, 50–71. [CrossRef]
4. Gibbard, P.L.; Head, M.J. The Quaternary Period. In *Geologic Time Scale 2020*; Gradstein, F.M., Ogg, J.G., Schmitz, M.D., Ogg, G.M., Eds.; Elsevier: Amsterdam, The Netherlands, 2020; pp. 1217–1255.
5. Iannucci, A.; Sardella, R. What does the "Elephant-*Equus*" event mean today? Reflections on mammal dispersal events around the Pliocene-Pleistocene boundary and the flexible ambiguity of biochronology. *Quaternary* **2023**, *6*, 16. [CrossRef]
6. Rightmire, G.P.; Margvelashvili, A.; Lordkipanidze, D. Variation among the Dmanisi hominins: Multiple taxa or one species? *Am. J. Phy. Anthrop.* **2019**, *168*, 481–495. [CrossRef] [PubMed]
7. Tappen, M.; Bukhsianidze, M.; Ferring, R.; Coil, R.; Lordkipanidze, D. Life and death at Dmanisi, Georgia: Taphonomic signals from the fossil mammals. *J. Hum. Evol.* **2022**, *171*, 103249. [CrossRef]
8. Muttoni, G.; Scardia, G.; Kent, D.V. Early hominins in Europe: The Galerian migration hypothesis. *Quat. Sci. Rev.* **2018**, *180*, 1–29. [CrossRef]
9. Falguères, C. The first human settlements out africa into Europe: A chronological perspective. *Quat. Sci. Rev.* **2020**, *247*, 106551. [CrossRef]
10. Azzaroli, A. Quaternary mammals and the "end-Villafranchian" dispersal event—A turning point in the history of Eurasia. *Palaeogeogr. Palaeoclimatol. Palaeoecol.* **1983**, *44*, 117–139. [CrossRef]
11. Azzaroli, A.; De Giuli, C.; Ficcarelli, G.; Torre, D. Late Pliocene to early mid-Pleistocene mammals in Eurasia: Faunal succession and dispersal events. *Palaeogeogr. Palaeoclimatol. Palaeoecol.* **1988**, *66*, 77–100. [CrossRef]
12. Iannucci, A.; Mecozzi, B.; Sardella, R. Beware of the "Wolf event"–Remarks on large mammal dispersals in Europe and the late Villafranchian faunal turnover. *Alp. Mediterr. Quat.* **2023**, *36*, 75–90. [CrossRef]
13. Iannucci, A. *Sus strozzii* (Suidae, Mammalia) from the historical locality of Quercia (Early Pleistocene, Italy). *Geobios* **2023**, *77*, 27–36. [CrossRef]
14. Forsyth Major, C.I. Studi sugli avanzi pleistocenici del genere *Sus*. *Atti Soc. Toscana Sci. Nat. P. V.* **1881**, *2*, 227.
15. Linnaeus, C. *Systema Naturae Per Regna Tria Naturae, Secundum Classes, Ordines, Genera, Species, Cum Characteribus, Differentiis, Synonymis, Locis*; Tomus, I., Editio Decima, R., Eds.; Laurentius Salvius: Stockholm, Sweden, 1758.
16. Faure, M.; Guérin, C. *Sus strozzii* et *Sus scrofa*, deux mammiferes artiodactyles, marqueurs des paleoenvironnements. *Palaeogeogr. Palaeoclimatol. Palaeoecol.* **1984**, *48*, 215–228. [CrossRef]
17. Iannucci, A.; Gasparik, M.; Sardella, R. First report of *Sus strozzii* (Suidae, Mammalia) from the Early Pleistocene of Hungary (Dunaalmás) and species distinction based on deciduous teeth. *Sci. Nat.* **2020**, *107*, 5. [CrossRef] [PubMed]

18. Martínez-Navarro, B.; Madurell-Malapeira, J.; Ros-Montoya, S.; Espigares, M.P.; Medin, T.; Hortola, P.; Palmqvist, P. The Epivillafranchian and the arrival of pigs into Europe. *Quat. Int.* **2015**, *389*, 131–138. [CrossRef]
19. Kahlke, R.-D. *Untermassfeld. A Late Early Pleistocene (Epivillafranchian) Fossil Site near Meiningen (Thuringia, Germany) and Its Position in the Development of the European Mammal Fauna*; British Archaeological Reports, International Series 1578; BAR Publishing: Oxford, UK, 2006.
20. Kahlke, R.-D. Late Early Pleistocene European large mammals and the concept of an Epivillafranchian biochron. *Cour. Forsch.-Inst. Senckenberg* **2007**, *259*, 265–278.
21. Kahlke, R.-D. Les communautés de grands mammifères du Pléistocène inférieur terminal et le concept d'un biochrone Épivillafranchien. *Quaternaire* **2009**, *20*, 415–427. [CrossRef]
22. Bellucci, L.; Sardella, R.; Rook, L. Large mammal biochronology framework in Europe at Jaramillo: The Epivillafranchian as a formal biochron. *Quat. Int.* **2015**, *389*, 84–89. [CrossRef]
23. Bona, F.; Sala, B. Villafranchian-Galerian mammal faunas transition in South-Western Europe. The case of the late Early Pleistocene mammal fauna of the Frantoio locality, Arda River (Castell'Arquato, Piacenza, northern Italy). *Geobios* **2016**, *49*, 329–347. [CrossRef]
24. Van der Made, J.; Rosell, J.; Blasco, R. Faunas from Atapuerca at the Early-Middle Pleistocene limit: The ungulates from level TD8 in the context of climatic change. *Quat. Int.* **2017**, *433*, 296–346. [CrossRef]
25. Cherin, M.; Alba, D.M.; Crotti, M.; Menconero, S.; Moullé, P.-É.; Sorbelli, L.; Madurell-Malapeira, J. The post-Jaramillo persistence of *Sus strozzii* (Suidae, Mammalia) in Europe: New evidence from the Vallparadís Section (NE Iberian Peninsula) and other coeval sites. *Quat. Sci. Rev.* **2020**, *233*, 106234. [CrossRef]
26. Van der Made, J.; Moyà-Solà, S. European Suinae (Artiodactyla) from the Late Miocene onwards. *Boll. Soc. Paleontol. Ital.* **1989**, *28*, 329–339.
27. Gliozzi, E.; Abbazzi, L.; Argenti, P.; Azzaroli, A.; Caloi, L.; Capasso Barbato, L.; Di Stefano, G.; Esu, D.; Ficcarelli, G.; Girotti, O.; et al. Biochronology of selected mammals, molluscs and ostracods from the middle Pliocene to the late Pleistocene in Italy. The state of the art. *Riv. Ital. Paleontol. Stratigr.* **1997**, *103*, 369–388. [CrossRef]
28. Guérin, C.; Faure, M. The Wild Boar (*Sus scrofa priscus*) from the Post-Villafranchian Lower Pleistocene of Untermaßfeld. In *Das Pleistozän von Untermaßfeld bei Meiningen (Thüringen). Teil 1.*; Kahlke, R.-D., Ed.; Monographien des Römisch-Germanisches Zentralmuseum: Mainz, Germany, 1997; pp. 375–383.
29. Lister, A.M.; Parfitt, S.A.; Owen, F.J.; Collinge, S.E.; Breda, M. Metric analysis of ungulate mammals in the early Middle Pleistocene of Britain, in relation to taxonomy and biostratigraphy: II: Cervidae, Equidae and Suidae. *Quat. Int.* **2010**, *228*, 157–179. [CrossRef]
30. Antoine, P.; Moncel, M.H.; Limondin-Lozouet, N.; Locht, J.L.; Bahain, J.J.; Moreno, D.; Voinchet, P.; Auguste, P.; Stoetzel, E.; Dabkowski, J.; et al. Palaeoenvironment and dating of the Early Acheulean localities from the Somme River basin (Northern France): New discoveries from the high terrace at Abbeville-Carrière Carpentier. *Quat. Sci. Rev.* **2016**, *149*, 338–371. [CrossRef]
31. Cherin, M.; Sorbelli, L.; Crotti, M.; Iurino, D.A.; Sardella, R.; Souron, A. New material of *Sus strozzii* (Suidae, Mammalia) from the Early Pleistocene of Italy and a phylogenetic analysis of suines. *Quat. Sci. Rev.* **2018**, *194*, 94–115. [CrossRef]
32. Iannucci, A.; Sardella, R.; Strani, F.; Mecozzi, B. Size shifts in late Middle Pleistocene to Early Holocene *Sus scrofa* (Suidae, Mammalia) from Apulia (southern Italy): Ecomorphological adaptations? *Hystrix* **2020**, *31*, 10–20. [CrossRef]
33. Luzón, C.; Yravedra, J.; Courtenay, L.A.; Saarinen, J.; Blain, H.A.; DeMiguel, D.; Viranta, S.; Azanza, B.; Rodrigez-Alba, J.J.; Herranz-Rodrigo, D.; et al. Taphonomic and spatial analyses from the early Pleistocene site of Venta Micena 4 (Orce, Guadix-Baza Basin, southern Spain). *Sci. Rep.* **2021**, *11*, 13977. [CrossRef]
34. Valli, A.M.F.; Caron, J.-B.; Debard, E.; Guérin, C.; Pastre, J.-F.; Argant, J. Le gisement paléontologique villafranchien terminal de Peyrolles (Issoire, Puy-de-Dôme, France): Résultats de nouvelles prospections. *Geodiversitas* **2006**, *28*, 297–317.
35. Bout, P.; Azzaroli, A. Stratigraphie et faune du Creux de Peyrolles près Perrier (Puy-de-Dôme). *Annal. Paloentol.* **1952**, *38*, 37–56.
36. Bout, P. *Le Villafranchien du Velay et du Basin Hydrographique Moyen et Supérieur de L'allier*; Imprimerie Jeanne d'Arc: Le Puy-en-Velay, France, 1960.
37. Couthures, J.; Pastre, J.F. Chronostratigraphie du Plio-PIéistocène d'Auvergne et du Velay: Nouveaux apports des datations radiométriques et du paléomagnétisme. *Quaternaire* **1983**, *20*, 9–18. [CrossRef]
38. Poidevin, J.L.; Cantagrel, J.M.; GUERPA. Un site unique du Plio-Pleistocene en Europe: Le pluteau de Perrier (Puy-de-Dôme)—Confrontation des donneés volcanologiques, stratigraphiques et paléontologiques. *Rev. Sc. Nat. d'Auvergne* **1984**, *50*, 87–95.
39. Nomade, S.; Pastre, J.-F.; Guillou, H.; Faure, M.; Guérin, C.; Delson, E.; Debard, E.; Voinchet, P.; Msessager, E. $^{40}Ar/^{39}Ar$ constraints on some French landmark Late Pliocene to Early Pleistocene large mammalian paleofaunas: Paleoenvironmental and paleoecological implications. *Quat. Geochronol.* **2014**, *21*, 2–15. [CrossRef]
40. Lo Bello, P. Géochronologie par la Méthode ^{39}Ar-^{40}Ar de Ponces Quaternaires Contaminées: Exemple des Ponces du Mont-Dore (Massif Central, France). Utilisation d'un Laser Continu Pour la Datation des Min´Eraux Individuels. Unpublished Ph.D. Thesis, Université de Nice, Nice, France, 1988.
41. Palombo, M.R.; Valli, A.M.F. Remarks on the biochronology of mammalian faunal complexes from the Pliocene to the Middle Pleistocene in France. *Geol. Romana* **2004**, *37*, 145–163.
42. Thouveny, N.; Taieb, M.; Bonnet, A. Etude magnétostratigraphique du remplissage du Pléistocene inférieur de la grotte de La Sartanette (Remoulins, Gard). *Bull. Soc. Geol. Fr.* **1984**, *26*, 1385–1388. [CrossRef]

43. Paquette, J.L.; Médard, E.; Poidevin, J.L.; Barbet, P. Precise dating of middle to late Villafranchian mammalian paleofaunae from the Upper Allier River valley (French Massif Central) using U-Pb geochronology on volcanic zircons. *Quat. Geochronol.* **2021**, *65*, 101198. [CrossRef]
44. Heintz, E.; Guérin, C.; Martin, P.; Prat, F. Principaux gisements Villafranchiens de France: Listes fauniques et biostratigraphie. *Mem. BRGM* **1974**, *78*, 411–417.
45. Iannucci, A.; Mecozzi, B.; Sardella, R.; Iurino, D.A. The extinction of the giant hyena *Pachycrocuta brevirostris* and a reappraisal of the Epivillafranchian and Galerian Hyaenidae in Europe: Faunal turnover during the Early–Middle Pleistocene Transition. *Quat. Sci. Rev.* **2021**, *272*, 107240. [CrossRef]
46. Cherin, M.; Breda, M.; Esattore, B.; Hart, V.; Turek, J.; Porciello, F.; Angeli, G.; Holpin, S.; Iurino, D.A. A Pleistocene Fight Club revealed by the palaeobiological study of the *Dama*-like deer record from Pantalla (Italy). *Sci. Rep.* **2022**, *12*, 13898. [CrossRef]
47. Soria, C.D.; Pacifici, M.; Di Marco, M.; Stephen, S.M.; Rondinini, C. COMBINE: A coalesced mammal database of intrinsic and extrinsic traits. *Ecology* **2021**, *102*, e03344. [CrossRef]
48. Hammer, Ø.; Harper, D.A.T.; Ryan, P.D. PAST: Paleontological statistics software package for education and data analysis. *Palaeontol. Electron.* **2001**, *4*, 4.
49. Mazza, P.P.A.; Bertini, A.; Magi, M. The late Pliocene site of Poggio Rosso (central Italy): Taphonomy and paleoenvironment. *Palaios* **2004**, *19*, 227–248. [CrossRef]
50. Barash, A.; Belmaker, M.; Bastir, M.; Soudack, M.; O'Brien, H.D.; Woodward, H.; Prendergast, A.; Barzilai, O.; Been, E. The earliest Pleistocene record of a large-bodied hominin from the Levant supports two out-of-Africa dispersal events. *Sci. Rep.* **2022**, *12*, 1721. [CrossRef] [PubMed]
51. Kahlke, R.-D. (Ed.) The Pleistocene of Untermassfeld—A synopsis of site origin, palaeobiodiversity, taphonomic characteristics, palaeoenvironment, chronostratigraphy, and significance in western Palaearctic faunal history. In *The Pleistocene of Untermassfeld near Meiningen (Thüringen, Germany)*; Part 5; Monographien des Römisch-Germanisches Zentralmuseum: Mainz, Germany, 2022; pp. 1635–1734.
52. Madurell-Malapeira, J.; Alba, D.M.; Espigares, M.-P.; Vinuesa, V.; Palmqvist, P.; Martínez-Navarro, B.; Moyà-Solà, S. Were large carnivorans and great climatic shifts limiting factors for hominin dispersals? Evidence of the activity of *Pachycrocuta brevirostris* during the Mid-Pleistocene Revolution in the Vallparadís Section (Vallès-Penedès Basin, Iberian Peninsula). *Quat. Int.* **2017**, *431*, 42–52. [CrossRef]
53. Tchernov, E. The age of the 'Ubeidiya Formation, an Early Pleistocene hominid site in the Jordan Valley, Israel. *Isr. J. Earth Sci.* **1987**, *36*, 3–30.
54. Martínez-Navarro, B.; Belmaker, M.; Bar-Yosef, O. The large carnivores from 'Ubeidiya (early Pleistocene, Israel): Biochronological and biogeographical implications. *J. Hum. Evol.* **2009**, *56*, 514–524. [CrossRef]
55. Belmaker, M. Insights from carnivore community composition on the paleoecology of early Pleistocene Eurasian sites: Implications for the dispersal of hominins out of Africa. *Quat. Int.* **2018**, *464*, 3–17. [CrossRef]
56. Kostopoulos, D.S.; Konidaris, G.E.; Amanatidou, M.; Chitoglou, K.; Fragkioudakis, E.; Gerakakis, N.; Giannakou, V.; Gkeme, A.; Kalaitzi, C.; Tsakalidis, C.; et al. The new fossil site Krimni-3 in Mygdonia Basin and the first evidence of a giant ostrich in the Early Pleistocene of Greece. *PalZ* **2023**, *97*, 147–161. [CrossRef]
57. De Giuli, C.; Masini, F. Late Villafranchian faunas in Italy: The Casa Frata local fauna (Upper Valdarno, Tuscany). *Palaeontogr. Ital.* **1986**, *74*, 1–9.
58. Napoleone, G.; Albanielli, A.; Azzaroli, A.; Bertini, A.; Magi, M.; Mazzini, M. Calibration of the Upper Valdarno Basin to the Plio-Pleistocene for correlating the Apennine continental sequences. *Ital. J. Quat. Sci.* **2003**, *16*, 131–166.
59. Iannucci, A.; Conti, J.; Curcio, F.; Iurino, D.A.; Mancini, M.; Mecozzi, B.; Strani, F.; Sardella, R. Middle Pleistocene mammal faunas of the area of Rome: Recent results and ongoing work on the MUST collection. *J. Mediterr. Earth Sci.* **2023**, *15*, 19–37. [CrossRef]
60. Duval, M.; Arnold, L.J.; Demuro, M.; Parés, J.M.; Campaña, I.; Carbonell, E.; de Castro, J.M.B. New chronological constraints for the lowermost stratigraphic unit of Atapuerca Gran Dolina (Burgos, N Spain). *Quat. Geochron.* **2022**, *71*, 101292. [CrossRef]
61. Iannucci, A.; Cherin, M.; Sardella, R. The lost hyena from Paciano (Umbria, Italy) reconsidered. *Alp. Mediterr. Quat.* **2022**, *35*, 105–117.
62. Masini, F.; Palombo, M.R.; Rozzi, R. A reappraisal of the early to middle Pleistocene Italian Bovidae. *Quat. Int.* **2013**, *288*, 45–62. [CrossRef]
63. Sorbelli, L.; Cherin, M.; Kostopoulos, D.S.; Sardella, R.; Mecozzi, B.; Plotnikov, V.; Prat-Vericat, M.; Azzarà, B.; Bartolini-Lucenti, S.; Madurell-Malapeira, J. Earliest bison dispersal in Western Palearctic: Insights from the *Eobison* record from Pietrafitta (Early Pleistocene, central Italy). *Quat. Sci. Rev.* **2023**, *301*, 107923. [CrossRef]
64. Siori, M.S.; Sala, B. The mammal fauna from the late Early Biharian site of Castagnone (Northern Monferrato, Piedmont, NW Italy). *Geobios* **2007**, *40*, 207–217. [CrossRef]
65. Guérin, C.; Dewolf, Y.; Lautridou, J.P. Révision d'un site paléontologique célèbre: Saint-Prest (Chartres, France). *Geobios* **2003**, *36*, 55–82. [CrossRef]
66. Owen, R. Description of teeth and portions of jaws of two extinct anthracotherioid quadrupeds (*Hyopotamus vectianus* and *Hyop. bovinus*) discovered by the Marchioness of Hastings in the Eocene deposits on the N. W. coast of the Isle of Wight: With an attempt to develop Cuvier's idea of the classification of pachyderms by the number of their toes. *Q. J. Geol. Soc. Lond.* **1848**, *4*, 103–141.
67. Gray, J.E. On the natural arrangement of vertebrose animals. London Medical Reposit. *Monthly J. Rev.* **1821**, *15*, 296–310.

68. Iannucci, A. New results on suids from the Early Pleistocense site of Untermassfeld. In *The Pleistocene of Untermassfeld near Meiningen (Thüringen, Germany)*; Part 5; Kahlke, R.-D., Ed.; Monographien des Römisch-Germanisches Zentralmuseum: Mainz, Germany, 2022; pp. 1339–1354.
69. Schaub, S. Die oberpliocaene Säugetierfauna von Senèze (Haute-Loire) und ihre verbreitungsgeschichtliche Stellung. *Eclogae Geol. Helv.* **1943**, *36*, 270–289.
70. Azzaroli, A. Revisione dei terreni fluvio-lacustri del Valdarno Superiore. V. Filogenesi e biologia di *Sus strozzii* e *Sus minor*. *Palaeontogr. Ital.* **1954**, *48*, 41–76.
71. Dubost, G.; Charron, F.; Courcoul, A.; Rodier, A. The Chinese water deer, *Hydropotes inermis*—A fast-growing and productive ruminant. *Mammal. Biol.* **2011**, *76*, 190–195. [CrossRef]
72. Frauendorf, M.; Gethöffer, F.; Siebert, U.; Keuling, O. The influence of environmental and physiological factors on the litter size of wild boar (*Sus scrofa*) in an agriculture dominated area in Germany. *Sci. Total Environ.* **2016**, *541*, 877–882. [CrossRef]
73. Pianka, E.R. On r- and K-Selection. *Am. Nat.* **1970**, *104*, 592–597. [CrossRef]
74. Tuomi, J. Mammalian reproductive strategies: A generalized relation of litter size to body size. *Oecologia* **1980**, *45*, 39–44. [CrossRef] [PubMed]
75. Stearns, S.C. Life history tactics: A review of the ideas. *Quart. Rev. Biol.* **1976**, *51*, 3–47. [CrossRef] [PubMed]
76. Reznick, D.; Bryant, M.J.; Bashey, F. r-and K-selection revisited: The role of population regulation in life-history evolution. *Ecology* **2002**, *83*, 1509–1520. [CrossRef]
77. MacArthur, R.H.; Wilson, E.O. An equilibrium theory of insular zoogeography. *Evolution* **1963**, *17*, 373–387. [CrossRef]
78. MacArthur, R.H.; Wilson, E.O. *The Theory of Island Biogeography*; Princeton University Press: Princeton, NJ, USA, 1967.
79. Bieber, C.; Ruf, T. Population dynamics in wild boar *Sus scrofa*: Ecology, elasticity of growth rate and implications for the management of pulsed resource consumers. *J. Appl. Ecol.* **2005**, *42*, 1203–1213. [CrossRef]
80. Geisser, H.; Reyer, H.-U. The influence of food and temperature on population density of wild boar *Sus scrofa* in the Thurgau (Switzerland). *J. Zool.* **2005**, *267*, 89–96. [CrossRef]
81. Keuling, O.; Baubet, E.; Duscher, A.; Ebert, C.; Fischer, C.; Monaco, A.; Podgórski, T.; Prevot, C.; Ronnenberg, K.; Sodeikat, G.; et al. Mortality rates of wild boar *Sus scrofa* L. in central Europe. *Eur. J. Wildl. Res.* **2013**, *59*, 805–814. [CrossRef]
82. Massei, G.; Kindberg, J.; Licoppe, A.; Gačić, D.; Šprem, N.; Kamler, J.; Baubet, E.; Hohmann, U.; Monaco, A.; Ozoliņš, J.; et al. Wild boar populations up, numbers of hunters down? A review of trends and implications for Europe. *Pest Manag. Sci.* **2015**, *71*, 492–500. [CrossRef]
83. Vetter, S.G.; Ruf, T.; Bieber, C.; Arnold, W. What is a mild winter? Regional differences in within-species responses to climate change. *PLoS ONE* **2015**, *10*, e0132178. [CrossRef] [PubMed]
84. Melletti, M.; Meijaard, E. *Ecology, Conservation and Management of Wild Pigs and Peccaries*; Cambridge University Press: Cambridge, UK, 2017.
85. White, T.D.; Harris, J.M. Suid evolution and correlation of African hominid localities. *Science* **1977**, *198*, 13–21. [CrossRef] [PubMed]
86. Cooke, H.B.S.; Wilkinson, A.F. Suidae and Tayassuidae. In *Evolution of African Mammals*; Maglio, V.J., Cooke, H.B.S., Eds.; Harvard University Press: Cambridge, UK, 1978; pp. 435–482.
87. Harris, J.M.; Cerling, T.E. Dietary adaptations of extant and Neogene African suids. *J. Zool.* **2002**, *256*, 45–54. [CrossRef]
88. Harris, J.M.; White, T.D. Evolution of the Plio-Pleistocene African Suidae. *Trans. Am. Philos. Soc.* **1979**, *69*, 1–128. [CrossRef]
89. Souron, A. Diet and ecology of extant and fossil wild pigs. In *Ecology, Conservation and Management of Wild Pigs and Peccaries*; Melletti, M., Meijaard, E., Eds.; Cambridge University Press: Cambridge, UK, 2017; pp. 29–38.
90. Yang, D.; Pisano, A.; Kolasa, J.; Jashashvili, T.; Kibii, J.; Cano, A.R.G.; Viriot, N.; Grine, F.E.; Souron, A. Why the long teeth? Morphometric analysis suggests different selectivse pressures on functional occlusal traits in Plio-Pleistocene African suids. *Paleobiology* **2022**, *48*, 655–676. [CrossRef]
91. Cherin, M.; Basilici, G.; Duval, M.; Shao, Q.; Sier, M.J.; Parés, J.M.; Gliozzi, E.; Mazzini, I.; Magri, D.; Di Rita, F.; et al. The dawn of the Late Villafranchian: Paleoenvironment and age of the Pantalla paleontological site (Italy; Early Pleistocene). *Quat. Sci. Rev.* **2023**, *316*, 108279. [CrossRef]
92. Strani, F.; Bellucci, L.; Iannucci, A.; Iurino, D.A.; Mecozzi, B.; Sardella, R. Palaeoenvironments of the MIS 15 site of Cava di Breccia—Casal Selce 2 (central Italian Peninsula) and niche occupation of fossil ungulates during Middle Pleistocene interglacials. *Hist. Biol.* **2022**, *34*, 555–565. [CrossRef]
93. Magurran, A.E.; Henderson, P.A. Explaining the excess of rare species in natural species abundance distributions. *Nature* **2003**, *422*, 714–716. [CrossRef]
94. Sorbelli, L.; Alba, D.M.; Cherin, M.; Moullé, P.É.; Brugal, J.P.; Madurell-Malapeira, J. A review on *Bison schoetensacki* and its closest relatives through the early-Middle Pleistocene transition: Insights from the Vallparadís Section (NE Iberian Peninsula) and other European localities. *Quat. Sci. Rev.* **2021**, *261*, 106933. [CrossRef]
95. Hermier, R.; Merceron, G.; Kostopoulos, D.S. The emblematic Eurasian Villafranchian antelope *Gazellospira* (Mammalia: Bovidae): New insights from the Lower Pleistocene Dafnero fossil sites (Northern Greece). *Geobios* **2020**, *61*, 11–29. [CrossRef]
96. Palombo, M.R. Discrete dispersal bioevents of large mammals in Southern Europe in the post-Olduvai Early Pleistocene: A critical overview. *Quat. Int.* **2017**, *431*, 3–19. [CrossRef]

97. Breda, M.; Marchetti, M. Systematical and biochronological review of Plio-Pleistocene Alceini (Cervidae; Mammalia) from Eurasia. *Quat. Sci. Rev.* **2005**, *24*, 775–805. [CrossRef]
98. Mazza, P.P.A. Poggio Rosso (Upper Valdarno, central Italy), a window on latest Pliocene wildlife. *Palaios* **2006**, *21*, 493–498. [CrossRef]
99. Moncel, M.-H.; Lemorini, C.; Eramo, G.; Fioretti, G.; Daujeard, C.; Curci, A.; Berto, C.; Hardy, B.; Pineda, A.; Rineau, V.; et al. A taphonomic and spatial distribution study of the new levels of the middle Pleistocene site of Notarchirico (670–695 ka, Venosa, Basilicata, Italy). *Archaeol. Anthropol. Sci.* **2023**, *15*, 106. [CrossRef]
100. Ollé, A.; Lombao, D.; Asryan, L.; Gárcia-Medano, P.; Arroyo, A.; Fernández-Marchena, J.L.; Yesilova, G.C.; Cáceres, I.; Huguet, R.; López-Polín, L.; et al. The earliest European Acheulean: New insights into the large shaped tools from the late Early Pleistocene site of Barranc de la Boella (Tarragona, Spain). *Front. Earth Sci.* **2023**, *11*, 1188663. [CrossRef]
101. Mayer, J.J.; Brisbin, I.E. *Wild Pigs: Biology, Damage, Control Techniques and Management*; Savannah River National Laboratory: Aiken, SC, USA, 2009.
102. Pazzaglia, F.; Barchi, M.R.; Buratti, N.; Cherin, M.; Pandolfi, L.; Ricci, M. Pleistocene calcareous tufa from the Ellera basin (Umbria, central Italy) as a key for an integrated paleoenvironmental and tectonic reconstruction. *Quat. Int.* **2013**, *292*, 59–70. [CrossRef]
103. Abbazzi, L.; Benvenuti, M.; Rook, L.; Masini, F. Biochronology of the Mugello intermontane basin (northern Apennines, Italy). *Il Quat.* **1995**, *8*, 5–10.
104. Fidalgo, D.; Rosas, A.; Madurell-Malapeira, J.; Pineda, A.; Huguet, R.; García-Tabernero, A.; Cáceres, I.; Olle, A.; Vallverdú, J.; Saladie, P. A review on the Pleistocene occurrences and palaeobiology of *Hippopotamus antiquus* based on the record from the Barranc de la Boella Section (Francolí Basin, NE Iberia). *Quat. Sci. Rev.* **2023**, *307*, 108034. [CrossRef]
105. Alberdi, M.T.; Palombo, M.R. The late Early to early Middle Pleistocene stenonoid horses from Italy. *Quat. Int.* **2013**, *288*, 25–44. [CrossRef]
106. De Giuli, C. Late Villafranchian faunas of Italy: The Selvella local fauna in the southern Chiana Valley-Umbria. *Palaeontogr. Ital.* **1986**, *74*, 11–50.
107. Geraads, D.; Guérin, C.; Faure, M. Les suidés du Pléistocène ancien d'Oubeidiyeh (Israël). *Mem. Travaux Cent. Rech. Fr. Jerus.* **1986**, *5*, 93–105.
108. Rink, W.J.; Bartoll, J.; Schwarcz, H.P.; Shane, P.; Bar-Yosef, O. Testing the reliability of ESR dating of optically exposed buried quartz sediments. *Radiat. Meas.* **2007**, *42*, 1618–1626. [CrossRef]
109. Freudenthal, M. Neogene vertebrates from the Gargano Peninsula, Italy. *Scripta Geol.* **1971**, *3*, 1–10.
110. De Giuli, C.; Masini, F.; Torre, D. The latest Villafranchian faunas in Italy: The Pirro Nord fauna (Apricena, Gargano). *Palaeontogr. Ital.* **1986**, *74*, 51–62.
111. Arzarello, M.; Marcolini, F.; Pavia, G.; Pavia, M.; Petronio, C.; Petrucci, M.; Rook, L.; Sardella, R. Evidence of earliest human occurrence in Europe: The site of Pirro Nord (southern Italy). *Naturwissenschaften* **2007**, *94*, 107–112. [CrossRef]
112. Arzarello, M.; Pavia, G.; Peretto, C.; Petronio, C.; Sardella, R. Evidence of an early Pleistocene hominin presence at Pirro Nord (Apricena, Foggia, southern Italy): P13 site. *Quat. Int.* **2012**, *267*, 56–61. [CrossRef]
113. Arzarello, M.; De Weyer, L.; Peretto, C. The first European peopling and the Italian case: Peculiarities and "opportunism". *Quat. Int.* **2016**, *393*, 41–50. [CrossRef]
114. Pavia, M.; Zunino, M.; Coltorti, M.; Angelone, C.; Arzarello, M.; Bagnus, C.; Bellucci, L.; Colombero, S.; Marcolini, F.; Peretto, C.; et al. Stratigraphical and palaeontological data from the Early Pleistocene Pirro 10 site of Pirro Nord (Puglia, south eastern Italy). *Quat. Int.* **2012**, *267*, 40–55. [CrossRef]
115. Petrucci, M.; Cipullo, A.; Martínez-Navarro, B.; Rook, L.; Sardella, R. The Late Villafranchian (Early Pleistocene) carnivores (Carnivora, Mammalia) from Pirro Nord (Italy). *Palaeontogr. Abt. A* **2013**, *298*, 113–145. [CrossRef]
116. Carpentieri, M.; Arzarello, M. For our world without sound: The opportunistic debitage in the Italian context—A methodological evaluation of the lithic assemblages of Pirro Nord, Cà Belvedere di Montepoggiolo, Ciota Ciara Cave and Riparo Tagliente. *J. Paleolithic Archaeol.* **2022**, *5*, 12. [CrossRef]
117. Stefanelli, D.; Mecozzi, B.; Breda, M.; Girone, A.; La Perna, R.; Maiorano, P.; Marino, M.; Sardella, R. *Paleodays 2023—XXIII Edizione Delle Giornate di Paleontologia Lecce, 7–9 Giugno 2023, Abstract Book*; Cherin, M., Collareta, A., Eds.; Rediscovering the Fossil Collection from the Late Early Pleistocene Site of Pirro Nord (Foggia, Southern Italy) at the Bari University; University of Salento: Lecce, Italy, 2023; p. 76.
118. Tsoukala, E.; Bonifay, M.-F. The Early Pleistocene Carnivores (Mammalia) from Ceyssaguet (Haute-Loire). *Paléo* **2004**, *16*, 193–242.
119. Brugal, J.P.; Argant, A.; Boudadi-Maligne, M.; Crégut-Bonnoure, E.; Croitor, R.; Fernandez, P.; Fourvel, J.-P.; Fosse, P.; Gaudelli, J.-L.; Labe, B.; et al. Pleistocene herbivores and carnivores from France: An updated overview of the literature, sites and taxonomy. *Ann. Paleontol.* **2020**, *106*, 102384. [CrossRef]
120. Koufos, G.D.; Kostopoulos, D.S. *The Plio-Pleistocene large mammal record of Greece: Implications for early human dispersals into Europe In Paleoanthropology of the Balkans and Anatolia*; Harvati, K., Roksandic, M., Eds.; Springer: Dordrecht, The Netherlands, 2016; pp. 269–280.
121. Konidaris, G.E.; Kostopoulos, D.S.; Maron, M.; Schaller, M.; Ehlers, T.A.; Aidona, E.; Marini, M.; Tourloukis, V.; Muttoni, G.; Koufos, G.D.; et al. Dating of the Lower Pleistocene vertebrate site of Tsiotra Vryssi (Mygdonia Basin, Greece): Biochronology, magnetostratigraphy, and cosmogenic radionuclides. *Quaternary* **2021**, *4*, 1. [CrossRef]

122. Alifieri, E. Dietary Adaptations of the Early Pleistocene Gerakarou Mammal Community (Mygdonia Basin, Greece). Unpublished Master's Thesis, Interinstitutional Program of Postgraduate Studies in Palaeontology-Geobiology, Aristotle University of Thessaloniki, Thessaloniki, Greece, 2021.
123. Beck, H.E.; Zimmermann, N.E.; McVicar, T.R.; Vergopolan, N.; Berg, A.; Wood, E.F. Present and future Köppen-Geiger climate classification maps at 1-km resolution. *Sci. Data* **2018**, *5*, 180214. [CrossRef]
124. Vekua, A. Die Wirbeltierfauna des Villafranchium von Dmanisi und ihre biostratigraphische Bedeutung. *Jahrb. RGZM* **1995**, *42*, 77–180.
125. Channell, J.E.T.; Singer, B.S.; Jicha, B.R. Timing of Quaternary geomagnetic reversals and excursions in volcanic and sedimentary archives. *Quat. Sci. Rev.* **2020**, *228*, 106114. [CrossRef]

quaternary

Review

A Review on the Latest Early Pleistocene Carnivoran Guild from the Vallparadís Section (NE Iberia)

Joan Madurell-Malapeira [1,2,*], Maria Prat-Vericat [3], Saverio Bartolini-Lucenti [1,3], Andrea Faggi [1], Darío Fidalgo [4], Adrian Marciszak [5] and Lorenzo Rook [1]

1 Earth Science Department, University of Florence, 50121 Florence, Italy; saverio.bartolinilucenti@unifi.it (S.B.-L.); andrea.faggi@unifi.it (A.F.); lorenzo.rook@unifi.it (L.R.)
2 Department of Geology, Faculty of Sciences, Universitat Autònoma de Barcelona, 08193 Cerdanyola del Vallès, Spain
3 Catalan Institute of Paleontology, Universitat Autònoma de Barcelona, Edifici ICTA-ICP, c/Columnes s/n, Campus de la UAB, 08193 Cerdanyola del Vallès, Spain; maria.prat@icp.cat
4 Department of Palaeobiology, Museo Nacional de Ciencias Naturales (CSIC), 28006 Madrid, Spain; dfidal01@ucm.es
5 Department of Palaeozoology, University of Wrocław, Sienkiewicza 21, 50-335 Wrocław, Poland; adrian.marciszak@uwr.edu.pl
* Correspondence: joan.madurellmalapeira@unifi.it

Abstract: The Vallparadís Section encompasses various geological layers that span a significant chronological range, extending from the latest Early Pleistocene to the early Middle Pleistocene, covering a timeframe from approximately 1.2 to 0.6 Ma. This period holds particular importance, as it coincides with a significant climatic transition known as the *Early–Middle Pleistocene Transition*, a pivotal phase in Quaternary climatic history. This transition, marked by the shift from a 41,000-year obliquity-driven climatic cycle to a 100,000-year precession-forced cyclicity, had profound effects on the Calabrian carnivorous mammal communities. Notably, the once diverse carnivore guild began to decline across Europe during this period, with their last documented occurrences coinciding with those found within the Vallparadís Section (e.g., *Megantereon* or *Xenocyon*). Concurrently, this period witnessed the initial dispersals of African carnivorans into the European landscape (e.g., steppe lions), marking a significant shift in the composition and dynamics of the region's carnivorous fauna.

Keywords: Early Pleistocene; carnivoran guild; Vallparadís Section; Epivillafranchian; Iberia

Citation: Madurell-Malapeira, J.; Prat-Vericat, M.; Bartolini-Lucenti, S.; Faggi, A.; Fidalgo, D.; Marciszak, A.; Rook, L. A Review on the Latest Early Pleistocene Carnivoran Guild from the Vallparadís Section (NE Iberia). *Quaternary* **2024**, 7, 40. https://doi.org/10.3390/quat7030040

Academic Editor: Juan Rofes

Received: 12 June 2024
Revised: 5 August 2024
Accepted: 19 September 2024
Published: 23 September 2024

1. Introduction

1.1. Paleoenviromental Background

Throughout the latest Pliocene and the Early Pleistocene, the taphocenosis of the Northern Palearctic underwent profound transformations in response to the discernible cooling trend and heightened seasonality, a pattern intensified around 3 Ma. Moreover, there is widespread acknowledgment of a shift towards aridification across Europe during the Pleistocene, marking a transition from the tropical–subtropical ecosystems characteristic of the Pliocene to the contemporary environments we recognize today [1–3]. These environmental shifts prompted significant adaptations in large mammal assemblages, favoring cursorial species better suited to open habitats, a transition that coincided with the dispersals of the first hominins out of Africa [4].

Within this period, marked by glacial–interglacial dynamics forced by obliquity cycles with a periodicity of approximately 41 ka, pivotal events shaped the global climatic structure as we understand it. Notable among these were the intensification of cold and warm phases around 1.8 Ma, reflecting the dynamic of Pleistocene climate fluctuations [2,5]. Furthermore, a non-linear phase of the glacial–interglacial cycles, characterized by significant amplitude and asymmetry, began approximately 1.4–1.2 Ma: *the Early–Middle Pleistocene Transition* [6–8].

The *Early–Middle Pleistocene Transition* (EMPT) refers to a critical period in Earth's history during the Pleistocene epoch, roughly spanning from around 1.2 to 0.7 Ma. This transition marks a significant shift in the dynamics of climatic and environmental conditions, as well as in the evolutionary trajectories of various species. Understanding the *EMPT* is crucial for unraveling the complex interactions between climate, ecology, and human evolution during this pivotal period in Earth's history. Specifically, in the North Mediterranean, the *EMPT* induced an increase in aridity and seasonality which could be related to the extinction of several groups of large mammals and the dispersal of new forms from Asia and Africa [4,6,8]

The end of the Early Pleistocene was characterized by even more drastic events, including a substantial glacial phase around 0.9 million years ago, known as the *'900 ka event'* [7,9,10], and a progressive transition towards forcing by precession cycles, with a periodicity of approximately 100 ka [11]. These transformative processes are intricately linked to shifts in global temperature and seasonality, often inferred through biological proxies within animal populations and communities [12–14].

1.2. The Early–Middle Pleistocene Transition Carnivoran Guild

Early Calabrian times were characterized mostly by open habitats [2], with evidence of aridity periods, exemplified on the site of Venta Micena by the presence of *Hemitragus* and *Praeovibos* [15]. Since the beginning of the Late Villafranchian (ca. 2.0 Ma), the carnivoran guild was dominated mostly by cursorial species adapted to open environments: the sabertoothed cats *Homotherium crenatidens* and *Megantereon adroveri*, the large pantherine *Panthera gombaszoegensis*, the feline *Viretailurus pardoides*, and the Mediterranean lynx *Lynx pardinus*. *Pachycrocuta brevirostris* emerged as the dominant large hyaena within these ecosystems, alongside canids such as *Xenocyon lycaonoides*, *Canis mosbachensis*, and *Vulpes alopecoides*. Additionally, a derived omnivorous form of large ursid, *Ursus etruscus*, was extensively recorded in Mediterranean Europe. Unfortunately, despite this rich carnivore diversity, the record of mustelids is relatively scarce; only badgers, represented by the extant species *Meles meles*, are abundantly recorded in Europe.

This situation started to change around 1.2 Ma with the record of more humid and forested habitats, precisely at the Vallparadís Section [16]. This trend has a break around the Jaramillo magnetostratigraphic subchron (MIS30), with another pulse of aridity recorded at the layer EVT12 of the Vallparadís Section but also in the Iberian site of Quibas [17,18], with new evidences of *Hemitragus* and *Praeovibos*. However, the increase in humidity and seasonality and the general harshening of the environmental conditions were perfectly recorded since the 0.9 Ma event. All the former changes had a significant impact on the carnivoran guild. The high diversity and intraspecific competition of Villafranchian species [19] led to the vanishing of the genus *Megantereon* around 1 Ma, followed later around MIS21 by *Xenocyon* and *Viretailurus* and the progressive decrease on the abundance of *Homotherium* and *P. gombaszoegensis*. These species were replaced by the African newcomers, *Panthera spelaea fossilis*, *Panthera pardus*, and *Crocuta crocuta*, who started being recorded as being present in several sites since 1.2 Ma, with the first leopard record from Vallonnet Cave [20].

1.3. The Vallparadís Section

The Vallparadís Section encompasses the paleontological sites of Cal Guardiola (CGR) and Vallparadís Estació (EVT), situated in the Vallès-Penedès Basin in Northeastern Iberia [21,22]. Over the course of emergency excavations conducted from 1997 to 2008, these sites yielded 32,000 vertebrate remains dating from the late Early to Middle Pleistocene. Both locations exhibit a distinctive depositional setting strongly influenced by the dynamics of an alluvial fan system and the geometry of the Miocene paleorelief.

Biochronological, magnetostratigraphic, and U-series–ESR evidence collectively suggests that the Vallparadís Section ranges from before the Jaramillo paleomagnetic subchron (approximately 1.2–1.1 Ma) to the early Middle Pleistocene (around 0.6 Ma) [21–24].

Analyzing the older part of the section (circa 1.2–1.1 Ma), pollen and wood studies from CGRD2 indicate a warm–temperate and humid ancient environment, likely a river or river–marsh ecosystem that supported diverse plant life, including aquatic macrophytes, deciduous trees, and grasses. The abundance of hippopotamuses that remains in this layer supports the interpretation of a fluvial primary depositional environment [25], while the high diversity of large ungulates suggests a landscape with both woodlands and more open, arid areas [16].

Meso- and microwear analyses conducted on a substantial sample of ungulate teeth from the section indicate a noteworthy shift in paleoenvironments since 0.9 Ma (MIS22). Initially dominated by open, dry grasslands with discernible seasonality (Layer EVT12, circa 1.0 Ma; MIS30), the environment transitioned to more humid woodlands, possibly with an even more pronounced seasonality (Layers EVT7 and CGRD7; circa 0.86 Ma; MIS21). These findings align with data from other contemporaneous Southern European sites [10]. Preliminary investigations into the stable isotopic signal from layers EVT12 and EVT7 also suggest a period of increased aridity during the Jaramillo subchron and more wooded environments during it. The study of enamel hypoplasia in hippopotamuses points in the same direction, highlighting increased seasonality in Europe during the Early Pleistocene [26].

2. Materials and Methods

The specimens reviewed in the present paper are housed at the Catalan Institute of Paleontology, Sabadell, Spain (ICP), and came from the Cal Guardiola and Vallparadís Estació emergency excavations. The code of the ICP collections is IPS. The detailed record of carnivoran species in each Vallparadís Section layer is displayed in Table 1.

Table 1. Occurrences of the different carnivoran species in each geological layer of the Vallparadís Section (see Supplementary Figure S1 for the stratigraphic scheme of the Vallparadís Section).

	Geological Layers of the Vallparadís Section					
	Cal Guardiola Local Section		Vallparadís Estació Local Section			
	CGRD2	CGRD7	EVT12	EVT10	EVT7	EVT3
Homotherium crenatidens		■				
Panthera fossilis				■	■	
Megantereon sp.			■			
Panthera gombaszoegensis		■			■	
Puma pardoides					■	
Lynx pardinus		■	■	■	■	■
Pachycrocuta brevirostris	■	■	■	■	■	
Canis (Xenocyon) lycaonoides			■		■	
Canis mosbachensis	■	■	■		■	
Vulpes alopecoides					■	
Vulpes vulpes						■
Ursus deningeri		■	■		■	
Meles meles			■		■	

All specimens discussed in the text have been personally studied by the authors in the last years and compared with late Early Pleistocene assemblages they also personally studied, including Dmanisi (Georgia), Upper Valdarno, Olivola, Pirro Nord, Collecurti and Cava Redicicoli (Italy), Venta Micena, Barranco León, Fuente Nueva 3, Incarcal Complex, La Boella and Bòvila Ordis (Spain), and Le Vallonnet (France).

Bootstrapping Cluster Analysis and Non-Metric Multidimensional Scaling

To evaluate the similarity of the carnivore guild in Vallparadís with that from other Early Pleistocene localities in Eurasia and Africa, we conducted a bootstrapping cluster analysis (BCA) on specific composition matrices of the selected assemblages (Supplementary Table S1). The twenty-three localities are fairly evenly distributed geographically (six in Africa, eight in Asia, and nine in Europe) and date between approximately 2.0 and 0.7 million years ago (see Supplementary Figure S2). Similar to previous studies using comparable analyses [27,28], we selected localities with well-established chronologies and a substantial number of recorded carnivorans to avoid biased results. The occurrence matrices compiled are based on published data ([9,21,23,28–44]), with some cases revised and updated.

BCA is a segmentation technique that is valuable for evaluating the stability of clustering outcomes, also utilized to detect notable statistical resemblances among Pleistocene large mammal communities (refer to [27,28,45,46]). The process commences with an initial grouping of the dataset via the unweighted pair group method with arithmetic mean (UPGMA) algorithm [47]. This initial grouping acts as a baseline clustering of the chosen localities based on their taxonomic similarity, enabling us to gauge the likelihood of each branch in the baseline using a permutation approach. Following this, a random subset of the original data is selected to perform a new UPGMA cluster analysis, leading to a new clustering arrangement. This procedure is repeated numerous times. The newly generated clustering is then compared to the baseline one. A similarity index (G^*) is computed between the baseline and the sample clustering, with values ranging from 0 (if the clusterings are entirely dissimilar) to 1 (if the clusters from the original and sampled data are identical). The G^* is subsequently contrasted with the expected similarity value (G°) under the null hypothesis that the sampled data are a true random subset of the original dataset. We performed the resampling procedure 1000 times, and in each instance, we conducted UPGMAs on the sampled data and determined similarity indices. If the likelihood that G^* is greater than or equal to G° surpasses the significance threshold ($p(G^\circ \leq G^*)$; $\alpha = 0.05$), the partitioning levels of the baseline cluster analysis are deemed robust (see [48]). We also explored ecological subdivisions within the analyzed guild, classifying carnivorans based on traditional dietary categories: hypocarnivores (diet containing less than 50% vertebrate meat), mesocarnivores (diet containing 50%–70% vertebrate meat), and hypercarnivores (diet containing more than 70% vertebrate meat), following [49,50]. Furthermore, we included an insectivore category (species specialized in consuming insects) and incorporated hunting strategies and dietary habits based on ecological categories proposed in the literature, e.g., [28,35,50–55].

Finally, we categorized the fossil carnivorans according to their habitat preference, particularly as taxa of open, mixed, closed, or aquatic environments. The same dataset was used to perform a non-metric multidimensional scaling (NMDS), an ordination method suitable for taxa occurrence matrices, as it uses distance matrices of the presence/absence of the taxa (but also relative abundances) to project the dataset in bi- or tridimensional spaces [56]. In the case at hand, NMDS on taxa occurrences was performed, taking into consideration the habitat preference per each site. To visualize the proportions of carnivoran abundance in the different sites, we used barplots of their ecological and environmental specifics. We used the software RStudio (v. 2023.12.1+402 'Ocean Storm' Release 4da58325ffcff29d157d9264087d4b1ab27f7204, 28 January 2024 [57]) in R environment (v. 4.3.2, [58]) to perform analyses and produce graphs. The BCA was obtained with the function *pvclust()* (*'pvclust'* v. 2.2-0 [59]), and for NMDS, the function *metaMDS()* (*'vegan'* package v. 2.6-4 [60]). The barplots were obtained with *barplot()*

('*graphics*' package v.4.3.2 [58]), and the plot of the NMDS was obtained with *ggplot()* ('*ggplot2*' package v.3.4.0; [61]).

3. Results

3.1. Felids

3.1.1. *Homotherium crenatidens*

Homotherium was a high-speed pursuit and slender predator adapted to open environments recorded in Eurasia, Africa, and America during the Plio-Pleistocene, and it became extinct around 20 ka [29,35,62–64]. In Europe, it is documented as having been present since the beginning of the Villafranchian period, around 3.0 Ma, with findings in locations such as Perrier-Les Etouaires, Saint Vallier, Senèze, Upper Valdarno, and the Incarcal complex [31,65–67].

Comparisons between *Homotherium*, other sabertoothed cats, and lions in terms of social behavior have been one of the most debated topics among Quaternary paleontologists in recent decades. While much attention has been given to the purported social behavior of *Smilodon*, particularly regarding its packs hunting similar to extant lions [68], the social dynamics of *Homotherium* remain less understood. A recent fossil DNA analysis suggests the social behavior of *Homotherium*; however, the extent of this behavior remains uncertain [63,64]. It is noteworthy that *Panthera spelaea* and *Homotherium* co-existed during more than half million years in Europe; however, the social behavior of both species was probably not similar to that of extant lionesses, according to available data [63,64]

Homotherium was only incidentally recorded at the Vallparadís Section, with a partial upper Canine from the layer CGRD7 [69], whereas it is very abundant, close to 500 specimens, in the Incarcal site of coeval chronology and only 100 km north [67].

3.1.2. *Panthera spelaea fossilis*

The decline in *Homotherium* records since the Early-to-Middle Pleistocene boundary is partially explained by the arrival of lions from Africa around the Jaramillo subchron, despite the fact that the sociality of both species was probably not very high the first European lions were larger and heavier built than the slender *Homotherium*. Three migration waves indicate the lion's arrival in Europe. The migration from Asia via Eastern Europe into Central Europe is documented by the remains from Bachatsk (western Siberia), dated at ca. 1.0 Ma [70], and later by the find from Kozi Grzbiet (750–700 ka, Poland [71]). The Southwestern Asia and Southeastern Europe route, via the Balkan Peninsula, is recorded at the Greek site Megalopolis–Marathousa [72,73] and Moldovan locality Sinjakovo 1 [74,75]. The earliest European appearance is documented by the Vallparadís Estació record and well supported by the oldest (750–700 ka) so-far known record from Pakefield [76]. Since then, lions were ubiquitously recorded during the Pleistocene of Europe and North America. *P. s. fossilis* remains were found in the Vallparadís Section in layers EVT10 and EVT7 (MIS30, MIS21; Figure 1W–P'), corresponding to at least two different individuals. The most complete one includes a radius, and a practically complete forepaw came from layer EVT10. Additionally, a femur, a tibia, and a third metatarsal of slender appearance were recovered from layer EVT7. These remains represent the oldest records of lions outside Africa and exemplify the dispersal of African carnivorans into Europe during the *Early–Middle Pleistocene Transition*. The Vallparadís record of *P. s. fossilis* fits well into the Mediterranean migration route of many large carnivores of African origin, like *P. pardus* from Le Vallonnet Cave (1.2 Ma; [20]) or *C. crocuta* from Trinchera Dolina 4–6 (850–800 ka) [77]. The arrival in Europe of these three large carnivores of African origin was an important event for the faunal assemblages.

This was a main break in the hitherto stable carnivore paleoguild and led to some changes in its structure. Most of some far dominant species, like *H. crenatidens* and *P. gombaszogensis*, survived but were removed from the apex predator position in carnivore paleoguild. Their previously compact and wide geographic range shrunk into isolated and restricted areas. *H. crenatidens* had noticeably decreased in size and especially in massiveness [62,78–80]. Those lesser, gracile homotheres appeared at the same time in different Eurasian areas. Lion pressure might play a significant role alongside the combination of climatic-induced changes in vegetation, exacerbated competition among large carnivores, and the increased pressure from *Homo* as a member of the predatory guild [81].

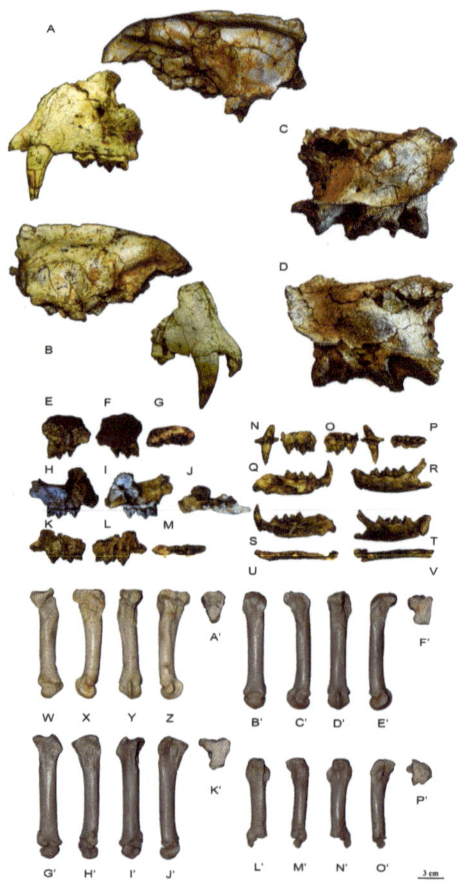

Figure 1. Felids recorded at the Vallparadís Section. *Panthera gombaszoegensis* cranium (EVT21072) from layer EVT7 in (**A**) left and (**B**) right lateral view; Neurocranium of *Megantereon* sp. (IPS125140) from layer EVT12 in (**C**) left and (**D**) right lateral views; maxillary with P3–P4 of *Lynx pardinus* from layer EVT12 (IPS84917) in (**E**) buccal, (**F**) lingual, and (**G**) occlusal views; maxillary with P3-P4 of *Lynx pardinus* from layer EVT12 (IPS84920) in (**H**) buccal, (**I**) lingual, and (**J**) occlusal views; maxillary with P3–P4 of *Lynx pardinus* from layer EVT12 (IPS84916) in (**K**) buccal, (**L**) lingual, and (**M**) occlusal views; maxillary fragment with C1–P4 of *Lynx pardinus* from layer EVT7 (IPS84915) in (**N**) buccal, (**O**) lingual, and (**P**) occlusal views; right hemi-mandible with c1–m1 of *Lynx pardinus* from EVT7 (IPS84914) in (**Q**) buccal, (**S**) lingual, and (**U**) occlusal views; left hemi-mandible with p3–m1 of *Lynx pardinus* from EVT12 (IPS60461) in (**R**) buccal, (**T**) lingual, and (**V**) occlusal views; right second metacarpal of *Panthera spelaea fossilis* from layer EVT10 (EVT14276) in (**W**) anterior, (**X**) lateral, (**Y**) posterior, (**Z**) medial, and (**A′**) proximal views; right third metacarpal of *Panthera spelaea fossilis* from layer EVT10 (EVT16012) in (**B′**) anterior, (**C′**) lateral, (**D′**) posterior, (**E′**) medial, and (**F′**) proximal views; right fourth metacarpal of *Panthera spelaea fossilis* from layer EVT10 (EVT14720) in (**G′**) anterior, (**H′**) lateral, (**I′**) posterior, (**J′**) medial, and (**K′**) proximal views; right fifth metacarpal of *Panthera spelaea fossilis* from layer EVT10 (EVT16013) in (**L′**) anterior, (**M′**) lateral, (**N′**) posterior, (**O′**) medial, and (**P′**) proximal views.

The early European lions were notable for their enormous size and robúst build. Kurtén [82] observed that these lions made their entry into Europe as giant forms, stating the following: 'The lion entered the European scene (...) with a gigantic form. (...) The Cromerian lion in Europe may be the largest felid that ever existed'. Subsequent studies revealed a general trend of decreasing body size in the *P. spelaea* lineage over time ([82–87]). Throughout the Middle Pleistocene, variations in size among *P. spelaea* were minimal, with specimens dating from around 900 to 300 ka being only marginally larger than those from 300 to 100 ka. Significant size reductions did not occur until approximately 50 ka, coinciding with a marked decline in genetic diversity among lion populations [88]. Between 45 and 15 ka, much smaller lions emerged, comparable in size to modern African lions, *P. leo*. This reduction in body size, along with the eventual extinction of the steppe lion, may have been an ecological response to increasingly harsh environmental conditions and a decrease in prey availability.

3.1.3. *Megantereon* sp.

Megantereon was a stoutly built ambush predator adapted for stalk prey in closed environments, with a body size of a small jaguar (ca. 100 Kg) [89]. This genus was present in Eurasia and Africa during the Plio-Pleistocene. In Europe, as with *Homotherium*, *Megantereon* has been identified in several sites dating back to 3.0 Ma [44]. It is typically characterized as an ambush predator, and recent studies on Asian fossil assemblages suggest reconsidering *Megantereon*'s habitat preference, indicating a potential adaptation to more open environments rather than dense forests [89,90].

Debate has long surrounded the species diversity within the genus *Megantereon*. Some researchers advocate for the existence of only two primary species in the Old World [89], while others argue for a higher level of diversity [91].

In Europe, *Megantereon* fossils have been unearthed at various sites, including Venta Micena, Pirro Nord, Argentario, Apollonia-1, and, more recently, at Untermassfeld and Vallparadís Section EVT12 [21–23,89].

As an efficient hunter, *Megantereon* likely employed its robust forelimbs to overpower prey, utilizing its sharp and elongated canines (ca. 10 cm in length) to bite, suffocate, and exsanguinate its victims.

As with *Homotherium*, this genus is scarcely recorded in the Vallparadís Section. Only a basicranial fragment from the layer EVT12 is here attributed to *Megantereon* (Figure 1C,D) [92]. The fragmentary nature of the specimen does not permit taxonomical considerations. However, together with the remains from the German site of Untermassfeld, this element represents the last occurrence of this genus in Europe at MIS31-30.

3.1.4. *Panthera gombaszoegensis*

Several researchers have proposed an African origin for *P. gombaszoegensis* [93,94], suggesting its dispersal into Europe around 2.0 Ma, linked to the presence of a large pantherine form in the Late Pliocene of Laetoli Upper Beds (circa 3.7 million years ago). Other scholars have attributed the first pantherine fossils from Laetoli to *P. leo* [95] or, more recently, to a new species, *P. principalis* [80]. Ultimately, [35,96] offered a more parsimonious hypothesis, noting that African pantherine specimens older than 2 Ma are not identifiable at the species level. We concur with the latter authors, supporting the notion that there is no direct connection between the African specimens and the European *P. gombaszoegensis* s.l. Since around 2.0 Ma, early forms of *P. gombaszoegensis* (or *P. toscana*) have been common in the Late Villafranchian faunas of Europe, evidenced by records from Olivola, Upper Valdarno, Tegelen, Pirro Nord, Untermassfeld, and up to the Middle Pleistocene of Chateau [93,94,97]. In fact, the relationship between the early *P. toscana* and *P. gombaszoegensis* remains unclear, with several anatomical differences distinguishing the two forms from different time periods. Recently, Chatar et al. [98] studied Belgian specimens from the Middle Pleistocene and identified more similarities with tigers than with jaguars. It is true that *P. gombaszoegensis* was originally described based on mandibular specimens, whose robust morphology can resemble the ones of jaguars. However, in terms of cranial morphology, *P. gombaszoegensis* is more like the Asian lineage of tiger–snow leopard than to American jaguars. The post-cranial morphology of this species is largely unknown. No partial or complete skeletons were recorded up to now; the most complete record comes from the Middle Pleistocene of Château [94], pointing to a stout and large felid of more than 100 kg, with a similar morphology of a basal pantherine like *Panthera pardus*.

In the Vallparadís Section, *P. gombaszoegensis*, also known as 'The Eurasian Pantherine', was found only in the post-Jaramillo layers CGRD7 and EVT7, with few fragmented postcranial remains [69], isolated teeth, and a relatively complete cranium (Figure 1A,B). The preliminary study of the cranium morphological traits indicates more similarities with Middle Pleistocene forms than to *P. toscana* and suggests a putative second dispersal from Asia rather than to a local gradual evolution of characters. However, the virtual absence of remains from the period 1.8–1.2 Ma precludes accurate comparisons.

3.1.5. *Viretailurus pardoides*

The Eurasian puma-like cats are exceedingly rare in the fossil record: only one cranium and very few postcranial bones were identified. Consequently, the anatomy and ecology of these middle-sized carnivores remain poorly understood, and their taxonomic classification is a subject of considerable debate.

Panthera schaubi was initially described by Viret [65] at the French site of Saint-Vallier, dating back approximately 2.1 Ma. Viret suggested that these materials exhibited dimensions and morphological characteristics typical of a small pantherine. However, in 1964, Hemmer demonstrated that the skull morphology from St. Vallier was distinctly non-pantherine [99]. Instead, it shares numerous features with the American puma, *Puma concolor*. Consequently, Hemmer assigned the St. Vallier puma-like cat remains to a new genus, *Viretailurus*. Following this, Kurtén and Crusafont [100] described carnivore remains from the Iberian site of La Puebla de Valverde, identifying some as *Panthera* cf. *schaubi*, stressing similarities with specimens from the English Red Crags identified by Owen as *Felis pardoides* [101].

The first puma-like cats from the Iberian Peninsula were first recorded at the Puebla de Valverde site, dating to 2.2 Ma [100]. In Cueva Victoria, some remains were also identified belonging to a puma-like cat. Initial descriptions by Pons-Moyà and Moyà-Solà [102] attributed the specimens to the genus *Jansofelis* sp. Nevertheless, more recent publications attributed these remains to *Puma pardoides* [103–105]. The most recent Iberian puma remains were described in the site of El Chaparral [106].

Evidence from Central Europe includes the German site of Untermassfeld (ca. 1.0 Ma) and the Stránská skála site in the Czech Republic (ca. 0.6 Ma). In Bulgaria, the site of Varshets yielded a humerus fragment, dated to the early part of MN17. Additionally, maxillary fragments from the Georgian site of Kvabebi, previously attributed to *Lynx issiodorensis*, were later ascribed to *P. pardoides*. More recently, Werdelin et al. [107] attributed some new specimens from Graunceanu to a puma-like cat.

To summarize, recent researchers agree on including this taxon in the Felinae subfamily instead of Pantherinae felids. As in the case of *P. gombaszoegensis*, the puma affinities of the mandible are clear, whereas the cranial affinities are more difficult to ascertain. Notwithstanding, the taxonomy of this taxon is outside the scope of this paper, and despite the fact that *Puma pardoides* is the most commonly accepted, we prefer to use *Viretailurus pardoides* until the phylogenetic relationships of this taxon can be clarified.

Viretailurus is only recorded in the layer EVT7 of the Vallparadís Section based on a partial corpus with p4 and a fragment of m1 [105]. No further considerations are needed, except to remark that this specimen can be the last citation of the taxon in Europe.

3.1.6. *Lynx pardinus*

Small-to-medium-sized felids normally included in *Lynx issiodorensis* have been recorded in Europe since the Pliocene in sites like Serrat d'en Vaquer or Cuevas de Alzamora [108,109]. Nevertheless, the taxonomic adscription of these early felines remains controversial because of the scanty record, and we prefer to include it in *Lynx* sp. Since the beginning of the Villafranchian, true lynxes have been recorded at several sites, such as Perrier-Les Etouaires, Saint Vallier, Olivola, or Pantalla [65,110–112].

According to some scholars, *L. issiodorensis* was the ancestor of both *Lynx lynx* and *Lynx pardinus* [108,112–114], with *L. pardinus* first recorded in the Eastern Iberian Peninsula at ca. 1.6 Ma [113]. However, the morphology of *Lynx issiodorensis* and their similarities with the Mediterranean *lynx* suggests a closer relationship and an early split of *L. lynx* from the *L. issiodorensis–L. pardinus* lineage. The virtually absent Pliocene record in Asia precludes us from ascertaining the origin of the Boreal lynx. Recent studies of Early Pleistocene specimens suggested the inclusion of the specimens from Vallonnet, Apollonia, and Pirro Nord in *Lynx pardinus*, attesting the wide expansion of this taxon in Mediterranean Europe in the latest Early Pleistocene. The ecological role of the first Mediterranean Lynx and when they started their dietary specialization on lagomorphs are two of the questions remaining to be answered in the following years [113].

In the Vallparadís Section, the remains of lynx were recovered from the layers CGRD7, EVT7, and EVT12 [114], including a wide variety of morphological forms and size classes. Interestingly, the remains from the Jaramillo layer, with evidence of aridity and harsh climate [17], are considerably larger than those unearthed from the post-Jaramillo layers, reaching the size of a small *Viretailurus*. The observed variability in dental morphology also points to reconsider most of the previous citations of *Lynx issiodorensis* in the latest Early Pleistocene (Figure 1E–V) [114].

3.2. *Hyaenids*

Pachycrocuta brevirostris

Pachycrocuta brevirostris is a prevalent carnivore in the late Early Pleistocene faunal assemblages of Europe. Nonetheless, the origin of the genus *Pachycrocuta* has been a subject of intense debate in recent decades. Palmqvist et al. [115] advocated for an African origin of this genus, while other researchers, such as Werdelin [116] and Liu et al. [117], have convincingly argued for an Asian origin. These latter researchers highlight that the earliest records in both Asia and Africa are nearly concurrent: approximately 4.0–3.5 Ma [118].

Regardless of its origin, *P. brevirostris* spread into Europe around 2.2 Ma, with its earliest likely record at the French site of Senèze (J.M.-M unpublished data), where it emerged as one of the most prevalent elements in European faunas, frequently the primary agent responsible for bone accumulations. Its widespread presence persisted until the late Early Pleistocene in Europe, up to 0.86 Ma [21,22,118]. The influence of this species on the European Pleistocene taphocenosis and its potential competition with early hominins has been a topic of significant discussion in recent years [22,118].

One of the more debated topics around *Pachycrocuta* was their putative social behavior and strict scavenging behavior, as favored by several authors [115]. The social behavior of extinct hyaenas was recently studied in several works by Vinuesa et al. [119,120], who suggested that the social behavior of the extant spotted hyaena is probably a recently acquired trait, and the most parsimonious scenario is interpreting most of the extinct hyaenas as mostly solitary animals.

The strict scavenger behavior of *Pachycrocuta* has been deduced based on its fossil accumulations and the morphological traits of the only known skeleton from the Middle Pleistocene of Zhoukoudian-1 in China [121]. First, it is difficult to ascertain if these accumulations were the product of strict scavenging behavior or a combination of scavenging and hunting, as in the extant spotted hyena. Secondly, there are no certainties that this skeleton is really from the same individual and not a composite one. Additionally, all the Middle Pleistocene Chinese *Pachycrocuta* are considerably larger and morphologically derived compared with their European counterparts from the Early Pleistocene.

Pachycrocuta was found in several layers of the Vallparadís Section from MIS35 to 21, including CGRD2, EVT12, EVT10, CGRD7, and EVT7. The most noteworthy specimen is a partial slender skeleton found in layer EVT7 (Figure 2A–B,D–V,X–Z). This skeleton displays several differences compared with the Zoukoudian-1 one. Firstly, the proportional shortening of the distal-limb elements (tibia and radius), which putatively support the scavenging behavior and cursorial locomotor behavior, is not accentuated in EVT7, being more like that of the brown hyaena. Secondly, the mean of body mass estimations for this skeleton is ca. 82 kg, much smaller than 140 kg, which was estimated for the Chinese specimens. Indeed, the EVT7 specimen is smaller than most of the *Crocuta spelaea* skeletons from the European Late Pleistocene and probably similar in size to the earliest European *Crocuta*. The in-depth study of this specimen will provide clues in the near future that we can use to discern the dietary behavior of this hotly debated taxon.

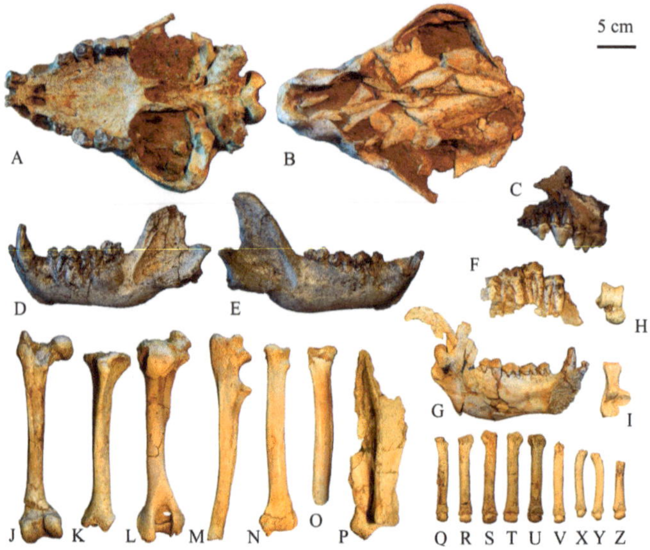

Figure 2. *Pachycrocuta brevirostris* remains from the Vallparadís Section. Partial cranium (EVT21863) from layer EVT7 in (**A**) ventral and (**B**) dorsal views; maxillary fragment with P2-P4 (IPS14521) from layer CGRD2 in (**C**) buccal view; left hemi-mandible with i3-m1 (EVT24222) from layer EVT12 in (**D**) buccal view; right hemi-mandible with i3–m1 (EVT24641) from layer EVT12 in (**E**) buccal view; fragment of right hemi-mandible with p3–m1 (EVT21980) from layer EVT7 in (**F**) buccal view; right hemi-mandible with i3–m1 (EVT23222) from layer EVT7 in (**G**) lingual view; right talus (EVT23130) from layer EVT7 in (**H**) anterior view; right calcaneum (EVT23066) from layer EVT7 in (**I**) anterior view; left femur (EVT23118) from layer EVT7 in (**J**) posterior view; right tibia (EVT23224) from layer EVT7 in (**K**) anterior view; right humerus (EVT23230) from layer EVT7 in (**L**) posterior view; right ulna (EVT23228) from layer EVT7 in (**M**) medial view; left radius (EVT23209) from layer EVT7 in (**N**) anterior view; right partial radius (EVT23213) from layer EVT7 in (**O**) anterior view; left scapula (EVT23226) from layer EVT7 in (**P**) dorsal view; right fifth metacarpal (EVT23233) from layer EVT7 in (**Q**) anterior view; left fourth metatarsal (EVT23144) from layer EVT7 in (**R**) anterior view; left fourth metacarpal (EVT21709) from layer EVT7 in (**S**) anterior view; right third metacarpal (EVT23184) from layer EVT7 in (**T**) anterior view; left third metacarpal (EVT23131) from layer EVT7 in (**U**) anterior view; left fifth metacarpal (EVT23060) from layer EVT7 in (**V**) anterior view; left fifth metatarsal (EVT23068) from layer EVT7 in (**X**) anterior view; right fifth metacarpal (EVT23027) from layer EVT7 in (**Y**) anterior view; indeterminate metacarpal (EVT23067) from layer EVT7 in (**Z**) anterior view.

3.3. Canids

3.3.1. *Xenocyon lycaonoides*

Adaptations for hypercarnivory behavior, characterized by a diet comprising 70% or more vertebrate meat, are relatively common in both fossil and contemporary Canidae lineages [122]. Molecular data suggest that the divergence between the extant species *Lycaon pictus* and *Cuon alpinus* from other wolf-like canids occurred between the Late Pliocene and the Early Pleistocene [123]. The sparse and scattered fossil records of large hypercarnivorous canids across the Old World add complexity to our understanding of the evolutionary history of these wild dogs. These hypercarnivorous dogs are well-documented during the Early Pleistocene in Europe, Asia, and Africa, though different taxa are attributed to them depending on the author [30,124–139].

Martínez-Navarro and Rook [128] proposed a hypothesis highlighting a gradual reduction in certain dental cusps, an increase in others, a tendency towards hypercarnivory behavior, and the loss of the first metacarpal (a unique feature of the modern *Lycaon pictus*). According to their hypothesis, all *Lycaon*-like dogs from the Early Pleistocene should be classified within the genus *Lycaon*, with three chronospecies: *Lycaon falconeri* for the earliest Eurasian forms of the Early Pleistocene, *Lycaon lycaonoides* for those from the latter part of the Early Pleistocene and the beginning of the Middle Pleistocene in Eurasia and Africa, and *Lycaon pictus* for the Middle–Late Pleistocene and contemporary African forms. However, other researchers [131,132,140] have argued that most or all of these forms should be placed in the genus *Xenocyon*, which is considered a sister genus to *Lycaon* and *Cuon*. Given the uncertainty regarding the generic classification of fossil wild-dog specimens and the scope of this paper, we prefer to refer to all large hypercarnivorous taxa with pronounced carnassial teeth and skull adaptations for hypercarnivory from the Late Pliocene and Early Pleistocene of the Old World as belonging to the genus *Xenocyon* until a more definitive hypothesis clarifies their taxonomy.

In the Vallparadís Section, remains attributable to *Xenocyon lycaonoides* were found in layers EVT12 and EVT7, corresponding to a complete foot and two hemi-mandibles, respectively. The large-sized hypercarnivorous canid *Xenocyon lycaonoides* was one of the most characteristic taxa of the second half of the Early Pleistocene ([141], FAD ca. 1.76 Ma [136]) persisting until the Middle Pleistocene age (LAD ca. 450–400 ka [139]).

The hemi-mandibles from layer EVT7 (Figure 3A–F) share several morphologic characteristics with other previously reported remains of *X. lycaonoides* form European assemblages [134].

The studied material exhibits minor differences compared to other specimens of the same species, specifically in the dimensions and sharpness of the main cusps of the m1 talonid and the m2 trigonid, as well as the reduction in the m1 and m2 entoconids. This variation is interpreted as a derived condition relative to the 'typical' *X. lycaonoides* material, indicating a small evolutionary step towards the hypercarnivory observed in this lineage [134]. This finding further supports the gradual nature of hunting-dog evolution. Additionally, the material studied represents one of the latest well chronologically constrained European records of the genus *Xenocyon* at 0.86 Ma, just before the Early–Middle Pleistocene boundary.

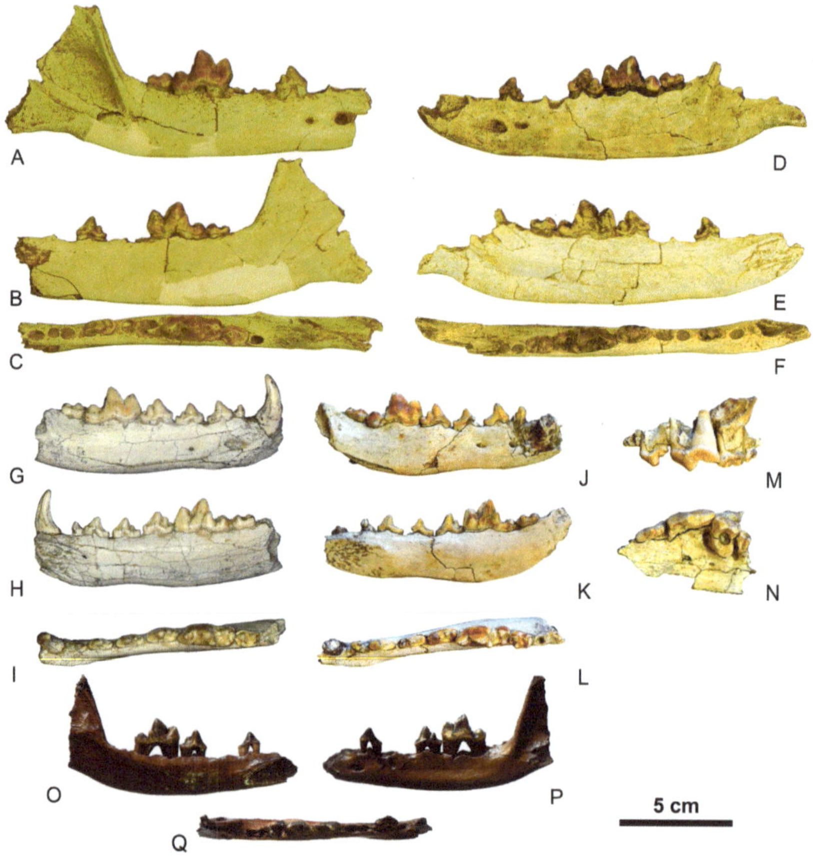

Figure 3. Canids recovered from the Vallparadís Section. Right hemi-mandible with p3, m1–m2 (EVT22049) of *Xenocyon lycaonoides* from layer EVT7 in (**A**) buccal, (**B**) lingual, and (**C**) occlusal view; left hemi-mandible with p2 and p4–m2 (EVT23434) of *Xenocyon lycaonoides* from layer EVT7 in (**D**) buccal, (**E**) lingual, and (**F**) occlusal view; right hemi-mandible with c1–m2 (EVT24342) of *Canis mosbachensis* from layer EVT12 in (**G**) buccal, (**H**) lingual, and (**I**) occlusal view; right hemi-mandible with p1–m2 (EVT13840) of *Canis mosbachensis* from layer EVT7 in (**J**) buccal, (**K**) lingual, and (**L**) occlusal view; left maxillary fragment with P3–M2 (EVT25504) of *Canis mosbachensis* from layer EVT12 in (**M**) buccal and (**N**) occlusal view; left hemi-mandible with p2, p4–m1 (IPS120168) of *Vulpes vulpes* from layer EVT3 in (**O**) lingual, (**P**) buccal, and (**Q**) occlusal view.

3.3.2. *Canis mosbachensis*

Historically, the appearance of *Canis* spp. marked the beginning of the biochronological event called the 'Wolf Event' around 2.0 Ma [142,143]. Recent discoveries pointed out an earlier European dispersal for canids. This evidence was based on scarce specimens attributed to *Canis* cf. *C. etruscus* from sites of Costa S. Giacomo FU (late Middle Villafranchian; [44]) and specimens of *Canis* sp. from the French site of Vialette [144]. The latter are dated back to the Early Villafranchian, around 3 Ma [144]. Therefore, these findings not only backdate the Wolf Event, but they show its diachronic nature (as noted by [145]), questioning its significance as a biochronological event [146]. In this sense, some authors [44,147] suggest using a different definition: the '*Pachycrocuta brevirostris* event', referring to the dispersal of the hyenid in Eurasia, would be a better designation of the faunal turnover around 2 Ma due to its ecological importance in Early Pleistocene habitats.

The earliest well-documented evidence of the genus *Canis* from the Early Pleistocene of Europe has been recorded in the Italian sites of Olivola (Val di Magra, Tuscany, Italy; ca. 2 Ma; [44]) and Poggio Rosso (Upper Valdarno, Tuscany, Italy; ca. 1.9–1.8 Ma; [2,148,149]), with three species, *Canis arnensis*, *Canis etruscus*, and *Xenocyon falconeri* [38,125,142,150,151].

The taxonomical state of *Canis* species since their appearance in Europe has been widely studied [125–127,134,150,152–156], and different species have been described across Europe on the base of morphology and size. Nevertheless, their reciprocal phylogenetic relationships, as well as in comparison with extant canids, are still highly debated.

In the Iberian Peninsula, the first accounted finding of canids is that of Fonelas P-1 (Guadix Basin, Granada, Spain; approximately 1.9–1.7 Ma). From this site, four canids taxa have been recovered: *Vulpes alopecoides*, *C. etruscus*, *C.* cf. *falconeri*, and a new small species, *C. accitanus* [156,157]. Furthermore, the Guadix–Baza Basin has three of most important Early Pleistocene European sites, spanning approximately 1.4–1.2 Ma [158–160]: Venta Micena (around 1.4 Ma), Barranco León (around 1.3 Ma), and Fuente Nueva 3 (around 1.2 Ma). Along with remains of many Late Villafranchian mammal species, in all three of the sites remains, of the medium-sized *C. mosbachensis* have been found [23]. These findings are among the most ancient records of this taxon in Europe, and from this moment on, it becomes very common in the late Early Pleistocene and through the Middle Pleistocene; see, among others, [30,41,126,161].

Besides its wide diffusion, the true phylogenetic relationship of *C. mosbachensis* with other modern and fossil species is still highly debated. This wolf-like, medium-sized species for a long time has been considered to be derived from *C. etruscus* and to have then originated the modern wolf (among others [125,145,151,162–168]). Nevertheless, Soergel [169] argued the possibility of a close relationship between *C. mosbachensis* and *C. arnensis*, rather than *C. etruscus*. Furthermore, some authors [151,154] suggested the presence of a different lineage of wolf-like dogs in Southern Europe, parallel to the one of Central Europe of *C. mosbachensis*. Bartolini-Lucenti et al. [170] reveals that, around 1 Ma, the medium-sized canid present in the Iberian Peninsula was very close to that of the coeval German site of Untermassfeld [126] more than a more closely derived form of *C. arnensis* (Figure 3G–N). In either case, the overall dimensions of *C. mosbachensis* are smaller than those of *C. lupus*, and it probably exhibited a more hypocarnivorous diet as compared with its extant counterpart.

3.3.3. *Vulpes alopecoides*

The genus *Vulpes*, which includes fossil foxes, has been present since the Late Miocene in North America and the Old World. However, the fossil record of these foxes is extremely sparse and fragmented, often attributed to various species with uncertain relationships. The Early and Middle Pleistocene records of *Vulpes* in Europe are particularly confusing and have been the subject of much debate. Four species have been described from this period: *Vulpes alopecoides* from Il Tasso (Upper Valdarno, Italy), *Vulpes praeglacialis* and *Vulpes praecorsac* from Villany (Hungary), and *Vulpes angustidens* from Hundsheim (Germany). The limited fossil evidence used to describe these species has led to considerable debate among scholars regarding their relationships to both extant foxes, such as *Vulpes vulpes* and *Vulpes lagopus*, and to each other.

Vulpes alopecoides has been reported from several Early Pleistocene sites across Europe, including Dafnero-1 in Greece [171], Dmanisi in Georgia [28,172], Fonelas-P1 in Spain [157], Kastritsi in Greece [171], La Puebla de Valverde in Spain [100], Makinia in Greece [171], Pirro Nord in Italy [30], Sesklon in Greece [171], St. Vallier in France [65], Upper Valdarno in Italy [34,150], Villarroya in Spain [173,174], and Volax in Greece [171]. In contrast, *Vulpes praeglacialis* has been found in later Early Pleistocene localities, such as Apollonia-1 in Greece [32], Atapuerca Trinchera Dolina 6 TD6 in Spain [127], Barranco Leon-5 in Spain [23], Cal Guardiola in Spain [23], Gombaszög/Gombasek in Slovakia [175], Caune de l'Arago in France [176], Deutsch Altenburg 2C in Austria [177], El Chaparral in Spain [23], Fuente

Nueva 3 in Spain [23], L'Escale in France [178], Le Vallonnet in France [20], Püspökfurdö-Betfia 2 in Romania [179], Venta Micena in Spain [23], and Villany 3–8 in Hungary [180].

The classification of these species has been further complicated by differing opinions among researchers. Kormos [181] suggested that *Vulpes praeglacialis* is closely related to the modern arctic fox, assigning it to the genus *Alopex* (an earlier name for *Vulpes lagopus*). Rabeder [177], studying material from Deutsch-Altenburg 2C, proposed that *Vulpes alopecoides*, *Vulpes praeglacialis*, and *Vulpes angustidens* from Hundsheim form a phyletic line leading to the contemporary *Vulpes vulpes*. This lineage is characterized by an increase in size and a slight reduction in molar size [129]. Rabeder also argued that *Vulpes praecorsac* is part of the lineage leading to *Vulpes corsac* but does not represent a direct ancestor–descendant relationship. He suggested that the arctic fox diverged from the *Vulpes* lineage during the Pliocene, predating the European fossil record and earlier than the interpretation by Wang et al. [182] regarding *Vulpes qiuzhudingi*.

Bartolini-Lucenti and Madurell-Malapeira [34] conducted a comparative study of the known variability in Pleistocene fox forms and their extant counterparts. They concluded that all Early Pleistocene fox remains should be classified under *Vulpes alopecoides*, as the observed tooth morphology variations among European Pleistocene taxa were less significant than those within each extant species.

In the Vallparadís Section, the remains of Early Pleistocene foxes are so scarce and limited to the post-Jaramillo layers CGRD7 and EVT7, where a partial hemi-mandible and a P4 were recovered, respectively, and attributed to *V. alopecoides* [34,69].

3.3.4. *Vulpes vulpes*

The red fox (*Vulpes vulpes)* first appeared in the Middle Pleistocene of Eurasia [115]. Pei [121,183] identified a few dental specimens from Locality 1 and several cranial fragments from Locality 3 of Zhoukoudian (ZKD) as '*V.* cf. *vulgaris*'. However, age calibration of these ZKD localities remains problematic. According to Li et al. [184], the *V. vulpes* remains from layers 8 to 11 of Locality 1, dated around 700–780 ka, are the oldest in the sequence [137]. Locality 3 is considered younger than Locality 1, dating to more than 400 ka [137,138]. Although Pei's [121] attribution should be revisited, the Locality 1 specimens might represent the earliest record of *V. vulpes* in the Old World.

More certain records of *V. vulpes* come from the French sites of Lunel-Viel and L'Escale, with the oldest European occurrence dated to MIS12 [178]. A questionable Middle Pleistocene occurrence of *V. vulpes* is reported from Thomas Quarry 1, level 'Grotte des Hominides' in Morocco, correlated to approximately 600 thousand years ago [185]. However, Geraads [185] noted that the weathered mandible with worn premolars and carnassial from this site is of doubtful age and significance, possibly even sub-fossil. Late Pleistocene sites such as Doukkala and 'Grotte de Gazelles' in Morocco provide more reliable evidence of the red fox's presence in North Africa [185].

Despite the taxonomic confusion, *V. alopecoides* has generally been considered the ancestor of the modern red fox, *Vulpes vulpes* [34,177]. The European Middle Pleistocene record of *V. vulpes* is extremely sparse, first appearing in the French localities of l'Escale (ca. 0.6 Ma) [178] and Caune de l'Arago (ca. 0.6–0.45 Ma) [176]. In the Iberian Peninsula, red foxes have been identified from Sima de los Huesos (MIS12, ca. 0.45 Ma) and Galería (ca. 0.3 Ma) in Atapuerca [77]. In Italy, *V. vulpes* likely appeared slightly later, around the MIS12-11 transition (ca. 0.43 Ma), at sites such as Malagrotta and Torre in Pietra [186–188].

The precise chronology of *V. vulpes* appearance and its phylogenetic relationship with the earlier late Early Pleistocene *V. alopecoides* have not been thoroughly investigated. In the Iberian Peninsula, the earliest records of *V. vulpes* are from the Middle Pleistocene sites of Trinchera Galería (ca. 0.3 Ma) and Sima de los Huesos (MIS12, ca. 0.45 Ma) within the Atapuerca complex [77]. Although the Atapuerca fox remains have not been described in detail, García [77] compared the Galería specimens metrically with living and fossil foxes, noting close similarities with the extant species.

At the Middle Pleistocene layers of the Vallparadís Section, *Vulpes vulpes* remains are rarely recorded but were previously described by Madurell-Malapeira et al. [189] (Figure 3O–Q). The *Vulpes* specimens from EVT3 (ca. 0.6 Ma) closely resemble those of the modern *V. vulpes*, though slightly larger, and can be differentiated from *V. alopecoides* based on dental morphology. Therefore, the EVT3 sample, dated to 0.6 Ma, likely represents one of the earliest records of *V. vulpes* in Western Europe, pending a more detailed revision of the similarly aged sample from the French site of L'Escale.

3.4. Ursids

Ursus deningeri

The cave bear's lineage has been recorded in Europe since the beginning of the Pleistocene (ca. 2.6 Ma), likely related to the intensification of the Northern Hemisphere glacial processes and the establishment of the 40 ka obliquity-forced cyclicity. The first species of this lineage, *Ursus etruscus*, is known from European localities such as Saint Vallier, Upper Valdarno, or Kuruksay [54,65]. Here, we support the traditional viewpoint of Kurtén [82], who first hypothesized a phylogenetic line for cave bears starting with the Early Pleistocene *U. etruscus*, followed by the Middle Pleistocene *Ursus deningeri* and finally in the Late Pleistocene for *Ursus spelaeus* s.l. Calabrian cave bears inhabited mixed environments of woodlands and/or wooded grasslands, where they fed on a broad variety of food items, with a substantial intake of meat and/or fish [54]. The trend in the derivation of morphological characters had an abrupt change since 1.2 Ma, when the first specimens of *Ursus deningeri* were recorded from Vallonnet cave as displaying clear speleoid characters and a large size [20]. This tendency towards herbivory is also documented in Untermassfeld and Gran Dolina (MIS31 and MIS21-19, respectively) [77].

In the Vallparadís Section, cave bears were abundantly recorded from MIS30 to MIS21 in layers EVT12, CGRD7, and EVT7. These specimens show significant intraspecific variability and sexual dimorphism, favoring our previous idea that the *Early–Middle Pleistocene Transition* associated with the climatic shifts was the driving force behind the final changes that conducted the cave bear's lineage to hyperherbivory behavior [21–23,54] (Figures 4 and 5).

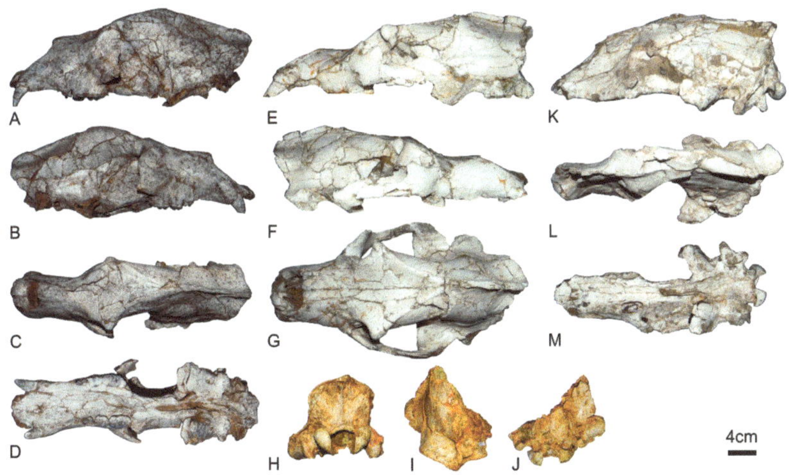

Figure 4. Crania of *Ursus deningeri* from the Vallparadís Section. Cranium with left and right C1, P4–M2 (EVT24876) from layer EVT12 in (**A**) left lateral, (**B**) right lateral, (**C**) dorsal, and (**D**) ventral views; cranium with left and right C1 and P4–M2 (EVT15872) from layer EVT7 in (**E**) left lateral, (**F**) right lateral, and (**G**) dorsal views; cranium with right m2 (EVT7116) from layer EVT7 in (**K**) left lateral, (**L**) dorsal, and (**M**) ventral views; basicranial fragment (IPS14951) from layer CGRD7 in (**H**) occipital, (**I**) dorsal, and (**J**) ventral views.

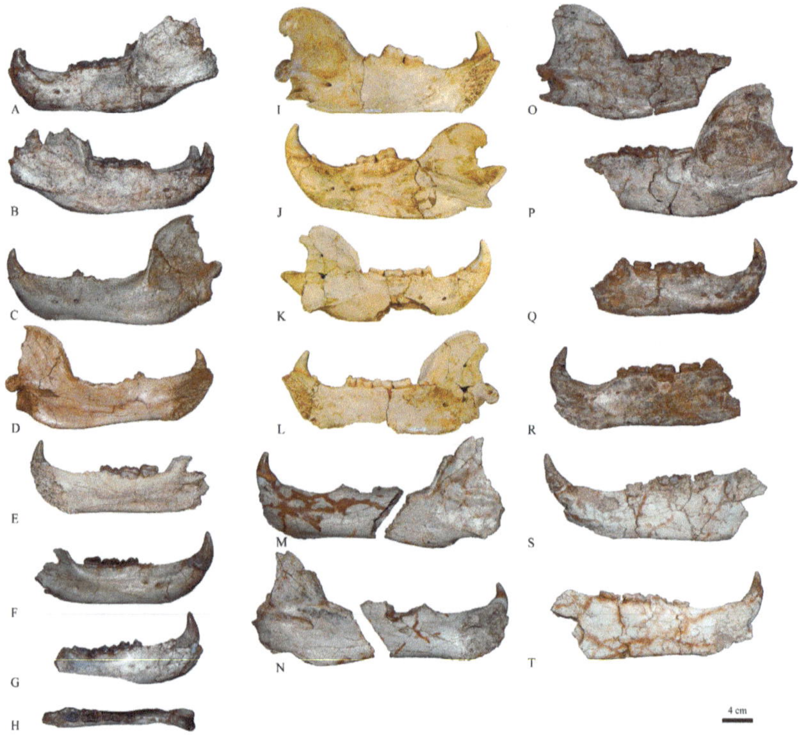

Figure 5. Mandibular remains of *Ursus deningeri* from the Vallparadís Section. Mandible with left and right c1–m3 (EVT24876) from layer EVT12 in (**A**) left buccal and (**B**) right buccal views; left hemi-mandible (EVT25680) from layer EVT12 in (**C**) buccal and (**D**) lingual views; right hemi-mandible with c1 and m1–m3 (EVT20710) from layer EVT12 in (**E**) lingual and (**F**) buccal views; partial right hemi-mandible with c1 and p4–m3 (EVT24876b) from layer EVT12 in (**G**) buccal and (**H**) occlusal views; mandible with left c1, p3, and m1–m2 and right c1 and p1–m3 (IPS14950) from layer CGRD7 in (**I**) left lingual, (**J**) left buccal, (**K**) right buccal, and (**L**) right lingual views; left hemi-mandible with c1 (EVT20234) from layer EVT7 in (**M**) buccal and (**N**) lingual views; left hemi-mandible with m1–m3 (EVT842) from layer EVT6 in (**O**) lingual and (**P**) buccal views; right hemi-mandible with c1–m3 (EVT920) from layer EVT6 in (**Q**) buccal and (**R**) lingual views; left hemi-mandible with c1–m2 (EVT11277) from layer EVT7 in (**S**) buccal and (**T**) lingual views.

In a recent study on the microwear and isotopic samples of the former specimens, Vizcaíno-Varo [190] found that the Vallparadís ursids are at a midpoint between the generalist omnivore behavior of *U. etruscus* and the hyperherbivory behavior of the Late Pleistocene *U. speleaus*. Additionally, the isotopic evidence shows that, probably around MIS21, these bears started hibernating and, consequently, habiting in karstic environments during winters, coinciding with the harshening of climatic conditions [190].

3.5. Mustelids

Meles meles

Badgers of the genus *Meles* have been reported from Eurasian localities since the Late Pliocene (3.5–3 Ma); their fossil record is, however, very scarce. This caused confusion among the taxonomy and prevented any good phylogenetic reconstruction of the genus. The first species to appear, almost simultaneously, are *Meles thorali* in Europe and *Meles chiai* in China [191–193]. Even if this synchronous appearance makes it difficult to hypothesize where the genus originated, the timeframe coincides with the progressive transition of northern hemisphere biocoenosis from a subtropical, predominantly wooded habitat to more open environments and mixed forests with a progressive increase in seasonality [8]. Until recently, the earliest known evidence of *Meles meles* was found at Fuente Nueva 3, dating back to around 1.2 million years ago. However, Marciszak et al. [194] reported several *M. meles* remains from Żabia Cave in Poland, a site estimated to be between 1.7 and 1.5 Ma. If this finding is verified, it will bridge the temporal gap between the latest records of *M. thorali* at Apollonia-1 (around 1.5 Ma) and the earliest European badgers. The presence of *M. meles* at Żabia Cave could indicate that extant European badgers likely evolved from an *M. thorali*-like ancestor in Asia and then migrated to Europe, where they gradually replaced *M. thorali*.

The European badger (*M. meles*) is the only mustelid recorded in the Vallparadís Section. Abundant cranial remains of this species were recovered from the Lower and Middle Units (layers EVT10, EVT11, and EVT12; and layers EVT6, EVT7, and EVT8; [23] and references therein). The *M. meles* specimens from Vallparadís display a large intraspecific variability, characteristic of the genus, often leading to great taxonomic confusion. This appears particularly evident when comparing the two most complete skulls of Vallparadís, as they have very different sizes. However, the robust skull with a large anteriorly placed infraorbital foramen and the distolingually placed metaconule of M1 are characteristic of *M. meles* and differentiate it from *M. thorali* (Figure 6).

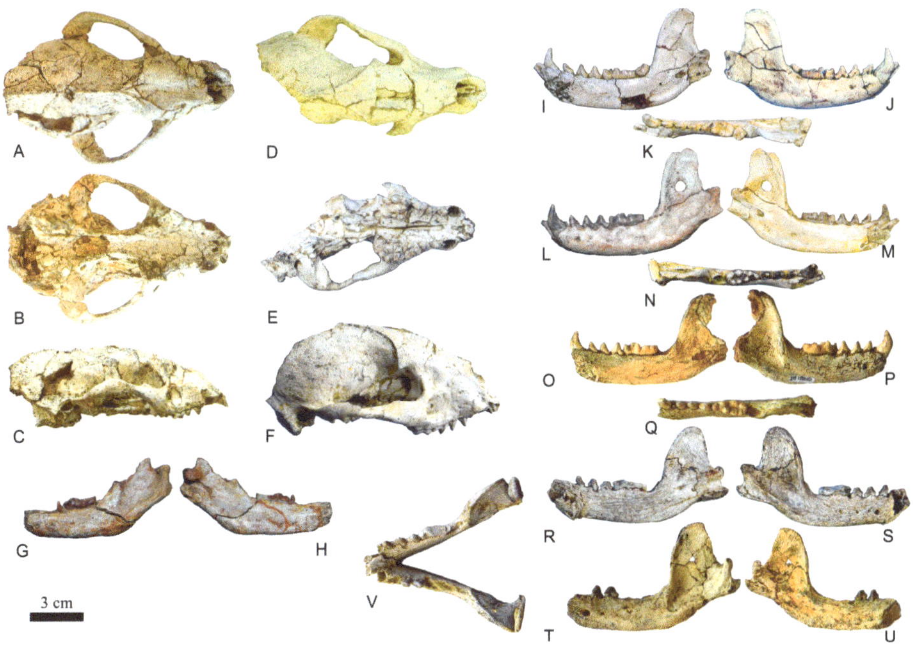

Figure 6. *Meles meles* remains recovered from the Vallparadís Section. Cranium with left and right p4–m1 (EVT25417) from layer EVT12 in (**A**) dorsal, (**B**) ventral, and (**C**) right lateral views; partial cranium with left and right p2–m1 (IPS94288) from layer EVT12 in (**D**) dorsal, (**E**) ventral, and (**F**) right lateral views; left partial hemi-mandible with m1 (IPS122011) from layer EVT7 in (**G**) buccal and (**H**) lingual views; right hemi-mandible with i1–m2 (IPS122008) from layer EVT12 in (**I**) lingual, (**J**) buccal, and (**K**) occlusal views; left hemi-mandible with i3–m1 (IPS122009) from layer EVT12 in (**L**) buccal, (**M**) lingual and (**N**) occlusal views; right hemi-mandible with c1–m1 (IPS122010) from layer EVT12 in (**O**) lingual, (**P**) buccal, and (**Q**) occlusal views; mandible with left p2 + p3 and right p1–m1 (IPS94288) from layer EVT12 in (**R**) right lingual, (**S**) right buccal, (**T**) left buccal, (**U**) left lingual, and (**V**) occlusal views.

It is unclear whether the emergence of the modern badger species reflects a general trend of progressive omnivorous adaptations within the lineage itself or is due to specific environmental conditions. Unfortunately, the few post-cranial remains of *Meles* recovered from Vallparadís (a radius and three phalanges) do not allow for inferences to be made on the evolution of the dietary behavior of this genus.

3.6. Vallparadís Guild Structure: Composition and Ecological Comparison with Other Early Pleistocene Guilds

The results of the BCA are reported in Figure 7A. The analysis clearly separates two large groups: a group of African sites (node 1) and one of all the Eurasian localities (node 3). The first group that branches out includes the East and North African guilds. Within this, the Shungura Formation of Omo, levels G–K, stems at the base of the cluster. The major difference with the other sites includes the presence of *Helogale* gr. *hirtula* and *Megantereon whitei*. Within the African cluster, two subgroupings branch from node 2 (percentage *p*-value = 70%): The first one (node 7) clusters the guilds of Koobi Fora Okote Fm. and that of Tighennif. The two localities have similar age and share several species (e.g., *Crocuta crocuta*, *Hyaena hyaena*, and *Panthera leo*). The percentage *p*-value (=71%) of this node is a testament of their own peculiarities. The second subcluster is that of node 9 and is well supported (percentage *p*-value = 94%). The subcluster includes East African localities with an age between ca. 1.9 and 1.2 Ma. All three localities share numerous taxa (*Panthera pardus*, *Pseudocivetta ingens*, *Crocuta crocuta*, and *Hyaena hyaena*), but the Olduvai Bed 1 equally differs from Koobi Fora KBS Fm. and Olduvai Bed 2 from a more diverse guild for exclusive carnivorans (e.g., *Xenocyon africanus* or *Otocyon recki*) and for occurrences shared only with one of the sites clustered together at node 14 (e.g., the *Lutra* in both Olduvai and the *Lupulella* in the coeval Bed 1 and KBS Fm.). The larger cluster composed of Eurasian localities (node 3) is further subdivided into numerous subgroups. Node 6 identifies the separation of Early–Middle Pleistocene Northeastern Chinese sites of Jinyuan Cave Upper levels and Zhoukoudian Locality 1. Their grouping is justified by some shared elements, e.g., *Canis variabilis* and *Vulpes chikushanensis*, and also for the peculiarity in comparison to other Asian sites. Nevertheless, the percentage *p*-value is not among the highest, although it is fairly high (*p*-value = 78%). The second group is that of the node 13 (percentage *p*-value = 98%) group, together with the Jinyuan Cave Lower levels and Nihewan Classic Fauna. As in the case of the cluster of node 6, chronological and geographic proximity might be the reason for the clusterization of these sites. They indeed share a large number of taxa (for a total of eleven species; see Supplementary Table S1). A similar reason might lie at the base of the grouping of node 11 (percentage *p*-value = 97%), that of *Gigantopithecus* Cave and of Gongwangling, two sites of Central–Southern China. The grouping is supported by the presence of taxa exclusive that are to these sites, e.g., *Ailuropoda melanoleuca* and *Panthera pardus*, or rarer taxa, like *Arctonyx* and *Felis sylvestris*. Node 8 denotes a larger set of European and Western Asian localities of Dmanisi and 'Ubeidiya. Trinchera Dolina TD6 is the first site stemming out from node 8: this position is due to the presence of clear European species (e.g., *Lynx pardinus*, *Canis mosbachensis*, and *Vulpes alopecoides*) but also new and rare occurrences for Early Pleistocene, like *Crocuta crocuta*. The grouping of Poggio Rosso and Casa Frata is well supported (percentage *p*-value = 94%) and justified by the close chronological and geographical position and their taxonomical composition (with six shared taxa, e.g., *Canis arnensis*, *Lynx issiodorensis*, and *Ursus etruscus*). The cluster of node 11 has 'Ubeidiya at its base. The Israeli site is characterized by the compresence of Eurasian and out-of-Africa taxa that are clearly affine with late Calabrian European sites (e.g., *Canis mosbachensis*, *Xenocyon lycaonoides*, and *Panthera gombaszoegensis*). The other localities are organized in two groups: one of earlier localities from Georgia and Southern Europe (node 16) and one of Epivillafranchian sites (node 18). In the first one, Dmanisi stems from the group made of Pirro Nord DE, Venta Micena, and Apollonia-1. At the level of carnivore guild, there is very little difference in terms of composition between the four sites: they share nine carnivores (Supplementary Table S1). This is particularly true especially for

Venta Micena and Apollonia-1, which share all the taxa (see Supplementary Table S1). A similar situation is that of the four Epivillafranchian sites (among them Vallparadís ones). These sites have six taxa in common (*Meles meles*, *Lynx pardinus*, *Pachycrocuta brevirostris*, *Canis mosbachensis*, *Xenocyon lycaonoides*, and *Ursus deningeri*), and two more are shared between EVT7/CGRD7, Vallonnet, and Untermassfeld (reaching the number of eight, respectively, nearly the 73%, the 62%, and the 73% of their record).

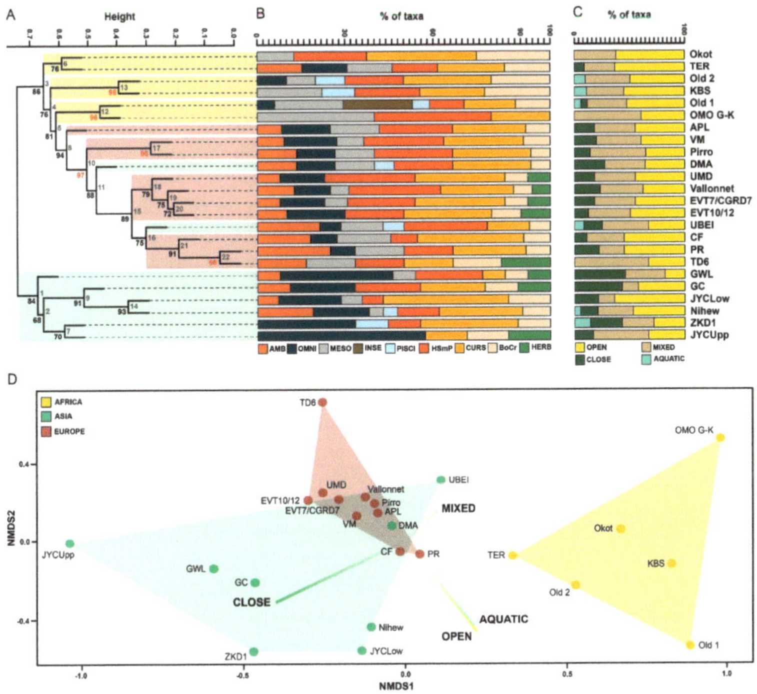

Figure 7. Bootstrapping cluster analysis and non-metric multidimensional scaling for Old World Early Pleistocene sites. (**A**) Dendrogram resulting from the bootstrapping cluster analysis on presence/absence of taxa in selected sites. (**B**) Histograms of the relative proportions of the ecological preference of the carnivorans in the analyzed guilds of the Old World, arranged according to a BCA based on ecological matrix of the sites. (**C**) Histograms of the relative proportions of the ecological preference of the carnivorans in the analyzed guilds of the Old World, arranged according to a BCA based on environmental matrix of the sites. (**D**) Non-metric multidimensional scaling plot on the environmental parameters of Old World Early Pleistocene sites. Abbreviations: AMB, hypercarnivorous ambush predators; BoCr, hypercarnivorous species with bone-cracking adaptations; CURS, hypercarnivorous cursorial predators; HERB, hypocarnivorous adaptations towards herbivory; HSmP, small-sized hypercarnivorous species; INSE, insectivorous species; MESO, mesocarnivores; OMNI, omnivorous, hypocarnivorous, and opportunistic feeders; PISC, hypercarnivores adapted to a piscivorous diet; APL, Apollonia-1 (Greece); CF, Casa Frata (Italy); DMA, Dmanisi (Georgia); EVT10/12, Vallparadís Estacio layers 10–12 (Spain); EVT7/CGRD7, Vallparadís Estacio layer 7/Cal Guardiola 7 (Spain); GWC, Gongwangling (China); JYCLow, Jinyuan Cave Lower fauna (China); JYCUpp, Jinyuan Cave Upper Fauna (China); KBS, KBS Member, Koobi Fora Formation (Kenya); Nihew, Xiashagou, Classic Nihewan (China); Okot, Okote Member, Koobi Fora Formation (Kenya); Old 1, Olduvai Bed I (Tanzania); Old 2, Olduvai Bed II (Tanzania); OMO G-K, Members GeK, Shungura Formation (Ethiopia); Pirro, Pirro Nord (Italy); PR, Poggio Rosso (Italy); TD6, Trincher Dolina 6 (Spain); UBEI, 'Ubeidiya (Israel); UMD, Untermassfeld (Germany); Vallonnet, Vallonnet (France); VM, Venta Micena (Spain); ZKD1, Zhoukoudian Locality 1 (China).

Figure 7C reports the result of NMDS on the occurrences reported in each site, taking into consideration the environmental preferences of the taxa. In the plot, the distribution of the localities reflects the similarity between them (i.e., the closer the sites, the more similar) and generally confirms the results of the BCA, as described above. In the graph, European localities are greatly associated with one another, despite their chronological differences. Among these, particularly the sites of Poggio Rosso and Casa Frata have carnivore guilds similar between them but fairly distinct from the group composed of late Early Pleistocene of Venta Micena, Pirro Nord, Vallonnet, and Vallparadís layers. The exception among the European sites is Trinchera Dolina TD6, which lies greatly separated along the NMDS Axis 2. The limited number of carnivorans and the occurrence of peculiar elements, chiefly *Crocuta crocuta* and *Cuon alpinus*, might explain this distance from the other European guilds. Asian localities occupy a wide space in the NMDS. For instance, the (nearly) coeval *Gigantopithecus* Cave, Nihewan Classic Fauna, and Jinyuan Cave Lower Fauna are separated in two positions, with the latter two much closer to each other than to the Southern Chinese locality of *Gigantopithecus* Cave. In turn, this site is closer to the Late Calabrian site of Gongwanling. Regarding temporal distinctions, Jinyuan Cave Upper Fauna is more distant from the Lower Fauna than from any other East Asian locality. This is possibly due to the arrival of new taxa like *Canis variabilis* and *Ursus deningeri*. The two sites of Dmanisi and 'Ubeidiya are located considerably far from the Eastern Asian ones and much closer to European sites. Dmanisi lies close to Apollonia-1 and Venta Micena, as their carnivoran guilds share numerous taxa (e.g., *Pachycrocuta brevirostris*, *Homotherium crenatidens*, *Panthera gombaszoegensis*, *Xenocyon lycaonoides*, and *Ursus etruscus*). 'Ubeidiya differs slightly from the European sites because of the mixture of African/Eurasian elements, such as *Crocuta crocuta*, *Vormela peregusna*, *Canis mosbachensis*, and *Ursus etruscus*. Far on the positive end of the NMDS 1 Axis, the African sites make up a separate and distinguishable group. Among them, we see that there is some kind of difference, as there is no evident clustering based on age or in terms of geographic or stratigraphic proximity. Tighennif is the most similar to European localities, given the number of European taxa recorded in the Algerian site. The high NMDS 2 Axis values of Shungura Fm. of Omo, levels G-K, might be explained by the presence of unique taxa (e.g., *Helogale*) not recorded in the other selected sites and the limited number of occurrences.

4. Discussion

4.1. The Vallparadís Section Carnivore Guild: Ecological and Taxonomical Comparison

The results of the BCA and the NMDS (Figure 7) show the different grouping of the carnivore guilds of Africa and Eurasia. The various guilds are correctly clustered into groups which testify the similarities but also the peculiarities between one another, for instance, the cluster of African guilds. These sites are separated by the other ones both taxonomically and ecologically (Figure 7), confirming previous results [28]. It is worth mentioning the differences in environmental preferences among African guilds: the large dominance of open habitat taxa in all guilds with different proportions in terms of aquatic and mixed habitat species. Another well-characterized grouping is that of *Gigantopithecus* Fauna and Gongwangling, as opposed to that of Nihewan Classic Fauna–Jinyuan Cave Lower Fauna and of Jinyuan Cave Upper Fauna–Zhoukoudian Loc. 1. This is evident in both BCA and in NDMS. Furthermore, the latter enhances the differences between clusters, adding the environmental parameters. The pattern of Asian localities might be the signal of the complete pattern of temporal and spatial turnover/latitudinal distribution and segregation that characterized and still characterize Asia today. Following the preferences of the carnivorans from the Asian sites supports a certain distinction in terms of habitats, with *Gigantopithecus* Fauna, Gongwangling, and Jinyuan Cave Upper Fauna dominated by a closed environment, in comparison to Nihewan Classic Fauna, Jinyuan Cave Lower Fauna, and Zhoukoudian Loc. 1, in which taxa of a more open environment dominate. The European localities are well clusterized together, with a clear affinity between Dmanisi and 'Ubeidiya, as reported and discussed in other previous works [27,28,33,41]. This clusterization and the position in the NMDS space suggest that a majority of the European localities were characterized by relatively mixed environments, with no dominance between open- or closed-environment taxa. Among the European locality, only the Upper Valdarno localities of Poggio Rosso and Casa Frata are marked by proportionally more abundant open habitat taxa, in agreement with current understanding of the environments of Upper Valdarno Basin around 1.9–1.8 Ma [31,40]. In comparison to other considered localities from Asia, Africa, and Europe, the carnivore guild from the Vallparadís Section is, unsurprisingly, close to other Western European Epivillafranchian localities, particularly Untermassfeld and Vallonnet. This is consistently supported by the taxonomic composition of their guilds, equally composed by common elements like *Meles meles*, *Lynx pardinus*, and *Ursus deningeri*. This is confirmed by both the BCA and the NMDS (Figure 7).

4.2. The Vallparadís Section Carnivorans in the Iberian Context

The latest Early Pleistocene (Epivillafranchian) is not abundantly recorded in the Iberian Peninsula, with few sites exhibiting low diversity of recorded carnivorans, precluding accurate comparisons with the Vallparadís Section guild. On the Northeastern Iberia, the sites of the Incarcal complex basically have a good record of two carnivoran species (*H. crenatidens* and *P. brevirostris*) correlated with MIS21 [23]. However, at the close Bòvila Ordis site, no carnivore was recorded in the layers with an approximate age of 1.2 Ma [23]. Further south, in the Francolí valley, the recently discovered site of Barranc de la Boella, with an approximate age of 0.9 Ma, records *P. gombaszoegensis*, Hyaenidae indet., *U. deningeri*, and *Canis* sp. Unfortunately, there are only a few poorly preserved specimens [25].

In the Southern Iberian Peninsula, the site of Húescar-1, with a debated chronology around 0.8 Ma, records very few specimens of *H. crenatidens*, *P. gombaszoegensis*, Hyaenidae indet., and *C. mosbachensis* [195]. In Cadiz, in the Chaparral site [106], well-preserved but scarce specimens of *V. pardoides*, *C. mosbachensis*, and *V. alopecoides* have been recovered. A similar low diversity of carnivorans was documented in the Quibas site [18] (MIS30), with scarce records of *Lynx pardinus* and few indeterminate canid fragments.

Particularly interesting in this discussion is the site of Cueva Negra del Estrecho del Rio Quípar, with an estimated chronology of 990–772 ka [196]. The former authors describe carnivoran remains attributable to *Ursus* sp., Mustelidae indet., *Crocuta* sp., and *Lynx* sp. Probably the most interesting in this assemblage is the putative presence of a spotted hyena, coeval with the earliest records of this genus outside Africa in Trinchera Dolina [77]. However, a quick examination of the remains attributed to *Crocuta* open serious doubts on former authors' attribution. *Crocuta* remains are not described in detail and only discussed based on a biometrical analysis of the lower p4 of a partial and poorly preserved right hemi-mandible [196]. The dimensions of the p4 fit perfectly with the specimens of the Vallparadís Section, securely attributed to *Pachycrocuta*. Additionally, the presence of mesial and distal accessory cuspulids on the p4 and the massive and high mandibular symphysis clearly favors its attribution to *P. brevirostris* [196].

A similar scenario is observed in the Early Pleistocene layers of the the Atapuerca site complex. Specifically, the layer TE9c of the Sima del Elefante site, with an estimated chronology of 1.2 Ma [197], evidences the presence of *P. gombaszoegensis*, *Lynx pardinus*, cf. *Pannonictis*, cf. *Baranogale*, *Mustela* sp., *C. mosbachensis*, *V. alopecoides*, and *U. deningeri* [197]. In the Trinchera Dolina TD6 layer [77], dated approximately 0.9 Ma, the recorded species includes *U. deningeri*, *C. crocuta*, *M. palerminea*, *L. pardinus*, *C. mosbachensis*, and *V. alopecoides* [77]. Other layers in the complex also include *H. crenatidens* and various carnivorans never described in detail. Despite the abundance of small mustelid species, which are extremely rare in the Iberian record, the most significant highlight of the Atapuerca site is the confirmed presence of the genus *Crocuta* in Early Pleistocene layers [77].

Finally, the most diverse carnivore guild on Iberian Early Pleistocene is found at the Cueva Victoria site [23], with an estimated chronology of 1.0–0.8 Ma; however, several doubts about the stratigraphical provenance of some specimens exist. The record includes

H. crenatidens, *M. adroveri*, *P. gombaszoegensis*, *V. pardoides*, *L. pardinus*, *P. brevirostris*, *U. deningeri*, *X. lycaonoides*, *C. mosbachensis*, and *V. alopecoides*. Additionally, an unpublished third metatarsal is clearly attributable to *P. s. fossilis*. This guild is the same as the one recorded in the Vallparadís Section; however the uncertainties about their stratigraphical scheme preclude detailed discussion on their implications.

To summarize, the carnivoran guild of the latest Iberian Early Pleistocene shows no significant changes compared to the previous Late Villafranchian guild recorded at sites like Venta Micena [23], at least until MIS30 (ca. 1.0 Ma). Since MIS30, African-origin carnivorans such as *P. s. fossilis* and *C. crocuta* started to being recorded in Iberia but only in two sites or complexes, probably due to the scarce and fragmentary record of the Iberian Epivillafranchian record. Despite no secure records of *P. pardus* during this period, the main causes of the disappearance of several species may include the competition with newcomer species and the environmental changes associated with the onset of the *Early–Middle Pleistocene Transition* (i.e., increase in seasonality, aridity, and dissymmetrical climatic cycles prompted by an increase in wood cover as compared with previous Late Villafranchian).

4.3. The Vallparadís Section's Carnivorans in the Eurasian Context

The scenario for the latest Early Pleistocene sites in Europe is similar to the Iberian one, characterized by several sites, predominantly in Mediterranean Europe, with poorly recorded carnivoran guilds. The European sites from the 1.2–0.8 Ma interval include Collecurti, Cava Redicciolì, Slivia, and Frantoio in Italy [198,199]; Apollonia-1 in Greece [32]; Blassac-la-Girondie, Bois-de-Riquet, Cagnes-sur-Mer, Ceyssaguet, Chagny, Rosières, Vallonnet, Soleilhac, Saint-Prest, Sartanette, Tour-de-Grimaldi, Trois Pigeons, and Durfort in France [39,200]; Untermassfeld in Germany [201]; Happisburgh in England [202]; Trilica in Montenegro [203]; Somssich Hill 2 in Hungary [204]; or Akhalkalaki in Georgia [205].

Among the former list, only Untermassfeld, Vallonnet, and Ceyssaguet exhibit a comparable or even most diverse carnivoran guild, which includes *H. crenatidens*, *M. adroveri*, *P. gombaszoegensis*, *A. pardinensis*, *V. pardoides*, *L. pardinus*, *F. sylvetris*, *P. brevirostris*, *U. deningeri*, *X. lycaonoides*, *C. mosbachensis*, *V. alopecoides*, and *M. meles*. Additionally, at Le Vallonnet, two lower molars of *P. pardus* were personally studied by the authors, supporting the taxonomic attribution made by [20].

Despite the different taxonomical attributions provided by different scholars, the Epivillafranchian carnivore guild of Europe displays homogeneity, except for the scarcely recorded African immigrants (lion, leopard, and spotted hyaena). However, several minor issues need to be addressed.

Firstly, since the beginning of the Pleistocene, *U. etruscus* was the only ursid recorded in European assemblages. This omnivorous specie, adapted to moderately wooded environments, was recorded at numerous sites. However, starting around 1.2 Ma and probably influenced by the climatic instability associated with the *Early–Middle Pleistocene Transition*, these forms progressively become stouter, larger, and more sexually dimorphic, starting their transition to a purely herbivorous diet [190]. *Ursus deningeri* was first recorded at the Vallonnet cave but also at the Vallparadís Section and Ceyssaguet. Alternatively, in Trinchera Dolina, Frantoio, and Untermassfeld, the potential same derived form has been cited under the attribution of *U. dolinensis* [77]. Secondly, in previous papers, one of us (J. M.-M.) hypothesized an Iberian origin for the Mediterranean lynx, first recorded at the Avenc Marcel cave (1.6 Ma) [113,114]. Recent research studies demonstrated that, by the end of Late Villafranchian and Epivillafranchian, *L. pardinus* was already distributed throughout all of Mediterranean Europe. This species was present at sites such as Pirro Nord, Vallonnet, Ceyssaguet, the Vallparadís Section, Cueva Victoria, or Untermassfeld, making it difficult to determine the precise geographical origin of the species. A more parsimonious hypothesis suggests a gradual transition from *L. issiodorensis*-like forms to smaller *L. pardinus*-like ones. Lastly, the evolutive history of the wild cat (*Felis sylvestris*) remains poorly understood. The presence of a mandible of this species in the approximately 2.0 Ma layers of the Italian Upper Valdarno complex is known. However, until the late Middle Pleistocene, it seems that this species did not exist in

Europe. Despite the exceedingly limited record, often consisting of one or two fragments, this evidence from the Epivillafranchian European sites pointed to a continuous presence of the wild cat in Europe.

In summary, the Calabrian carnivore guild remained stable throughout the Late Villafranchian but began to decline with the disappearance of several Villafranchian-character species during the arid phase associated with MIS30. This period is precisely recorded at the Vallparadís Section, which also marks the first appearance of lions and the last occurrence of dirk-toothed cats (layers EVT10-12). The various layers of the Vallparadís Section have become a key reference for studying these changes in the carnivoran guild. Future stable isotope and microwear analyses will provide further insights into the impact of the Early–Middle Pleistocene Transition on European taphocenosis [190].

5. Conclusions

The *Early–Middle Pleistocene Transition* in the Northern Hemisphere marked a shift from the predominantly obliquity-forced cyclicity to the strongly asymmetric and severely cold precession-forced cyclicity characteristic of the Middle and Late Pleistocene. These changes, along with increased aridity and seasonality, profoundly affected the European late Early Pleistocene taphocenosis. The Vallparadís Section, with its multiple layers comprising a 12 m thick sequence, is currently the only European site where the impact on large mammal assemblages during this period can be studied in detail. This period roughly coincides with the transition from Oldowan to Acheulian culture in Western Europe.

In this study, we detail the carnivoran guild composition of the Vallparadís Section and its changes over time, comparing it with other Iberian and European sites. The results show the stability of most of the guild during the Calabrian Early Pleistocene, with the arrival of several newcomers of African origin (e.g., steppe lions). These long-surviving Calabrian species are recorded until the interglacial stage MIS31 in layers EVT7 and CGRD7 of the section (e.g., *Megantereon* or *Xenocyon*).

Supplementary Materials: The following supporting information can be downloaded at https://www.mdpi.com/article/10.3390/quat7030040/s1, Figure S1: Geographical location of the Vallparadís Section within the Iberian Peninsula and the city of Terrassa. Additionally, composite stratigraphic section with the layer of precedence of the studied specimens; Figure S2: Map of the Old World showing the considered localities in the present study. Number: 1 = Vallparadís Section (Iberia); 2 = Apollonia-1 (Greece); 3 = Casa Frata (Italy); 4 = Dmanisi (Georgia); 5 = Gongwangling (China); 6 = Jinyuan Cave Lower Fauna (China); 7 = Jinyuan Cave Upper Fauna (China); 8 = KBS Member, Koobi Fora Formation (Kenya); 9 = Xiashagou, Classic Nihewan (China); 10 = Okote Member, Koobi Fora Formation (Kenya); 11 = Olduvai Bed I (Tanzania); 12 = Olduvai Bed II (Tanzania); 13 = Members G-K, Shungura Formation (Ethiopia); 14 = Pirro Nord (Italy); 15 = Poggio Rosso (Italy); 16 = Trinchera Dolina 6 (Spain); 17 = Ubeidiya (Israel); 18 = Untermassfeld (Germany); 19 = Vallonnet (France); 20 = Venta Micena (Spain); 21 = Zhoukoudian Locality 1 (China); Table S1: Database of species occurrences per selected sites used in the analysis. Abbreviations: APL, Apollonia-1 (Greece); CF, Casa Frata (Italy); DMA, Dmanisi (Georgia); EVT10/12, Vallparadís Estacio layers 10–12 (Spain); EVT7/CGRD7, Vallparadís Estacio layer 7/Cal Guardiola 7 (Spain); GWC, Gongwangling (China); JYCLow, Jinyuan Cave Lower Fauna (China); JYCUpp, Jinyuan Cave Upper Fauna (China); KBS, KBS Member, Koobi Fora Formation (Kenya); Nihew, Xiashagou, Classic Nihewan (China); Okot, Okote Member, Koobi Fora Formation (Kenya); Old 1, Olduvai Bed I (Tanzania); Old 2, Olduvai Bed II (Tanzania); OMO G-K, Members GeK, Shungura Formation (Ethiopia); Pirro, Pirro Nord (Italy); PR, Poggio Rosso (Italy); TD6, Trinchera Dolina 6 (Spain); UBEI, 'Ubeidiya (Israel); UMD, Untermassfeld (Germany); Vallonnet, Vallonnet (France); VM, Venta Micena (Spain); ZKD1,. Zhoukoudian Locality 1 (China). References: [9,21,23,28–44].

Author Contributions: Conceptualization, J.M.-M.; methodology, S.B.-L. and D.F.; software, S.B.-L.; formal analysis, S.B.-L. and D.F.; investigation, all co-authors; resources, J.M.-M.; data curation, J.M.-M.; writing—review and editing, all co-authors. All authors have read and agreed to the published version of the manuscript.

Funding: Research on the Vallparadís Section was funded by the Departament de Cultura de la Generalitat de Catalunya under the research project 'Evolució dels ecosistemes dels Pirineus Orientals i àrees adjaçents durant el Pleistocè: 2022–2025' for which J.M.-M. is the principal investigator. D.F. was supported by the Ayuda del Programa de Formación de Profesorado Universitario (FPU20/03,389) and is a Ph.D. student at the Programa de Doctorado en Biología at the Universidad Complutense de Madrid.

Data Availability Statement: All data included in this paper are available upon corresponding author upon reasonable request.

Acknowledgments: The present manuscript is part of the PhD thesis of MPV, under the Biodiversity Ph.D. program of the Universitat Autònoma de Barcelona (UAB). The authors thank the Servei d'Arqueologia i Paleontologia de la Generalitat de Catalunya for the help on the Vallparadís Section Research.

Conflicts of Interest: The authors declare no conflicts of interest.

References

1. Bertini, A. Pliocene to Pleistocene Palynoflora and Vegetation in Italy: State of the Art. *Quat. Int.* **2010**, *225*, 5–24. [CrossRef]
2. Bertini, A.; Magi, M.; Mazza, P.P.A.; Fauquette, S. Impact of Short-Term Climatic Events on Latest Pliocene Land Settings and Communities in Central Italy (Upper Valdarno Basin). *Quat. Int.* **2010**, *225*, 92–105. [CrossRef]
3. Magri, D.; Di Rita, F.; Aranbarri, J.; Fletcher, W.; González-Sampériz, P. Quaternary Disappearance of Tree Taxa from Southern Europe: Timing and Trends. *Quat. Sci. Rev.* **2017**, *163*, 23–55. [CrossRef]
4. Leroy, S.A.G.; Arpe, K.; Mikolajewicz, U. Vegetation Context and Climatic Limits of the Early Pleistocene Hominin Dispersal in Europe. *Quat. Sci. Rev.* **2011**, *30*, 1448–1463. [CrossRef]
5. Blain, H.; Agustí, J.; Lordkipanidze, D.; Rook, L.; Del, M. Paleoclimatic and Paleoenvironmental Context of the Early Pleistocene Hominins from Dmanisi (Georgia, Lesser Caucasus) Inferred from the Herpetofaunal Assemblage. *Quat. Sci. Rev.* **2014**, *105*, 136–150. [CrossRef]
6. Head, M.J.; Gibbard, P.L. *Early–Middle Pleistocene Transitions: The Land–Ocean Evidence*; Geological Society of London: London, UK, 2005; ISBN 978-1-86239-495-7.
7. Clark, P.U.; Archer, D.; Pollard, D.; Blum, J.D.; Rial, J.A.; Brovkin, V.; Mix, A.C.; Pisias, N.G.; Roy, M. The Middle Pleistocene Transition: Characteristics, Mechanisms, and Implications for Long-Term Changes in Atmospheric pCO_2. *Quat. Sci. Rev.* **2006**, *25*, 3150–3184. [CrossRef]
8. Head, M.J.; Gibbard, P.L. Early–Middle Pleistocene Transitions: Linking Terrestrial and Marine Realms. *Quat. Int.* **2015**, *389*, 7–46. [CrossRef]
9. Kahlke, R.D.; García, N.; Kostopoulos, D.S.; Lacombat, F.; Lister, A.M.; Mazza, P.P.A.; Spassov, N.; Titov, V.V. Western Palaearctic Palaeoenvironmental Conditions during the Early and Early Middle Pleistocene Inferred from Large Mammal Communities, and Implications for Hominin Dispersal in Europe. *Quat. Sci. Rev.* **2011**, *30*, 1368–1395. [CrossRef]
10. Strani, F.; DeMiguel, D.; Alba, D.M.; Moyà-Solà, S.; Bellucci, L.; Sardella, R.; Madurell-Malapeira, J. The Effects of the "0.9 Ma Event" on the Mediterranean Ecosystems during the Early-Middle Pleistocene Transition as Revealed by Dental Wear Patterns of Fossil Ungulates. *Quat. Sci. Rev.* **2019**, *210*, 80–89. [CrossRef]
11. Mudelsee, M.; Schulz, M. The Mid-Pleistocene Climate Transition: Onset of 100 Ka Cycle Lags Ice Volume Build-up by 280 Ka. *Earth Planet. Sci. Lett.* **1997**, *151*, 117–123. [CrossRef]
12. Blain, H.-A.; Lozano-Fernández, I.; Agustí, J.; Bailon, S.; Granda, L.; Patrocinio, M.; Ortiz, E.; Ros-Montoya, S.; Jiménez-Arenas, J.M.; Toro-Moyano, I.; et al. Refining upon the Climatic Background of the Early Pleistocene Hominid Settlement in Western Europe: Barranco Le on and Fuente Nueva-3 (Guadix-Baza Basin, SE Spain). *Quat. Sci. Rev.* **2016**, *144*, 132–144. [CrossRef]
13. Crippa, G.; Angiolini, L.; Bottini, C.; Erba, E.; Felletti, F.; Frigerio, C.; Hennissen, J.; Leng, M.; Petrizzo, M.R.; Raffi, I.; et al. Seasonality Fluctuations Recorded in Fossil Bivalves during the Early Pleistocene: Implications for Climate Change. *Palaeogeogr. Palaeoclimatol. Palaeoecol.* **2016**, *446*, 234–251. [CrossRef]
14. Saarinen, J.; Oksanen, O.; Zliobaite, I.; Fortelius, M.; DeMiguel, D.; Azanza, B.; Bocherens, H.; Luzón, C.; Solano, J.; Yravedra, J.; et al. Pliocene to Middle Pleistocene Climate History in the Guadix-Baza Basin, and the Environmental Conditions of Early Homo Dispersal in Europe. *Quat. Sci. Rev.* **2021**, *268*, 107132. [CrossRef]
15. Rodríguez-Gómez, G.; Palmqvist, P.; Ros-Montoya, S.; Espigares, M.; Martínez-Navarro, B. Resource Availability and Competition Intensity in the Carnivore Guild of the Early Pleistocene Site of Venta Micena (Orce, Baza Basin, SE Spain). *Quat. Sci. Rev.* **2017**, *143*, 69–83. [CrossRef]

16. Postigo-Mijarra, J.; Burjachs, F.; Manzaneque, F.; Morla, C. A Palaeoecological Interpretation of the Lower-Middle Pleistocene Cal Guardiola Site (Terrassa, Barcelona, NE Spain) from the Comparative Study of Wood and Pollen. *Rev. Palaeobot. Palynol.* **2007**, *146*, 247–264. [CrossRef]
17. Sorbelli, L.; Alba, D.; Cherin, M.; Moullé, P.-E.; Brugal, J.-P.; Madurell Malapeira, J. A Review on Bison Schoetensacki and Its Closest Relatives through the Early-Middle Pleistocene Transition: Insights from the Vallparadís Section (NE Iberian Peninsula) and Other European Localities. *Quat. Sci. Rev.* **2021**, *261*, 106933. [CrossRef]
18. Piñero, P.; Agustí, J.; Laborda, C.; Duval, M.; Zhao, J.; Blain, H.A.; Furió, M.; Laplana, C.; Rosas, A.; Sevilla, P. Quibas-Sima: A Unique 1 Ma-Old Vertebrate Succession in Southern Iberian Peninsula. *Quat. Sci. Rev.* **2022**, *283*, 107469. [CrossRef]
19. Rodríguez, J.; Mateos, A.; Hertler, C.; Palombo, M.R. Modelling Human Presence and Environmental Dynamics during the Mid-Pleistocene Revolution: New Approaches and Tools. *Quat. Int.* **2015**, *393*, 19–23. [CrossRef]
20. Moulle, P.-E. Les Grand Mammifères Du Pléistocène Inférieur de La Grotte Du Vallonnet (Roquegrune-Cap-Martin, Alpes Maritimes): Étude Paléontologie Des Carnivores, Équidé, Suidé et Bovidés. Ph.D. Thesis, Muséum national d'histoire naturelle, Paris, France, 1992.
21. Madurell-Malapeira, J.; Minwer-Barakat, R.; Alba, D.M.; Garcés, M.; Gómez, M.; Aurell-Garrido, J.; Ros-Montoya, S.; Moyà-Solà, S.; Berástegui, X. The Vallparadís Section (Terrassa, Iberian Peninsula) and the Latest Villafranchian Faunas of Europe. *Quat. Sci. Rev.* **2010**, *29*, 3972–3982. [CrossRef]
22. Madurell Malapeira, J.; Alba, D.; Espigares, M.; Vinuesa Vinuesa, V.; Palmqvist, P.; Martínez-Navarro, B.; Moyà-Solà, S. Were Large Carnivorans and Great Climatic Shifts Limiting Factors for Hominin Dispersals? Evidence of the Activity of *Pachycrocuta Brevirostris* during the Mid-Pleistocene Revolution in the Vallparadís Section (Vallès-Penedès Basin, Iberian Peninsula). *Quat. Int.* **2017**, *431*, 42–52. [CrossRef]
23. Madurell-Malapeira, J.; Ros-Montoya, S.; Espigares, M.P.; Alba, D.M.; Aurell-Garrido, J. Villafranchian Large Mammals from the Iberian Peninsula: Paleobiogeography, Paleoecology and Dispersal Events. *J. Iber. Geol.* **2014**, *40*, 167–178. [CrossRef]
24. Minwer-Barakat, R.; Madurell-Malapeira, J.; Alba, D.M.; Aurell-Garrido, J.; De Esteban-Trivigno, S.; Moyà-Solà, S. Pleistocene Rodents from the Torrent de Vallparadís Section (Terrassa, Northeastern Spain) and Biochronological Implications. *J. Vertebr. Paleontol.* **2011**, *31*, 849–865. [CrossRef]
25. Fidalgo, D.; Rosas, A.; Madurell-Malapeira, J.; Pineda, A.; Huguet, R.; García-Tabernero, A.; Cáceres, I.; Ollé, A.; Vallverdú, J.; Saladie, P. A Review on the Pleistocene Occurrences and Palaeobiology of Hippopotamus Antiquus Based on the Record from the Barranc de La Boella Section (Francolí Basin, NE Iberia). *Quat. Sci. Rev.* **2023**, *307*, 108034. [CrossRef]
26. Fidalgo, D.; Rosas, A.; Bartolini-Lucenti, S.; Boisserie, J.-R.; Pandolfi, L.; Martínez-Navarro, B.; Palmqvist, P.; Rook, L.; Madurell-Malapeira, J. Increase on Environmental Seasonality through the European Early Pleistocene Inferred from Dental Enamel Hypoplasia. *Sci. Rep.* **2023**, *13*, 16941. [CrossRef] [PubMed]
27. Bartolini-Lucenti, S.; Cirilli, O.; Pandolfi, L.; Bernor, R.L.; Bukhsianidze, M.; Carotenuto, F.; Lordkipanidze, D.; Tsikaridze, N.; Rook, L. Zoogeographic Significance of Dmanisi Large Mammal Assemblage. *J. Hum. Evol.* **2022**, *163*, 103125. [CrossRef] [PubMed]
28. Bartolini-Lucenti, S.; Madurell-Malapeira, J.; Martínez-Navarro, B.; Cirilli, O.; Pandolfi, L.; Rook, L.; Bushkhianidze, M.; Lordkipanidze, D. A Comparative Study of the Early Pleistocene Carnivore Guild from Dmanisi (Georgia). *J. Hum. Evol.* **2022**, *162*, 103108. [CrossRef] [PubMed]
29. Jin, C.; Wang, Y.; Liu, J.; Ge, J.; Zhao, B.; Liu, J.; Zhang, H.; Shao, Q.; Gao, C.; Zhao, K.; et al. Late Cenozoic Mammalian Faunal Evolution at the Jinyuan Cave Site of Luotuo Hill, Dalian, Northeast China. *Quat. Int.* **2021**, *577*, 15–28. [CrossRef]
30. Petrucci, M.; Cipullo, A.; Martínez-Navarro, B.; Rook, L.; Sardella, R. The Late Villafranchian (Early Pleistocene) Carnivores (Carnivora, Mammalia) from Pirro Nord (Italy). *Palaeontogr. Abt. A Palaozool. Stratigr.* **2013**, *298*, 113–145. [CrossRef]
31. Rook, L.; Croitor, R.; Delfino, M.; Ferretti, M.P.; Gallai, G.; Pavia, M. The Upper Valdarno Plio-Pleistocene Vertebrate Record: An Historical Overview, with Notes on Palaeobiology and Stratigraphic Significance of Some Important Taxa. *Ital. J. Geosci.* **2013**, *132*, 104–125. [CrossRef]
32. Koufos, G. New Material and Revision of the Carnivora, Mammalia from the Lower Pleistocene Locality Apollonia 1, Greece. *Quaternary* **2018**, *1*, 6. [CrossRef]
33. Konidaris, G.E. Guilds of Large Carnivorans during the Pleistocene of Europe: A Community Structure Analysis Based on Foraging Strategies. *Lethaia* **2022**, *55*, 1–18. [CrossRef]
34. Bartolini-Lucenti, S.; Madurell-Malapeira, J. Unraveling the Fossil Record of Foxes: An Updated Review on the Plio-Pleistocene *Vulpes* Spp. from Europe. *Quat. Sci. Rev.* **2020**, *236*, 106296. [CrossRef]
35. Werdelin, L.; Peigné, S. Carnivora. In *Cenozoic Mammals of Africa*; Werdelin, L., Sanders, W., Eds.; University of California Press: Berkeley, CA, USA, 2010.
36. *Fossil Mammals of Asia*; Wang, X.; Flynn, L.J.; Fortelius, M. (Eds.) Columbia University Press: New York, NY, USA, 2013; ISBN 978-0-231-15012-5.
37. Geraads, D. Pleistocene Carnivora (Mammalia) from Tighennif (Ternifine), Algeria. *Geobios* **2016**, *49*, 445–458. [CrossRef]
38. Bartolini-Lucenti, S.; Rook, L. A Review on the Late Villafranchian Medium-Sized Canid *Canis Arnensis* Based on the Evidence from Poggio Rosso (Tuscany, Italy). *Quat. Sci. Rev.* **2016**, *151*, 58–71. [CrossRef]
39. Palombo, M.R.; Valli, A.M.F. Remarks on the Biochronology of Mammalian Faunal Complexes from the Pliocene to the Middle Pleistocene in France. *Geol. Romana* **2003**, *37*, 145–163.

40. Mazza, P.; Bertini, A.; Magi, M. The Late Pliocene Site of Poggio Rosso (Central Italy): Taphonomy and Paleoenvironment. *Palaios* **2004**, *19*, 227–248. [CrossRef]
41. Martínez-Navarro, B.; Belmaker, M.; Bar-Yosef, O. The Large Carnivores from 'Ubeidiya (Early Pleistocene, Israel): Biochronological and Biogeographical Implications. *J. Hum. Evol.* **2009**, *56*, 514–524. [CrossRef]
42. De Giuli, C.; Masini, F. Late Villafranchian Faunas in Italy: The Casa Frata Local Fauna (Upper Valdarno, Tuscany). *Palaeontogr. Ital.* **1986**, *74*, 1–9.
43. Arribas, A.; Garrido, G.; Viseras, C.; Soria, J.M.; Pla, S.; Solano, J.G.; Garcés, M.; Beamud, E.; Carrión, J.S. A Mammalian Lost World in Southwest Europe during the Late Pliocene. *PLoS ONE* **2009**, *4*, e7127. [CrossRef]
44. Rook, L.; Martínez-Navarro, B. Villafranchian: The Long Story of a Plio-Pleistocene European Large Mammal Biochronologic Unit. *Quat. Int.* **2010**, *219*, 134–144. [CrossRef]
45. Carotenuto, F.; Tsikaridze, N.; Rook, L.; Lordkipanidze, D.; Longo, L.; Condemi, S.; Raia, P. Venturing out Safely: The Biogeography of Homo Erectus Dispersal out of Africa. *J. Hum. Evol.* **2016**, *95*, 1–12. [CrossRef] [PubMed]
46. Raia, P.; Carotenuto, F.; Meloro, C.; Piras, P.; Barbera, C.; Kotsakis, T. More than Three Million Years of Community Evolution. The Temporal and Geographical Resolution of the Plio-Pleistocene Western Eurasia Mammal Faunas. *Palaeogeogr. Palaeoclimatol. Palaeoecol.* **2009**, *276*, 15–23. [CrossRef]
47. Sokal, R.R.; Michener, C.D. A Statistical Method for Evaluating Systematic Relationships. *Univ. Kans. Sci. Bull.* **1958**, *38*, 1409–1438.
48. Pillar, V.D. How Sharp Are Classifications? *Ecology* **1999**, *80*, 2508–2516. [CrossRef]
49. Crusafont-Pairó, M.; Truyols-Santonja, J. A Biometric Study of the Evolution of Fissiped Carnivores. *Evolution* **1956**, *10*, 314–332. [CrossRef]
50. Van Valkenburgh, B. Carnivore Dental Adaptations and Diet: A Study of Trophic Diversity within Guilds. In *Carnivore*; Gittleman, J.L., Ed.; Springer: Berlin/Heidelberg, Germany, 1989.
51. Van Valkenburgh, B.; Koepfli, K.-P. Cranial and Dental Adaptations to Predation in Canids. *Symp. Zool. Soc. Lond.* **1993**, *65*, 15–37.
52. Popowics, T.E. Postcanine Dental Form in the Mustelidae and Viverridae (Carnivora: Mammalia). *J. Morphol.* **2003**, *256*, 322–341. [CrossRef]
53. Flower, L.O.H.; Schreve, D.C. An Investigation of Palaeodietary Variability in European Pleistocene Canids. *Quat. Sci. Rev.* **2014**, *96*, 188–203. [CrossRef]
54. Medin, T.; Martínez-Navarro, B.; Madurell-Malapeira, J.; Figueirido, B.; Kopaliani, G.; Rivals, F.; Kiladze, G.; Palmqvist, P.; Lordkipanidze, D. The Bears from Dmanisi and the First Dispersal of Early Homo out of Africa. *Sci. Rep.* **2019**, *9*, 17752. [CrossRef]
55. Estes, R.D. *The Behavior Guide to African Mammals: Including Hoofed Mammals, Carnivores, Primates*; University of California Press: Berkeley, CA, USA, 2012.
56. Kruskal, J.B. Nonmetric Multidimensional Scaling: A Numerical Method. *Psychometrika* **1964**, *29*, 115–129. [CrossRef]
57. R Studio Team. *RStudio Team RStudio: Integrated Development Environment for R.*; R Studio Team: Boston, MA, USA, 2024.
58. R Core Team. *R: A Language and Environment for Statistical Computing*; R Core Team: Vienna, Austria, 2024.
59. Suzuki, R.; Terada, Y.; Shimodaira, H. *Pvclust: Hierarchical Clustering with P-Values via Multiscale Bootstrap Resampling 2019*; Comprehensive R Archive Network: Tokio, Japan, 2019.
60. Oksanen, J.; Blanchet, F.G.; Friendly, M.; Kindt, R.; Legendre, P.; McGlinn, D.; Minchin, P.; O'Hara, R.; Simpson, G.; Solymos, P.; et al. Vegan: Community Ecology Package. Ordination Methods, Diversity Analysis and Other Functions for Community and Vegetation Ecologists, Version 2.4-3; Comprehensive R Archive Network: 2017. Available online: https://cran.r-project.org/ (accessed on 10 June 2024).
61. Wickham, H.; Chang, W.; Wickham, M.H. Package 'Ggplot2'. Create Elegant Data Visualisations Using the Grammar of Graphics; Version 2; Comprehensive R Archive Network: 2016.
62. Hemmer, H. Notes on the Ecological Role of European Cats (Mammalia: Felidae) of the Last Two Million Years. In *Miscelánea en Homenaje a Emiliano Aguirre*; Baquedano Pérez, E., Rubio Jara, S., Aguirre, E., Eds.; Museo Arqueológico Regional: Madrid, España, 2004; Volume 2, pp. 214–232. ISBN 84-451-2654-7.
63. Antón, M. *Sabertooth*; Indiana University Press: Bloomington, IN, USA, 2013.
64. Barnett, R.; Westbury, M.V.; Sandoval-Velasco, M.; Vieira, F.G.; Jeon, S.; Zazula, G.; Martin, M.D.; Ho, S.Y.W.; Mather, N.; Gopalakrishnan, S.; et al. Genomic Adaptations and Evolutionary History of the Extinct Scimitar-Toothed Cat, Homotherium Latidens. *Curr. Biol.* **2020**, *30*, 5018–5025.e5. [CrossRef] [PubMed]
65. Viret, J. Le lœss à bancs durcis de Saint-Vallier (Drôme), et sa faune de mammifères villafranchiens (avec une analyse granulométrique et une analyse pollinique). *Nouv. Arch. Musèum D'Histoide Nat. Lyon* **1954**, *4*, 3–67. [CrossRef]
66. Ballesio, R. Monographie d'un Machairodus Du Gisement Villafranchien de Senèze : Homotherium Crenatidens Fabrini. *Trav. Lab. Géologie Fac. Sci. Lyon Nouv. Série* **1963**, *9*, 1–129.
67. Sorbelli, L.; Rufí, I.; Prat-Vericat, M.; Madurell-Malapeira, J. El Complex d'Incarcal: Quan els dents de sabre dominaven el comtat de Besalú. *Quad. Assem. D'Estudis* **2022**, *3*, 63–72.
68. Carbone, C.; Maddox, T.; Funston, P.J.; Mills, M.G.L.; Grether, G.F.; Van Valkenburgh, B. Parallels between Playbacks and Pleistocene Tar Seeps Suggest Sociality in an Extinct Sabretooth Cat, Smilodon. *Biol. Lett.* **2009**, *5*, 81–85. [CrossRef] [PubMed]
69. Madurell-Malapeira, J.; Alba, D.M.; Moyà-Solà, S. Carnivora from the Late Early Pleistocene of Cal Guardiola (Terrassa, Vallès-Penedès Basin, Catalonia, Spain). *J. Paleontol.* **2009**, *83*, 969–974. [CrossRef]

70. Sotnikova, M.V.; Foronova, I.V. First Asian Record of Panthera (Leo) Fossilis (Mammalia, Carnivora, Felidae) in the Early Pleistocene of Western Siberia, Russia. *Integr. Zool.* **2014**, *9*, 517–530. [CrossRef]
71. Marciszak, A.; Lipecki, G.; Pawłowska, K.; Jakubowski, G.; Ratajczak-Skrzatek, U.; Zarzecka-Szubińska, K.; Nadachowski, A. The Pleistocene Lion *Panthera Spelaea* (Goldfuss, 1810) from Poland—A Review. *Quat. Int.* **2021**, *605–606*, 213–240. [CrossRef]
72. Sickenberg, O. Eine Säugenerfauna Des Eferen Bihariums Aus Dem Becken von Megalopolis (Peloponnes, Griechenland). *Ann. Géologiques Pays Helléniques* **1976**, *27*, 25–73.
73. Tsoukala, E.; Chatzopoulou, K.; Rabeder, G.; Pappa, S.; Nagel, D.; Withalm, G. Paleontological and stratigraphical research in Loutra Arideas Bear Cave (Almopia Speleopark, Pella, Macedonia, Greece). *Sci. Ann. Sch. Geol. Aristotle Univ. Thessalon.* **2006**, *98*, 41–67.
74. David, A.I.; Tatarinov, K.A.; Svistun, V.I. *Khishnye, Hobotnye I Kopytnye Rannego Pleistocena Yugo–zapada SSSR [Carnivores, Proboscidians and Ungulates of Early Pleistocene at Southwest USSR]*, Shtiintsa: Kishinev, Moldova, 1990.
75. Krakhmalnaya, T.V. Carnivore remains from Late Pleistocene and Holocene deposits in the Ukraine. In *The Holocene History of the European Vertebrate Fauna*; Benecke, N., Ed.; Modern aspects of research; Verlag Marie Leidorf GmbH: Nordrhein-Westfalen, Germany, 1999.
76. Lewis, M.; Pacher, M.; Turner, A. The larger Carnivora of the West Runton Freshwater Bed. *Quat. Int.* **2010**, *228*, 116–135. [CrossRef]
77. García, N. *Osos Y Otros Carnívoros De La Sierra De Atapuerca*; Fundacion Oso de Asturias: Oviedo, Spain, 2003.
78. Hemmer, H. Pleistozäne Katzen Europas—Eine Übersicht. *Cranium* **2003**, *20*, 6–22.
79. Hemmer, H. The story of the cave lion—Panthera leo spelaea (Goldfuss, 1810)—A review. *Quaternaire* **2010**, *4*, 201–208.
80. Hemmer, H. The identity of the "lion", Panthera principialis sp. nov., from the Pliocene Tanzanian site of Laetoli and its significance for molecular dating the pantherine phylogeny, with remarks on Panthera shawi (Broom, 1948), and a revision of Puma incurva (Ewer, 1956), the Early Pleistocene Swartkrans "leopard" (Carnivora, Felidae). *Palaeobiodiversity Palaeoenvironments* **2023**, *103*, 465–487.
81. Serangeli, J.; Kolfschoten, T.V.; Starkovich, B.M.; Verheijen, I. The European saber-toothed cat (Homotherium latidens) found in the "Spear Horizon" at Schöningen (Germany). *J. Hum. Evol.* **2015**, *89*, 172–180. [CrossRef]
82. Kurtén, B. *Pleistocene Mammals of Europe*; Weidenfeld and Nicolson: London, UK, 1968.
83. Schütt, G.; Hemmer, H. Zur Evolution des Löwen (*Panthera leo* L.) im europäischen Pleistozän. *Neues Jahrb. Für Geol. Paläontologie Monatsh.* **1978**, *18*, 228–255.
84. Argant, A.; Argant, J.; Jeannet, M.; Erbajeva, M. The big cats of the fossil site Château Breccia Northern Section (Saône-et-Loire, Burgundy, France), stratigraphy, palaeoenvironment, ethology and biochronological dating. *Cour. Forschungsinstitut Senckenberg* **2007**, *259*, 121–140.
85. Marciszak, A.; Schouwenburg, C.; Gornig, W.; Lipecki, G.; Mackiewicz, P. Morphometric comparison of Panthera spelaea (Goldfuss, 1810) from Poland with the lion remains from Eurasia over the last 700 ka. *Quat. Sci. Rev.* **2019**, *223*, 105950. [CrossRef]
86. Puzachenko, A.Y.; Argant, A.; Baryshnikov, G.F.; Boeskorov, G.G.; Kirillova, I.V.; Klimovsky, A.I.; Kosintsev, P.A.; Marciszak, A.; Plotnikov, V.V.; Sabol, M.; et al. Distribution history of the cave lion (*Panthera spelaea* (Goldfuss, 1810)). *Earth Hist. Biodivers.* **2024**, *1*, 100006. [CrossRef]
87. Prat-Vericat, M.; Marciszak, A.; Rufí, I.; Sorbelli, L.; Llenas, M.; Lucenti, S.B.; Madurell-Malapeira, J. Middle Pleistocene Steppe Lion Remains from Grotte de la Carrière (Têt Valley, Eastern Pyrenees). *J. Mamm. Evol.* **2022**, *29*, 547–569. [CrossRef]
88. Barnett, R.; Shapiro, B.; Barnes, I.; Ho, S.Y.; Burger, J.; Yamaguchi, N.; Higham, T.F.; Wheeler, H.T.; Rosendahl, W.; Sher, A.V.; et al. Phylogeography of lions (*Panthera leo* ssp.) reveals three distinct taxa and a late Pleistocene reduction in genetic diversity. *Mol. Ecol.* **2008**, *18*, 1668–1677. [CrossRef] [PubMed]
89. Palmqvist, P.; Torregrosa, V.; Pérez-Claros, J.A.; Martínez-Navarro, B.; Turner, A. A re-evaluation of the diversity of *Megantereon* (Mammalia, Carnivora, Machairodontinae) and the problem of species identification in extinct carnivores. *J. Vertebr. Paleontol.* **2007**, *27*, 160–175. [CrossRef]
90. Zhu, M.; Jiangzuo, Q.; Qin, D.; Jin, C.; Sun, C.; Wang, Y.; Yan, Y.; Liu, J. First discovery of *Megantereon* skull from southern China. *Hist. Biol.* **2021**, *33*, 3413–3422. [CrossRef]
91. Werdelin, L.; Lewis, M.E. Species identification in *Megantereon*: A reply to Palmqvist. *J. Paleontol.* **2002**, *76*, 931–933. [CrossRef]
92. Tura-Poch, C.; Bartolini-Lucenti, S.; Jiangzuo, Q.; Prat-Vericat, M.; Martínez-Navarro, B.; Rook, L.; Madurell-Malapeira, J. The disappearance of European dirk toothed-cats. *Paleoworld* **2024**, *in press*. [CrossRef]
93. Hemmer, H.; Kahlke, R.D.; Vekua, A.K. *Panthera onca georgica* ssp. nov. from the Early Pleistocene of Dmanisi (Republic of Georgia) and the phylogeography of jaguars (Mammalia, Carnivora, Felidae). *Neues Jahrb. Geol. Paläontologie Abh.* 2010; 115–127.
94. Argant, A.; Argant, J. The *Panthera gombaszogensis* story: The contribution of the Château breccia (Saône-et-Loire, Burgundy, France). *Quaternaire* **2011**, *4*, 247–269.
95. Turner, A. The evolution of the guild of larger terrestrial carnivores during the Plio-Pleistocene in Africa. *Geobios* **1990**, *23*, 349–368. [CrossRef]
96. Manthi, F.K.; Brown, F.H.; Plavcan, M.J.; Werdelin, L. Gigantic lion, *Panthera leo*, from the Pleistocene of Natodomeri, eastern Africa. *J. Paleontol.* **2018**, *92*, 305–312. [CrossRef]
97. Jiangzuo, Q.; Liu, J. First record of the Eurasian jaguar in southern Asia and a review of dental differences between pantherine cats. *J. Quat. Sci.* **2020**, *35*, 817–830. [CrossRef]

98. Chatar, N.; Michaud, M.; Fischer, V. Not a jaguar after all? Phylogenetic affinities and morphology of the Pleistocene felid Panthera gombaszoegensis. *Pap. Palaeontol.* **2022**, *8*, e1464. [CrossRef]
99. Hemmer, H. Studien an "Panthera schaubi" Viret aus dem Villafranchien von Saint-Vallier (Drôme). *Neues Jahrb. Geol. Palaontol. Abh.* **1964**, *122*, 324–336.
100. Kurtén, B.; Crusafont-Pairó, M. Villafranchian carnivores (Mammalia) from La Puebla de Valverde (Teruel, Spain). *Comm. Biologicae* **1977**, *85*, 1–39.
101. Owen, R. *A History of British Fossil Mammals, and Birds*; John Van Voorst: London, UK, 1846; 560p.
102. Pons-Moyà, J.; Moyà-Solà, S. La fauna de carnívors del Pleistoceno medio (Mindel) de la cueva Victoria (Cartagena, España). *Acta Geológica Hispánica* **1978**, *2*, 54–58.
103. Carbonell, E.; Escalera, J.E.; Solà, S.M.; Moyà, J.P.; de Villalta, J.F. Cueva Victoria (Murcia, España): Lugar de ocupación humana más antiguo de la Península Ibérica. *Endins Publicació D'Espeleologia* **1981**, *8*, 47–57.
104. Blain, H.; Bailon, S.; Agustí, J. Amphibians and squamate reptiles from the latest early Pleistocene of Cueva Victoria (Murcia, southeastern Spain, SW Mediterranean): Paleobiogeographic and paleoclimatic implications, *Geol. Acta* **2008**, *6*, 345–361.
105. Madurell-Malapeira, J.; Aurell-Garrido, J.; Alba, D.M.; Moyà-Solà, S. The Iberian record of the puma-like cat Puma pardoides (Owen, 1846)(Carnivora, Felidae). *Comptes Rendus Palevol* **2010**, *9*, 55–62. [CrossRef]
106. Pacheco, F.G.; Santiago, A.; Gutiérrez, J.M.; López-García, J.M.; Blain, H.A.; Cuenca-Bescós, G.; van der Made, J.; Cáceres, I.; García, N. The early pleistocene paleontological site in the sierra del chaparral (villaluenga del rosario, cádiz, southwestern spain). *Quat. Int.* **2011**, *243*, 92–104. [CrossRef]
107. Werdelin, L.; Dragusin, V.; Robu, M.; Petculescu, A.; Popescu, A.; Curran, S.; Terhune, C. Carnivora from the early pleistocene of Grăunceanu (Olteţ River Valley, Dacian Basin, Romania). *Riv. Ital. Paleontol. Stratigr.* **2023**, *129*, 457–476. [CrossRef] [PubMed]
108. Werdelin, L. The evolution of lynxes. *Ann. Zool. Fenn.* **1981**, *1*, 37–71.
109. Montoya, P.; Morales, J.; Sendra, J. Una mandiíbula de *Lynx issiodorensis* (Croizet y Jobert, 1828) (Carnivora, Mammalia) en el Plioceno inferior de Cuevas de Almanzora (Almeriía, España). *Span. J. Palaeontol.* **2001**, *16*, 125–131. [CrossRef]
110. Ficcarelli, G.; Torre, D. Phyletic relationships between Lynx group issiodorensis and Lynx pardina. *Boll. Soc. Paleontol. Ital.* **1977**, *16*, 197–202.
111. Cherin, M.; Iurino, D.A.; Sardella, R. New well-preserved material of *Lynx issiodorensis valdarnens* is (Felidae, Mammalia) from the Early Pleistocene of Pantalla (central Italy). *Boll. Soc. Paleontol. Ital.* **2013**, *54*, 103–111.
112. Mecozzi, B.; Sardella, R.; Boscaini, A.; Cherin, M.; Costeur, L.; Madurell-Malapeira, J.; Pavia, M.; Profico, A.; Iurino, D.A. The tale of a short-tailed cat: New outstanding Late Pleistocene fossils of Lynx pardinus from southern Italy. *Quat. Sci. Rev.* **2021**, *262*, 106840. [CrossRef]
113. Boscaini, A.; Madurell-Malapeira, J.; Llenas, M.; Martínez-Navarro, B. The origin of the critically endangered Iberian lynx: Speciation, diet and adaptive changes. *Quat. Sci. Rev.* **2015**, *123*, 247–253. [CrossRef]
114. Boscaini, A.; Alba, D.M.; Beltran, J.F.; Moyà-Solà, S.; Madurell-Malapeira, J. Latest Early Pleistocene remains of *Lynx pardinus* (Carnivora, Felidae) from the Iberian Peninsula: Taxonomy and evolutionary implications. *Quat. Sci. Rev.* **2016**, *143*, 96–106. [CrossRef]
115. Palmqvist, P.; Martínez-Navarro, B.; Pérez-Claros, J.A.; Torregrosa, V.; Figueirido, B.; Jiménez-Arenas, J.M.; Espigares, M.P.; Ros-Montoya, S.; De Renzi, M. The giant hyena Pachycrocuta brevirostris: Modelling the bone-cracking behavior of an extinct carnivore. *Quat. Int.* **2011**, *243*, 61–79. [CrossRef]
116. Werdelin, L. *Pachycrocuta* (hyaenids) from the Pliocene of east Africa. *Paläontologische Z.* **1999**, *73*, 157–165. [CrossRef]
117. Liu, J.; Liu, J.; Zhang, H.; Wagner, J.; Jiangzuo, Q.; Song, Y.; Liu, S.; Wang, Y.; Jin, C. The giant short-faced hyena *Pachycrocuta brevirostris* (Mammalia, Carnivora, Hyaenidae) from Northeast Asia: A reinterpretation of subspecies differentiation and intercontinental dispersal. *Quat. Int.* **2021**, *577*, 29–51. [CrossRef]
118. Espigares, M.P.; Martínez-Navarro, B.; Palmqvist, P.; Ros-Montoya, S.; Toro, I.; Agustí, J.; Sala, R. Homo vs. Pachycrocuta: Earliest evidence of competition for an elephant carcass between scavengers at Fuente Nueva-3 (Orce, Spain). *Quat. Int.* **2013**, *295*, 113–125. [CrossRef]
119. Vinuesa, V.; Madurell-Malapeira, J.; Fortuny, J.; Alba, D.M. The endocranial morphology of the Plio-Pleistocene bone-cracking hyena Pliocrocuta perrieri: Behavioral implications. *J. Mamm. Evol.* **2015**, *22*, 421–434. [CrossRef]
120. Vinuesa, V.; Iurino, D.A.; Madurell-Malapeira, J.; Liu, J.; Fortuny, J.; Sardella, R.; Alba, D.M. Inferences of social behavior in bone-cracking hyaenids (Carnivora, Hyaenidae) based on digital paleoneurological techniques: Implications for human–carnivoran interactions in the Pleistocene. *Quat. Int.* **2016**, *413*, 7–14. [CrossRef]
121. Pei, W.C. On the carnivora from locality 1 of Choukoutien. *Palaeontol. Sin. Ser. C* **1934**, *8*, 1–217.
122. Valkenburgh, B.V. *Iterative Evolution of Hypercarnivory in Canids (Mammalia: Carnivora): Evolutionary Interactions Among Sympatric Predators*; Cambridge University Press Stable: Cambridge, UK, 1991; Volume 17, pp. 340–362.
123. Chavez, D.E.; Gronau, I.; Hains, T.; Kliver, S.; Koepfli, K.P.; Wayne, R.K. Comparative Genomics Provides New Insights into the Remarkable Adaptations of the African Wild Dog (Lycaon Pictus). *Sci. Rep.* **2019**, *9*, 8329. [CrossRef] [PubMed]
124. Kretzoi, M. Die Raubtiere von Gombaszög Nebst Einer Übersicht Der Gesamtfauna. *Ann. Musei Natl. Hung. Pars Mineral. Geol. Palaeontol.* **1938**, *31*, 88–157.
125. Torre, D. I Cani Villafranchiani Della Toscana. *Palaeontogr. Ital.* **1967**, *53*, 113–138.

126. Sotnikova, M.V. Remains of Canidae from the Lower Pleistocene Site of Untermassfeld. In *The Pleistocene of Untermassfeld near Meiningen (Thuringia)*; Kahlke, R.-D., Ed.; Monographs of the RGZM. Propylaeum: Heidelberg, Germany, 2001; Volume 40, pp. 607–632.
127. Rook, L. The Plio-Pleistocene Old World *Canis* (*Xenocyon*) Ex Gr. *Falconeri. Boll. Soc. Paleontol. Ital.* **1994**, *33*, 71–82.
128. Martínez-Navarro, B.; Rook, L. Gradual Evolution in the African Hunting Dog Lineage Systematic Implications. *Comptes Rendus Palevol* **2003**, *2*, 695–702. [CrossRef]
129. Koizumi, A. First Record of the Plio-Pleistocene Hypercarnivorous Canid, Canis (Xenocyon) Falconeri from Tama River. *Quat. Res.* **2003**, *42*, 105–111. [CrossRef]
130. Werdelin, L.; Lewis, M.E. Plio-Pleistocene Carnivora of Eastern Africa: Species Richness and Turnover Patterns. *Zool. J. Linn. Soc.* **2005**, *144*, 121–144. [CrossRef]
131. Moullé, P.E.; Lacombat, F.; Echassoux, A. Apport Des Grands Mammifères de La Grotte Du Vallonnet (Roquebrune-Cap-Martin, Alpes-Maritimes, France) à La Connaissance Du Cadre Biochronologique de La Seconde Moitié Du Pléistocène Inférieur d'Europe. *Anthropologie* **2006**, *110*, 837–849. [CrossRef]
132. Tedford, R.H.; Wang, X.; Taylor, B.E. Phylogenetic Systematics of the North American Fossil Caninae (Carnivora: Canidae). *Bull. Am. Mus. Nat. Hist.* **2009**, *325*, 1–218. [CrossRef]
133. Hartstone-Rose, A.; Werdelin, L.; De Ruiter, D.J.; Berger, L.R.; Churchill, S.E. The Plio-Pleistocene Ancestor of Wild Dogs, Lycaon Sekowei n. Sp. *J. Paleontol.* **2010**, *84*, 299–308. [CrossRef]
134. Madurell-Malapeira, J.; Rook, L.; Martínez-Navarro, B.; Alba, D.M.; Aurell-Garrido, J.; Moyà-Solà, S. The Latest European Painted Dog. *J. Vertebr. Paleontol.* **2013**, *33*, 1244–1249. [CrossRef]
135. Wang, X.; Li, Q.; Xie, G. Earliest Record of Sinicuon in Zanda Basin, Southern Tibet and Implications for Hypercarnivores in Cold Environments. *Quat. Int.* **2015**, *355*, 3–10. [CrossRef]
136. Bartolini-Lucenti, S.; Madurell-Malapeira, J.; Palmqvist, P.; Rook, L. The First Hunting Dog from Dmanisi : Comments of Social Behaviour in Canidae and Hominins. *Sci. Rep.* **2021**, *11*, 13501. [CrossRef]
137. Jiangzuo, Q.; Wang, Y.; Song, Y.; Liu, S.; Jin, C.; Liu, J. Middle Pleistocene *Xenocyon Lycaonoides* Kretzoi, 1938 in Northeastern China and the Evolution of Xenocyon-Lycaon Lineage. *Hist. Biol.* **2022**, 1–13. [CrossRef]
138. Bartolini-Lucenti, S.; Spassov, N. Cave Canem! The Earliest *Canis* (*Xenocyon*) (Canidae, Mammalia) of Europe: Taxonomic Affinities and Paleoecology of the Fossil Wild Dogs. *Quat. Sci. Rev.* **2022**, *276*, 107315. [CrossRef]
139. Marciszak, A.; Gornig, W.; Szynkiewicz, A. Carnivores from Draby 3 (Central Poland): The Latest Record of *Lycaon lycaonoides* (Kretzoi, 1938) and the Final Accord in the Long History of Ancient Faunas. *Quat. Int.* **2023**, *674*, 62–86. [CrossRef]
140. Tong, H.; Hu, N.; Wang, X. New Remains of Canis Chihliensis (Mammalia, Carnivora) from Shanshenmiaozhou, a Lower Pleistocene Site in Yangyuan, Hebei. *Vertebr. PalAsiatica* **2012**, *50*, 335–360.
141. Rook, L.; Bartolini-Lucenti, S.; Cirilli, O.; Delfino, M.; Ferretti, M.P. Pando Vertebrate Records. In *Reference Module in Earth Systems and Environmental Sciences*; Elsevier Inc.: Amsterdam, The Netherlands, 2023; pp. 1–11.
142. Azzaroli, A. Quaternary Mammals and the "End-Villafranchian" Dispersal Event—A Turning Point in the History of Eurasia. *Palaeogeogr. Palaeoclimatol. Palaeoecol.* **1983**, *44*, 117–139. [CrossRef]
143. Azzaroli, A.; De Giuli, C.; Ficcarelli, G.; Torre, D. Late Pliocene to Early Mid-Pleistocene Mammals in Eurasia: Faunal Succession and Dispersal Events. *Palaeogeogr. Palaeoclimatol. Palaeoecol.* **1988**, *66*, 77–100. [CrossRef]
144. Lacombat, F.; Abbazzi, L.; Ferretti, M.P.; Martínez-Navarro, B.; Moullé, P.E.; Palombo, M.R.; Rook, L.; Turner, A.; Valli, A.M.F. New Data on the Early Villafranchian Fauna from Vialette (Haute-Loire, France) Based on the Collection of the Crozatier Museum (Le Puy-En-Velay, Haute-Loire, France). *Quat. Int.* **2008**, *179*, 64–71. [CrossRef]
145. Sotnikova, M.; Rook, L. Dispersal of the Canini (Mammalia, Canidae: Caninae) across Eurasia during the Late Miocene to Early Pleistocene. *Quat. Int.* **2010**, *212*, 86–97. [CrossRef]
146. Sardella, R.; Palombo, M.R. The Pliocene-Pleistocene Boundary: Which Significance for the so Called "Wolf Event"? Evidences from Western Europe. *Quaternaire* **2007**, *18*, 65–71. [CrossRef]
147. Martínez-Navarro, B. Early Pleistocene Faunas of Eurasia and Hominin Dispersals. In *Out of Africa I: The First Hominin Colonization of Eurasia*; Fleagle, J.G., Shea, J.J., Grine, F.E., Baden, A.L., Leakey, R.E., Eds.; Vertebrate Paleobiology and Paleoanthropology; Springer Science+Business Media: Dordrecht, The Netherlands, 2010; pp. 207–224. ISBN 978-90-481-9035-5.
148. Napoleone, G.; Albianelli, A.; Azzaroli, A.; Bertini, A.; Magi, M.; Mazzini, M. Calibration of the Upper Valdarno Basin to the Plio-Pleistocene for Correlating the Apennine Continental Sequences. *Alp. Mediterr. Quat.* **2003**, *16*, 131–166.
149. Fidolini, F.; Ghinassi, M.; Magi, M.; Papini, M.; Sagri, M. The Plio-Pleistocene Fluvio-Lacustrine Upper Valdarno Basin (Central Italy): Stratigraphy and Basin Fill Evolution. *Ital. J. Geosci.* **2013**, *132*, 13–32. [CrossRef]
150. Del Campana, D. Cani Pliocenici Della Toscana. *Palaeontogr. Ital.* **1913**, *19*, 189–254.
151. Rook, L.; Torre, D. The Latest Villafranchian and Early Galerian Small Dogs of the Mediterrenean Area. *Acta Zool. Cracoviensia* **1996**, *39*, 427–434.
152. Thenius, E. Die Caniden (Mammalia) Aus Dem Altquartär von Hundsheim (Niederösterreich) Nebst Bemer Kungen Zur Stammesgeschichte Der Gattung Cuon. *Neues Jahrb. Geol. Paläontologie Abh.* **1954**, *99*, 230–286.
153. Kurtén, B. A History of Coyote-like Dogs (Canidae, Mammalia). *Acta Zool. Fenn.* **1974**, *140*, 2197–2199. [CrossRef]
154. Rook, L. *I Cani Dell'Eurasia Dal Miocene Superiore al Pleistocene Medio. Ph.D. Thesis, Università consorziate: Modena, Italy; Bologna, Italy; Firenze, Italy;* Roma "La Sapienza"—V Ciclo: Firenze, Italy, 1993.

155. Koufos, G.D.; Kostopoulos, D.S. New Carnivore Material from the Plio-Pleistocene of Macedonia (Greece) with a Description of a New Canid. *Münchner Geowiss. Abhlungen* **1997**, *34*, 33–63.
156. Garrido, G.; Arribas, A. *Canis Accitanus* Nov. Sp., a New Small Dog (Canidae, Carnivora, Mammalia) from the Fonelas P-1 Plio-Pleistocene Site (Guadix Basin, Granada, Spain). *Geobios* **2008**, *41*, 751–761. [CrossRef]
157. Garrido, G. El Registro de *Vulpes Alopecoides* (Forsyth-Major, 1877), *Canis Etruscus* Forsyth-Major, 1877 y *Canis* Cf. Falconeri Forsyth-Major, 1877 (Canidae, Carnivora, Mammalia) En Fonelas P-1 (Cuenca de Guadix, Granada). *Vertebr. Plioceno Super. Termin. Suroeste Eur. Fonelas P 1 Y Proy. Fonelas.* **2008**, *10*, 159–186.
158. Duval, M.; Falguères, C.; Bahain, J.J.; Grün, R.; Shao, Q.; Aubert, M.; Hellstrom, J.; Dolo, J.M.; Agusti, J.; Martínez-Navarro, B.; et al. The Challenge of Dating Early Pleistocene Fossil Teeth by the Combined Uranium Series-Electron Spin Resonance Method: The Venta Micena Palaeontological Site (Orce, Spain). *J. Quat. Sci.* **2011**, *26*, 603–615. [CrossRef]
159. Duval, M.; Falguères, C.; Bahain, J.-J. Age of the Oldest Hominin Settlements in Spain: Contribution of the Combined U-Series/ESR Dating Method Applied to Fossil Teeth. *Quat. Geochronol.* **2012**, *10*, 412–417. [CrossRef]
160. Toro-Moyano, I.; Martínez-Navarro, B.; Agustí, J.; Souday, C.; Bermúdez de Castro, J.M.; Martinón-Torres, M.; Fajardo, B.; Duval, M.; Falguères, C.; Oms, O.; et al. The Oldest Human Fossil in Europe, from Orce (Spain). *J. Hum. Evol.* **2013**, *65*, 1–9. [CrossRef] [PubMed]
161. Ghezzo, E.; Berté, D.F.; Sala, B. The Revaluation of Galerian Canidae, Felidae and Mustelidae of the Cerè Cave (Verona, Northeastern Italy). *Quat. Int.* **2014**, *339–340*, 76–89. [CrossRef]
162. Torre, D. Affinità Dentali Del Cane Della Grotta Di "L'Escale. " *Riv. Ital. Paleontol. Stratigr.* **1974**, *80*, 147–156.
163. Torre, D. The Ruscinian and the Villafranchian Dogs of Europe. *Boll. Della Soc. Paleontol. Ital.* **1979**, *18*, 162–165.
164. Sotnikova, M.V.; Baigusheva, V.S.; Titov, V.V. Carnivores of the Khapry faunal assemblage and their stratigraphic implications. *Stratigr. Geol. Correl.* **2002**, *10*, 375–390.
165. Musil, R. Die Caniden Der Stránská Skála. *Anthropos* **1972**, *20*, 107–112.
166. Martin, R. Trois Nouvelles Especes de Caninae (Canidae, Carnivora) Des Gisements Plio-Villafranchiens d'Europe. *Doc. Lab. Géologie Fac. Sci. Lyon* **1973**, *57*, 87–96.
167. Sotnikova, M.V. *Carnivores Mammals of the Pliocene–Early Pleistocene: Stratigraphic Implications*; Nakua: Moscow, Russia, 1989.
168. Argant, A. Carnivores Quaternaires de Bourgogne. *Trav. Doc. Lab. Géologie Lyon* **1991**, *115*, 3–301.
169. Soergel, W. Die Säugetierfauna Des Altdiluvialen Tonlagers von Jockgrim in Der Pfalz. *Z. Dtsch. Geol. Ges.* **1925**, *77*, 405–438.
170. Bartolini-Lucenti, S.; Alba, D.M.; Rook, L.; Moyà-Solà, S.; Madurell-Malapeira, J. Latest Early Pleistocene Wolf-like Canids from the Iberian Peninsula. *Quat. Sci. Rev.* **2017**, *162*, 12–25. [CrossRef]
171. Koufos, G.D. The Villafranchian Carnivoran Guild of Greece: Implications for the Fauna, Biochronology and Paleoecology. *Integr. Zool.* **2014**, *9*, 444–460. [CrossRef]
172. Krijgsman, W.; Tesakov, A.; Yanina, T.; Lazarev, S.; Danukalova, G.; Van Baak, C.G.C.; Agustí, J.; Alçiçek, M.C.; Aliyeva, E.; Bista, D.; et al. Quaternary Time Scales for the Pontocaspian Domain: Interbasinal Connectivity and Faunal Evolution. *Earth Sci. Rev.* **2019**, *188*, 1–40. [CrossRef]
173. De Villalta-Comella, J.F. Contribución al Conocimiento de La Fauna de Mamíferos Fósiles Del Plioceno de Villaroya. *Bol. Inst. Geol. Y Min. España* **1952**, *64*, 1–204.
174. Pueyo, E.L.; Muñoz, A.; Laplana, C.; Parés, J.M. The Last Appearance Datum of Hipparion in Western Europe: Magnetostratigraphy along the Pliocene–Pleistocene Boundary in the Villarroya Basin (Northern Spain). *Int. J. Earth Sci.* **2016**, *105*, 2203–2220. [CrossRef]
175. García, N.; Arsuaga, J.L. Carnivores from the Early Pleistocene Hominid-Bearing Trinchera Dolina 6 (Sierra de Atapuerca, Spain). *J. Hum. Evol.* **1999**, *37*, 415–430. [CrossRef]
176. Moigne, A.M.; Palombo, M.R.; Belda, V.; Heriech-Briki, D.; Kacimi, S.; Lacombat, F.; de Lumley, M.A.; Moutoussamy, J.; Rivals, F.; Quilès, J.; et al. Les Faunes de Grands Mammifères de La Caune de l'Arago (Tautavel) Dans Le Cadre Biochronologique Des Faunes Du Pléistocène Moyen Italien. *Anthropologie* **2006**, *110*, 788–831. [CrossRef]
177. Rabeder, G. Die Carnivoren (Mammalia) Aus Dem Altpleistozän Mit Beiträgen Zur Systematik Einiger Musteliden Und Caniden. *Beiträge Paläontologie Von Osterr.* **1976**, *1*, 5–119.
178. Bonifay, M.-F. Carnivores Quaternaires Du Sud-Est de La France. *Mémoires Muséum Natl. D'Histoire Nat.* **1971**, *21*, 49–380.
179. Terzea, E. Biochronology of the Pleistocene Deposits at Betfia (Bihor, Romania). *Acta Zool. Cracoviensia* **1996**, *39*, 531–540.
180. Jánossy, D. Pleistocene Vertebrate Faunas of Hungary. In *Developments in Palaeontology and Stratigraphy*; Elsevier Science Publishers: Amsterdam, New York, NY, USA, 1986; ISBN 978-0-444-99526-1.
181. Kormos, N. Die Füchse Des Ungarischen Oberpliozan. *Folia Zool. Hydrobiol.* **1932**, *4*, 167–188.
182. Wang, X.; Tseng, Z.J.; Li, Q.; Takeuchi, G.T.; Xie, G. From "third Pole" to North Pole: A Himalayan Origin for the Arctic Fox. *Proc. R. Soc. Biol. Sci.* **2014**, *281*, 20140893. [CrossRef]
183. Jiangzuo, Q.; Liu, J.; Wagner, J.; Dong, W.; Chen, J. Taxonomical revision of fossil *Canis* in Middle Pleistocene sites of Zhoukoudian, Beijing, China and a review of fossil records of *Canis mosbachensis variabilis* in China. *Quat. Int.* **2018**, *482*, 93–108. [CrossRef]
184. Li, Y.; Zhang, Y.; Wu, X.; Ao, H.; An, Z.; An, Z. Mammalian Evolution in Asia Linked to Climate Changes. In *Late Cenozoic Climate Change in Asia*; Developments in Paleoenvironmental Research, Ed.; Springer Science+Business Media: Dordrecht, The Netherlands, 2014; pp. 435–490. ISBN 978-94-007-7816-0.
185. Geraads, D. A Revision of the Fossil Canidae (Mammalia) of North-Western Africa. *Palaeontology* **2011**, *54*, 429–446. [CrossRef]

186. Petronio, C.; Di Stefano, G.; Kotsakis, T.; Salari, L.; Marra, F.; Jicha, B.R. Biochronological framework for the late Galerian and early-middle Aurelian Mammal Ages of peninsular Italy. *Geobios* **2019**, *53*, 35–50. [CrossRef]
187. Capasso Barbato, L.; Minieri, M.R. Nuovi resti di carnivori del Pleistocene medio dei dintorni di Roma. *Geol. Romana* **1987**, *26*, 1–15.
188. Caloi, L.; Palombo, M.R. Anfibi, rettili e mammiferi di Torre del Pagliaccetto (Torre in Pietra, Roma) in Torre in Pietra, Roma. *Quaternaria* **1978**, *20*, 315–428.
189. Madurell-Malapeira, J.; Lucenti, S.B.; Alba, D.M. Middle Pleistocene fox from the Vallparadís Section (Vallès-Penedès Basin, NE Iberian Peninsula) and the earliest records of the extant red fox. *Riv. Ital. Paleontol. Stratigr.* **2021**, *127*, 179–187.
190. Vizcaíno-Varo, V. Paleoecology and Dietary Habits of Cave Bear Lineage through Stable Isotopes Analyses ($\delta^{13}C$ and $\delta^{18}O$) and Dental Microwear on Ursus Deningeri from Vallparadís Section (Terrassa, Vallès-Penedès Basin, NE Iberian Peninsula). Unpublished. Master's Thesis, Universitat Rovira I Virgili, Tarragona, Spain, 2023.
191. Madurell-Malapeira, J.; Santos-Cubedo, A.; Marmi, J. Oldest European Occurence of *Meles* (Mustelidae, Carnivora) from the Middle Pliocene (MN16) of Almenara-Casablanca-4 Karstic Site (Castelló, Spain). *J. Vertebr. Paleontol.* **2009**, *29*, 961–965. [CrossRef]
192. Jiangzuo, Q.G.; Liu, J.Y.; Wagner, J.; Chen, J. Taxonomical Revision of "*Arctonyx*" Fossil Remains from the Liucheng Gigantopithecus Cave (South China) by Means of Morphotype and Morphometrics, and a Review of Late Pliocene and Early Pleistocene *Meles* Fossil Records in China. *Palaeoworld* **2018**, *27*, 282–300. [CrossRef]
193. Faggi, A.; Bartolini-Lucenti, S.; Madurell-Malapeira, J.; Abramov, A.V.; Puzachenko, A.Y.; Jiangzuo, Q.; Peiran, L.; Rook, L. Quaternary Eurasian Badgers: Intraspecific Variability and Species Validity. *J. Mamm. Evol.* **2024**, *31*, 3. [CrossRef]
194. Marciszak, A.; Ambros, D.; Hilpert, B. Mustelids from Sackdilling Cave (Bavaria, Germany) and Their Biostratigraphic Significance. *Geobios* **2021**, *68*, 83–107. [CrossRef]
195. Ros-Montoya, S.; Palombo, M.R.; Espigares, M.P.; Palmqvist, P.; Martinez-Navarro, B. The mammoth from the archaeo-paleontological site of Huéscar-1: A tile in the puzzling question of the replacement of Mammuthus meridionalis by Mammuthus trogontherii in the late Early Pleistocene of Europe. *Quat. Sci. Rev.* **2018**, *197*, 336–351. [CrossRef]
196. Walker, M.J.; Uriarte, M.H.; Jimenez, A.L.; Martínez, M.L.; Lerma, I.M.; van der Made, J.; Duval, M.; Grün, R. Cueva Negra del Estrecho del Río Quípar: A dated late Early Pleistocene Palaeolithic site in southeastern Spain. *J. Paleolit. Archaeol.* **2020**, *3*, 816–855. [CrossRef]
197. Huguet, R.; Vallverdú, J.; Rodríguez-Álvarez, X.P.; Terradillos-Bernal, M.; Bargalló, A.; Lombera-Hermida, A.; Menéndez, L.; Modesto-Mata, M.; Van der Made, J.; Soto, M.; et al. Level TE9c of Sima del Elefante (sierra de Atapuerca, Spain): A comprehensive approach. *Quat. Int.* **2017**, *433*, 278–295. [CrossRef]
198. Palombo, M.R.; Sardella, R.; Novelli, M. Carnivora dispersal in Western Mediterranean during the last 2.6 Ma. *Quat. Int.* **2008**, *179*, 176–189. [CrossRef]
199. Bona, F.; Sala, B. Villafranchian-Galerian mammal faunas transition in South-western Europe. The case of the late early Pleistocene mammal fauna of the Frantoio locality, Arda River (Castell'Arquato, Piacenza, northern Italy). *Geobios* **2016**, *49*, 329–347. [CrossRef]
200. Brugal, J.P.; Argant, A.; Boudadi-Maligne, M.; Crégut-Bonnoure, E.; Croitor, R.; Fernandez, P.; Fourvel, J.B.; Fosse, P.; Guadelli, J.L.; Labe, B.; et al. Pleistocene herbivores and carnivores from France: An updated overview of the literature, sites and taxonomy. *Ann. Paléontologie* **2020**, *106*, 102384. [CrossRef]
201. Kahlke, R.D. The early Pleistocene (Epivillafranchian) faunal site of Untermassfeld (Thuringia, central Germany) synthesis of new results. *Eraul* **2000**, *92*, 123–138.
202. Parfitt, S.A.; Ashton, N.M.; Lewis, S.G.; Abel, R.L.; Coope, G.R.; Field, M.H.; Gale, R.; Hoare, P.G.; Larkin, N.R.; Lewis, M.D.; et al. Early Pleistocene human occupation at the edge of the boreal zone in northwest Europe. *Nature* **2010**, *466*, 229–233. [CrossRef] [PubMed]
203. Vislobokova, I.A.; Agadjanian, A.K. On the history of Early—Middle Pleistocene mammal faunas of the Central Balkans. *Paleontol. J.* **2016**, *50*, 187–201. [CrossRef]
204. Gasparik, M.; Pazonyi, P. The macromammal remains and revised faunal list of the Somssich Hill 2 locality (late Early Pleistocene, Hungary) and the Epivillafranchian faunal change. *Fragm. Palaeontol. Hung.* **2018**, *35*, 153–178. [CrossRef]
205. Tappen, M.; Adler, D.S.; Ferring, C.R.; Gabunia, M.; Vekua, A.; Swisher III, C.C. Akhalkalaki: The taphonomy of an Early Pleistocene locality in the Republic of Georgia. *J. Archaeol. Sci.* **2002**, *29*, 1367–1391. [CrossRef]

quaternary

MDPI

Article

Carrying Capacity, Available Meat and the Fossil Record of the Orce Sites (Baza Basin, Spain)

Guillermo Rodríguez-Gómez [1,2,*], M. Patrocinio Espigares [3], Bienvenido Martínez-Navarro [4,5,6], Sergio Ros-Montoya [3], Antonio Guerra-Merchán [3], Jesús A. Martín-González [7], Isidoro Campaña [3], Alejandro Pérez-Ramos [3], Alejandro Granados [3], José Manuel García-Aguilar [3], María Dolores Rodríguez-Ruiz [8] and Paul Palmqvist [3]

[1] Departamento de Geodinámica, Estratigrafía y Paleontología, Universidad Complutense de Madrid, C/José Antonio Novais 12, 28040 Madrid, Spain
[2] Centro UCM-ISCIII de Evolución y Comportamiento Humanos, Avd/Monforte de Lemos, 5, Pabellón 14, 28029 Madrid, Spain
[3] Departamento de Ecología y Geología, Universidad de Málaga, Campus Universitario de Teatinos, 29071 Málaga, Spain; mpespigares@uma.es (M.P.E.); sergiorosm@uma.es (S.R.-M.); antguerra@uma.es (A.G.-M.); i.campana.lozano@uma.es (I.C.); alejoperezramos5@gmail.com (A.P.-R.); a.granados@uma.es (A.G.); chemacyanos@msn.com (J.M.G.-A.); ppb@uma.es (P.P.)
[4] IPHES-CERCA, Institut Català de Paleoecologia Humana i Evolució Social, C/Marcel.lí Domingo s/n, Campus Sescelades, Edifici W3, 43007 Tarragona, Spain; bienvenido.martinez@icrea.cat
[5] Area de Prehistoria, Universitat Rovira i Virgili (URV), Avda. Catalunya 35, 43002 Tarragona, Spain
[6] ICREA, Pg. Lluís Companys 23, 08010 Barcelona, Spain
[7] Departamento de Matemáticas y Computación, Universidad de Burgos, Plaza Misael Bañuelos s/n, 09001 Burgos, Spain; jamartin@ubu.es
[8] Departamento de Química Inorgánica, Cristalografía y Mineralogía, Universidad de Málaga, Campus Universitario de Teatinos, 29071 Málaga, Spain; mdrodriguez@uma.es
* Correspondence: grodriguezgomez@ucm.es

Citation: Rodríguez-Gómez, G.; Espigares, M.P.; Martínez-Navarro, B.; Ros-Montoya, S.; Guerra-Merchán, A.; Martín-González, J.A.; Campaña, I.; Pérez-Ramos, A.; Granados, A.; García-Aguilar, J.M.; et al. Carrying Capacity, Available Meat and the Fossil Record of the Orce Sites (Baza Basin, Spain). *Quaternary* **2024**, *7*, 37. https://doi.org/10.3390/quat7030037

Academic Editors: Raffaele Sardella, Alessio Iannucci, George Konidaris, Joan Madurell Malapeira and Dimitris S. Kostopoulos

Received: 31 May 2024
Revised: 17 August 2024
Accepted: 23 August 2024
Published: 27 August 2024

Abstract: The Early Pleistocene sites of Orce in southeastern Spain, including Fuente Nueva-3 (FN3), Barranco León (BL) and Venta Micena (VM), provide important insights into the earliest hominin populations and Late Villafranchian large mammal communities. Dated to approximately 1.4 million years ago, FN3 and BL preserve abundant Oldowan tools, cut marks and a human primary tooth, indicating hominin activity. VM, approximately 1.6 million years old, is an outstanding site because it preserves an exceptionally rich assemblage of large mammals and predates the presence of hominins, providing a context for pre-human conditions in the region. Research suggests that both hominins and giant hyenas were essential to the accumulation of skeletal remains at FN3 and BL, with secondary access to meat resources exploited by saber-toothed felids. This aim of this study aims to correlate the relative abundance of large herbivores at these sites with their estimates of Carrying Capacity (CC) and Total Available Biomass (TAB) using the PSEco model, which incorporates survival and mortality profiles to estimate these parameters in paleoecosystems. Our results show: (i) similarities between quarries VM3 and VM4 and (ii) similarities of these quarries with BL-D (level D), suggesting a similar formation process; (iii) that the role of humans would be secondary in BL-D and FN3-LAL (Lower Archaeological Level), although with a greater human influence in FN3-LAL due to the greater presence of horses and small species; and (iv) that FN3-UAL (Upper Archaeological Level) shows similarities with the expected CC values for FN3/BL, consistent with a natural trap of quicksand scenario, where the large mammal species were trapped according to their abundance and body mass, as there is a greater presence of rhinos and mammoths due to the greater weight per unit area exerted by their legs. Given the usefulness of this approach, we propose to apply it first to sites that have been proposed to function as natural traps.

Keywords: prey biomass; large mammals; taphonomy; Early Pleistocene; Western Europe; Venta Micena; Fuente Nueva 3; Barranco León

1. Introduction

Reconstructing the ecological conditions under which past species evolved is of great interest, because it allows us to interpret how different aspects affected their evolution. Resource availability is among the parameters that most influence the distribution and survival of species. Animal resources played a key role in the survival of the human lineage since at least 2.5 million years ago (Ma) [1] and were particularly critical for Pleistocene hominins in Europe [2–4], although for some human species it has been suggested that they also consumed high amounts of plant resources [5–11]. Even today, most hunter-gatherer societies rely heavily on animal resources [12–14]. Many researchers have argued that large mammals were the primary source of meat and fat for hominins in the Pleistocene [3,11,15–22]. In this way, the study of the ecological framework in which hominin groups evolved can shed light on their subsistence strategies and patterns of resource exploitation, which are key to the study of human evolution [23–25].

In order to assess the ecological conditions of past large mammal communities, Carrying Capacity and meat availability can be used as indices useful for characterizing ecosystems and paleoecosystems (e.g., [26–32], among others). The term Carrying Capacity is widely used in the life sciences and is applied at a variety of levels of analysis, from molecular to ecological studies (see [33]). In present-day ecosystems, Carrying Capacity has been used to assess the status of large mammal populations and to guide management and conservation efforts [27–29]. Typically, these studies use the third meaning of Carrying Capacity as defined by Sayre [33], where K is the intrinsic population growth limit after long-term environmental stability. Applied to paleosynecology, it is useful for establishing predator–prey relationships derived from the obtaining of accurate biomass estimates [32,34–37]. In this study, as proposed by Coe et al. [26], we will consider Carrying Capacity (CC) as synonymous with the prey biomass species that can be sustained over time in an ecosystem. The concept of CC is sometimes used as a synonym for the availability of meat resources in ecosystems (e.g., [38,39]). However, CC (i.e., the total biomass of an ecosystem) cannot be used by the secondary consumers without leading the ecosystem to collapse. For this reason, we support the use of the term 'available meat' to refer to the biomass fraction of the ecosystem that can be sustainably used by the secondary consumers (see [30,37,40–45]). Therefore, we use the concept of available meat as the primary consumer biomass potentially available to the secondary consumers, which was termed Total Available Biomass (TAB) by Rodríguez-Gómez et al. [40].

In present-day ecosystems, CC estimates for large mammals are calculated from observed densities and average population weights, sometimes using three-quarters of adult female weight [27–29] and in other cases using adult female weight directly [31]. These approaches are not easy to apply in the case of paleoecosystems (see [37]) and alternatives have been sought, sometimes using average adult body masses (e.g., [35,46–52]; but see [37]) or estimating ecosystem Carrying Capacity from climatic parameters (e.g., [32,36,38,39]). Survival profiles are another alternative used to infer average population body mass [37], which allows for obtaining values similar to the Hatton et al. [31] approach using mean adult female body mass for current ecosystems. These survival profiles provide information on the proportion of individuals in a population who survive by age group and thus make it possible to obtain mortality profiles (i.e., the proportion of individuals who die by age group in the population). From these mortality profiles, if conditions of stability and stationarity are met (see [53,54]), it is then possible to estimate the long-term meat resources available in a paleoecosystem [30,37,40–45,55,56]. Leslie-Lewis matrices or the Weibull model are both useful tools for obtaining survival and mortality profiles, thus allowing for the estimation of age structures for fossil populations and modeling their dynamics [37,54,55,57]. We developed the PSEco model from estimates of survival and mortality profiles of herbivore species, which makes it possible to estimate ecosystem CC, TAB and secondary consumer biomass [45]. Results from this model on predator–prey biomass relationships resemble those found in current African ecosystems [45], supporting the use of this methodology.

Located about 150 km northeast of the city of Granada (SE Spain), in the eastern sector of the Guadix-Baza Depression and within the Baza Basin, the Orce sites are of great interest for the study of human evolution in Europe, since they provide information on the ecological conditions encountered by the first human settlers in Western Europe, as well as on the period preceding the first hominin dispersal in this area. The archaeological sites of Fuente Nueva-3 (FN3) and Barranco León (BL) preserve skeletal remains of 18 large mammalian species [58–61] (13 large prey, Table 1), as well as anthropic markings associated with Oldowan lithics (i.e., Mode 1) [22,62–68]. Using a combined approach based on biostratigraphy, magnetostratigraphy and electron spin resonance (ESR), the ages of BL and FN3 were estimated in 1.43 ± 0.38 Ma and 1.19 ± 0.21 Ma, respectively [66,69]. Additionally, cosmogenic nuclides yielded an age estimate of 1.50 ± 0.31 Ma for FN3 [70]. Currently, the oldest human fossil in Western Europe is a deciduous molar tooth of BL dated at ~1.4 Ma [66]. A marshy environment is associated with the excavated levels of BL, with the exception of level D, which has fluvial features and contains the majority of the archaeological assemblage [66]. The D level is subdivided into two layers, D1 and D2 and the time of its deposition was very short, which excludes it from being a palimpsest [66,71]. Two archaeological levels are grouped for the six layers of the fertile strata of FN3: a Lower Archaeological Level (LAL) (layers 1–3) and an Upper Archaeological Level (layers 4–6) [22,64,72,73], which do not differ in the composition of their faunal assemblages (see Table 1), except for the abundance of megaherbivores, and probably represent two very close temporal events. Further from Orce is also a key paleontological site for studying Early Pleistocene large mammal communities, Venta Micena (VM), which is slightly older biostratigraphically (1.6–1.5 Ma) than FN3 and BL and has no evidence of human presence (i.e., there are no lithic tools, anthropic marks on bones or human fossil remains from this site). However, it presents a huge amount of fossil remains from an excavation area of ~400 m^2 in several quarries of Venta Micena (e.g., VM2, VM3 and VM4), with more than 24,000 skeletal remains showing excellent preservation and low levels of weathering in more than 90% of the remains, which has allowed the identification of 21 species (14 large prey, Table 1). The role of scavenging hyenas was essential in the accumulation of these skeletal remains in the vicinity of their denning sites during a period of several years [74–76]. In contrast, Granados et al. [77] proposed a different interpretation of the Venta Micena paleontological layer, based on isotopic and geochemical studies. Their findings suggested that this layer was formed by a series of episodes of upward and downward expansion of a shallow lake that originated in the vicinity of the deposits. However, in a small sample excavated in quarry VM4, ~350 m away from VM3, Luzón et al. [78] proposed that a different carnivore than in VM3 participated with a key role in bone accumulation, arguing based on taphonomic differences between the two quarries (e.g., a lower weathering rate in VM4, more abundant articulated remains than in VM3 and less intensity of fractures and gnawing marks by the hyenas). In a later paper, however, Palmqvist et al. [79] demonstrated that the differences between these two quarries were not due to the intervention of different carnivores as bone accumulating agencies, but to differences in the time elapsed until the skeletal remains were covered by limestone sediments (this period was shorter in VM4, which explains the better preservation of the remains in this quarry compared to VM3).

Table 1. Minimum number of individuals (MNI) data from [22,68,78–80] of large herbivores identified in the faunal assemblages of the Orce sites (Granada, Spain) analyzed in this study, together with the relative percentage of abundance for each species. Abbreviations: VM: Venta Micena; BL: Barranco León; FN3: Fuente Nueva-3; UAL: Upper Archeological Level of FN3; LAL: Lower Archeological Level of FN3. * Remains showing gnaw marks made by a porcupine. [a] Palmqvist et al. [79]; [b] Espigares [80]; [c] Luzón et al. [78]; [d] Espigares et al. [22]; [e] Yravedra et al. [68].

Herbivore Species	VM3 [a,b]	VM4 [a,c]	FN3-UAL [d,e]	FN3-LAL [d,e]	BL-D [d]
Ammotragus europaeus					1 (2.1%)
Bison sp.	51 (17.3%)	3 (7.1%)	8 (11.9%)	3 (8.1%)	4 (8.3%)

Table 1. *Cont.*

Herbivore Species	VM3 [a,b]	VM4 [a,c]	FN3-UAL [d,e]	FN3-LAL [d,e]	BL-D [d]
Bovidae indet. (cf. *Rupicapra*)	1 (0.3%)				
Hemibos cf. *gracilis*	1 (0.3%)	1 (2.4%)	1 (1.5%)	1 (2.7%)	
Hemitragus albus	16 (5.5%)	3 (7.1%)	4 (6.0%)	4 (10.8%)	3 (6.3%)
Praeovibos sp.	1 (0.3%)				
Soergelia minor	20 (6.9%)	1 (2.4%)			
Cervidae indet. (cf. *Capreolus*)	1 (0.3%)				
Metacervocerus rhenanus	33 (11.4%)	6 (14.3%)	4 (6.0%)	5 (11.6%)	8 (16.7%)
Praemegaceros cf. *verticornis*	56 (19.3%)	7 (16.7%)	5 (7.5%)	2 (4.7%)	5 (10.4%)
Hippopotamus antiquus	5 (1.7%)	1 (2.4%)	7 (10.4%)	1 (2.3%)	5 (10.4%)
Equus altidens	91 (31.4%)	10 (23.8%)	5 (7.5%)	14 (32.6%)	3 (6.3%)
Equus sp.		2 (4.8%)	9 (13.4%)	5 (11.6%)	13 (27.1%)
Equus suessenbornensis			2 (3.0%)	2 (4.7%)	2 (4.2%)
Stephanorhinus hundsheimensis	7 (2.4%)	5 (11.9%)	8 (11.9%)	2 (4.7%)	2 (4.2%)
Mammuthus meridionalis	5 (1.7%)	2 (4.8%)	13 (19.4%)	3 (7.0%)	1 (2.1%)
Hystrix sp.	2 (0.7%)	1 * (2.4%)	1 * (1.5%)	1 * (2.3%)	1 (2.1%)
Total	290	42	67	43	48

The Orce sites have provided relevant information on the late Early Pleistocene large mammals that inhabited southern Europe, on their paleocommunities and also on the human groups that inhabited the region (e.g., [34,81–88]). Their richness and uniqueness have allowed us to model and compare the paleoecological conditions that preceded and followed the first hominin arrival in Western Europe [41,42]. Thus, these sites offer unique opportunities for paleoecological analyses. We have conducted several paleoecological studies on the Orce sites, analyzing both CC and TAB [25,41,42,45,65,79,87]. However, we have not yet addressed the comparison of CC and TAB values between VM, BL and FN3. PSEco, with its estimates of survival and mortality profiles, provides optimal scenarios for the large herbivore species, with conditions of stability and stationarity, averaging the temporal fluctuations of their populations [30,54]. Thus, our PSEco proposal can be considered as an ideal situation and thus be used to analyze how the faunal assemblages deviate from the situation represented by this model and why, as we did in a preliminary analysis with the BL and FN3 records [65]. Assuming that the estimates of CC and TAB obtained from PSEco can be considered under optimal ecological conditions, our main objective in this study is to analyze whether the records from these sites can be correlated with the representation that herbivorous mammals had in the paleocommunity and/or with the meat that carnivores and humans could obtain from them and to compare the assemblages of VM with those of BL and FN3. For doing so, we estimate the CC and TAB of the large herbivore community of VM, BL and FN3 and compare them with the relative abundances of species at these sites.

2. Materials and Methods

Based on the materials and methods used in previous analyses, this study aims to reconstruct the paleocommunities of large herbivorous mammals preserved at the Orce sites, focusing on Venta Micena (VM), Barranco León (BL) and Fuente Nueva-3 (FN3) (Table 1), specifically in quarries 3 and 4 of Venta Micena (VM3 and VM4, respectively), the Upper and Lower Archaeological Levels of FN3 (FN3-UAL and FN3-LAL, respectively) and level D of Barranco León (BL-D). These faunal assemblages contain at least 11 species of large herbivores (>10 kg), meeting the criterion of recording at least 8 prey species proposed by Rodríguez-Gómez et al. [43] to identify those Pleistocene faunal assemblages with high conservation completeness. In the analysis of CC and TAB, we used the same species as those of the previous analysis [37,45] and considered FN3 and BL together because they had the same faunal assemblages [22,37,41,45]. In addition to the information derived from the faunal lists, we used Rodríguez-Gómez et al. [45] estimates of prey life history trait values on the basis of their modern analogues (Table 2). We reconstructed their survival and mortality profiles using these values. We used survival profiles to estimate the average body mass of herbivore populations and CC (see [37]). Mortality profiles allowed for the estimation of TAB for the secondary consumers in the paleoecosystems (see [30,40,41,45,55]). The life history traits included are adult and neonate body mass (ABM and NBM, respectively), age at first birth (or age of sexual maturity + pregnancy) (AFB), number of litters per year (LY), litter size (LS) and longevity (L). Estimates of mean adult body mass (ABM) were based on metric measurements from fossil bones and teeth [37] (Table 2).

Table 2. Herbivore life history traits derived from comparison with living analogues data taken from [45]. Abbreviations: ABM (adult body mass, in kg); AFB (age at first birth, in years); LS (litter size); LY (litters per year); NBM (neonate body masses, in kg); L (longevity, in years); D (density, in ind./km^2). Densities were estimated using Damuth's [89] equation (see text).

Order	Family	Species	ABM	AFB	LS	LY	NBM	L	D
Artiodactyla	Bovidae	*Ammotragus europaeus*	135	2.00	1.19	1.00	4.95	19.2	2.38
Artiodactyla	Bovidae	*Bison* sp.	450	2.62	1.00	0.91	25.79	25.0	0.92
Artiodactyla	Bovidae	Bovidae indet. (cf. *Rupicapra*)	25	2.00	1.19	1.00	1.89	19.2	9.03
Artiodactyla	Bovidae	*Hemibos* cf. *gracilis*	300	2.50	1.00	0.96	14.50	22.4	1.27
Artiodactyla	Bovidae	*Hemitragus albus*	75	2.00	1.19	1.00	3.28	19.2	3.79
Artiodactyla	Bovidae	*Praeovibos* sp.	315	2.00	1.19	1.00	9.96	19.2	1.22
Artiodactyla	Bovidae	*Soergelia minor*	225	2.00	1.19	1.00	7.45	19.2	1.59
Artiodactyla	Cervidae	Cervidae indet. (cf. *Capreolus*)	25	2.86	1.00	1.10	1.82	20.8	9.03
Artiodactyla	Cervidae	*Metacervocerus rhenanus*	95	2.86	1.00	1.10	5.57	20.8	3.14
Artiodactyla	Cervidae	*Praemegaceros* cf. *verticornis*	400	2.86	1.00	1.10	18.63	20.8	1.01
Artiodactyla	Hippopotamidae	*Hippopotamus antiquus*	3200	4.00	1.00	0.52	40.20	54.5	0.20
Perissodactyla	Equidae	*Equus altidens*	350	3.50	1.00	0.67	30.70	38.8	1.12
Perissodactyla	Equidae	*Equus suessenbornensis*	565	3.50	1.00	0.67	54.69	38.8	0.77
Perissodactyla	Rhinocerotidae	*Stephanorhinus hundsheimensis*	1000	6.75	1.00	0.36	41.75	47.0	0.49
Proboscidea	Elephantidae	*Mammuthus meridionalis*	6000	11.25	1.13	0.24	101.00	65.0	0.12
Rodentia	Hystricidae	*Hystrix refossa*	20	1.46	1.51	1.51	0.31	20.0	10.77
Rodentia	Hystricidae	*Hystrix* sp.	15	1.46	1.51	1.51	0.31	20.0	13.52

In order to estimate CC or prey biomass of paleocommunities, in this study, we followed the approach of Rodríguez-Gómez et al. [37]. In the case of estimating the amount of TAB provided by the prey community, we applied the PSEco model [45]. Both the estimation of CC by means of survival profiles and the estimation of TAB by means of mortality profiles require values of the average body mass of individuals at different ages and population densities. These approaches take into account the proportion of subadults, an aspect that is relevant for CC estimates (see [37]) or for analyses aimed at estimating the exploitation of meat resources by large predators, since body size is one of the most fundamental parameters in prey selection [74,90–92]. For the estimation of body mass of prey species at distinct age intervals, we followed the proposal of Zullinger et al. [93]:

$$M(t) = ABM * e^{-e^{-K(t-I)}}, \quad (1)$$

where ABM is the asymptotic body mass (that is, the adult body mass in g), M (t) is the mass at age t, K is the growth rate constant (days^{-1}) and I is the age at turning point (days). K refers to the adult body mass according to the equation:

$$log(K) = -0.901 - 0.302 * log(ABM) \quad (2)$$

Mean mass values for each age interval were estimated as the arithmetic mean of the two most extreme values within each age interval. To estimate density values, we used the equation developed by Damuth [89] for European mixed temperate forests:

$$log(D) = -0.79 * log(ABM) + 4.33; r^2 = 0.94 \quad (3)$$

where D is the density of the population (ind./km^2) and ABM is expressed in g.

To estimate CC for the Orce paleocommunities, we used the approach of Rodríguez-Gómez et al. [37], which takes into account the relative proportion of individuals (l_i) and body mass (M_i) in each age interval:

$$B = \sum_{i=1}^{n} l_i * M_i * D,\ i = 1, \ldots, n. \quad (4)$$

l_i was calculated from survival profiles and age structures obtained from the Weibull model [45,53,54]:

$$l_i = \frac{X_i}{\sum_{i=1}^{n} X_i},\ i = 1, \ldots n, \quad (5)$$

with X_i being the number of individuals in each age interval. The biomass that each age interval contributes to the total biomass of the population is derived from the proportion of individuals and body mass per age interval and population density. By adding the biomass of all species, the paleoecosystem CC is calculated.

For TAB, we used PSEco (see [45]), which estimates the number of individuals that could die annually without causing their populations to collapse using mortality profiles of large herbivores of a community (see [30,53,54]). PSEco uses the faunal lists (Table 1), the values of life history traits (Table 2) and the prey species densities in the paleoecosystems to estimate the amount of prey biomass that can be extracted from a paleoecosystem on an annual basis. We used an equation similar to equation 4, substituting l_i for the proportion of deaths between ages (d_i) ($d_x = l_x - l_{x+1}$):

$$Output\ Biomass = \sum_{i=1}^{n} d_i * M_i * D,\ i = 1, \ldots, n. \quad (6)$$

The total output biomass (TBO) is the sum of the biomass output of all species in the community results, as presented by Rodríguez-Gómez et al. [40]. From the TBO, PSEco employs a "wastage factor" that estimates the percentage of biomass that is not used by the secondary consumers (e.g., skin, horns and bones) (see [30,40,44,53,54]). The result is the biomass available for the secondary consumers (in kg/km^2*year^{-1} and kcal/km^2*year^{-1}), which is called TAB (Total Available Biomass) in the PSEco model [45]. Since the Weibull model provides many mortality profiles, PSEco selects only the extreme values corresponding to maximum and minimum subadult mortalities [53,54], where TAB-min is the available biomass corresponding to maximum subadult mortality (and minimum adult mortality) and TAB-MAX refers to the reverse situation.

For both CC and TAB, we divided biomasses into six classes: Class 1: 10–45 kg; Class 2: 45–90 kg; Class 3: 90–180 kg; Class 4: 180–360 kg; Class 5: 360–1000 kg; Class 6: >1000 kg. Thus, we can consider the importance of body size as the most significant parameter in prey selection [74,90–92] as well as the contribution of each population to the different body size classes, which in turn allows to compare values and patterns between CC and TAB.

Based on the results of CC and TAB, we estimated the relative percentage of each family in each index. We grouped the species into families to facilitate and simplify the analyses. For the fossil record, we used the minimal number of individuals (MNI) values to estimate relative percentages (Table 1). We are aware of the tendency of MNIs to underestimate common species and overestimate rare ones when considering the MNI (see [75]). However, this allows us to consider the presence of some species that are represented by scarce fossil or ichnofossil remains in the record. The comparison between the relative proportions of families in the CC, TAB and fossil record was performed using the χ^2 test.

3. Results

Tables 3 and 4 show the values obtained for the Carrying Capacity (CC) and available meat (TAB) provided by the different species of primary consumers of the faunal assemblages of Venta Micena (VM), Barranco León (BL) and Fuente Nueva-3 (FN3). These values are averages of those obtained from the survival and mortality profiles. The CC of Venta Micena was 3813 kg/km^2*year and that of Fuente Nueva-3 and Barranco León was 3535 kg/km^2*year, while the TAB values were 468 and 387 kg/km^2*year, respectively. With these values, CC was 8 and 9 times higher than TAB in these sites, respectively, and the relative percentage of TAB to CC were 12.28% and 10.96%, respectively. These percentages can be interpreted as the annual turnover rates of these faunal sets that could be consumed by the secondary consumers without causing ecosystem degradation. For both VM (Table 3) and FN3/BL (Table 4), the species that contributed the most biomass to the CC was *Mammuthus meridionalis* (534 kg/km^2*year in both cases), while the species that contributed the least biomass to the ecosystems was the porcupine (*Hystrix*) (135 and 138 kg/km^2*year, respectively). The opposite situation occurred with the values of TAB, where the species that provided the most meat to the environment was the porcupine (42 kg/km^2*year) and the one providing the least was the mammoth (18 kg/km^2*year). This is because species with low reproductive rates, such as the megaherbivores, cannot tolerate high rates of mortality, an aspect that species with high reproductive rates, such as the porcupine, can tolerate (see Table 2) [45]. As can be seen in Tables 3 and 4, most species contribute biomass to three size classes for both CC and TAB, with the exception of porcupines, bovidae indet. (cf. *Rupicapra*) and Cervidae indet. (cf. *Capreolus*) from VM, which only contribute to the first class (10–45 kg), and goats *Hemitragus albus* and horses *Equus altidens*, which both contribute to two classes.

Figure 1 shows the estimated CC and TAB values for VM, FN3 and BL distributed across the six size classes provided by PSEco [45]. As mentioned above, we grouped the species values into families to simplify and facilitate comparisons. Differences in biomass distribution patterns can be observed between VM and FN3/BL for both CC and TAB. In terms of CC, two blocks can be distinguished for both VM and FN3/BL, formed by the first three size classes and the last three ones, with the second block standing out in the two assemblages with the highest biomass (Figure 1A,B). The largest differences between the two assemblages are found in class 1 (10–45 kg) and class 4 (180–360 kg), with class 1 in VM contributing more biomass than classes 2 (45–90 kg) and 3 (90–180 kg) and 4 more than 5 (360–1000 kg) and 6 (>1000 kg) (Figure 1A). This is not the case for FN3/BL (Figure 1B). In the case of TAB, classes 1 and 4 are also prominent in VM (Figure 1C). However, the distribution in FN3/BL was more homogeneous, except for class 6, which had a lower value (Figure 1D). Comparing the figures of CC with those of TAB, it is possible to observe the differences in the biomass contribution of megaherbivores to one or the other index, playing a major role in CC and a very secondary one in TAB (Figure 1). The families Bovidae and Cervidae contribute to all size classes except the one in excess of 1000 kg (class 6), where the main contribution is made by the members of the families Hippopotamidae and Elephantidae (Figure 1). Equids in VM are only represented by *E. altidens* and contribute biomass to two size classes. In contrast, two species (*E. altidens* and *E. suessenbornensis*) are present in FN3/BL and contribute to four size classes.

Figures 2 and 3 show the proportions of the different families according to the estimates of CC, TAB and minimum number of individuals (MNI) obtained for each site (Table 1). In VM, bovids and cervids stand out in the estimates of CC, TAB and in the relative abundance of the VM3 and VM4 quarries. In these quarries, equid remains are also relevant, which is not the case for CC and TAB. The χ^2 tests show significant differences between the CC and TAB values with relative abundance in the VM quarries, but no significant differences between the two quarries (χ^2 = 11.709; p =0.068795), supporting the suggestions of Palmqvist et al. [79] that these assemblages have a similar composition. In terms of CC and TAB values, VM3 and VM4 show higher proportions of cervids and equids and lower representations of bovids and porcupines. VM4 shows proportions of rhino *Stephanorhinus hundsheimensis* close to CC and mammoths close to TAB values. Although the χ^2 analysis shows no differences between the proportions of the different families in the VM3 and VM4 assemblages, it can be seen that VM3 has lower proportions of mammoth, rhinoceros and porcupine and higher proportions of bovids than VM4 (Figure 2C,D). If we distribute the proportion of the different species in three body mass categories (small: 10–90 kg; medium-large: 90–600 kg; megaherbivores: >600 kg) (Table 5), we can see that in both VM3 and VM4 there is a greater proportion of medium-to-large species, although in

VM4 megaherbivores reach almost 20% of the assemblage while in VM3 they are 6%. In the case of BL and FN3, bovids and cervids also stand out in the values of CC, TAB and in the fossil record, as recorded by MNI counts, as well as equids in the latter. The χ^2 tests show no significant differences between the CC values and those of the UAL level of FN3 ($\chi^2 = 4.6274$; $p = 0.5924$). Regarding CC and TAB values, FN3-LAL and BL-D show higher estimates for equids and lower values for bovids, porcupines and rhinoceros (Figure 3). FN3-LAL shows cervid values close to CC and rhinoceros close to TAB values and it is noteworthy that the representation of equids is 49%. In the case of BL-D, values close to CC are observed for hippo *Hippopotamus antiquus* and close to TAB for rhinoceros (Figure 3). For both CC and TAB, FN3-UAL has higher values for equids and mammoths, lower values for bovids and porcupines and similar values for cervids and hippos (Figure 3). Regarding the distribution by body size categories (Table 5), very close values can be observed for the CC of the FN3/BL and FN3-UAL records, where the biomass is concentrated in medium-large species (>50%) and megaherbivores (>35%). In FN3-LAL and BL-D, most species are of medium to large size (~75%), with megaherbivores being more relevant in BL-D (17%), while in FN3-LAL both small species and megaherbivores are similarly relevant (12% and 14%, respectively) (Table 5).

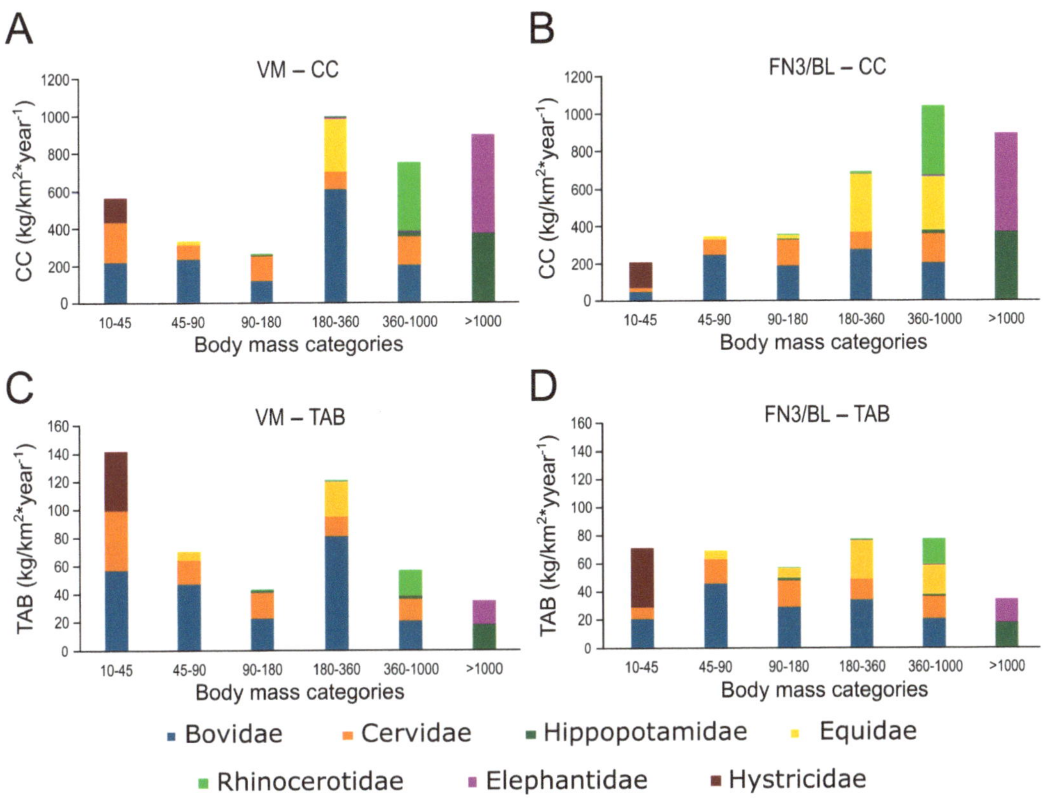

Figure 1. Histograms of Carrying Capacity (CC, in kg/km^2*year^{-1}) for Venta Micena (**A**) and Fuente Nueva-3/Barranco León (**B**) together with the values of Total Available Biomass (TAB, in kg/km^2*year^{-1}) by body mass categories (10–45 kg: Class 1; 45–90 kg: Class 2; 90–180 kg: Class 3; 180–360 kg: Class 4; 360–1000 kg: Class 5; >1000 kg: Class 6) for both paleocommunities ((**C**,**D**), respectively).

Table 3. Estimated values of Carrying Capacity (CC, in kg/km2*year^{-1}) and Total Available Biomass (TAB, in kg/km^2*year^{-1}) from the PSEco model for the Venta Micena (VM) faunal assemblage by body mass classes (Class 1: 10–45 kg; Class 2: 45–90 kg; Class 3: 90–180 kg; Class 4: 180–360 kg; Class 5: 360–1000 kg; Class 6: >1000 kg).

CC	*Bison* sp.	Bovidae Indet. (cf. *Rupicapra*)	*Hemibos* cf. *gracilis*	*Hemitragus albus*	*Praeovibos* sp.	*Soergelia minor*	Cervidae Indet. (cf. *Capreolus*)	*Metacervocerus rhenanus*	*Praemegaceros* cf. *verticornis*	*Hippopotamus antiquus*	*Equus altidens*	*Stephanorhinus hundsheimensis*	*Mammuthus meridionalis*	*Hystrix* sp.	
Class 1	0.00	178.31	0.00	24.70	0.00	20.00	184.29	22.68	0.00	0.00	0.00	0.00	0.00	135.34	
Class 2	18.85	0.00	18.68	178.51	18.75	0.00	0.00	60.96	18.62	0.00	18.15	0.00	0.00	0.00	
Class 3	0.00	0.00	28.29	0.00	27.79	60.51	0.00	139.91	0.00	4.06	0.00	6.06	0.00	0.00	
Class 4	60.10	0.00	212.66	0.00	186.96	146.18	0.00	0.00	91.17	0.00	287.27	13.21	2.03	0.00	
Class 5	201.58	0.00	0.00	0.00	0.00	0.00	0.00	0.00	153.58	21.56	0.00	373.10	4.91	0.00	
Class 6	0.00	0.00	0.00	0.00	0.00	0.00	0.00	0.00	0.00	368.07	0.00	0.00	526.60	0.00	
Total	280.53	178.31	259.63	203.20	233.50	226.69	184.29	223.55	263.37	393.68	305.42	392.37	533.54	135.34	CC = 3813.42
TAB	*Bison* sp.	Bovidae Indet.(cf. *Rupicapra*)	*Hemibos* cf. *gracilis*	*Hemitragus albus*	*Praeovibos* sp.	*Soergelia minor*	Cervidae Indet. (cf. *Capreolus*)	*Metacervocerus rhenanus*	*Praemegaceros* cf. *verticornis*	*Hippopotamus antiquus*	*Equus altidens*	*Stephanorhinus hundsheimensis*	*Mammuthus meridionalis*	*Hystrix* sp.	
Class 1	0.00	38.13	0.00	10.26	0.00	8.31	34.19	8.55	0.00	0.00	0.00	0.00	0.00	42.21	
Class 2	4.35	0.00	6.43	28.97	7.31	0.00	0.00	10.39	6.58	0.00	6.52	0.00	0.00	0.00	
Class 3	0.00	0.00	4.92	0.00	5.59	11.76	0.00	18.76	0.00	1.51	0.00	0.94	0.00	0.00	
Class 4	7.99	0.00	26.44	0.00	26.57	19.82	0.00	0.00	14.03	0.00	24.83	1.11	0.21	0.00	
Class 5	20.40	0.00	0.00	0.00	0.00	0.00	0.00	0.00	15.90	1.68	0.00	18.65	0.27	0.00	
Class 6	0.00	0.00	0.00	0.00	0.00	0.00	0.00	0.00	0.00	17.60	0.00	0.00	17.03	0.00	
Total	32.74	38.13	37.78	39.24	39.47	39.88	34.19	37.71	36.52	20.79	31.35	20.70	17.50	42.21	TAB = 468.21

Table 4. Estimated values of Carrying Capacity (CC, in kg/km^2*year^{-1}) and Total Available Biomass (TAB, in kg/km^2*year^{-1}) from the PSEco model for the Fuente Nueva-3 (FN3) and Barranco León (BL) faunal assemblages by body mass classes (Class 1: 10%–45 kg; Class 2: 45–90 kg; Class 3: 90–180 kg; Class 4: 180–360 kg; Class 5: 360–1000 kg; Class 6: >1000 kg).

CC	*Ammotragus europaeus*	*Bison* **sp.**	*Hemibos* cf. *gracilis*	*Hemitragus albus*	*Metacervocerus rhenanus*	*Praemegaceros* cf. *verticornis*	*Hippopotamus antiquus*	*Equus altidens*	*Equus suessenbornensis*	*Stephanorhinus hundsheimensis*	*Mammuthus meridionalis*	*Hystrix refossa*	
Class 1	23.48	0.00	0.00	24.70	22.68	0.00	0.00	0.00	0.00	0.00	0.00	138.30	
Class 2	30.05	18.85	18.68	178.51	60.96	18.62	0.00	18.15	0.00	0.00	0.00	0.00	
Class 3	157.94	0.00	28.29	0.00	139.91	0.00	4.06	0.00	19.09	6.06	0.00	0.00	
Class 4	0.00	60.10	212.66	0.00	0.00	91.17	0.00	287.27	24.70	13.21	2.03	0.00	
Class 5	0.00	201.58	0.00	0.00	0.00	153.58	21.56	0.00	286.43	373.10	4.91	0.00	
Class 6	0.00	0.00	0.00	0.00	0.00	0.00	368.07	0.00	0.00	0.00	526.60	0.00	
Total	211.46	280.53	259.63	203.20	223.55	263.37	393.68	305.42	330.21	392.37	533.54	138.30	CC = 3535.26
TAB	*Ammotragus europaeus*	*Bison* **sp.**	*Hemibos* cf. *gracilis*	*Hemitragus albus*	*Metacervocerus rhenanus*	*Praemegaceros* cf. *verticornis*	*Hippopotamus antiquus*	*Equus altidens*	*Equus suessenbornensis*	*Stephanorhinus hundsheimensis*	*Mammuthus meridionalis*	*Hystrix refossa*	
Class 1	10.44	0.00	0.00	10.26	8.55	0.00	0.00	0.00	0.00	0.00	0.00	41.98	
Class 2	6.19	4.35	6.43	28.97	10.39	6.58	0.00	6.52	0.00	0.00	0.00	0.00	
Class 3	24.37	0.00	4.92	0.00	18.76	0.00	1.51	0.00	6.86	0.94	0.00	0.00	
Class 4	0.00	7.99	26.44	0.00	0.00	14.03	0.00	24.83	2.90	1.11	0.21	0.00	
Class 5	0.00	20.40	0.00	0.00	0.00	15.90	1.68	0.00	20.43	18.65	0.27	0.00	
Class 6	0.00	0.00	0.00	0.00	0.00	0.00	17.60	0.00	0.00	0.00	17.03	0.00	
Total	41.00	32.74	37.78	39.24	37.71	36.52	20.79	31.35	30.19	20.70	17.50	41.98	TAB = 387.49

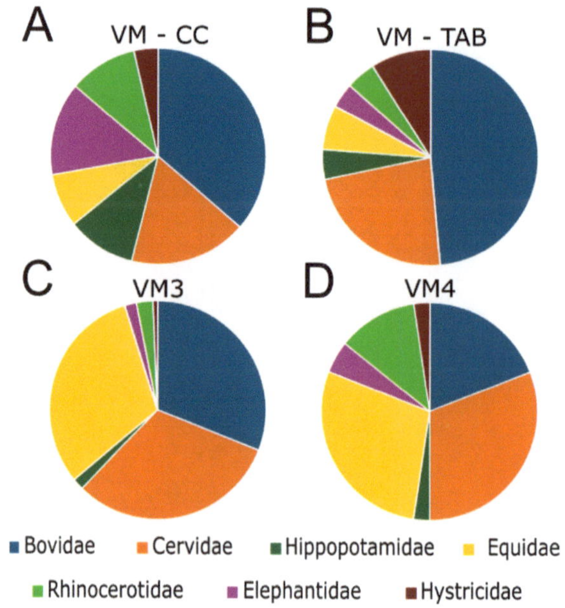

Figure 2. Carrying Capacity (CC) (**A**) and Total Available Biomass (TAB) (**B**) estimated for the Venta Micena (VM) site. The relative abundances of herbivores in the VM3 (**C**) and VM4 (**D**) quarries, estimated from values of minimum number of individuals (MNI) from [78,79], is also shown.

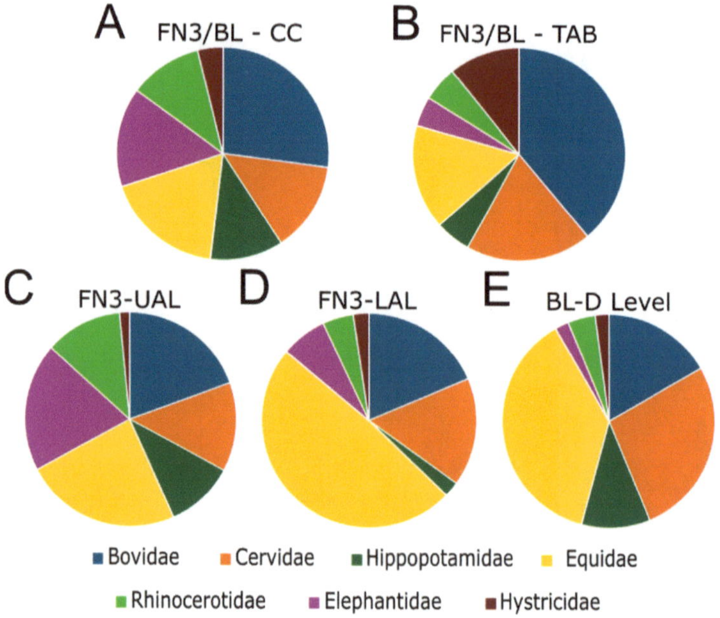

Figure 3. Carrying Capacity (CC) (**A**) and Total Available Biomass (TAB) (**B**) for Fuente Nueva-3 (FN3) and Barranco León (BL). The relative abundance of herbivores in the Upper Archeological Level (UAL-FN3) (**C**) and Lower Archeological Level of FN3 (LAL-FN3) (**D**) and the BL-D Level (**E**), estimated from the minimum number of individuals (MNI) values of [22,68], is depicted.

Table 5. Percentages of species by gross body size classes in the CC and TAB estimates for Venta Micena (VM-CC and VM-TAB, respectively) and Fuente Nueva-3 and Barranco León (FN3/BL-CC and FN3/BL-TAB, respectively) sites, in the fossil record of the Venta Micena 3 (VM3) and Venta Micena 4 (VM4) quarries, in the UAL and LAL levels of Fuente Nueva-3 (FN3–UAL and FN3–LAL, respectively) and in the D level of Barranco León (BL-D). The species were divided into three size categories: small (10–90 kg), medium-large (90–600 kg) and megaherbivores (>600 kg). Species included in the small size class: Bovidae indet. (cf. *Rupicapra*), Cervidae indet. (cf. *Capreolus*), *Hemitragus albus* and *Hystrix* sp. Species included in the medium-large size class: *Ammotragus europaeus*, *Bison* sp., *Hemibos* cf. *gracilis*, *Praeovibos* sp., *Soergelia minor*, *Metacervocerus rhenanus*, *Praemegaceros* cf. *verticornis*, *Equus altidens*, *Equus suessenbornensis* and *Equus* sp. Species included in the megaherbivores size class: *Hippopotamus antiquus*, *Stephanorhinus hundsheimensis* and *Mammuthus meridionalis*.

	Small Size	Medium-Large Size	Megaherbivores
VM-CC	18	47	35
VM-TAB	33	55	13
VM3	7	87	6
VM4	10	71	19
FN3/BL-CC	10	53	37
FN3/BL-TAB	21	64	15
FN3–UAL	7	51	42
FN3–LAL	12	74	14
BL-D Level	8	75	17

4. Discussion

As used in studies of ecology and paleoecology, Carrying Capacity (CC) represents the optimal and ideal conditions of a population and/or community, being the maximum growth it could have in the long term (see [33]). As previously proposed by Palmqvist et al. [65], in this study, we used these optimal conditions to evaluate and interpret the fossil assemblages, especially in communities rich enough to be considered as completely preserved in the fossil record. To estimate these optimal conditions, we used the PSEco model, which provided values of CC and Total Available Biomass or meat (TAB), as well as the ecological densities of carnivores and humans that could sustain the ecosystems [45]. For those paleocommunities of large mammals from Orce and the Sierra de Atapuerca, PSEco produced results for predator–prey biomass ratios comparable to those found in modern African ecosystems [45], which are close to steady conditions [31]. These results validated our approach as a means to estimate these ecological indices (i.e., CC and TAB).

The CC results indicate similar values for the VM and FN3/BL faunal assemblages, although VM (3813 kg/km^2*year) would have a slightly higher value than FN3/BL (3535 kg/km^2*year) (Tables 3 and 4), as we advocated in various previous studies ([37,42,45,94]; however, see [95]). These values are similar to those presented by Nairobi National Park (Kenya), Serengeti National Park (Tanzania), the Savuti area of Chobe National Park (Botswana) and Kruger National Park (South Africa), including for megaherbivores [31]. In terms of TAB, there are greater differences between VM and FN3/BL, with VM being more productive than FN3/BL due to the composition of the prey species and their differences in reproductive rates (see [41,42,45,54]). When the distribution patterns of CC and TAB are analyzed, there emerge differences between both communities (Figure 1). Following Ripple and Van Valkenburgh [96], Rodríguez-Gómez et al. [43] suggested that the differences between VM and FN3/BL in the patterns of TAB distribution among size classes could be due to top-down forces, because the body size classes with higher values of TAB (45–90, 180–360, 360–1000 kg) correspond to those of the preferred prey of the three top predators of FN3/BL (wild dog *Lycaon lycaonoides* as well as saber-tooths *Homotherium latidens* and *Megantereon whitei*) [71]. As shown in Figure 1D, the distribution of TAB is similar in classes 1 (10–45 kg), 2 (45–90 kg), 4 (180–360 kg) and 5 (360–1000 kg). The pattern of TAB distribution between size classes in VM shows more inequality between these classes than in FN3/BL, with classes 1 and 4 standing out (Figure 1C). This situation could have led to a greater competition for meat in VM, as more biomass was concentrated in specific size classes, rather than being more homogeneously distributed among the size classes (Figure 1C), as shown in FN3/BL (Figure 1D). This concentration of resources could favor the specialist carnivores, while the homogeneous distribution could favor the generalist ones, who could exploit a wider spectrum of resources. The compositional differences in the guild of secondary consumers between VM and FN3/BL are mainly due to the presence of the European jaguar *Panthera* cf. *gombaszoegensis* in VM and of *Homo* sp. in FN3/BL, since the jaguar was a specialist flesh-eating hypercarnivore and humans were an omnivorous and more generalist species that probably behaved more as a scavenger than as a predator. In this way, this dietary difference could support our proposal. However, in a previous study, Rodríguez-Gómez et al. [42] argued that competition was lower in VM than in FN3/BL due to the greater TAB of VM. Therefore, they did not relate the presence of humans in FN3 and BL to the presence of greater amounts of resources compared to VM, a site that does not preserve any evidence of the presence of humans [97], but rather to climatic or biogeographic barriers to hominin dispersal that posed a delay on the settlement of Western Europe until about 1.4 Ma (i.e., the age estimated for FN3/BL) [88]. Nevertheless, it is possible that conditions in FN3 and BL favored human presence due to a more homogeneous distribution of resources among size classes. This suggestion would argue against the proposal that the distribution of TAB in FN3/BL was driven by top-down forces. It will be interesting to further investigate this line of work in future studies to reach a conclusion.

Using our CC and TAB results as a reference for the analysis of the fossil record from Orce, it is observed that VM3 and VM4 are very different from the expected values (Figure 2). The giant hyenas, the biological agents involved in the formation of these assemblages, mainly focused on medium and large-sized species (between 90 and 600 kg) (Table 5), while small-sized species and megaherbivores were less relevant for them. This is more pronounced in VM3 than in VM4, where megaherbivores make up 6% of the total, while in VM4 they reach 19%. Nevertheless, according to our analysis, VM3 and VM4 do not show significant differences in the family composition of their assemblages (Figure 2C,D). This supports the suggestion of Palmqvist et al. [79] that there are no differences in the bone accumulation and modification patterns of the agent involved in these accumulations, *P. brevirostris*, only in the length of time that the skeletal remains were exposed to hyena consumption and weathering before the assemblage was capped by sedimentation of micritic limestone. The differences observed in VM3 and VM4 with respect to the expected values according to CC and TAB may be due to the coincidence of the activity of different carnivores in the selection and consumption of prey, since *P. brevirostris* scavenged the prey hunted selectively by saber-toothed predators (*Homotherium latidens* and *Megantereon whitei*) and wild dogs (*Lycaon lycaonoides*), being a strict scavenger of solitary habits as opposed to the present spotted hyena, which acts more like an active hunter [76,86,98].

Significant differences with the expected values of CC and TAB are observed for BL-D. However, it is interesting to note that no significant differences are observed between BL-D and the VM quarries, VM3 (χ^2 = 11.403; p = 0.0767) and VM4 (χ^2 = 12.215; p = 0.0573). Based on these results, we could argue that the factors involved in site formation for VM, together with the conservation biases involved in the accumulation of bone remains at this site [76], may have been similar to those involved in forming BL-D. It should be noted, however, that there is evidence of the presence of humans at BL-D [22,62,63,66]. The limited characteristics of the stone tools from FN3 and BL, including the small size of the flint and limestone flakes, have led to the assumption that access to large and megaherbivore prey through hunting would have been difficult for these hominins [65,86], which led researchers to consider that the optimal strategy of these humans at BL would be that of a strict and generalist scavenger [41]. It is most likely that these hominins played a role similar to that of *P. brevirostris* at VM, although the hyena was better equipped for aggressive scavenging (i.e., kleptoparasitism). Table 5 shows that the presence of megaherbivores in BL-D (17%, with hippos at 10%) is similar to that of VM4 (19%) and also to that expected in FN3/BL (15%) according to the TAB values, but is somewhat further from the TAB value derived for VM (13%). This suggests that the available meat resources of megaherbivores were exploited, but that the small-sized species were exploited less intensively than expected (Table 5). Depending on body size, carnivorous species are limited in the exploitation of certain resources [90], with larger carnivores having the ability and need to exploit larger prey [99]. Thus, these results appear to be consistent with the proposed *Megantereon*-hominin-*Pachycrocuta* model for FN3 (see discussions in [65,68]), which proposes that humans had secondary acquisition of prey remains from saber-tooths and other top predators. Saber-tooth cats would kill prey and consume only a part of the carcass (i.e., the softer tissues), leaving a large amount of meat and all the nutrients in the bones intact for scavenging by humans and, subsequently, by the giant hyenas. Saber-tooths would not focus on small prey; rather, they would take advantage of the meat resources available in the young megaherbivores, but their preferred prey would be medium and large ungulates (90–600 kg), as deduced by Palmqvist et al. [34,84,85] based on isotopic analyses of fossil bone collagen in VM3.

A similar case to that of BL-D can be observed for FN3-LAL, with very close values of medium-to-large species and megaherbivores (Table 5), with the relative percentages of megaherbivores very close to those provided by the TAB estimates (Figure 3 and Table 5). However, FN3-LAL shows significant differences with the other assemblages analyzed in the proportion represented by the different families, with equids showing a high representation (Figure 3D). In terms of prey size, FN3-LAL has a greater presence of small-sized species than the other fossil assemblages analyzed (Table 5). On the other hand, horses are an animal resource used by humans throughout the Pleistocene and Assaf [100] has recently argued that horses were an essential resource for Early Pleistocene humans due to the nutritional value of their meat, especially bone marrow, which is superior to that of other ungulates (see [100] and references therein). Given the importance of human presence at this level of FN3, as indicated by the finding of abundant manuports (i.e., limestone cobbles used by the hominins for breaking bones and accessing the medullary cavities; see [22,64]), both the greater number of small-sized species and horses may be a reflection of higher activity of humans in the FN3-LAL compared to other levels.

Regarding FN3-UAL, this is the only faunal assemblage that does not show significant differences with the expected value of CC, with a high value of megaherbivores (42%) as well as medium and large species (51%) and a low abundance of small species (7%) (Table 5). In comparison with the FN3-LAL level, Espigares et al. [64] and Palmqvist et al. [101] have suggested that this level is a death trap associated with quicksand due to its geological composition (two-thirds of layer 5 of LAL are fine and very fine sands and the remaining third is composed of silts and clays, which were combined with the oligosaline waters of the Orce paleolake). The functioning of this level as a quicksand trap for herbivores is consistent with the fact that no significant differences are observed with the values expected by CC (Figure 3A,D). As can be seen in Figure 3, FN3-UAL gives slightly higher values than expected for mammoths and rhinos in the CC scenario, which could be explained by the high weight per unit area supported by their legs, which posed a risk of entrapment ([101], see Figure 8). These results support that this methodology may allow the interpretation of sites that functioned as traps for the fauna recorded. This would be expected if these traps showed no relevant bias, trapping the species according to their abundance in the ecosystems, which is what CC measures. The only exception would be the largest species, like elephants, which are scarce in the ecosystem but tend to be overrepresented in the natural traps given the elevated weight supported by their feet, which results in a greater risk of entrapment. It will be interesting to confirm these results by applying the methodology to other sites with conditions similar to those of the UAL level of FN3.

Given the usefulness that CC values can have in evaluating sites that originally functioned as traps, could TAB be useful in evaluating the faunal composition of some sites? Our TAB results have allowed us to discuss megaherbivore values at BL-D and their consistency with the subsistence strategies of humans and the giant hyenas, as well as deviations from equid values at FN3-LAL. For this part, it is an interesting index to make interpretative proposals on the sites, but it is a more complex index to analyze than CC, because it is difficult to find similar and compatible scenarios for FN3-LAL, as that in the case of the traps considered for CC. Primarily, predators select prey based on body size [74,90–92], so large carnivores must prey on large prey species for reasons of energy efficiency [99]. Thus, sites where large felids played a relevant role, such as the Orce sites, do not show a high abundance of small prey because they were not frequently preyed upon by these predators. In present-day African ecosystems, lions (*Panthera leo*) and spotted hyenas (*Crocuta crocuta*) prey on different species, most of which weigh less than 600 kg [102,103] and their prey selection overlaps significantly [104]. In Kruger National Park, for example, spotted hyenas prey primarily on wildebeest (*Connochaetes taurinus*, 163 kg), buffalo (*Syncerus caffer*, 520 kg), Burchell's zebra (*Equus quagga burchellii*, 220 kg), greater kudu (*Tragelaphus strepsiceros*, 170 kg) and impala (*Aepyceros melampus*, 44 kg) [105] and for lions, buffalo, wildebeest and zebra, with a distinction between males and females, with males preferring buffalo and females wildebeest and zebra [106]. According to the CC values of Hatton et al. [31], buffalo, zebra and wildebeest account for approximately 30% of the total prey biomass or 67% if megaherbivores are not considered. In the case of Kalahari (South Africa), the main prey for lions and hyenas is the gemsbok (*Oryx gazella*, 162 kg), with small prey such as springbok (*Antidorcas marsupialis*, 32 kg) or porcupines (*Hystrix*, <20 kg) also taken in the absence of large ungulates [107–111]. CC estimates for the Kalahari are very low at 258 kg/km^2*year [31], with these species accounting for >75% of CC. In Serengeti National Park (Tanzania), spotted hyenas hunt mainly wildebeest, Thomson's gazelle (*Eudorcas thomsonii*, 18 kg) and zebra [112–114] and seven species account for 90% of the total meat intake of lions: buffalo, kongoni (*Alcelaphus buselaphus*, 125 kg), Thomson's gazelle (*Gazella thomsonii*, 18 kg), topi (*Damaliscus lunatus*, 108 kg), warthog (*Phacochoerus aethiopicus*, 55 kg), wildebeest and zebra [115]. In Serengeti National Park, total CC is approximately 2875 kg/km^2*year and 2158 kg/km^2*year excluding megafauna, with these species accounting for 82% of the CC [31]. Spotted hyenas have a preferred prey range of 56 to 182 kg, with the mode at 102 kg (Hayward 2006). This range may be slightly higher for lions, with wildebeest being the preferred prey in parts of South Africa and East Africa [27]. Nevertheless, both spotted hyenas and lions may prey on young individuals of megafauna, such as elephants, giraffes, hippopotamuses and rhinoceroses [116,117], but Hayward [104] argued that hyenas tend to avoid buffalo, plains zebras and giraffes. Given the above, we can generally expect hyenas and lions to hunt prey between 50 and 200 kg, which tend to be the most abundant, with megaherbivores being a less exploited resource, as well as species under 50 kg, except in poor resource conditions, as in the Kalahari [31,107–111]. Therefore, if we focus on the prey consumed by spotted hyenas and lions, we will have a very biased view of the resources that could be exploited from the ecosystem and, in turn, many difficulties in relating them to TAB values. This could be overcome by combining different carnivore species with different preferred prey sizes (e.g., leopard, cheetah, wild dog, etc.), which could provide a ratio of prey remains that would approach the values of meat available in the ecosystems. However, it would be very difficult to find an accumulation of remains of different predators preserved at the same site and that the prey species are represented according to resource availability. For this reason, this approach makes it difficult to obtain relative values for fossil sites similar to those based on TAB values. Another possibility could be provided by generalist species that could exploit all the resources of the environment with a similar preference for all potential prey, both as scavengers and as dominant predators of the ecosystem. It will be interesting to evaluate sites where humans exerted a major role in the paleoecosystem after the inception of the Acheulean (i.e., mode 2) techno-culture, being the main accumulating agent, in order to interpret their ability to exploit the resources of the ecosystems in which they lived.

5. Conclusions

Knowing the ecological conditions under which humans evolved during the Pleistocene is of great interest in the study of human evolution, and indices such as Carrying Capacity (CC) and Total Available Biomass (TAB) are useful to infer aspects of these conditions in the paleoecosystems. These indices represent optimal conditions for the communities studied and therefore serve as a reference for the evaluation and interpretation of faunal assemblages. Our results lead us to propose that (i) the distribution of TAB in FN3/BL could have been more beneficial for humans than in VM being a species with a generalist behavior; (ii) the differences observed between the fossil assemblages and CC and TAB values are due to the intervention of different secondary consumers; (iii) humans may have been more important in shaping the faunal assemblage of FN3-LAL than in the one of BL-D because the former shows a higher abundance of small herbivores and equids, both of which were essential resources for humans in the Early Pleistocene; and (iv) the relative abundances of taxa from FN3-UAL show consistency with the interpretation that this level acted as a quicksand trap, showing a proportion of taxa similar to those present in the ecosystem under CC conditions, with a slightly higher abundance of rhinoceroses and mammoths than expected. Considering all of the above, it can be argued that the approach of this work, using CC and TAB values to evaluate the fossil faunal assemblages, is useful with its application in sites rich in fauna such as those of Orce. It will be interesting to use this methodology and working perspective in other sites to test its usefulness and sites considered natural traps could be an ideal next step.

Author Contributions: G.R.-G. wrote the main manuscript text and all co-authors made contributions. All authors have read and agreed to the published version of the manuscript.

Funding: Funding for this research has been provided by the Spanish Ministry of Science, Innovation and University (Refs.: CGL2016-78577-P, CGL2016-80975-P, PGC2018-093925-B-C31, PGC2018-093925-B-C33, PID2019-111185GB-I00, PID2021-122355NB-C31), Junta de Andalucía (Refs: UMA18-FEDERJA-188, P18-FR-3193), Generalitat de Catalunya (Ref.: 2021SGR 01238 (AGAUR)), Universidad de Málaga (Refs.: B1-2022_13, B1-2020_24) and by research group RNM-146 of Junta de Andalucía. This research has been authorized by the Consejería de Cultura of the Junta de Andalucía. G. Rodríguez-Gómez enjoys a postdoctoral contract "Atracción de Talento Investigador César Nombela" (Ref. 2023-T1/PH-HUM-29222) co-funded by the Comunidad de Madrid and the Universidad Complutense de Madrid. I. Campaña and A. Pérez-Ramos are beneficiaries of posdoctoral grants from Junta de Andalucía. This work has also been supported by the Madrid Government (Comunidad de Madrid-Spain) under the Multiannual Agreement with Universidad Complutense de Madrid in the line Research Incentive for Young PhDs, in the context of the V PRICIT (Regional Programme of Research and Technological Innovation) (Ref. PR27/21–004).

Data Availability Statement: The results of this work are presented in this paper and are complemented by those presented in the publication of Rodriguez-Gomez et al. [48].

Acknowledgments: We are grateful for the thoughtful and insightful comments provided by three anonymous reviewers and the editors of the special issue 'Mammals Biochronology and Paleoecology of the Euro-Mediterranean Quaternary'. In addition, G.R.-G. would like to thank Mireia Parera and Javier Moreno for making this work possible by their time.

Conflicts of Interest: The authors declare that they have no known competing financial interests or personal relationships that could have appeared to influence the work reported in this paper.

Abbreviations

BL-D	Barranco León-Level D
CC	Carrying Capacity
FN3-LAL	Fuente Nueva 3-Lower Archaeologial Level
FN3-UAL	Fuente Nueva 3-Upper Archaeological Level
MNI	minimum number of individuals
PSEco	Paleosynecological model

TAB	Total Available Biomass
VM3	Venta Micena quarry 3
VM4	Venta Micena quarry 4

References

1. Ferraro, J.V.; Plummer, T.W.; Pobiner, B.L.; Oliver, J.S.; Bishop, L.C.; Braun, D.R.; Ditchfield, P.W.; Seaman, J.W., III; Binetti, K.M.; Seaman, J.W., Jr.; et al. Earliest Archaeological Evidence of Persistent Hominin Carnivory. *PLoS ONE* **2013**, *8*, e62174. [CrossRef] [PubMed]
2. Speth, J.D. Early Hominid Subsistence Strategies in Seasonal Habitats. *J. Archaeol. Sci.* **1987**, *14*, 13–29. [CrossRef]
3. Roebroeks, W. Hominid behaviour and the earliest occupation of Europe: An exploration. *J. Hum. Evol.* **2001**, *41*, 437–461. [CrossRef] [PubMed]
4. Hublin, J.J.; Richards, M.P. *The Evolution of Hominin Diets. Integrating Approaches to the Study of Palaeolithic Subsistence*; Spinger: Leipzig, Germany, 2009; p. 288.
5. El Zaatari, S.; Grine, F.E.; Ungar, P.S.; Hublin, J.J. Ecogeographic variation in Neandertal dietary habits: Evidence from occlusal molar microwear texture analysis. *J. Hum. Evol.* **2011**, *61*, 411–424. [CrossRef] [PubMed]
6. Henry, A.G.; Brooks, A.G.; Piperno, D.R. Microfossils in calculus demonstrate consumption of plants and cooked foods in Neanderthal diets (Shanidar III, Iraq; Spy I and II, Belgium). *Proc. Natl. Acad. Sci. USA* **2011**, *108*, 486–491. [CrossRef]
7. Hardy, K.; Buckley, S.; Collins, M.J.; Estalrrich, A.; Brothwell, D.; Copeland, L.; García-Tabernero, A.; García-Vargas, S.; De La Rasilla, M.; Lalueza-Fox, C.; et al. Neanderthal medics? Evidence for food, cooking, and medicinal plants entrapped in dental calculus. *Naturwissenschaften* **2012**, *99*, 617–626. [CrossRef]
8. Sistiaga, A.; Mallol, C.; Galván, B.; Summons, R.E. The Neanderthal Meal: A New Perspective Using Faecal Biomarkers. *PLoS ONE* **2014**, *9*, e101045. [CrossRef]
9. Estalrrich, A.; El Zaatari, S.; Rosas, A. Dietary reconstruction of the El Sidrón Neandertal familial group (Spain) in the context of other Neandertal and modern hunter-gatherer groups. A molar microwear texture analysis. *J. Hum. Evol.* **2017**, *104*, 13–22. [CrossRef]
10. Power, R.C.; Salazar-García, D.C.; Rubini, M.; Darlas, A.; Harvati, K.; Walker, M.; Hublin, J.J.; Henry, A.G. Dental calculus indicates widespread plant use within the stable Neanderthal dietary niche. *J. Hum. Evol.* **2018**, *119*, 27–41. [CrossRef]
11. Salazar-García, D.C.; Power, R.C.; Rudaya, N.; Kolobova, K.; Markin, S.; Krivoshapkin, A.; Henry, A.G.; Richards, M.P.; Viola, B. Dietary evidence from Central Asian Neanderthals: A combined isotope and plant microremains approach at Chagyrskaya Cave (Altai, Russia). *J. Hum. Evol.* **2021**, *156*, 102985. [CrossRef]
12. Cordain, L.; Miller, J.B.; Eaton, S.B.; Mann, N.; Holt, S.H.; Speth, J.D. Plant-animal subsistence ratios and macronutrient energy estimations in worldwide hunter-gatherer diets. *Am. J. Clin. Nutr.* **2000**, *71*, 682–692. [CrossRef]
13. Marlowe, F.W. Hunter-Gatherers and Human Evolution. *Evol. Anthropol.* **2005**, *14*, 54–67. [CrossRef]
14. Fiorenza, L.; Benazzi, S.; Estalrrich, A.; Kullmer, O. Diet and cultural diversity in Neanderthals and modern humans from dental macrowear analyses. In *Dental Wear in Evolutionary and Biocultural Contexts*; Academic Press: Cambridge, MA, USA, 2020; pp. 39–72. [CrossRef]
15. Binford, L.R. *Constructing Frames of Reference. An Analytical Method for Archaeological Theory Building Using Ethnographic and Environmental Data Sets*, 1st ed.; University of California Press: Berkeley, CA, USA, 2001; p. 624.
16. Gaudzinski, S.; Roebroeks, W. Adults only. Reindeer hunting at the Middle Palaeolithic site Salzgitter Lebenstedt, Northern Germany. *J. Hum. Evol.* **2000**, *38*, 497–521. [CrossRef]
17. Bocherens, H.; Drucker, D.G.; Billiou, D.; Patou-Mathis, M.; Vandermeersch, B. Isotopic evidence for diet and subsistence pattern of the Saint-Césaire I Neanderthal: Review and use of a multi-source mixing model. *J. Hum. Evol.* **2005**, *49*, 71–87. [CrossRef] [PubMed]
18. Richards, M.P.; Trinkaus, E. Isotopic evidence for the diets of European Neanderthals and early modern humans. *Proc. Natl. Acad. Sci. USA* **2009**, *106*, 16034–16039. [CrossRef] [PubMed]
19. Saladié, P.; Huguet, R.; Díez, C.; Rodríguez-Hidalgo, A.; Cáceres, I.; Vallverdú, J.; Rosell, J.; Bermúdez de Castro, J.M.; Carbonell, E. Carcass transport decisions in *Homo antecessor* subsistence strategies. *J. Hum. Evol.* **2011**, *61*, 425–446. [CrossRef]
20. Speth, J.D. *The Paleoanthropology and Archaeology of Big-Game Hunting*; Springer: Berlin/Heidelberg, Germany, 2010; p. 268.
21. Smith, G.M. Neanderthal megafaunal exploitation in Western Europe and its dietary implications: A contextual reassessment of La Cotte de St Brelade (Jersey). *J. Hum. Evol.* **2015**, *78*, 181–201. [CrossRef]
22. Espigares, M.P.; Palmqvist, P.; Guerra-Merchán, A.; Ros-Montoya, S.; García-Aguilar, J.M.; Rodríguez-Gómez, G.; Serrano, F.J.; Martínez-Navarro, B. The earliest cut marks of Europe: A discussion on hominin subsistence patterns in the Orce sites (Baza basin, SE Spain). *Sci. Rep.* **2019**, *9*, 15408. [CrossRef]
23. Kahlke, R.D.; Gaudzinski, S. The blessing of a great flood: Differentiation of mortality patterns in the large mammal record of the Lower Pleistocene fluvial site of Untermassfeld (Germany) and its relevance for the interpretation of faunal assemblages from archaeological sites. *J. Archaeol. Sci.* **2005**, *32*, 1202–1222. [CrossRef]
24. Stringer, C.B.; Finlayson, J.C.; Barton, R.N.E.; Fernández-Jalvo, Y.; Cáceres, I.; Sabin, R.C.; Rhodes, E.J.; Currant, A.P.; Rodríguez-Vidal, J.; Giles-Pacheco, F.; et al. Neanderthal exploitation of marine mammals in Gibraltar. *Proc. Natl. Acad. Sci. USA* **2008**, *105*, 14319–14324. [CrossRef]

25. Palmqvist, P.; Rodríguez-Gómez, G.; Figueirido, B.; García-Aguilar, J.M.; Pérez-Claros, J.A. On the ecological scenario of the first hominin dispersal out of Africa. *L'anthropologie* **2022**, *126*, 102998. [CrossRef]
26. Coe, M.J.; Cumming, D.H.; Phillipson, J. Biomass and Production of Large African Herbivores in Relation to Rainfall and Primary Production. *Oecologia* **1976**, *22*, 341–354. [CrossRef]
27. Schaller, G.B. *The Serengeti Lion: A Study of Predator-Prey Relations*; The University of Chicago Press: Chicago, IL, USA; London, UK, 1972; p. 518.
28. Owen-Smith, N. *Megaherbivores: The Influence of Very Large Body Size on Ecology*; Cambridge University Press: Cambridge, UK, 1988; p. 382.
29. Hayward, M.W.; O'Brien, J.; Kerley, G.I.H. Carrying capacity of large African predators: Predictions and tests. *Biol. Conserv.* **2007**, *139*, 219–229. [CrossRef]
30. Rodríguez-Gómez, G.; Rodríguez, J.; Martín-González, J.A.; Goikoetxea, I.; Mateos, A. Modeling trophic resource availability for the first human settlers of Europe: The case of Atapuerca TD6. *J. Hum. Evol.* **2013**, *64*, 645–657. [CrossRef]
31. Hatton, I.A.; McCann, K.S.; Fryxell, J.M.; Davies, T.J.; Smerlak, M.; Sinclair, A.R.E.; Loreau, M. The predator-prey power law: Biomass scaling across terrestrial and aquatic biomes. *Science* **2015**, *349*, aac6284-1. [CrossRef] [PubMed]
32. Rodríguez, J.; Blain, H.-A.; Mateos, A.; Martín-González, J.A.; Cuenca-Bescós, G.; Rodríguez-Gómez, G. Ungulate carrying capacity in Pleistocene Mediterranean ecosystems: Evidence from the Atapuerca sites. *Palaeogeogr. Palaeoclimatol. Palaeoecol.* **2014**, *393*, 122–134. [CrossRef]
33. Sayre, N.F. The Genesis, History, and Limits of Carrying Capacity. *Ann. Am. Assoc. Geogr.* **2008**, *98*, 120–134. [CrossRef]
34. Palmqvist, P.; Gröcke, D.R.; Arribas, A.; Fariña, R.A. Paleoecological reconstruction of a lower Pleistocene large mammal community using biogeochemical (δ^{13}C, δ^{15}N, δ^{18}0, Sr: Zn) and ecomorphological approaches. *Paleobiology* **2003**, *29*, 205–229. [CrossRef]
35. Barnosky, A.D. Megafauna biomass tradeoff as a driver of Quaternary and future extinctions. *Proc. Natl. Acad. Sci. USA* **2008**, *105*, 11543–11548. [CrossRef]
36. Rodríguez, J.; Mateos, A. Carrying capacity, carnivoran richness and hominin survival in Europe. *J. Hum. Evol.* **2018**, *118*, 72–88. [CrossRef]
37. Rodríguez-Gómez, G.; Palmqvist, P.; Martínez-Navarro, B.; Martín-González, J.A.; Bermúdez de Castro, J.M. Mean body size estimation in large mammals and the computation of biomass in past ecosystems: An application to the Pleistocene sites of Orce and Sierra de Atapuerca (Spain). *C. R. Palevol.* **2022**, *21*, 207–233. [CrossRef]
38. Vidal-Cordasco, M.; Ocio, D.; Hicker, T.; Marín-Arroyo, A.B. Ecosystem productivity affected the spatiotemporal disappearance of Neanderthals in Iberia. *Nat. Ecol. Evol.* **2022**, *6*, 1644–1657. [CrossRef] [PubMed]
39. Vidal-Cordasco, M.; Terlato, G.; Ocio, D.; Marín-Arroyo, A. Neanderthal coexistence with *Homo sapiens* in Europe was affected by herbivore carrying capacity. *Sci. Adv.* **2023**, *9*, eadi4099. [CrossRef] [PubMed]
40. Rodríguez-Gómez, G.; Mateos, A.; Martín-González, J.A.; Blasco, R.; Rosell, J.; Rodríguez, J. Discontinuity of Human Presence at Atapuerca during the Early Middle Pleistocene: A Matter of Ecological Competition? *PLoS ONE* **2014**, *9*, e101938. [CrossRef]
41. Rodríguez-Gómez, G.; Palmqvist, P.; Rodríguez, J.; Mateos, A.; Martín-González, J.A.; Espigares, M.P.; Ros-Montoya, S.; Martínez-Navarro, B. On the ecological context of the earliest human settlements in Europe: Resource availability and competition intensity in the carnivore guild of Barranco Leon-D and Fuente Nueva-3 (Orce, Baza Basin, SE Spain). *Quat. Sci. Rev.* **2016**, *143*, 69–83. [CrossRef]
42. Rodríguez-Gómez, G.; Palmqvist, P.; Ros-Montoya, S.; Espigares, M.P. Resource availability and competition intensity in the carnivore guild of the Early Pleistocene site of Venta Micena (Orce, Baza Basin, SE Spain). *Quat. Sci. Rev.* **2017**, *164*, 154–167. [CrossRef]
43. Rodríguez-Gómez, G.; Rodríguez, J.; Martín-González, J.A.; Mateos, A. Evaluating the impact of *Homo*-carnivore competition in European human settlements during the early to middle Pleistocene. *Quat. Res.* **2017**, *88*, 129–151. [CrossRef]
44. Rodríguez-Gómez, G.; Cassini, G.H.; Palmqvist, P.; Bargo, M.S.; Toledo, N.; Martín-González, J.A.; Muñoz, N.A.; Kay, R.F.; Vizcaíno, S.F. Testing the hypothesis of an impoverished predator guild in the Early Miocene ecosystems of Patagonia: An analysis of meat availability and competition intensity among carnivores. *Palaeogeogr. Palaeoclimatol. Palaeoecol.* **2020**, *554*, 109805. [CrossRef]
45. Rodríguez-Gómez, G.; Martín-González, J.A.; Espigares, M.P.; Bermúdez de Castro, J.M.; Martínez-Navarro, B.; Arsuaga, J.L.; Palmqvist, P. From meat availability to hominin and carnivore biomass: A paleosynecological approach to reconstructing predator-prey biomass ratios in the Pleistocene. *Quat. Sci. Rev.* **2024**, *328*, 108474. [CrossRef]
46. Bermúdez de Castro, J.M.; Díez Fernández-Lomana, J.C.; Mosquera Martínez, M.; Nicolás Checa, M.E.; Pérez Pérez, A.; Rodríguez Méndez, J.; Sánchez Marco, A. El nicho ecológico de los homínidos del Pleistoceno Medio de Atapuerca. *Complutum* **1995**, *6*, 9–56.
47. Fariña, R.A. Trophic relationships among Lujanian mammals. *Evol. Theory* **1996**, *11*, 125–134.
48. Vizcaíno, S.F.; Fariña, R.A.; Zárate, M.A.; Bargo, M.S.; Schultz, P. Palaeoecological implications of the mid-Pliocene faunal turnover in the Pampean Region (Argentina). *Palaeogeogr. Palaeoclimatol. Palaeoecol.* **2004**, *213*, 101–113. [CrossRef]
49. Vizcaíno, S.F.; Bargo, M.S.; Kay, R.F.; Fariña, R.A.; Di Giacomo, M.; Perry, J.M.G.; Prevosti, F.J.; Toledo, N.; Cassini, G.H.; Fernicola, J.C. A baseline paleoecological study for the Santa Cruz Formation (late–early Miocene) at the Atlantic coast of Patagonia, Argentina. *Palaeogeogr. Palaeoclimatol. Palaeoecol.* **2010**, *292*, 507–519. [CrossRef]

50. Marín Arroyo, A.B. *Arqueozoología en el Cantábrico Oriental Durante la Transición Pleistoceno/Holoceno*; Editorial Universidad de Cantabria, Colección Difunde: Santander, Spain, 2010; p. 686.
51. Meloro, C.; Clauss, M. Predator-prey biomass fluctuations in the Plio-Pleistocene. *Palaios* **2012**, *27*, 90–96. [CrossRef]
52. Palombo, M.R. To what extent could functional diversity be a useful tool in inferring ecosystem responses to past climate changes? *Quat. Int.* **2016**, *413 Pt B*, 15–31. [CrossRef]
53. Martín-González, J.A.; Mateos, A.; Rodríguez-Gómez, G.; Rodríguez, J. A parametrical model to describe a stable and stationary age structure for fossil populations. *Quat. Int.* **2016**, *413*, 69–77. [CrossRef]
54. Martín-González, J.A.; Rodríguez-Gómez, G.; Palmqvist, P. Survival profiles from linear models versus Weibull models: Estimating stable and stationary population structures for Pleistocene large mammals. *J. Archaeol. Sci. Rep.* **2019**, *25*, 370–386. [CrossRef]
55. Rodríguez-Gómez, G.; Pérez-Fernández, E.; Fernández, P.; Arsuaga, J.L.; Díez, C.; Arceredillo, D. Palaeoecology of the Southern chamois from Valdegoba Cave (Burgos, Spain) and its exploitation by the Neanderthals. *Lethaia* **2022**, *55*, 1–25. [CrossRef]
56. Wilson, O.E.; Parker, A.K. Low predator competition indicates occupation of macro-predatory niches by giant Miocene reptiles at La Venta, Colombia. *Palaeogeogr. Palaeoclimatol. Palaeoecol.* **2023**, *632*, 111843. [CrossRef]
57. Pérez-Pérez, A.; Bravo-Cuevas, V.M.; Fernández, P. Population dynamics of *Equus conversidens* (Perissodactyla, Equidae) from the late Pleistocene of Hidalgo (central Mexico): Comparison with extant and fossil equid populations. *J. South Am. Earth Sci.* **2021**, *106*, 103100. [CrossRef]
58. Moullé, P.E.; Echassoux, A.; Martínez-Navarro, B. *Ammotragus europaeus*: Une nouvelle espèce de Caprini (Bovidae, Mammalia) du Pléistocène inférieur à la grotte du Vallonnet (France). *C. R. Palevol* **2004**, *3*, 663–673. [CrossRef]
59. Martínez-Navarro, B.; Palmqvist, P.; Madurell, J.; Ros-Montoya, S.; Espigares, M.P.; Torregrosa, V.; Pérez-Claros, J.A. La Fauna de Grandes mamíferos de Fuente Nueva-3 y Barranco León-5: Estado de la Cuestión. In *Ocupaciones Humanas en el Pleistoceno Inferior y Medio de la Cuenca de Guadix-Baza*; Toro, I., Martínez-Navarro, B., Agustí, J., Eds.; Junta de Andalucía, Consejería de Cultura, Arqueología Monografías: Sevilla, Spain, 2010; pp. 197–236.
60. Madurell-Malapeira, J.; Martínez-Navarro, B.; Ros-Montoya, S.; Espigares, M.P.; Toro, I.; Palmqvist, P. The earliest European badger (*Meles meles*), from the Late Villafranchian site of Fuente Nueva 3 (Orce, Granada, SE Iberian Peninsula). *C. R. Palevol* **2011**, *10*, 609–615. [CrossRef]
61. Ros-Montoya, S.; Bartolini-Lucenti, S.; Espigares, M.P.; Palmqvist, P.; Martínez- Navarro, B. First review of lyncodontini material (Mustelidae, Carnivora, Mammalia) from the lower Pleistocene archaeo-palaeontological sites of Orce (southeastern Spain). *Riv. Ital. Paleontol. Stratigr.* **2021**, *127*, 33–47. [CrossRef]
62. Barsky, D.; Celiberti, V.; Cauche, D.; Grégire, S.; Levègue, F.; de Lumley, H.; Toro-Moyano, I. Raw material discernment and technological aspects of the Barranco León and Fuente Nueva 3 stone assemblages (Orce southern Spain). *Quart. Int.* **2010**, *223–224*, 201–219. [CrossRef]
63. Barsky, D.; Vergès, J.-M.; Sala, R.; Menéndez, L.; Toro-Moyano, I. Limestone percussion tools from the late Early Pleistocene sites of Barranco León and Fuente Nueva 3 (Orce, Spain). *Phil. Trans. R. Soc. B* **2015**, *370*, 20140352. [CrossRef]
64. Espigares, M.P.; Palmqvist, P.; Rodríguez-Ruiz, M.D.; Ros-Montoya, S.; Pérez-Ramos, A.; Rodríguez-Gómez, G.; Guerra-Merchán, A.; García-Aguilar, J.M.; Granados, A.; Campaña, I.; et al. Sharing food with hyenas: A latrine of *Pachycrocuta brevirostris* in the Early Pleistocene assemblage of Fuente Nueva-3 (Orce, Baza Basin, SE Spain). *Archaeol. Anthropol. Sci.* **2023**, *15*, 81. [CrossRef]
65. Palmqvist, P.; Rodríguez-Gómez, G.; Martínez-Navarro, B.; Espigares, M.P.; Figueirido, B.; Ros-Montoya, S.; Guerra-Merchán, A.; Granados, A.; García-Aguilar, J.M.; Pérez-Claros, J.A. Déjà vu: On the use of meat resources by sabretooths, hominins, and hyaenas in the Early Pleistocene site of Fuente Nueva 3 (Guadix-Baza Depression, SE Spain). *Archaeol. Anthropol. Sci.* **2023**, *15*, 17. [CrossRef]
66. Toro-Moyano, I.; Martínez-Navarro, B.; Agustí, J.; Souday, C.; Bermúdez de Castro, J.M.; Martinón-Torres, M.; Fajardo, B.; Duval, M.; Falguères, C.; Oms, O.; et al. The oldest human fossil in Europe, from Orce (Spain). *J. Hum. Evol.* **2013**, *65*, 1–9. [CrossRef]
67. Titton, S.; Oms, O.; Barsky, D.; Bargalló, A.; Serrano-Ramos, A.; García-Solano, J.; Sánchez-Bandera, C.; Yravedra, J.; Blain, H.-A.; Toro-Moyano, I.; et al. Oldowan stone knapping and percussive activities on a raw material reservoir deposit 1.4 million years ago at Barranco León (Orce, Spain). *Archaeol. Anthropol. Sci.* **2021**, *13*, 108. [CrossRef]
68. Yravedra, J.; Solano, J.A.; Courtenay, L.A.; Saarinen, J.; Linares-Matás, G.; Luzón, C.; Serrano-Ramos, A.; Herranz-Rodrigo, D.; Cámara, J.M.; Ruiz, A.; et al. Use of meat resources in the Early Pleistocene assemblages from Fuente Nueva 3 (Orce, Granada, Spain). *Archaeol. Anthropol. Sci.* **2021**, *13*, 213. [CrossRef]
69. Duval, M.; Falguères, C.; Bahain, J.-J.; Grün, R.; Shao, Q.; Aubert, M.; Dolo, J.-M.; Agusti, J.; Martínez-Navarro, B.; Palmqvist, P.; et al. On the limits of using combined U-series/ESR method to date fossil teeth from two Early Pleistocene archaeological sites of the Orce area (Guadix-Baza basin, Spain). *Quat. Res.* **2012**, *77*, 481–482. [CrossRef]
70. Álvarez, C.; Parés, J.M.; Granger, D.; Duval, M.; Sala, R.; Toro, I. New magnetostratigraphic and numerical age of the Fuente Nueva-3 site (Guadix-Baza Basin, Spain). *Quat. Int.* **2015**, *389*, 224–234. [CrossRef]
71. Arribas, A.; Palmqvist, P. The first human dispersal to Europe: Remarks on the archaeological and palaeoanthropological record from Orce (Guadix-Baza basin, southeastern Spain). *J. Hum. Evol.* **2002**, *17*, 55–78. [CrossRef]
72. Turq, A.; Martínez-Navarro, B.; Palmqvist, P.; Arribas, A.; Agustí, J.; Rodríguez-Vidal, J. Le Plio-Pléistocène de la région d'Orce, province de Grenada, Espagne: Bilan et perspectivas de Recherche. *Paléo* **1996**, *8*, 161–204. [CrossRef]

73. Martínez-Navarro, B.; Turq, A.; Agustí, J.; Oms, O. Fuente Nueva-3 (Orce, Granada, Spain) and the first human occupation of Europe. *J. Hum. Evol.* **1997**, *33*, 611–620. [CrossRef] [PubMed]
74. Palmqvist, P.; Martínez-Navarro, B.; Arribas, A. Prey selection by terrestrial carnivores in a Lower Pleistocene paleocommunity. *Paleobiology* **1996**, *22*, 514–534. [CrossRef]
75. Arribas, A.; Palmqvist, P. Taphonomy and palaeoecology of an assemblage of large mammals: Hyaenid activity in the lower Pleistocene site at Venta Micena (Orce, Guadix-Baza Basin, Granada, Spain). *Geobios* **1998**, *31*, 3–4. [CrossRef]
76. Palmqvist, P.; Arribas, A. Taphonomic decoding of the paleobiological information locked in a lower Pleistocene assemblage of large mammals. *Paleobiology* **2001**, *27*, 512–530. [CrossRef]
77. Granados, A.; Oms, O.; Anadón, P.; Ibáñez-Insa, J.; Kaakinen, A.; Jiménez-Arenas, J.M. Geochemical and sedimentary constraints on the formation of the Venta Micena Early Pleistocene site (Guadix-Baza basin, Spain). *Sci. Rep.* **2021**, *11*, 22437. [CrossRef]
78. Luzón, C.; Yravedra, J.; Courtenay, L.A.; Saarinen, J.; Blain, H.-A.; DeMiguel, D.; Viranta, S.; Azanza, B.; Rodríguez-Alba, J.J.; Herranz-Rodrigo, D.; et al. Taphonomic and spatial analyses from the Early Pleistocene site of Venta Micena 4 (Orce, Guadix-Baza Basin, southern Spain). *Sci. Rep.* **2021**, *11*, 13977. [CrossRef]
79. Palmqvist, P.; Espigares, M.P.; Pérez-Claros, J.A.; Figueirido, B.; Guerra-Merchán, A.; Ros-Montoya, S.; Rodríguez-Gómez, G.; García-Aguilar, J.M.; Granados, A.; Martínez-Navarro, B. Déjà vu: A reappraisal of the taphonomy of quarry VM4 of the Early Pleistocene site of Venta Micena (Baza Basin, SE Spain). *Sci. Rep.* **2022**, *12*, 705. [CrossRef]
80. Espigares Ortiz, M.P. Análisis y Modelización del Contexto Sedimentario y los Atributos Tafonómicos de los Yacimientos Pleistocénicos del Borde Nororiental de la Cuenca de Guadix-Baza. Ph.D. Thesis, Departamento de Estratigrafía y Paleontología, Universidad de Granada, Granada, Spain, 2010.
81. Martínez-Navarro, B.; Palmqvist, P. Presence of the African Machairodont *Megantereon whitei* (Broom, 1937) (Felidae, carnivora, Mammalia) in the Lower Pleistocene site of Venta Micena (Orce, Granada, Spain), with some considerations on the origin, evolution and dispersal of the genus. *J. Archaeol. Sci.* **1995**, *22*, 569–582. [CrossRef]
82. Turner, A.; Antón, M. The giant hyaena, *Pachycrocuta brevirostris* (Mammalia, Carnivora, Hyaenidae). *Geobios* **1996**, *29*, 455–468. [CrossRef]
83. Palmqvist, P.; Arribas, A.; Martínez Navarro, B. Ecomorphological study of large canids from the lower Pleistocene of southern Spain. *Lethaia* **1999**, *32*, 75–88. [CrossRef]
84. Palmqvist, P.; Pérez-Claros, J.A.; Janis, C.M.; Figueirido, B.; Torregrosa, V.; Gröcke, D.R. Biogeochemical and ecomorphological inferences on prey selection and resource partitioning among mammalian carnivores in an Early Pleistocene community. *Palaios* **2008**, *23*, 724–737. [CrossRef]
85. Palmqvist, P.; Pérez-Claros, J.A.; Janis, C.M.; Gröcke, D.R. Tracing the ecophysiology of ungulates and predator–prey relationships in an early Pleistocene large mammal community. *Palaeogeogr. Palaeoclimatol. Palaeoecol.* **2008**, *266*, 95–111. [CrossRef]
86. Palmqvist, P.; Martínez-Navarro, B.; Pérez-Claros, J.A.; Torregrosa, V.; Figueirido, B.; Jiménez-Arenas, J.M.; Espigares, M.P.; Ros-Montoya, S.; de Renzi, M. The giant hyena *Pachycrocuta brevirostris*: Modelling the bone-cracking behavior of an extinct carnivore. *Quat. Int.* **2011**, *243*, 61–79. [CrossRef]
87. Palmqvist, P.; Rodríguez-Gómez, G.; Bermúdez de Castro, J.M.; García-Aguilar, J.M.; Espigares, M.P.; Figueirido, B.; Ros-Montoya, S.; Granados, A.; Serrano, F.J.; Martínez-Navarro, B.; et al. Insights on the Early Pleistocene Hominin Population of the Guadix-Baza Depression (SE Spain) and a Review on the Ecology of the First Peopling of Europe. *Front. Ecol. Evol.* **2022**, *10*, 881651. [CrossRef]
88. Martínez-Navarro, B.; Madurell-Malapeira, J.; Ros-Montoya, S.; Espigares, M.P.; Medin, T.; Hortolà, P.; Palmqvist, P. The Epivillafranchian and the arrival of pigs into Europe. *Quat. Int.* **2015**, *389*, 131–138. [CrossRef]
89. Damuth, J. Population density and body size in mammals. *Nature* **1981**, *290*, 699–700. [CrossRef]
90. Carbone, C.; Mace, G.M.; Roberts, S.C.; Macdonald, D.W. Energetic constraints on the diet of terrestrial carnivores. *Nature* **1999**, *402*, 286–288. [CrossRef] [PubMed]
91. Radloff, F.G.T.; Toit, J.T.D. Large predators and their prey in a southern African savanna: A predator's size determines its prey size range. *J. Anim. Ecol.* **2004**, *73*, 410–423. [CrossRef]
92. Ercoli, M.D.; Prevosti, F.J.; Forasiepi, A.M. The structure of the mammalian predator guild in the Santa Cruz Formation (Late Early Miocene). *J. Mamm. Evol.* **2014**, *21*, 369–381. [CrossRef]
93. Zullinger, E.M.; Ricklefs, R.E.; Redford, K.H.; Mace, G.M. Fitting sigmoidal equations to mammalian growth curves. *J. Mammal.* **1984**, *65*, 607–636. [CrossRef]
94. Palmqvist, P.; Rodríguez-Gómez, G.; Espigares, M.P. Climate and environmental conditions in the Guadix-Baza Depression during the first hominin dispersal in Western Europe: Comment on Saarinen et al. (2021). *Quat. Sci. Rev.* **2022**, *297*, 107731. [CrossRef]
95. Saarinen, J.; Oksanen, O.; Žliobaitė, I.; Fortelius, M.; DeMiguel, D.; Azanza, B.; Bocherens, H.; Luzón, C.; Solano-García, J.; Yravedra, J.; et al. Pliocene to Middle Pleistocene climate history in the Guadix-Baza Basin, and the environmental conditions of early *Homo* dispersal in Europe. *Quat. Sci. Rev.* **2021**, *268*, 107132. [CrossRef]
96. Ripple, W.J.; Van Valkenburgh, B. Linking top-down forces to the Pleistocene megafaunal extinctions. *BioScience* **2010**, *60*, 516–526. [CrossRef]
97. Martínez-Navarro, B. The skull of Orce: Parietal bones or frontal bones? *J. Hum. Evol.* **2002**, *42*, 265–270. [CrossRef]

98. Coca-Ortega, C.; Pérez-Claros, J.A. Characterizing ecomorphological patterns in hyenids: A multivariate approach using postcanine dentition. *PeerJ* **2019**, *6*, e6238. [CrossRef]
99. Carbone, C.; Teacher, A.; Rowcliffe, J.M. The costs of carnivory. *PLoS Biol.* **2007**, *5*, 363–368. [CrossRef]
100. Assaf, E. Set in Stone: Human–Horse Relations as Embodied in Shaped Stone Balls. *Archaeologies* **2024**, *20*, 64–105. [CrossRef]
101. Palmqvist, P.; Campaña, I.; Granados, A.; Martínez-Navarro, B.; Pérez-Ramos, A.; Rodríguez-Gómez, G.; Guerra-Merchán, A.; Ros-Montoya, S.; Rodríguez-Ruiz, M.D.; García-Aguilar, J.M.; et al. The late Early Pleistocene site of Fuente Nueva-3 (Guadix-Baza Depression, SE Spain): A hyena latrine developed on a quicksand trap for megaherbivores? *J. Iber. Geol.* **2024**, 1–27. [CrossRef]
102. Mills, G.; Hofer, H. *Hyaenas. Status Survey and Conservation Action*; IUCN/SSC Hyaena Specialist Group: Gland, Switzerland; IUCN; Cambridge, UK, 1998; 154p.
103. Haas, S.K.; Hayssen, V.; Krausman, P.R. Panthera leo. Mamm. *Species* **2005**, *762*, 1–11. [CrossRef]
104. Hayward, M.W. Prey preferences of the spotted hyaena (*Crocuta crocuta*) and degree of dietary overlap with the lion (*Panthera leo*). *J. Zool.* **2006**, *270*, 606–614. [CrossRef]
105. Henschel, J.R.; Skinner, J.D. The diet of spotted hyaenas' *Crocuta crocuta* in Kruger National Park. *Afr. J. Ecol.* **1990**, *28*, 69–82. [CrossRef]
106. Funston, P.J.; Mills, G.L.; Biggs, H.C.; Richardson, P.R.K. Hunting by male lions: Ecological influences and socioecological implications. *Anim. Behav.* **1998**, *56*, 1333–1345. [CrossRef] [PubMed]
107. Mills, M.G.L. Prey selection and feeding habits of the large carnivores in the Southern Kalahari. *Koedoe* **1984**, *27*, 281–294. [CrossRef]
108. Mills, M.G.L. *Kalahari Hyaenas: The Comparative Behavioural Ecology of Two Species*; Allen and Unwin Hyman: London, UK, 1990; p. 304.
109. Eloff, F.C. Lion predation in the Kalahari Gemsbok National Park. *J. South Afr. Wildl. Mgmt. Ass.* **1973**, *3*, 59–64.
110. Eloff, F.C. Food ecology of the Kalahari lion *Panthera leo vernayi*. *Koedoe* **1984**, *27*, 249–258. [CrossRef]
111. Turner, A.; Antón, M. *The Big Cats and Their Fossil Relatives*; Columbia University Press: New York, NY, USA, 1997; p. 233.
112. Kruuk, H. *The Spotted Hyena. A Study of Predation and Social Behavior*; The University of Chicago Press: Chicago, IL, USA, 1972; p. 335.
113. Hofer, H.; East, M.L. The commuting system of Serengeti spotted hyaenas: How a predator copes with migratory prey. I. Social organization. *Anim. Behav.* **1993**, *46*, 547–557. [CrossRef]
114. Hofer, H.; East, M.L. Population dynamics, population size, and the commuting system of Serengeti spotted hyaenas. In *Serengeti II: Dynamics, Conservation and Management of an Ecosystem*; Sinclair, A.R.E., Arcese, P., Eds.; University of Chicago Press: Chicago, IL, USA, 1995; pp. 332–363.
115. Scheel, D.; Packer, C. Variation in predation by lions: Tracking a movable feast. In *Serengeti II: Dynamics, Conservation and Management of an Ecosystem*; Sinclair, A.R.E., Arcese, P., Eds.; University of Chicago Press: Chicago, IL, USA, 1995; pp. 299–314.
116. Pienaar, U.D.V.; Riche, E.; Roux, C.S. The use of drugs in the management and control of large carnivorous mammals. *Koedoe* **1969**, *12*, 177–183. [CrossRef]
117. Stander, P.E. The ecology of lions and conflict with people in north-eastern Namibia. Proceedings of a Symposium on Lions and Leopards as Game Ranch Animals, Onderstepoort, South Africa, 24–25 October 1997; van Heerden, J., Ed.; Wildlife Group of the South African Veterinary Association: Onderstepoort, South Africa, 1997; pp. 10–17.

quaternary

MDPI

Review

An Updated Review of The Quaternary Hippopotamus Fossil Records from the Iberian Peninsula

Darío Fidalgo [1,*], Joan Madurell-Malapeira [2,3], Roberta Martino [4,5], Luca Pandolfi [6] and Antonio Rosas [1]

1 Department of Palaeobiology, Museo Nacional de Ciencias Naturales (CSIC), st/José Gutiérrez Abascal 2, 28006 Madrid, Spain; arosas@mncn.csic.es
2 Earth Science Department, University of Florence, 50121 Firenze, Italy; joan.madurell@uab.cat
3 Department of Geology, Campus de la UAB, Universitat Autònoma de Barcelona, Edifici C, 08193 Cerdanyola del Vallès, Barcelona, Spain
4 Department of Earth Sciences, GeoBioTec, School of Science and Technology, FCT-NOVA, Campus de Caparica, Universidade NOVA de Lisboa, 2829-516 Caparica, Portugal; roberta.aska@gmail.com
5 Museu da Lourinhã, R. João Luís Moura, 95, 2530-158 Lourinhã, Portugal
6 Department of Science, University of Basilicata, viale dell'Ateneo Lucano 10, 85100 Potenza, Italy; luca.pandolfi@unibas.it
* Correspondence: dfidal01@ucm.es

Abstract: This work presents a comprehensive review of the Quaternary fossil records of hippopotamuses from the Iberian Peninsula, unveiling biogeographical insights of global significance. The results presented herein include the inference of a delayed arrival of *Hippopotamus* (*Hippopotamus antiquus*) populations onto the Iberian Peninsula compared to other European Mediterranean regions, with an estimated age of ca. 1.7 Ma, in contrast to 2.1–2.2 Ma elsewhere. Moreover, we hypothesize the possibility of a short-lived coexistence between *H. antiquus* and *Hippopotamus amphibius*, close to the extinction of the former taxon (ca. 0.45 Ma). The local extinction of all hippopotamus populations on the Iberian Peninsula between MIS 5 and 3 is suggested here, mirroring proposals made for the Italian peninsula. Notable aspects of this fossil record include the abundance of specimens, previously undocumented anatomical elements, and partially complete individuals with articulated body segments. The remains analyzed herein also present different ontogenetic stages and sexual dimorphism. Moreover, the presence of specimens displaying paleopathologies provides valuable insights into ethological and paleoecological studies. The exceptional record of at least three events of human exploitation of hippopotamus stands out, with this being a rarity in the broader context of the archaeological and paleontological record of the European continent.

Keywords: Pleistocene; Mediterranean; Spain; Portugal; biogeography; *Hippopotamus antiquus*; *Hippopotamus amphibius*

check for updates

Citation: Fidalgo, D.; Madurell-Malapeira, J.; Martino, R.; Pandolfi, L.; Rosas, A. An Updated Review of The Quaternary Hippopotamus Fossil Records from the Iberian Peninsula. *Quaternary* **2024**, 7, 4. https://doi.org/10.3390/quat7010004

Academic Editor: Juan Rofes

Received: 29 October 2023
Revised: 1 December 2023
Accepted: 18 December 2023
Published: 10 January 2024

1. Introduction

Despite their putative restrictive environmental necessities, fossil records suggest that *Hippopotamus* inhabited continental Europe for at least 2 million years during the Quaternary period [1–3]. According to the current data, changes in the African and European ecosystems as well as in their interconnecting areas allowed the passage and the following establishment of hippopotamus populations between these zones, forming a complex phylogeographical context [4]. In order to unravel these paleobiological networks, approaches at different scales are necessary, from a global approach that interrelates existing regional data to the evaluation of specific cases that provide a geographical and chronological basis for biological information. In the case of hippopotamuses, there has been an inertia in this type of work in the last few decades. Since the end of the last century, different studies focused on hippopotamus occurrences have been carried out. These works include Stuart and Gibbard [5] for the United Kingdom, Pandolfi and Petronio [6] and Martino and

Pandolfi [2] for the Italian peninsula, Athanassiou [7] for Greece, and Martino et al. [8] for the Portuguese record.

During the Quaternary, at least two hippopotamus dispersals into continental Europe are inferred, both from taxa belonging to the genus *Hippopotamus* [9]. *Hippopotamus antiquus* is a species erected on specimens from the Upper Valdarno (Tuscany, Italy) ([2] and references therein). The first appearance datum (FAD) of *H. antiquus* can probably be traced back to the Middle–Late Villafranchian boundary, probably before 2 Ma [2,10]. Its origin is still disputed, although it seems closely related to the large African species *Hippopotamus gorgops* [11]. *H. antiquus* has a wide paleogeographic distribution range, extending from the Iberian Peninsula to Georgia and reaching the United Kingdom and the entire Italian Peninsula [2,5,8,12]. The last records of this taxon in Europe are referred to in the mid-Middle Pleistocene and are rather conflicting [8], with the taxon *Hippopotamus tiberinus* (=*H.* ex gr. *H. antiquus* in [13]) being considered to encompass the last forms of this lineage or it could represent a new interaction with African populations (see [4,14]).

More limited are the fossil records of *Hippopotamus amphibius*, and its dispersal towards the European continent is still poorly documented [8]. However, it is generally considered that between 0.5–0.4 Ma, specimens belonging to this species were present in Europe [15]. This chronological range would allow a presumed co-occurrence of *H. amphibius* with *H. antiquus* in areas such as the Italian peninsula [2]. Even so, the specific discrimination of most of the skeletal elements of both species is complex. Some complete skulls and postcranial material of *H. amphibius* were collected from the English locality of Barrington [16], and a skull assigned to *H. amphibius* was described from the Italian locality of Tor di Quinto [15,17]. The last recorded occurrences of this taxon go as far as MIS 5-MIS 3 in Italy [6], without a clear assessment of the conditions that led to its extinction across Europe. Probably in Greece, *H. amphibius* survived until ca 30 ka since its presence is reported from Dyrós Cave [7]. An analysis of the fossil and archaeological records of the Levantine corridor shows that the last populations of hippopotamuses persisted in this area until historical times [18].

The long paleontological and archaeological research traditions of the Iberian Peninsula have favored a rigorous method for excavation and contextualization of the fossil and archaeological records. This, together with the presence of large basins with Quaternary sedimentary deposits and areas with karstic activity [19], creates very favorable conditions to house a fossil record of great interest. A better understanding of the spatiotemporal distribution of hippopotamuses on the Iberian Peninsula is necessary to improve the paleoenvironmental interpretation of their habitats at times of coexistence with human populations and to better understand how their ecosystems changed. Under these premises, this paper compiles, re-evaluates, and provides new data on the fossil record of hippopotamuses on the Iberian Peninsula during the Quaternary. Special attention is paid to paleobiological parameters, also contextually including the relevant archaeological information.

2. Materials and Methods

In order to carry out an exhaustive update of the Iberian fossil records of hippopotamuses, a four-stage methodology was applied:

1. Compilation and consultation of all of the available bibliographies on hippopotamus fossils on the Iberian Peninsula and the sites where they have been found. The key references for each case are listed in Table S1.
2. Direct consultation with the researchers responsible for excavations on the findings. Data from this type of consultation are referenced as "pers. comm.".
3. Review of previously published historical collections and unpublished material deposited in numerous Iberian institutions. Museo Nacional de Ciencias Naturales (MNCN), Museo Geominero (IGME), Museo Arqueológico y Paleontológico de Madrid (MAR), Museo Arqueológico Nacional (MAN), Museo de San Isidro (MSI), Catedral de Astorga, Museo de la Rinconada, Institut Català de Paleoecologia Humana

i Evolució Social (IPHES), Institut Català de Paleontologia (ICP), Museu Arqueològic Comarcal de Banyoles, Museu de Ciències Naturals de Barcelona, Museu del Seminari Conciliar de Barcelona, Museu d'Arqueologia de Catalunya, Museu d'Història de València, Museo Paleontològico de Elche, Museo de Arqueología de Múrcia, Museo Arqueológico de Cartagena, Museo Arqueológico y Etnológico de Granada, and Museu Geológico de Lisboa (MG).

4. Collaboration in the evaluation of new finds. Barranc de la Boella, Vallparadís Estació, Cal Guardiola, Incarcal Complex (I, II and V), Las Jarillas, Bòbila Ordis, Barranco León D, Fuente Nueva-3, and Venta Micena.

The large volume of fossil material evaluated (either directly, bibliographically or by personal communication) cannot be described in detail in this paper. Nevertheless, in the following sections, some of the specimens and fossil assemblages with the greatest paleobiological, archaeological, or heritage value will be highlighted. In turn, the fossil assemblages are summarized in the data associated with the locality or archaeo-paleontological level of the find (Table S1). These sites and geological layers will be treated as cases for the analysis of the fossil records.

The taxonomic determinations of the hippopotamus specimens included were reviewed, accepting those determinations that are sufficiently justified and based on discriminating anatomical elements. In this work, we considered the taxonomy proposed by Petronio [9]; we therefore accepted the validity of *H. antiquus* (=*H. major*) and *H. amphibius*, and we avoided using *H. tiberinus*, *H.* ex gr. *antiquus*, or *H. incognitus*.

The most current geochronology and stratigraphic framework proposed for each find included has been respected. Geochronologic ages are given in thousand or million years before the present, abbreviated as "ka" or "Ma", respectively.

3. Results

The review carried out in this work highlights the appraisal of at least forty-eight different localities or deposits with the occurrence of hippopotamuses on the Iberian Peninsula. Most of these localities are distributed parallel to the Mediterranean coast, although there are records from the center of the peninsula, the western part of the Atlantic coast, and a few specimens from the northern part of the peninsula (Figure 1).

A total of twenty-five deposits with hippopotamuses have been dated to the Early Pleistocene, while fifteen belong to the Middle Pleistocene and only two belong to the Late Pleistocene (Figure 2). Six of the cases in our dataset could not be assigned to a more precise chronology than the undifferentiated Pleistocene. The earliest record found on the Iberian Peninsula, the Mencal-9 site, is associated with a chronology of ca. 1.7 Ma (Figure 3j). On the other hand, the last records of hippopotamuses on the peninsula are ascribed to the Late Pleistocene, between MIS 5 and MIS 3, at the sites of Cueva del Toll and Sima de las Palomas de Cabezo Gordo (Figure 3k). If we focus on the taxa considered, the oldest specimens reliably determined as *H. antiquus* are found at Venta Micena (1.6–1.4 Ma), and the latest are found at Condeixa (ca. 0.45 Ma). According to the records from Solana de Zamborino and Las Jarillas, *H. amphibius* may have made its appearance in the records 0.48–0.4 Ma ago, surviving until the Late Pleistocene (Figure 2).

If we evaluate the geographic and chronological data of all hippopotamus occurrences, we can observe a wide distribution area of hippopotamuses in the Early Pleistocene that would extend from the southwest of the peninsula (Algoz) and the eastern Pyrenees (Incarcal Complex and Bòbila Ordis) to the center–north of the peninsula (Atapuerca Complex), including the entire Mediterranean coast (Figure 1). During the Middle Pleistocene, this extension remained constant, increasing the records in the center of the peninsula and in the central part of the Atlantic coast. At the end of the Middle Pleistocene and during the Late Pleistocene, hippopotamus records were much more limited and only located on the Mediterranean coast (Figure 1).

Figure 1. Schematic map of the Iberian Peninsula with the localities where *Hippopotamus* specimens have been found. In the lower part of the map, the minimum geographic distribution range that hippopotamus populations would reach in each sub-epoch according to their fossil records is highlighted.

Among the most important features of these fossil records, there is a great number of hippopotamus specimens from sites such as Vallparadís Estació, with more than 3000 fossils. The studied fossil assemblages contain several complete or partial skulls and mandibles (e.g., the Cal Guardiola Lower Unit, La Mina, Bòbila Ordis, the Vallparadís Estació Upper Unit, Incarcal-I, or Condeixa; Figure 3a–i) and partial skeletons (e.g., the Cal Guardiola Lower Unit or Incarcal-I; Figure 3n) including anatomical element such as the hyoid bone (e.g., Incarcal-I), together with specimens belonging to different age classes and sexes (e.g., the Vallparadís Section; Figure 3a–d). In addition to valuable data for the autoecological analysis of the hippopotamus populations, at some sites, we found information on the biotic (plant and faunal) and abiotic context in which these populations lived (e.g., Barranc de la Boella [10]). In some of the localities, direct interaction of other organisms with hippopotamuses was detected, such as marks on the bones of exploitation by carnivores (e.g., Barranco León [20] or the Vallparadís Section [21]; Figure 3m) or cutmarks by humans (Barranco León, Fuente Nueva-3 [20], and Cueva de Bolomor [22]; Figure 3l).

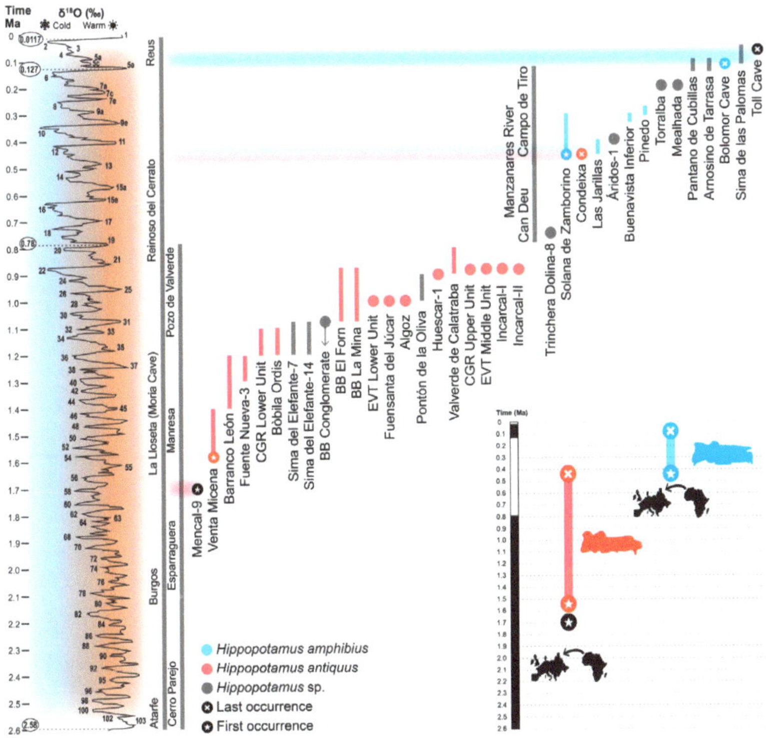

Figure 2. Chronological representation of *Hippopotamus* finds from the Iberian Peninsula and their taxonomic approximation. On the right, the estimated biochronological range of each species on the Iberian Peninsula is indicated with respect to its estimate for Europe as a whole.

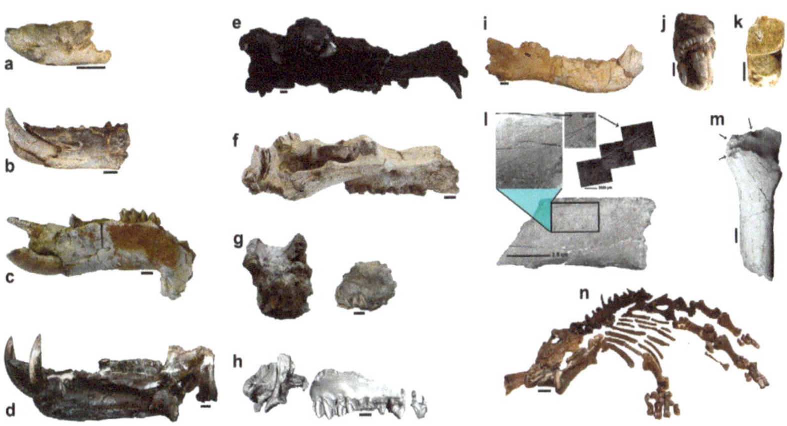

Figure 3. Some of the most outstanding finds of hippopotamuses on the Iberian Peninsula: (**a**–**d**) ontogenetic series of jaws of *Hippopotamus antiquus* from the Vallparadís Section, (**a**) infant jaw from Vallparadís Estació (EVT; IPS99003), (**b**) juvenile jaw from EVT (IPS127115), (**c**) young adult

jaw from EVT (IPS127243), and (**d**) adult jaw from Cal Guardiola (IPS14513; (**e**) skull of *H. antiquus* from Cal Guardiola (IPS14960); (**f**) skull of *H. antiquus* compressed dorso-ventrally found in the Vallparadís Estació layer EVT7 (IPS127242); (**g**) reconstruction of the skull of *H. antiquus* found in Bòbila Ordis; (**h**) reconstruction of the skull of *H. antiquus* found in Barranc de la Boella [10]; (**i**) mandible of *H. antiquus* found in Condeixa, assigned to the last recorded presence of the species on the Iberian Peninsula [8]; (**j**) fragment of a molar from Mencal-9, assigned to the oldest recorded presence of *Hippopotamus* on the Iberian Peninsula (Arribas, pers. comm.); (**k**) incisor fragment determined as *Hippopotamus amphibius* found in the Sima de las Palomas de Cabezo Gordo, assigned to one of the last recorded presences of hippopotamuses on the Iberian Peninsula (Walter, pers. comm.); (**l**) cut mark reported on a bone of *H. antiquus* at the locality of Fuente Nueva (modified from [20]); (**m**) femur of *H. antiquus* with marks of exploitation by hyenas found at Fuente Nueva (modified from [20]); and (**n**) partial skeleton of *H. antiquus* found at Cal Guardiola. Scale bar (**a–i**) 3 cm, (**j–k**) 1 cm and (**n**) 10 cm.

4. Discussion

4.1. Phylogenetic and Biogeographical Relationships of Hippopotamus of the Iberian Peninsula

Considering the data obtained, the oldest chronological record of specimens assignable to the genus *Hippopotamus* on the Iberian Peninsula was found at the Mencal-9 site. Despite the fragmentary nature of the remains found at Mencal-9 (two fragmented incisors (M9-2009-0012 and M9-2009-0059) and a fragmented molar [Figure 3j]), their taxonomic attribution to the genus *Hippopotamus* is reliable (Garrido and Arribas, pers. comm.). Chronologically, the Mencal-9 deposits have been dated to ca. 1.7 Ma by magnetostratigraphic and biochronological techniques [23] (Garrido and Arribas, pers. comm.). Some interesting specimens originally assigned to *H. antiquus* were also found in Venta Micena layers dated ca. 1.6–1.4 Ma [24]. Possibly, the remains from these two localities represent some of the earliest populations of *H. antiquus* that would have entered the Iberian Peninsula slightly later than in other Mediterranean areas: ca. 2 Ma in Greece (Elis [25]), ca. 2.2–2.1 Ma in Italy (e.g., Coste San Giacomo [26]), or ca. 2.2–2.1 Ma in France (Senèze [27]; J.M.-M. unpublished data). According to the hypothesis of van der Made et al. [14], these populations could correspond to an initial influx of African forms related to the controversial taxon *Hippopotamus kaisensis* (referred to as *H. antiquus* in Europe). Subsequently, these populations would have disappeared, leading to a period without hippopotamuses in Europe, which would have probably ended ca. 1.4 Ma ago with the arrival of populations of *Hippopotamus gorgops* (referred to as *Hippopotamus tiberinus* in Europe). However, the relatively close chronologies of the Barranco León site (ca.1.4–1.2 Ma), where hippopotamus fossils have also been found [28], and the general scarcity of fossil sites from this time span (1.6–1.4 Ma) on the Iberian Peninsula [29] do not particularly support the previously mentioned hypothesis.

The number of hippo sites increased enormously during the so-called '*Early-Middle Pleistocene Transition*', from the end of the Jaramillo subchron to the end of the Early Pleistocene. In this chronological timespan, *H. antiquus* populations extended from the southwesternmost tip of the peninsula (Algoz [30]) to its northeastern boundary (Incarcal Complex and Bòbila Ordis [31,32]). This increase in the fossil records of hippopotamuses is also shared in other areas of Europe, such as the Italian peninsula [2], where fossil assemblages, such as Collecurti, with more than five hundred specimens of this taxon were described by Mazza and Ventra [33]. With the transition to the Middle Pleistocene, the abundance of fossil sites with hippopotamus specimens decreased again. Indeed, Mazza [4] recognized a phenotypic trend in *H. antiquus* populations from ca. 1 Ma onwards that described as a new species *Hippopotamus tiberinus*. The validity of this taxon is disputed but a discussion on this topic is beyond the scope of this paper. However, it should be noted that van der Made et al. [34] noted a certain affinity between the specimens found in level TD8 from Gran Dolina (Atapuerca) and the specimens determined by Mazza [4] as *H. tiberinus*.

The last reliable occurrences of *H. antiquus* in Europe were recorded in the mid-Middle Pleistocene. Martino et al. [8] re-evaluated the specimens collected from Condeixa dated to ca. 0.45 Ma and confirmed their attribution to *H. antiquus*. If we consider as the earliest occurrence of *H. amphibius* in Europe the record from Tor di Quinto (most likely Cava Montanari [17]) dated ca. 0.55–0.45 Ma [15], a brief overlap of both taxa in Europe seems likely. On the Iberian Peninsula, there is no record of *H. amphibius* sufficiently informative to be sure of the chronological range of its presence. However, specimens from Solana de Zamborino (0.48–0.3 Ma [35]) and Las Jarillas (0.45–0.4 Ma [36]) have been determined as *H. amphibius*. If these determinations are confirmed, we should consider the possibility of the coexistence of *H. antiquus* and *H. amphibius* in a limited geographical area as the south and west of the Iberian Peninsula.

During the second half of the Middle Pleistocene, there are few sites where hippopotamuses are present, but unfortunately, their record is rather fragmentary. Despite this, the known finds point to the greatest geographical extension of this taxon, from the central-west (Condeixa and Melhada [8]) to the northeast (Can Deu [37]) of the peninsula. The group of fluvial terrace sites on the central and southern peninsula also highlight a constant presence of *H.* cf. *amphibius* populations until the end of the Middle Pleistocene (Áridos-1 [38], Campo de Tiro [39], Buenavista Inferior [39], or Las Jarillas [36]). A record with a slightly higher abundance of this taxon was found at the end of the Middle Pleistocene (MIS 5e) in the Bolomor Cave site [40] (Fernández Peris and Blasco, pers. comm.), although a detailed paleobiological study of these materials has not yet been carried out. These time periods coincide with those proposed for the findings of the largest accumulation of *H. amphibius* specimens at the Barrington locality (UK, ca. 0.13–0.115 Ma [16]). Faure [41], based on the English specimens, erected the disputed taxon Hippopotamus *incognitus*, listed as invalid by Petronio [9].

The local extinction of *H. amphibius* is poorly recorded on the Iberian Peninsula. The most recent findings of this taxon point to its last occurrences around MIS 4. Even so, these findings present a very limited stratigraphic and chronological contextualization. The specimens cited by Donner and Kurtén [42] at the Toll Cave site, and dated to MIS 4, have not been located in subsequent revisions (Blasco, pers. comm.). On the other hand, the specimen cited at the Sima de las Palomas de Cabezo Gordo site was found thirty years ago in the Upper Cuttin, without any stratigraphic context (Walker, pers. comm.). Estimation of the original stratigraphic position of the specimen has allowed us to assess a chronological range for its deposition between MIS 5 and MIS 3 [43]. These last local presences' chronologies also approximate the timing of the extinction of *H. amphibius* populations across Europe. Although specimens from the Italian site of Grotta Romanelli dated to MIS 3 were considered the last occurrence of hippopotamuses in continental Europe by Martino and Pandolfi [2], recently, Pieruccini et al. [44] re-dated the IUS3 to MI5e.

4.2. Paleoecological Considerations

Although the niche parameters of *H. antiquus* are still under discussion [10,45,46], the need for permanent bodies of water for its survival [47] and the presence of more or less open areas with grasslands [46] seem a clear requirement. Although these inferences are not very precise, they provide information on the minimum climatic parameters for its arrival in Europe and, specifically, for its dispersal across the Iberian Peninsula. The extension of the distribution area from Africa to the Iberian Peninsula in the middle of the Early Pleistocene raises two possible scenarios: (1) the presence of sufficient water bodies and an orographically traversable terrain for its passage through the Levantine corridor and the crossing of the entire European continent or (2) the reconsideration of a possible aquatic entry from North Africa through the western Mediterranean.

Once the populations of *H. antiquus* were established on the Iberian Peninsula, a continuous period of more than one million years of presence in this area was observed. The stability of these populations on a large part of the peninsula is surprising, as climatic conditions were progressively becoming harsher. These changes, known as the '*Early-Middle Pleistocene Transition*', were characterized by a gradual increase in the severity and duration of glacial cycles, along with an increase in seasonality and aridity [48,49]. Equally surprising is the possibility of a very brief coexistence of *H. antiquus* and *H. amphibius* in a restricted geographical area in the mid-Middle Pleistocene. This phenomenon has been previously proposed for the Italian peninsula as well [2]. More information on the niche differences between the two hippopotamuses is needed to assess the possible role of *H. amphibius* in the extinction of *H. antiquus*. The same can be said regarding the environmental conditions that may have led to a shift from the European to the extant hippopotamus, as it has been hypothesized that both taxa had the same dietary and climatic constraints.

The Middle Pleistocene is characterized by the presence of rapid cyclicity between glacial and interglacial intervals [50]. The arrival into the Northern Hemisphere of *H. amphibius*, classically associated with warm and wet conditions, is remarkable [47]. Even more surprising is its persistence in Europe until the Late Pleistocene, with an apparent continuous record on both the Italian peninsula [2] and the Iberian Peninsula. As has already been proposed for other taxa (e.g., [51,52]; among others), the population mobility dynamics of hippopotamuses during the Middle Pleistocene could correspond to a push–pull mechanism, lingering in glacial intervals in these warmer peninsulas and dispersing to more northern areas during interglacial intervals (e.g., UK [5,16,47]). However, a better understanding of the Central European and British records is crucial for the evaluation of these biogeographical hypotheses. Similarly, a detailed study of the state of health of the last well-recorded populations of *H. amphibius* in Europe (e.g., Barrington [16] or Bolomor [22]) would allow us to refine the causes of the continent-wide disappearance of a group of mammals that inhabited this area for more than two million years.

In particular, the Iberian Quaternary hippopotamus records offer exceptional specimens and fossil assemblages of great importance in paleobiological studies. Localities such as Incarcal-I (ca. 0.86 Ma; MIS21 [31,53]) have anatomical elements with an extremely limited representation in fossil records. In the same locality, together with the Vallparadís Section fossil assemblage (1.2–0.86 Ma; [21,54,55]), partially complete individuals have been excavated, with most of their anatomical elements articulated (Figure 3n). At the same time, at these sites, we found individuals characterized by different ontogenetic stages and belonging to both sexes (Figure 3a–d). Another highlight of this fossil record is the presence of different skeletal pathologies (e.g., dental enamel hypoplasias, malocclusion, stress deformities, or joint problems) that are being studied in a global bioclimatic context [49].

Most of the fossil assemblages with hippopotamus presence on the Iberian Peninsula are the result of interspecific interactions (e.g., accumulation by carnivores [56]) or show evidence of this type of event (e.g., feeding by hyenas [21,57]; Figure 3m). The peculiar hippopotamus lifestyle brings a particular value to the characterization of interspecific relationships, allowing us to infer the behavior of carnivores around water bodies. In contrast, there is very little evidence of human exploitation of hippopotamuses across the European continent [58] (Figure 3l). Of this scarce evidence, only one reference is outside the Iberian Peninsula, in the Greek locality of Marathousa [58]. In particular, a limited number of cut marks identified on specimens from the sites of Barranco León [59] and Fuente Nueva-3 [20] allow us to propose human access to hippopotamus carcasses in the Early Pleistocene. More evident seems to be the accumulation of hippopotamus remains in the locality of Bolomor Cave at the end of the Middle Pleistocene [22]. Further elaboration on this topic is beyond the scope of this paper and will be dealt with in more detail in future works.

5. Conclusions

The review and detailed study of the Quaternary fossil record of hippopotamuses from the Iberian Peninsula has allowed us to highlight several biogeographical data of global importance: (a) we infer the arrival of *Hippopotamus antiquus* populations to the Iberian Peninsula later than to the rest of the European Mediterranean areas (ca. 1.7 Ma vs. 2.1–2.2 Ma);

(b) the possibility of a brief coexistence between *H. antiquus* and *Hippopotamus amphibius* on the peninsula in chronologies close to the extinction of the populations of the former taxon (ca. 0.45 Ma); and (c) the idea of a local extinction of all hippopotamus populations in this area between MIS 5 and MIS 3 is considered, mostly coinciding with the proposals published for the Italic peninsula. The biogeographical proposals considered have been evaluated in a paleoecological and paleoclimatic context, opening the possibility of new paleoenvironmental reconstructions.

Together with these data, the potential of the Iberian record for the progress of paleozoological and paleoecological studies of the European Quaternary, including aspects related to human evolution, has been highlighted: (a) the abundance of records, the presence of poorly-known anatomical elements and the occurrence of partially complete individuals displaying articulated body segments; (b) a good representation of intraspecific diversity, with individuals of different ontogenetic stages and assignable to both sexes; (c) the existence of specimens with paleopathologies particularly relevant for paleoecological reconstructions; (d) the exceptional presence of at least three events of hippopotamus exploitation by humans, scarcely documented on the whole European continent.

Supplementary Materials: The following supporting information can be downloaded at: https://www.mdpi.com/article/10.3390/quat7010004/s1, Table S1: Quaternary hippopotamus fossil record from the Iberian Peninsula. References [60–85] are cited in the Supplementary Materials

Author Contributions: Conceptualization, D.F.; methodology, D.F. and J.M.-M.; software, D.F.; validation, D.F., J.M.-M., R.M. and L.P.; formal analysis, D.F.; investigation, D.F. and R.M.; resources, D.F., J.M.-M. and R.M.; data curation, D.F.; writing—original draft preparation, D.F.; writing—review and editing, D.F., J.M.-M., R.M., L.P. and A.R.; visualization, D.F.; supervision, J.M.-M., L.P. and A.R.; project administration, D.F.; funding acquisition, D.F., J.M-M., R.M. and A.R. All authors have read and agreed to the published version of the manuscript.

Funding: The research activity of D.F. and A.R. was funded by Ministerio de Ciencia e Innovación, PID2021-122356NB-I00. D.F. was supported by the Ayuda del Programa de Formación de Profesorado Universitario (FPU20/03389) and is a Ph.D. student at the Programade Doctorado en Biología at the Universidad Complutense de Madrid. R.M. benefited from a GeoBioTec NOVA grant UIDB/04035/2020 and a Ph.D. fellowship 2021.08458.BD provided by the Fundação para a Ciência e Tecnologia.

Data Availability Statement: All data included in this paper are available in the Supplementary Materials.

Acknowledgments: We would especially like to thank the editors and the three anonymous reviewers for their helpful comments, which significantly improved this paper. We would also like to thank all of the curators of the assessed collections included in this review for facilitating the work of collecting information and consulting the specimens: Susana Fraile (Museo Nacional de Ciencias Naturales), Silvia Menéndez (Museo Geominero (IGME)), Cesar Laplana (Museo Arqueológico y Paleontológico de la Comunidad de Madrid), Eduardo Galán (Museo Arqueológico Nacional), Alberto González (Museo de San Isidro), personal del Museo de la Catedral de Astorga, personal del Museo de la Rinconada, Núria Ibáñez (Institut Català de Paleoecologia Humana i Evolució Social), Josep M. Robles (Institut Català de Paleontologia), Andrea Ferrer (Museu Arqueològic Comarcal de Banyoles), Isidro Toro and Bienvenido Martínez-Navarro (Museo Arqueológico y Etnológico de Granada), and José Antonio Moita and Ruben Dias (Museu Geológico de Lisboa). Likewise, this work would not have been possible without the information provided by the researchers who dedicated personal communications to us: Guiomar Garrido (Universidad Internacional de la Rioja), Alfonso Arribas (Instituto Geológico y Minero de España), Ruth Blasco (Institut Català de Paleoecologia Humana i Evolució Social), Josep Fernández Peris (Museu de Prehistòria de València), and Michael Walker (Universidad de Murcia).

Conflicts of Interest: The authors declare no conflicts of interest.

References

1. Fidalgo, D.; Galli, E.; Madurell-Malapeira, J.; Rosas, A. Earliest Pleistocene European hippos: A review. *Comun. Geol.* **2021**, *108*, 69–73.
2. Martino, R.; Pandolfi, L. The Quaternary Hippopotamus records from Italy. *Hist. Biol.* **2022**, *34*, 1146–1156. [CrossRef]
3. Iannucci, A.; Mecozzi, B.; Sardella, R. Beware of the "Wolf event"–Remarks on large mammal dispersals in Europe and the late Villafranchian faunal turnover. *Alp. Mediterr. Quat.* **2023**, *36*, 75–90.
4. Mazza, P. Interrelations between pleistocene hippopotami of Europe and Africa. *Bollet. Soc. Paleont. Ital.* **1991**, *30*, 153–186.
5. Stuart, A.J.; Gibbard, P.L. Pleistocene occurrence of *Hippopotamus* in Britain. *Quartärpaläontologie* **1986**, *6*, 209–218.
6. Pandolfi, L.; Petronio, C. A brief review of the occurrences of Pleistocene *Hippopotamus* (Mammalia, Hippopotamidae) in Italy. *Geol. Croat.* **2015**, *68*, 313–319. [CrossRef]
7. Athanassiou, A. The fossil record of continental hippopotamids (Mammalia: Artiodactyla: Hippopotamidae) in Greece. In *Fossil Vertebrates of Greece Vol. 2: Laurasiatherians, Artiodactyles, Perissodactyles, Carnivorans, and Island Endemics*; Vlachos, E., Ed.; Springer: Cham Switzerland, 2022; pp. 281–299.
8. Martino, R.; Ríos, M.I.; Mateus, O.; Pandolfi, L. Taxonomy, chronology, and dispersal patterns of Western European Quaternary hippopotamuses: New insight from Portuguese fossil material. *Quat. Int.* **2023**, *674–675*, 121–137. [CrossRef]
9. Petronio, C. Note on the taxonomy of Pleistocene hippopotamuses. *Ibex* **1995**, *3*, 53–55.
10. Fidalgo, D.; Rosas, A.; Madurell-Malapeira, J.; Pineda, A.; Huguet, R.; García-Tabernero, A.; Cáceres, I.; Ollé, A.; Vallverdú, J.; Saladie, P. A review on the Pleistocene occurrences and palaeobiology of *Hippopotamus antiquus* based on the record from the Barranc de la Boella Section (Francolí Basin, NE Iberia). *Quat. Sci. Rev.* **2023**, *307*, 108034. [CrossRef]
11. Mazza, P. New evidence on the Pleistocene hippopotamuses of western Europe. *Geol. Rom.* **1995**, *31*, 61–241.
12. Vekua, A.K. The lower Pleistocene mammalian fauna of Ahkalkalaki (southern Georgia, USSR). *Palaeontogr. Ital.* **1986**, *74*, 63–96.
13. Mazza, P.; Bertini, A. Were Pleistocene hippopotamuses exposed to climate driven body size changes? *Boreas* **2013**, *42*, 194–209. [CrossRef]
14. van der Made, J.; Sahnouni, M.; Boulaghraief, K. *Hippopotamus gorgops* from El Kherba (Algeria) and the Context of its Biogeography. In Proceedings of the II MAP, Burgos, Spain, 15–16 April 2015; pp. 135–169.
15. Caloi, L.; Palombo, M.R.; Petronio, C. Resti cranici di *Hippopotamus antiquus* (= *H. major*) e *Hippopotamus amphibius* conservati nel Museo di Paleontologia dell'Università di Roma. *Geol. Rom.* **1980**, *19*, 91–119.
16. Reynolds, S.H. A Monograph on the British Pleistocene Mammalia VOL. III, Part I. Hippopotamus. Pages 1–38; plates I–VI. *Monogr. Palaeontogr. Soc.* **1922**, *74*, 1–38. [CrossRef]
17. Mecozzi, B.; Iannucci, A.; Mancini, M.; Tentori, D.; Cavasinni, C.; Conti, J.; Yuri Messina, M.; Sarra, A.; Sardella, R. Reinforcing the idea of an early dispersal of *Hippopotamus amphibius* in Europe: Restoration and multidisciplinary study of theskull from the Middle Pleistocene of Cava Montanari (Rome, central Italy). *PLoS ONE* **2023**, *18*, e0293405. [CrossRef] [PubMed]
18. Bytinski-Salz, H. Recent findings of *Hippopotamus* in Israel. *Isr. J. Zool.* **1965**, *14*, 38–48.
19. Silva, P.G.; Bardají, T.; Roquero García-Casal, E.; Baena Preysler, J.; Cearreta, A.; Rodríguez Pascua, M.A.; Rosas, A.; Zazo, C.; Goy, J.L. El periodo cuaternario: La historia geológica de la Prehistoria. *Cuat. Geomorf.* **2017**, *31*, 113–154.
20. Espigares, M.P. Análisis y Modelización del Contexto Sedimentario y los Atributos Tafonómicos de los Yacimientos Pleistocénicos del Borde Nororiental de la Cuenca de Guadix-Baza. Ph.D. Thesis, Universidad de Granada, Granada, Spain, 2010.
21. Madurell-Malapeira, J.; Alba, D.M.; Espigares, M.P.; Vinuesa, V.; Palmqvist, P.; Martínez-Navarro, B.; Moya-Sola, S. Were large carnivorans and great climatic shifts limiting factors for hominin dispersals? Evidence of the activity of Pachycrocuta brevirostris during the Mid-Pleistocene Revolution in the Vallparadís Section (Vallès-Penedès Basin, Iberian Peninsula). *Quat. Int.* **2017**, *431*, 42–52. [CrossRef]
22. Blasco, R.; Peris, J.F. From tortoises to elephants: The impact of elephants in the broad spectrum diet at Bolomor Cave (MIS 9–5 Spain). In *Contributions in Paleoanthropology Band 1: Human-Elephant Interactions: From Past to Present*; Konidaris, G.E., Barkai, R., Tourloukis, V., Harvati, K., Eds.; Tuebingen Paleoanthropology Book Series: Tuebingen, Germany, 2021; pp. 363–385.
23. Pla-Pueyo, S.; Viseras, C.; Soria, J.M.; Tent-Manclús, J.E.; Arribas, A. A stratigraphic framework for the pliocene-pleistocene continental sediments of the Guadix basin (betic Cordillera, S. Spain). *Quat. Int.* **2011**, *243*, 16–32. [CrossRef]
24. Alberdi, M.T.; Ruiz Bustos, A. Descripción y significado biostratigráfico y climático del Equus e Hippopotamus. en el yacimiento de Venta Micena (Granada). *Estud. Geol.* **1985**, *41*, 251–261. [CrossRef]
25. Reimann, C.K.; Strauch, F. Ein Hippopotamus-Schädel aus dem Pliozän von Elis (Peloponnes, Griechenland). *Neues Jahrbuch Geol. Palaontol.-Abhandlungen* **2008**, *249*, 203–222. [CrossRef]
26. Bellucci, L.; Bona, F.; Corrado, P.; Magri, D.; Mazzini, I.; Parenti, F.; Scardia, G.; Sardella, R. Evidence of late Gelasian dispersal of African fauna at Coste San Giacomo (Anagni Basin, central Italy): Early Pleistocene environments and the background of early human occupationin Europe. *Quat. Sci. Rev.* **2014**, *96*, 72–85. [CrossRef]
27. Mazza, P.; Rustioni, M. The fossil bear from Senèze (Southern France). *Rend. Lincei.* **1994**, *5*, 17–26. [CrossRef]
28. Martínez-Navarro, B.; Palmqvist, P.; Madurell-Malapeira, J.; Ros-Montoya, S.; Espigares, M.P.; Torregrosa, V.; Pérez-Claros, J.A. La fauna de grandes mamíferos de Fuente Nueva-3 y Barranco León-5: Estado de la cuestión. In *Ocupaciones Humanas en el Pleistoceno Inferior y Medio de la Cuenca de Guadix-Baza*; Toro, I., Martínez-Navarro, B., Agustí, J., Eds.; Junta de Andalucía, Consejería de Cultura: Sevilla, Spain, 2010; pp. 197–236.

29. Madurell-Malapeira, J.; Ros-Montoya, S.; Espigares, M.P.; Alba, D.M.; Aurell-Garrido, J. Villafranchian large mammals from the Iberian Peninsula: Paleobiogeography, paleoecology and dispersal events. *J. Iber. Geol.* **2014**, *40*, 167–178. [CrossRef]
30. Antunes, M.T.; Azzaroli, A.; Faure, M.; Guérin, C.; Mein, P. Mammifères pleistocènes de Algoz, en Algarve: Une révision. *Ciênc. Terra.* **1986**, *8*, 73–86.
31. Galobart, À.; Ros, X.; Maroto, J.; Vila, B. Descripción del material de hipopótamo (*Hippopotamus antiquus* Desmarest, 1822) de los yacimientos del Pleistoceno inferior de Incarcal (Girona, NE de la Península Ibérica). *Paleont. Evol.* **2003**, *34*, 153–173.
32. Agustí, J.; Leroy, S.A.; Lozano-Fernández, I.; Julià, R. Joint vegetation and mammalian records at the early Pleistocene sequence of Bòbila Ordis (Banyoles-Besalú Basin, NE Spain) and their bearing on early hominin occupation in Europe. *Paleobiodivers. Paleoenviron.* **2018**, *98*, 653–662. [CrossRef]
33. Mazza, P.P.A.; Ventra, D. Pleistocene debris-flow deposition of the hippopotamus-bearing Collecurti bonebed (Macerata, Central Italy): Taphonomic and paleoenvironmental analysis. *Palaeogeogr. Palaeoclimatol. Palaeoecol.* **2011**, *310*, 296–314. [CrossRef]
34. van der Made, J.; Rosell, J.; Blasco, R. Faunas from Atapuerca at the Early- Middle Pleistocene limit: The ungulates from level TD8 in the context of climatic change. *Quat. Int.* **2017**, *433*, 296–346. [CrossRef]
35. Álvarez-Posada, C.; Parés, J.M.; Sala, R.; Viseras, C.; Pla-Pueyo, S. New magnetostratigraphic evidence for the age of Acheulean tools at the archaeo-palaeontological site "Solana del Zamborino" (Guadix–Baza Basin, S Spain). *Sci. Rep.* **2017**, *7*, 13495. [CrossRef]
36. Baena, R.; Cunha, P.P.; Guerrero, I.C.; Fernández, J.J.; Cortés, M.; Silva, P.G.; Simón, M.; Murray, A.S.; Buylaert, J.P. El pleistoceno medio y superior en la secuencia ceneral de las terrazas del Guadalquivir (sgtg): Nuevas dataciones por luminiscencia de la t12 y t13 en la Rinconada (SW de España). In Proceedings of the XV Reunión Nacional Cuaternario, Bilbao, Spain, 1–5 July 2019; pp. 369–372.
37. Gómez-Alba, J. Catálogo razonado de los vertebrados fósiles de España del Museo de Geología de Barcelona (1882–1982). In *Treballs del Museu de Geologia de Barcelona*; Ajuntament de Barcelona: Museu de Geologia: Barcelona, Spain, 1997; pp. 1–296.
38. Blain, H.A.; Santonja, M.; Pérez-González, A.; Panera, J.; Rubio-Jara, S. Climate and environments during Marine Isotope Stage 11 in the central Iberian Peninsula: The herpetofaunal assemblage from the Acheulean site of Áridos-1, Madrid. *Quat. Sci. Rev.* **2014**, *94*, 7–21. [CrossRef]
39. Sese, C.; Soto, E. Vertebrados del Pleistoceno de Madrid. *Flora Fauna* **2000**, *6*, 216–243.
40. Blasco, R.; Rosell, J.; Peris, J.F.; Cáceres, I.; Vergès, J.M. A new element of trampling: An experimental application on the Level XII faunal record of Bolomor Cave (Valencia, Spain). *J. Archaeol. Sci.* **2008**, *35*, 1605–1618. [CrossRef]
41. Faure, M. *Hippopotamus incognitus* nov. sp., un hippopotame (Mammalia, Artiodactyla) du Pléistocène d'Europe occidentale. *Geobios* **1984**, *17*, 427–437. [CrossRef]
42. Donner, J.J.; Kurtén, B. The floral and faunal succession of "Cueva del Toll", Spain. *E G Quat. Sci. J.* **1958**, *9*, 72–82. [CrossRef]
43. Trinkaus, E.; Walker, M. *The People of Palomas: Neandertals from the Sima de las Palomas del Cabezo Gordo, Southeastern Spain*; Texas A&M University Press: College Station, TX, USA, 2017; p. 320.
44. Pieruccini, P.; Forti, L.; Mecozzi, B.; Iannucci, A.; Yu, T.L.; Shen, C.C.; Bona, F.; Lembo, G.; Muttillo, B.; Sardella, R.; et al. Stratigraphic reassessment of Grotta Romanelli sheds light on Middle-Late Pleistocene palaeoenvironments and human settling in the Mediterranean. *Sci. Rep.* **2022**, *12*, 13530. [CrossRef]
45. Palmqvist, P.; Rodríguez-Gomez, G.; Figueirido, B.; García-Aguilar, J.M.; Perez-Claros, J.A. On the ecological scenario of the first hominin dispersal out of Africa. *Anthropologie* **2022**, *126*, 102998. [CrossRef]
46. Kierdorf, U.; Kahlke, R.D. Pathological findings on remains of hippopotamids from the Early Pleistocene site of Untermassfeld. In *The Pleistocene of Untermassfeld near Meiningen (Thüringen, Germany). Part 4*; Kahlke, R.D., Ed.; Senckenberg: Frankfurt, Germany, 2020; Volume 40, pp. 1251–1272.
47. Adams, N.F.; Candy, I.; Schreve, D.C. An Early Pleistocene hippopotamus from Westbury Cave, Somerset, England: Support for a previously unrecognized temperate interval in the British Quaternary record. *J. Quat. Sci.* **2022**, *37*, 28–41. [CrossRef]
48. Head, M.J.; Gibbard, P.L. Early–Middle Pleistocene transitions: Linking terrestrial and marine realms. *Quat. Int.* **2015**, *389*, 7–46. [CrossRef]
49. Fidalgo, D.; Rosas, A.; Bartolini-Lucenti, S.; Boisserie, J.-R.; Pandolfi, L.; Martínez-Navarro, B.; Palmqvist, P.; Rook, L.; Madurell Malapeira, J. Increase on environmental seasonality through the European Early Pleistocene inferred from dental enamel hypoplasia. *Sci. Rep.* **2023**, *13*, 16941. [CrossRef]
50. Berends, C.J.; Köhler, P.; Lourens, L.J.; Van de Wal, R.S.W. On the cause of the Mid-Pleistocene transition. *Rev. Geophys.* **2021**, *59*, e2020RG000727. [CrossRef]
51. Fidalgo, D.; Rosas, A.; Estalrrich, A.; García-Tabernero, A.; Pastor, F.J.; Pineda, A.; Huguet, R.; Cáceres, I.; Ollé, A.; Vallverdú, J.; et al. First presence of Macaca sylvanus at the late early Pleistocene of barranc de la Boella (La Mina locality, Francolí basin, NE Iberia). *J. Hum. Evol.* **2023**, *179*, 103368. [CrossRef] [PubMed]
52. Boscaini, A.; Madurell-Malapeira, J.; Llenas, M.; Martínez-Navarro, B. The origin of the critically endangered Iberian lynx: Speciation, diet and adaptive changes. *Quat. Sci. Rev.* **2015**, *123*, 247–253. [CrossRef]
53. Duval, M.; Martínez-Navarro, B.; Oms, O.; Rufí, I.; Sorbelli, I.; Prat-Vericat, M.; Madurell-Malapeira, J. The Early Pleistocene fossil assemblage from Incarcal-I: First numerical dating results. In Proceedings of the XXI Inqua Congress, Rome, Italy, 13–20 July 2023.
54. Madurell-Malapeira, J. Estudi de les Restes d'*Hippopotamus antiquus* (Hippopotamidae, Mammalia) del Jaciment del Pleistocè Inferior de Cal Guardiola (Terrassa). Master's Thesis, Universidad Autònoma de Barcelona, Barcelona, Spain, 2006.

55. Madurell-Malapeira, J.; Minwer-Barakat, R.; Alba, D.M.; Garcés, M.; Gómez, M.; Aurell-Garrido, J.; Ros-Montoya, S.; Moyà-Solà, S.; Berástegui, X. The Vallparadís section (Terrassa, Iberian Peninsula) and the latest Villafranchian faunas of Europe. *Quat. Sci. Rev.* **2010**, *29*, 3972–3982. [CrossRef]
56. Espigares, M.P.; Palmqvist, P.; Rodríguez-Ruiz, M.D.; Ros-Montoya, S.; Pérez-Ramos, A.; Rodríguez-Gómez, G.; Guerra-Merchán, A.; García-Aguilar, J.M.; Granados, A.; Campaña, I.; et al. Sharing food with hyenas: A latrine of Pachycrocuta brevirostris in the Early Pleistocene assemblage of Fuente Nueva-3 (Orce, Baza Basin, SE Spain). *Archaeol. Anthropol. Sci.* **2023**, *15*, 1–29. [CrossRef]
57. Pineda, A.; Saladié, P.; Huguet, R.; Cáceres, I.; Rosas, A.; García-Tabernero, A.; Estalrrich, A.; Mosquera, M.; Ollé, A.; Vallverdú, J. Coexistence among large predators during the Lower Paleolithic at the site of La Mina (Barranc de la Boella, Tarragona, Spain). *Quat. Int.* **2015**, *388*, 177–187. [CrossRef]
58. Konidaris, G.; Tourloukis, V.; Boni, G.; Athanassiou, A.; Giusti, D.; Thompson, N.; Syrides, G.; Panagopoulou, E.; Karkanas, P.; Harvati, K. Marathousa 2: A New Middle Pleistocene Locality in the Megalopolis Basin (Greece) with Evidence of Hominin Exploitation of Megafauna (Hippopotamus). *PaleoAnthropology* **2023**, *2023*, 34–55.
59. Yravedra, J.; Solano, J.A.; Herranz-Rodrigo, D.; Linares-Matás, G.J.; Saarinen, J.; Rodríguez-Alba, J.J.; Titton, S.; Serrano-Ramos, A.; Courtenay, L.A.; Mielgo, C.; et al. Unravelling hominin activities in the zooarchaeological assemblage of Barranco León (Orce, Granada, Spain). *J. Paleolit. Archaeol.* **2022**, *5*, 6. [CrossRef]
60. Alonso Herrero, E.; Fernández-Martínez, E.; Rodríguez-Fernández, L.R.; Matías Rodríguez, R. *Guía Geológica Visual de León*; Editorial Celarayn: León, Spain, 2004; p. 274.
61. Rosas, A.; Huguet, R.; Pérez-González, A.; Carbonell, E.; Bermúdez de Castro, J.M.; Vallverdú, J.; van der Made, J.; Allue, E.; Garcia, N.; Martinez-Pérez, R.; et al. The "Sima del Elefante" cave site at Atapuerca (Spain). *Estud. Geol.* **2006**, *62*, 327–348. [CrossRef]
62. Rodríguez, J.; Burjachs, F.; Cuenca-Bescós, G.; García, N.; van der Made, J.; Pérez González, A.P.; Blain, H.-A.; Expósito, I.; López-García, J.M.; García Antón, M.; et al. One million years of cultural evolution in a stable environment at Atapuerca (Burgos, Spain). *Quat. Sci. Rev.* **2011**, *30*, 1396–1412. [CrossRef]
63. Van der Made, J. Ungulados de Gran Dolina, Atapuerca: Nuevos datos e interpretaciones. *Atapuerca Evol. Hum.* **1998**, 97–110.
64. Aguirre, E. Remarques sur la stratigraphie et la paléontologie du bassin de Granada (Espagne). *Comptes Rendus Hebd. Seances L Acad. Des Sci.* **1958**, *246*, 2140–2142.
65. Sánchez, B.; Pesquero, M.D.; Fraile, S.; Salesa, M.J. Las colecciones de Vertebrados fósiles del Museo Nacional de Ciencias Naturales (CSIC): Aportación del Profesor Emiliano Aguirre a la Paleontología española. *Zona Arqueol.* **2004**, *4*, 527.
66. Crusafont, M. El cuaternario español y su fauna de mamíferos: Ensayo de síntesis. *Rev. Fac. Cienc.* **1960**, *2*, 181–199.
67. Bech, J. Nuevo hallazgo de *Hippopotamus ampphibius major* en Banyoles (Gerona). *Acta Geológica Hispánica* **1970**, *5*, 51–53.
68. Julià, R. Nuevos datos sobre la posición cronoestratigráfica de los materiales cuaternarios de la cuenca lacustre de Banyoles-Besalú (Girona). *Acta Geol. Hisp.* **1977**, *12*, 55–59.
69. Faure, M. Les hippopotamidés du Pléistocène ancien d'Oubeidiyeh (Israèl). *Mem. Travaux Cent. Rech. Fr. Jerus.* **1986**, *5*, 107–142.
70. Ruiz-Bustos, A.; Fernández, J.; Morales, J.; Rodríguez-Fernández, J.; Vera, J.A. Bioestratigrafia de los materiales Plio-Cuaternarios del borde norte de la Depresion de Granada. *Estud. Geol.* **1990**, *46*, 277–290. [CrossRef]
71. Cardoso, J.L. *Contribuição Para o Conhecimento dos Grandes Mamíferos do Plistocénico Superior de Portugal*; OEIRAS: Oeiras, Portugal, 1993; pp. 171–567.
72. Estraviz, Q. Quaternary Fossil Vertebrates from Continental Portugal: Paleobiodiversity, Revision of Specimens and New Localities. Master Thesis, Universidade Nova de Lisboa, Lisboa, Portugal, May 2019.
73. Mazo, A.V.; Pérez-González, A.; Aguirre, E. Las faunas pleistocenas de Fuensanta del Jucar y El Provencio y su significado en la evolucion del Cuaternario de la Llanura manchega. *Bol. Geol. Min.* **1990**, *101*, 404–418.
74. Mazo, A.V. Los hipopótamos del Pleistoceno medio de Huescar-1 (Granada). *Trab. Neog. Cuantern.* **1989**, *1*, 317–325.
75. Alberdi, M.T.; Alcalá, L.; Azanza, B.; Cerdeño, E.; Mazo, A.V.; Morales, J.; Sesé, C. *Consideraciones Bioestratigráficas Sobre la Fauna de Vertebrados Fósiles de la Cuenca de Guadix-Baza (Granada, España)*; CSIC—Museo Nacional de Ciencias Naturales (MNCN): Madrid, Spain, 1989; Volume 11, pp. 347–355.
76. Domingo, M.S.; Alberdi, M.T.; Sánchez Chillón, B.; Cerdeño, E. La fauna cuaternaria de la Cornisa Cantábrica en las colecciones del Museo Nacional de Ciencias Naturales. *MUNIBE* **2005**, *57*, 5–30.
77. Mariezkurrena-Gastearena, K. Hallazgos de macromamíferos poco frecuentes en yacimientos arqueológicos y paleontológicos del Pleistoceno de la región cantábrica. *Kobie Ser. Paleoanthropol.* **2011**, *30*, 83–110.
78. Baena Escudero, R.; Fernández Caro, J.J.; Guerrero Amador, I.; Posada Simeón, J.C. La Terraza compleja del río Gualdaquivir en "Las Jarillas" (La Rinconada, Sevilla. SW de España): Cronoestratiagrafía, industria lítica y macro-fauna asociada. *Cuat. Geomorf.* **2014**, *28*, 107–125.
79. Soto, E. Estudio paleontologico. In *El yacimiento Achelense de Pinedo (Toledo)*; Querol, M.A., Santonja, M., Eds.; Ministerio de Cultura: Madrid, Spain, 1979; pp. 39–42.
80. López-Recio, M.; Silva, P.G.; Roquero, E.; Cunha, P.P.; Tapias, F.; Alcaraz-Castaño, M.; Baena, J.; Cuartero, F.; Morín, J.; Torres, T.; et al. Geocronología de los yacimientos achelenses de Pinedo y Cien Fanegas (Valle del Tajo) e implicaciones en la evolución fluvial en el entorno de Toledo (España). *Estud. Geol.* **2015**, *71*, e029. [CrossRef]
81. Nolte, E. Miscelánea Arqueológica. *Kobie* **1977**, *7*, 9–27.

82. Vilaseca, S. *Reus y su Entorno en la Prehistoria*; Asociación de Estudios Reusenses: Reus, Spain, 1973; pp. 1–282.
83. Jiménez-Arenas, J.M.; Santonja, M.; Botella, M.; Palmqvist, P. The oldest handaxes in Europe: Fact or artefact? *J. Archaeol. Sci.* **2011**, *38*, 3340–3349. [CrossRef]
84. Pineda, A.; Saladié, P. The Middle Pleistocene site of Torralba (Soria, Spain): A taphonomic view of the Marquis of Cerralbo and Howell faunal collections. *Archaeol. Anthropol. Sci.* **2019**, *11*, 2539–2556. [CrossRef]
85. Bonadonna, F.P.; Alberdi, M.T. *Equus stenonis* Cocchi as a biostratigraphical marker in the Neogene-Quaternary of the Western Mediterranean basin: Consequence on Galerian-Villafranchian chronostratigraphy. *Quat. Sci. Rev.* **1987**, *6*, 55–66. [CrossRef]

MDPI AG
Grosspeteranlage 5
4052 Basel
Switzerland
Tel.: +41 61 683 77 34

Quaternary Editorial Office
E-mail: quaternary@mdpi.com
www.mdpi.com/journal/quaternary

www.ingramcontent.com/pod-product-compliance
Lightning Source LLC
Chambersburg PA
CBHW041408210926
51850CB00013B/3216
* 9 7 8 3 7 2 5 8 3 7 5 2 6 *